Cracking the Code

A Practical Guide to the NEC®

ROB ZACHARIASON
Minnesota State Community and Technical College
Moorhead, Minnesota

Publisher
The Goodheart-Willcox Company, Inc.
Tinley Park, IL
www.g-w.com

BASED ON THE 2023 NATIONAL ELECTRICAL CODE®

Copyright © 2024
by
The Goodheart-Willcox Company, Inc.

All rights reserved. No part of this work may be reproduced, stored, or transmitted in any form or by any electronic or mechanical means, including information storage and retrieval systems, without the prior written permission of
The Goodheart-Willcox Company, Inc.

Library of Congress Control Number: 2022946966

ISBN 978-1-63776-705-4

1 2 3 4 5 6 7 8 9 – 24 – 28 27 26 25 24 23

The Goodheart-Willcox Company, Inc. Brand Disclaimer: Brand names, company names, and illustrations for products and services included in this text are provided for educational purposes only and do not represent or imply endorsement or recommendation by the author or the publisher.

The Goodheart-Willcox Company, Inc. Safety Notice: The reader is expressly advised to carefully read, understand, and apply all safety precautions and warnings described in this book or that might also be indicated in undertaking the activities and exercises described herein to minimize risk of personal injury or injury to others. Common sense and good judgment should also be exercised and applied to help avoid all potential hazards. The reader should always refer to the appropriate manufacturer's technical information, directions, and recommendations; then proceed with care to follow specific equipment operating instructions. The reader should understand these notices and cautions are not exhaustive.

The publisher makes no warranty or representation whatsoever, either expressed or implied, including but not limited to equipment, procedures, and applications described or referred to herein, their quality, performance, merchantability, or fitness for a particular purpose. The publisher assumes no responsibility for any changes, errors, or omissions in this book. The publisher specifically disclaims any liability whatsoever, including any direct, indirect, incidental, consequential, special, or exemplary damages resulting, in whole or in part, from the reader's use or reliance upon the information, instructions, procedures, warnings, cautions, applications, or other matter contained in this book. The publisher assumes no responsibility for the activities of the reader.

The Goodheart-Willcox Company, Inc. Internet Disclaimer: The Internet resources and listings in this Goodheart-Willcox Publisher product are provided solely as a convenience to you. These resources and listings were reviewed at the time of publication to provide you with accurate, safe, and appropriate information. Goodheart-Willcox Publisher has no control over the referenced websites and, due to the dynamic nature of the Internet, is not responsible or liable for the content, products, or performance of links to other websites or resources. Goodheart-Willcox Publisher makes no representation, either expressed or implied, regarding the content of these websites, and such references do not constitute an endorsement or recommendation of the information or content presented. It is your responsibility to take all protective measures to guard against inappropriate content, viruses, or other destructive elements.

Image Credits. Front cover: Bertl123 via Getty Images; Part 1 image: sockagphoto/Shutterstock.com; Part 2 image: Sashkin/Shutterstock.com

Preface

The *National Electrical Code* is a model code written in technical, legal language. To work in the electrical trades, it is critical that you learn the *NEC*'s structure, how to navigate it, and how to identify and use tables correctly. Because the *NEC* is revised every three years, it is vital that you understand how the information is organized rather than memorizing locations that might change. *Cracking the Code* was written to help you understand how to use the *NEC*. This text will teach you how to navigate and find information by focusing on the structure and organization of the *NEC*.

Part 1 of this text will break down the structure of the *NEC*, instructing on the layout and focusing on each chapter of the *NEC*, one at a time. Part 2 of this text delves into applying the knowledge learned in Part 1, providing you an opportunity to apply your knowledge of the *NEC* to find the necessary information to complete a task.

Extensive practice problems apply and reinforce *NEC* knowledge. End-of-unit problems start out as simple multiple-choice questions that reinforce the material learned within the unit and advance to open-ended questions that build off what was learned in previous units.

About the Author

Rob Zachariason has worked in the electrical industry for over 30 years. He is currently an instructor in the Electrical Technology program at Minnesota State Community and Technical College in Moorhead, Minnesota. He was formerly an instructor for the Dakotas JATC in Fargo, North Dakota. When not in school, Mr. Zachariason is the working owner of Rob Zachariason Electric. Mr. Zachariason has also worked on several electrical textbooks as well as instructor supplements and videos.

Mr. Zachariason has a diploma in Construction Electricity from Northwest Technical College and a bachelor's degree in Operations Management from Minnesota State University. He holds a Master Electrician license in both North Dakota and Minnesota and is a member of the International Brotherhood of Electrical Workers, the National Electrical Contractors Association, the International Association of Electrical Inspectors, the National Fire Protection Association, and the Minnesota State College Faculty union.

Mr. Zachariason lives in Fargo with his wife, Brandi. They have three daughters, Lauren, Kate, and Julia.

Reviewers

The author and publisher wish to thank the following industry and teaching professionals for their valuable input into the development of *Cracking the Code*.

Mark Andreason
Eastern Arizona College
Thatcher, Arizona

Erik Bade
Madison Area Technical College
Madison, Wisconsin

James Carmack
Kansas City Kansas Community College
Kansas City, Kansas

Terrence Caston
North Lake College
Dallas, Texas

Daniel Choate
Wytheville Community College
Wytheville, Virginia

Salvatore Ferrara
Electrical Training Center
Long Island, New York

Polly Friendshuh
Dunwoody College of Technology
Minneapolis, Minnesota

Jeff Hamilton
Terra Community College
Fremont, Ohio

Frank Holcomb
Tennessee College of Applied Technology-Paris
Paris, Tennessee

Brian Khairullah
Southern California Edison Training Center
Yucaipa, California

Byron Krull
Northwest Iowa Community College
Sheldon, Iowa

David Krzyston
State University of New York at Delhi
Delhi, New York

Ron McGary
Industrial Systems Technology Consulting and Training
Albany, Georgia

Daniel Neff
Palm Beach State College
Lake Worth, Florida

Eric Newcomer
Pennsylvania College of Technology
Williamsport, Pennsylvania

Ryan Northup
Chariho Career & Technical Center
Richmond, Rhode Island

David Robinson
Los Angeles Trade Technical College
Los Angeles, California

Barbara Salazar
El Paso Community College
El Paso, Texas

Bryan Schroder
Northcentral Technical College
Wausau, Wisconsin

Rebecca Velasquez
Salt Lake Community College
Salt Lake City, Utah

Chris Wiggins
Central Piedmont Community College
Charlotte, North Carolina

Matthew Wilkinson
Madison Area Technical College
Madison, Wisconsin

Josh Wilson
Inspector with North Dakota
West Fargo, North Dakota

Bart Wood
Southwest Wisconsin Technical College
Fennimore, Wisconsin

Acknowledgments

The author and publisher would like to thank the following companies, organizations, and individuals for their contribution of resource material, images, or other support in the development of *Cracking the Code*.

Bloom Energy
Convoy Solutions

Cooper Lighting
Eaton

Qualtek Electronics Corporation

The *National Electrical Code*®

The most informative and authoritative body of information concerning electrical wiring installation in the United States, and perhaps the world, is the *National Electrical Code*® (*NEC*). This code establishes a set of rules, regulations, and criteria for the installation of electrical equipment. Compliance with these methods will result in a safe installation.

The *NEC* is drafted by a team of experts assembled for this purpose by the National Fire Protection Association (NFPA). This team is formally called the *National Electrical Code* committee. They revise and update the *NEC* every three years. It is imperative that anyone installing electrical wiring obtains and studies the *NEC*. Articles and sections of the *NEC* are referred to throughout this text. Although certain portions, tables, and examples are directly quoted from its text, there is enough useful information in the *NEC* that not having it available would be a tremendous hindrance.

The latest edition of the *National Electrical Code* can be purchased from the National Fire Protection Association by visiting their website. Online access to the *National Electrical Code* and over 1,400 NFPA codes and standards is available with an NFPA LiNK® subscription. NFPA LiNK includes all current editions as well as a library of legacy codes and standards going back five editions. Subscribers also have early access to newly released editions before the printed book is available for purchase. To learn more about NFPA LiNK, visit the NFPA website at www.nfpa.org.

NFPA 70®, National Electrical Code®, NEC®, and NFPA LiNK® are registered trademarks of the National Fire Protection Association, Quincy, MA.

Features of the Textbook

The instructional design of this textbook includes student-focused learning tools to help you succeed. This visual guide highlights these features.

Unit Opening Materials

Each unit opener contains a list of learning objectives, a list of technical terms, and a brief introduction.
Learning Objectives clearly identify the knowledge and skills to be gained when the unit is completed.
Key Terms list the key words to be learned in the unit.
Introduction provides an overview and preview of the unit content.

Additional Features

Additional features are used throughout the body of each unit to further learning and knowledge.
Examples provide step-by-step sample problems that demonstrate how to perform calculations.
Procedures are highlighted throughout the textbook to provide clear instructions for various processes. **Pro Tips** provide advice and guidance that is especially applicable for on-the-job situations.
Navigation Tables highlight all the sections contained within some of the most-used articles of the *NEC*. This type of overview, which is not provided in the *NEC*, helps students understand the type of information contained in those articles.

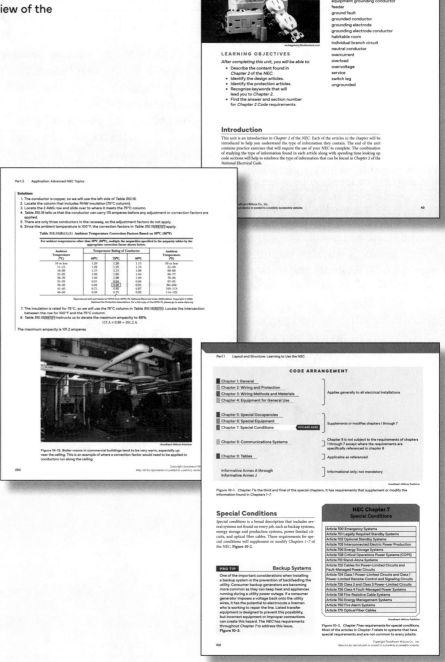

vi

Illustrations

Illustrations have been designed to clearly and simply communicate the specific topic. Illustrations and photographic images have been carefully chosen and designed to reinforce the text's content. Excerpts from the *NEC* are provided to give students real examples and visuals.

End-of-Unit Content

End-of-unit material provides an opportunity for review and application of concepts. **NEC Keywords** list numerous topics and words that should lead the user to a particular chapter of the *NEC*. A concise **Summary** provides an additional review tool and reinforces key learning objectives. This helps you focus on important concepts presented in the text. **Know and Understand** questions enable you to demonstrate knowledge, identification, and comprehension of unit material. **Apply and Analyze** questions extend learning and require you to select an answer by finding the correct section within the *NEC*. **Critical Thinking** questions are open ended and require you to find the answer and correct section within the *NEC*. These questions will begin testing you on the content contained within that unit and become cumulative, basing questions on everything learned up to that point within the text.

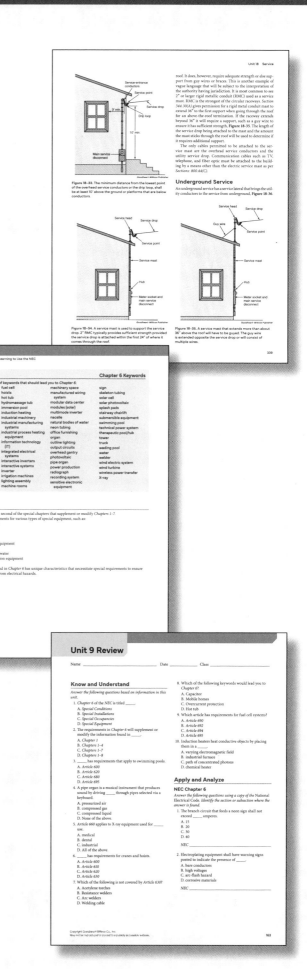

vii

TOOLS FOR STUDENT AND INSTRUCTOR SUCCESS

Student Tools

Student Text
Cracking the Code is a write-in text that focuses on the information, structure, and organization of the *National Electrical Code*.

G-W Digital Companion
For digital users, e-flash cards and vocabulary exercises allow interaction with content to create opportunities to increase achievement.

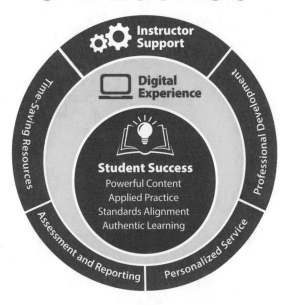

Instructor Tools

LMS Integration
Integrate Goodheart-Willcox content within your Learning Management System for a seamless user experience for both you and your students. EduHub® LMS-ready content in Common Cartridge® format facilitates single sign-on integration and gives you control of student enrollment and data. With a Common Cartridge integration, you can access the LMS features and tools you are accustomed to using and G-W course resources in one convenient location—your LMS.

G-W Common Cartridge provides a complete learning package for you and your students. The included digital resources help your students remain engaged and learn effectively:

- **Digital Textbook**
- **Drill and Practice** vocabulary activities

When you incorporate G-W content into your courses via Common Cartridge, you have the flexibility to customize and structure the content to meet the educational needs of your students. You may also choose to add your own content to the course.

For instructors, the Common Cartridge includes the Online Instructor Resources. QTI® question banks are available within the Online Instructor Resources for import into your LMS. These prebuilt assessments help you measure student knowledge and track results in your LMS gradebook. Questions and tests can be customized to meet your assessment needs.

Online Instructor Resources

- The **Instructor Resources** provide instructors with time-saving preparation tools such as answer keys, editable lesson plans, and other teaching aids.
- **Instructor's Presentations for PowerPoint®** are fully customizable, richly illustrated slides that help you teach and visually reinforce the key concepts from each unit.
- Administer and manage assessments to meet your classroom needs using **Assessment Software with Question Banks**, which include hundreds of matching, completion, multiple choice, and short answer questions to assess student knowledge of the content in each unit.

See www.g-w.com/cracking-code-2024 for a list of all available resources.

Professional Development

- Expert content specialists
- Research-based pedagogy and instructional practices
- Options for virtual and in-person Professional Development

Brief Contents

PART 1
Layout and Structure: Learning to Use the NEC

FOUNDATION
- Unit 1 NEC History and Revision Process 3
- Unit 2 Structure of the NEC. 11
- Unit 3 Preparing to Use the NEC. 21

GENERAL CHAPTERS
- Unit 4 NEC Chapter 1 35
- Unit 5 NEC Chapter 2 49
- Unit 6 NEC Chapter 3 73
- Unit 7 NEC Chapter 4 91

SPECIAL CHAPTERS
- Unit 8 NEC Chapter 5 117
- Unit 9 NEC Chapter 6 143
- Unit 10 NEC Chapter 7 167
- Unit 11 NEC Chapter 8 185
- Unit 12 NEC Chapter 9 and Informative Annexes. 197

PART 2
Application: Advanced NEC Topics

- Unit 13 Branch Circuits. 217
- Unit 14 Box Fill and Pull Box Calculations 241
- Unit 15 Raceway Fill Calculations 263
- Unit 16 Conductor Ampacity. 281
- Unit 17 Overcurrent Protection 305
- Unit 18 Service 325
- Unit 19 Grounding and Bonding. 349
- Unit 20 Dwelling Unit Cooking Appliance Calculations 379
- Unit 21 Dwelling Service Calculations 395

Contents

PART 1
Layout and Structure: Learning to Use the NEC

FOUNDATION

1 NEC History and Revision Process 3
  Introduction 3
  History 4
  National Fire Protection Association 4
  Changing the National Electrical Code 4
  Code-Change Process 5

2 Structure of the NEC 11
  Introduction 11
  Outline Format 12
  Layout 12
  National Electrical Code Revisions 16

3 Preparing to Use the NEC 21
  Introduction 21
  Article 90 22
  Finding Information 25
  Tools to Aid the Code User 27

GENERAL CHAPTERS

4 NEC Chapter 1 35
  Introduction 35
  Chapter 1, General 36
  Article 100, Definitions 36
  Article 110, General Requirements
    for Electrical Installations 37

5 NEC Chapter 2 49
  Introduction 49
  Chapter 2, Wiring and Protection 50
  Design Articles 50
  Protection Articles 58

6 NEC Chapter 3 73
  Introduction 73
  Chapter 3, Wiring Methods
    and Materials 74
  Common Format 74
  Wiring Methods and Materials 75

7 NEC Chapter 4 91
  Introduction 91
  Equipment Chapter 92
  Equipment Connection Articles 92
  Control Articles 94
  Consume Articles 100
  Create, Alter, and Store Articles 106

SPECIAL CHAPTERS

8 NEC Chapter 5 117
  Introduction 117
  Special Occupancies 118
  Hazardous Locations 118
  Gatherings and Entertainment 128
  Manufactured Structures and
    Agricultural Buildings 131
  Article 590, Temporary Installations 137

9 NEC Chapter 6 143
  Introduction 143
  Special Equipment 144

10 NEC Chapter 7 167
  Introduction 167
  Special Conditions 168
  Energy Backup and Storage Systems . . . 169
  Autonomous Power Production
    Systems 174

Power-Limited Circuits, Control Circuits,
   and Managed Power Systems 174
Article 770, Optical Fiber Cables 178

11 NEC Chapter 8 185
Introduction 185
Communications Systems 186

12 NEC Chapter 9 and Informative Annexes 197
Introduction 197
Chapter 9, Tables 198
Informative Annexes 204

PART 2
Application: Advanced NEC Topics

13 Branch Circuits 217
Introduction 217
Branch Circuits 218
Ratings 220
Dwelling Branch Circuits 224
Non-Dwelling Branch Circuits 234

14 Box Fill and Pull Box Calculations 241
Introduction 241
Electrical Boxes 242
Box Fill Calculations—NEC
   Section 314.16 244
Pull and Junction Boxes
   (1,000 Volts and Under)—NEC
   Section 314.28 252

15 Raceway Fill Calculations 263
Introduction 263
Raceway Fill Definitions 264
Circular Raceways 266
Conductors 267
Circular Raceway Fill 268
Noncircular Raceways 275

16 Conductor Ampacity 281
Introduction 281
General Conductor Information 282
Conductor Ampacities 285

17 Overcurrent Protection 305
Introduction 305
Article 240, Overcurrent Protection 306
Overcurrent Protection Devices
   (OCPD) 308
Overcurrent Protection Devices 315
Load 315
Sizing Overcurrent Devices 316

18 Service 325
Introduction 325
Service 326
Service Equipment 327
Service Disconnect 328
Wiring Methods 334
Types of Services 336

19 Grounding and Bonding 349
Introduction 349
Grounding Fundamentals 350
Grounding Electrode System 352
Equipment Grounding and
   Equipment Grounding Conductors 361
Bonding 366

20 Dwelling Unit Cooking Appliance Calculations 379
Introduction 379
Definitions 380
Household Cooking Equipment
   Demand Factors 380
Household Cooking Appliance
   Neutral Calculation 385
Sizing Branch Circuit Conductors 385
Branch Circuit Overcurrent Protective
   Device (OCPD) 387

21 Dwelling Service Calculations 395
Introduction 395
Article 220 396
General Requirements 396
Standard Method. 398
Minimum Size 405
Optional Method 405
Minimum Size 408

Appendix. 417
Math Review 418
Why Math? 418
Calculators .418
Whole Numbers. 419
Fractions 420
Reading a Ruler 421
Decimal Fractions 422
Converting Fractions to Decimals 423
Converting Decimals to Fractions 424
Equations 424
Area Measure 424
Volume Measure 426
Exponents and Roots 428
Working with Right Angles. 428
Computing Averages 429
Percent and Percentage 429
Ohm's Law Calculations 429
Service Calculation–Standard
  Method Template. 431
Service Calculation–Optional
  Method Template. 432

Glossary 433
Index of *NEC* References. 441
Index 446

Feature Contents

NEC KEYWORDS
NEC Chapter 2 Keywords	67
Chapter 3 Keywords	85
Chapter 3 Keywords	112
Chapter 5 Keywords	138
Chapter 6 Keywords	162
Chapter 7 Keywords	180
Chapter 8 Keywords	191

PRO TIP
NFPA 70E	5
NEC Style Manual	6
National Electrical Code Development Process	8
Code Hierarchy	12
Code Changes	17
NEC Handbook	18
Article 90	22
Adopting the NEC	24
Article 110 Requirements	39
Referencing the Correct NEC Section	40
Confined Spaces	42
Conductor Terminology	51
Article 300	78
Equipment of Article 312	79
Nonmetallic-Sheathed Cable	80
Receptacle Configurations	97
Doorbell Transformers	106
Table of Contents–Chapter 5	118
Hazardous Locations—Class vs Zone	122
Table of Contents–Chapter 6	144
Backup Systems	168
Class I, II, III Hazardous locations vs Class 1, 2, 3 Remote-Control and Power-Limited Circuits	175
Circular Mils	202
Voltage Drop	221
Bathroom Fans	229
Grounded Conductor in Switch Boxes	232
Finding the Correct Section	242
Show Your Work	251
Cable Protection	273
Conductor vs Cable	283
Motors	307
Branch Circuit Overcurrent Devices	308
Turning On a Breaker	311
Utility Locating Services	341
Grounded Conductor vs Grounding Conductor	350
Ground Fault on Ungrounded Branch Circuit	362
Insulated Equipment Grounding Conductor	364
Swimming Pools	368
Review NEC Structure	381
Instructional Programs	383
Commercial Cooking Equipment Calculations	385
Motor Values	397
Appliance Voltage Ratings	400
Temperate and Overcurrent Devices	403
Panelboards	405

PROCEDURE
Identifying Keywords	25
Finding a Code Requirement Using the Table of Contents	26
Finding a Code Requirement Using the Index	27
Maximum Number of Conductors	269
Raceway Area	270
Conductor Area	271
Finding the Correct Section	271
Bare Conductor Area	275
Conductor Ampacity—Finding the Correct Section	299
Steps To Size Overcurrent Protection Devices	318

PART 1
Layout and Structure: Learning to Use the NEC

sockagphoto/Shutterstock.com

FOUNDATION
Unit 1　NEC History and Revision Process
Unit 2　Structure of the NEC
Unit 3　Preparing to Use the NEC

GENERAL CHAPTERS
Unit 4　NEC Chapter 1
Unit 5　NEC Chapter 2
Unit 6　NEC Chapter 3
Unit 7　NEC Chapter 4

SPECIAL CHAPTERS
Unit 8　NEC Chapter 5
Unit 9　NEC Chapter 6
Unit 10　NEC Chapter 7
Unit 11　NEC Chapter 8
Unit 12　NEC Chapter 9

UNIT 1
NEC History and Revision Process

sockagphoto/Shutterstock.com

LEARNING OBJECTIVES

After completing this unit, you will be able to:

- Describe the history of the *National Electrical Code*.
- Explain the importance of the *National Electrical Code* being a consensus standard.
- List the steps to change the *National Electrical Code*.
- List the steps in the public input stage of the code-change process.
- List the steps in the public comment stage of the code-change process.
- Describe the reason for a tentative interim amendment (TIA)

KEY TERMS

code making panel (CMP)
consensus standard
first draft report
first revision
National Fire Protection Association (NFPA)
NEC style manual
NFPA standards council
public comment
public input
second draft report
second revision
technical correlating committee (TCC)
tentative interim amendment (TIA)

Introduction

The *National Electrical Code* is an electrical standard that has been governing electrical work since the late 1800s. The process for making changes to the *National Electrical Code* (*NEC*) is an extensive process that takes the perspective of many into consideration before changes are implemented. This unit gives a general overview of the process of updating the *National Electrical Code*.

History

In the late 1800s, the electrical industry was expanding, as were electricity-related fires. At that time, there were not many electrical standards. This led to many unsafe installations. Insurance companies, which were paying financial claims for many of these fires, were among the first to get involved in creating electrical standards. These standards were intended to ensure installations were safe from electrical hazards.

One of the first electrical standards was put into place in 1881 by the New York Board of Fire Underwriters. This standard consisted of five sections and was only one page long. Although the New York Board of Fire Underwriters were among the first, they were not alone in writing electrical standards. To create rules that were more universal, the National Electrical Light Association called a meeting to develop a standard that would be consistent and would have the ability to be widely adopted. The result of that meeting and the subsequent work was the first edition of the *National Electrical Code*, published in 1897 by the National Fire Protection Association (NFPA). See **Figure 1-1**.

Reproduced with permission of NFPA from NFPA 70, National Electrical Code, 2023 edition. Copyright © 2022, National Fire Protection Association. For a full copy of the NFPA 70, please go to www.nfpa.org

Figure 1-2. The NFPA 70 is the *National Electrical Code*.

Since 1897, the *NEC* has had over 50 editions and has expanded from a small booklet that can fit in your pocket to a book printed on full-size paper that is nearly 1000 pages long. The most current edition is the 2023 *National Electrical Code*, **Figure 1-2**.

National Fire Protection Association

The *National Electrical Code* is published by the **National Fire Protection Association (NFPA)**. The NFPA oversees the process of creating and revising the *NEC*. The official title of the *National Electrical Code* is NFPA 70. It is just one of many standards published by the NFPA. In addition to codes and standards, the National Fire Protection Association is involved in public education, outreach and advocacy, training, and research.

The NFPA is a self-funded, nonprofit organization that welcomes new members. NFPA membership comes with many benefits, including access to NFPA publications, resources, technical support, and discounts on NFPA products. Students qualify for a discounted membership rate.

Changing the National Electrical Code

The *NEC* is a ***consensus standard***. A consensus standard is a standard that takes the viewpoints of a group of people

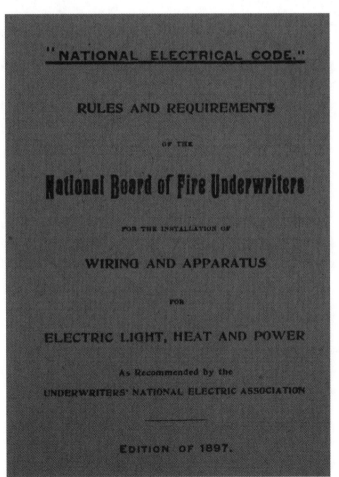

Goodheart-Willcox Publisher

Figure 1-1. The first edition of the *National Electrical Code* was published in 1897. It was the first widely accepted electrical standard.

PRO TIP — NFPA 70E

The NFPA 70E is titled *Standard for Electrical Safety in the Workplace*. The NFPA 70E has a title that is very similar to the NFPA 70 (*NEC*), but it is not the same document, and it has a vastly different purpose. The *NEC* is concerned with completed installations being free from electrical hazards, while the NFPA 70E is concerned with worker safety while installing or working on electrical systems. The NFPA 70E is a consensus standard that may be adopted or used in part or entirely by employer's safety programs. Many OSHA requirements are derived from information found in the NFPA 70E. See **Figure 1-3**.

Reproduced with permission of NFPA from NFPA 70, National Electrical Code, 2023 edition. Copyright © 2022, National Fire Protection Association. For a full copy of the NFPA 70, please go to www.nfpa.org

Figure 1-3. The NFPA 70E is the *Standard for Electrical Safety in the Workplace*. It contains practices and procedures to provide employee safety while working around electrical hazards.

with varied backgrounds into consideration during the standards development process. This prevents interested parties from pushing through changes just to benefit themselves personally or financially. Being a consensus standard, the *NEC* welcomes input and participation from anyone who is interested throughout the code-change process. The *NEC* uses code-making panels to make decisions on the proposed changes.

Code Making Panels (CMPs)

The *NEC* currently has 18 *code-making panels (CMPs)*. The NFPA also refers to the CMPs as technical committees (TCs). Each CMP is responsible for a specific portion of the *NEC*, **Figure 1-4**. The code-making panels are made up of groups of people with varied interests and backgrounds. This helps prevent individuals or industries from pushing through a change that may be for disingenuous reasons. Individuals that are appointed to CMPs are highly skilled in their area of expertise. Each member of a CMP is representing an industry or organization and will fall under one of the predetermined classifications, **Figure 1-5**.

The CMP members are volunteers and are not paid by the NFPA. Being a part of a code-making panel is a significant time commitment. Most of the members get involved due to their concern for safety and their desire to help improve the industry. Each CMP meets two times throughout the code-change process. Members also spend countless hours reviewing submitted proposals.

Technical Correlating Committee (TCC)

Technical correlating committee (TCC) members are the elite subject matter experts of the *NEC*. They continually look at the actions of the code-making panels to ensure they will not create a situation where one part of the *NEC* conflicts with another. They also ensure the code panel work complies with other NFPA standards and the *NEC* style manual.

Code-Change Process

There are four primary steps in the process to change the *NEC*.
1. Public Input Stage
2. Public Comment Stage
3. NFPA Technical Meeting
4. NFPA Standards Council

Public Input Stage

The public input stage is the first stage in the process of revising the *NEC*. It begins immediately after the current edition of the *NEC* is published. The public input stage begins with people submitting *public input*. Public input is a proposed change to the current edition of the *NEC*. Most of the individuals who submit public input are directly involved in the electrical industry, but anyone other than NFPA employees can send a submission. Submissions are completed through the NFPA's website. The code-change process is lengthy, so there is a limited window of time to submit public input. The schedule for all stages of the code-change process is published on the NFPA website.

Public input may suggest a new requirement, edit existing language, or delete a portion of the *Code*. Suggestions should be carefully thought out and offer reasons to support the change. Public input suggestions should use proper language and follow the *NEC* structure. It is more likely that the

Code-Making Panel No. 10

Articles 215, 225, 230, 235, 240, 242, 245

Nathan Philips, *Chair*
Integrated Electronic Systems, OR [IM]
Rep. National Electrical Contractors Association

Richard P. Anderson, Jr., Fluor Corporation, TX [U]
   Rep. Associated Builders & Contractors
Kevin S. Arnold, Eaton's Bussmann Business, MO [M]
Scott A. Blizard, American Electrical Testing Company, Inc., MA [IM]
   Rep. InterNational Electrical Testing Association
Randy Dollar, Siemens Industry, GA [M]
   Rep. National Electrical Manufacturers Association
Ed Koepke, Nidec Motor Corporation, MO [M]
Derrick Robey, Lyondellbasell, TX [U]

Roy K. Sparks, III, Eli Lilly and Company, IN [U]
   Rep. American Chemistry Council
Steven E. Townsend, General Motors Company, MI [U]
   Rep. IEEE-IAS/PES JTCC
Christopher R. Vance, National Grid, NY [UT]
   Rep. Electric Light & Power Group/EEI
David A. Williams, Delta Charter Township, MI [E]
   Rep. International Association of Electrical Inspectors
Danish Zia, UL LLC, NY [RT]

Alternates

Paul D. Barnhart, UL LLC, NC [RT]
   (Alt. to Danish Zia)
Joseph R. Chandler, Independent Electrical Contractors-Dallas, TX [IM]
   (Voting Alt.)
Anthony Dawes, DTE Energy, MI [UT]
   (Alt. to Christopher R. Vance)
Thomas A. Domitrovich, Eaton Corporation, MO [M]
   (Alt. to Kevin S. Arnold)
Richard E. Lofton, II, IBEW Local Union 280, OR [L]
   (Voting Alt.)

Elva Diane Lynch, Sturgeon Electric Company, Inc., CO [IM]
   (Alt. to Nathan Philips)
Alan Manche, Schneider Electric, KY [M]
   (Alt. to Randy Dollar)
David Morrissey, American Electrical Testing Company, Inc., MA [IM]
   (Alt. to Scott A. Blizard)
Douglas C. Smith, West Coast Code Consultants (WC3), UT [E]
   (Alt. to David A. Williams)
Peter R. Walsh, Teaticket Technical Associates, LLC, MA [U]
   (Alt. to Steven E. Townsend)

Jeffrey S. Sargent, NFPA Staff Liaison

Reproduced with permission of NFPA from NFPA 70, National Electrical Code, 2023 edition. Copyright © 2022, National Fire Protection Association. For a full copy of the NFPA 70, please go to www.nfpa.org

Figure 1-4. The first few pages of the *National Electrical Code* have all the code-making panels, their members, and the portions of the code they are responsible for. The classification and organization the panel members represent are written next to their name.

change will continue through the process if the submission is properly written.

After the public input closing date, the NFPA divides the public input among the appropriate code-making panels. The proposals are distributed to the CMP members ahead of the face-to-face meeting so they have a chance to review the information. Task groups are sometimes formed to address complex issues before the meeting.

First Draft Meeting

The first draft meeting is a public meeting at which the CMP members address public input. The panel members vote on all public input proposals that have been submitted. The code-making panels may take the following actions for each public input item.

- Accept the public input as written. No panel explanation is required.
- Accept an edited version of the public input. This is done if the CMP members feel the proposal is a good idea, or they like a portion of it. The CMP rewrites the public input and accepts the edited version. A panel explanation as to why the changes were made must be given.
- Reject the public input. If the proposal is rejected, the CMP must provide an explanation as to why it was rejected.

> **PRO TIP — NEC Style Manual**
>
> The **NEC style manual** is a writing tool for the *National Electrical Code*. Its purpose is to ensure the language and structure of the *NEC* are explicit, consistent, and promote uniform interpretation. It describes and gives examples of the specific language that is to be used, how the outline format is to be structured, units of measurement, standard electrical terms and their spelling, and much more. The *NEC* style manual is available to download from the NFPA website.

Committee Membership Classifications[1,2,3,4]

The following classifications apply to Committee members and represent their principal interest in the activity of the Committee.

1. **M** *Manufacturer:* A representative of a maker or marketer of a product, assembly, or system, or portion thereof, that is affected by the standard.
2. **U** *User:* A representative of an entity that is subject to the provisions of the standard or that voluntarily uses the standard.
3. **IM** *Installer/Maintainer:* A representative of an entity that is in the business of installing or maintaining a product, assembly, or system affected by the standard.
4. **L** *Labor:* A labor representative or employee concerned with safety in the workplace.
5. **RT** *Applied Research/Testing Laboratory:* A representative of an independent testing laboratory or independent applied research organization that promulgates and/or enforces standards.
6. **E** *Enforcing Authority:* A representative of an agency or an organization that promulgates and/or enforces standards.
7. **I** *Insurance:* A representative of an insurance company, broker, agent, bureau, or inspection agency.
8. **C** *Consumer:* A person who is or represents the ultimate purchaser of a product, system, or service affected by the standard, but who is not included in (2).
9. **SE** *Special Expert:* A person not representing (1) through (8), and who has special expertise in the scope of the standard or portion thereof.

Reproduced with permission of NFPA from NFPA 70, National Electrical Code, 2023 edition. Copyright © 2022, National Fire Protection Association. For a full copy of the NFPA 70, please go to www.nfpa.org

Figure 1-5. The committee membership classification description is in the back of the *National Electrical Code*, after the index. One classification not included in the image that some *NEC* code-making panels may have is "UT," which stand for utility. Utility representatives are on code-making panels that are responsible for portions of the code that may impact the serving utilities.

For public input to be accepted and continue as a potential change, it must pass with a simple majority vote. A simple majority vote is where the number of votes in support of a change exceeds the number of votes rejecting it.

After the first draft meeting, the CMPs develop ***first revisions***. A first revision is an edited portion of the *Code* rewritten to include the accepted public input.

First Draft Ballot

After the first draft meeting, CMP members are given some time to consider, discuss, and research the first revisions. After a designated period, a voting ballot is sent to each member. The first revisions must receive a 2/3 vote on the ballot for it to continue as accepted change.

First Draft Report

As the public input stage is completed, the ***first draft report*** is published. It contains the first revisions, the public input, and the CMP actions and statements. This is the first opportunity for the general public to see what the *NEC* will look like if all the initial code-making panel actions are upheld.

Public Comment Stage

The ***public comment*** stage of the code-change process is much like the public input stage. It begins with the general public sending in public comments about the first draft report. The public comments may only address first revisions and actions taken on public inputs. If a comment introduces new material, it will appear in the next revision cycle as public input. The public comments may support or oppose panel actions. After the period to accept comments has ended, the NFPA collates the comments and sends them to the appropriate code-making panels.

Second Draft Meeting

The CMPs hold a ***second draft meeting*** to act on and provide a response to the public comments. The CMP will use the first revisions and the public comments to put together the ***second revisions***. Each of the proposed changes in the second revision must receive a simple majority vote at the meeting to pass. A second revision is the edited portion of the *Code* rewritten to reflect the changes from the second draft meeting. It shows how a specific section of the *NEC* will look if the change is to continue as accepted.

Second Revision Ballot

After the second draft meeting, panel members are given some time to consider, discuss, and research the second revisions. After a designated period, a ballot is sent to each of the code-panel members so they can vote again on all the second revisions. Each revision must receive a 2/3 vote on the ballot for it to continue as an accepted change.

Second Draft Report

As the public comment stage is completed, the second draft report is published. It contains the second revisions, public comments, and the CMP actions and responses.

NFPA Technical Meeting

The NFPA technical meeting is held during the annual NFPA Conference and Expo. Every third year, when the meeting follows the comment stage of the code-change process, the

NFPA membership has the opportunity discuss, debate, and vote on certified amending motions.

A certified amending motion provides the next opportunity to amend or reverse the code-making panel actions. For the motion to become a certified amending motion, a notice of intent to make a motion (NITMAM) must be filed. A NITMAM must be filed by the published deadline and be approved by the motions committee. Upon motion committee approval, it becomes a certified amending motion.

Certified amending motions need a simple majority vote to pass. Motions that pass are returned to the respective code-making panel for further review.

NFPA Standards Council

The last opportunity to reverse a code-making panel action is by appealing to the NFPA standards council. The standards council is a group of individuals that oversee NFPA standards development activities. They ensure compliance with NFPA rules and regulations and serve as the appeals body for standards development matters. The standards council will look at the appeals and hold hearings if necessary. After deciding on any appeals, the standards council issues the next edition of the *NEC*.

Tentative Interim Amendment (TIA)

Occasionally it becomes necessary to make an immediate change to an *NEC* requirement. In this situation, the standards council will issue a tentative interim amendment (TIA). A tentative interim amendment is a revision that is implemented between *Code* cycles. TIAs are often issued in response to a change in the *NEC* that results in an unforeseen problem. Tentative interim amendments will only be in place until the next edition of the *NEC* is published. TIAs will show up as a public input in the upcoming first stage of the code-change process.

PRO TIP — **National Electrical Code Development Process**

There are four primary steps in the process of revising the *National Electrical Code*. Each step has a process of its own with important dates that must be adhered to. The NFPA website publishes all the dates for the upcoming code cycle. See **Figure 1-6**.

The National Electrical Code Development Process

Last edition published

Step 1—Public Input Stage	Step 2—Public Comment Stage	Step 3—NFPA Technical Meeting	Step 4—Council Appeals and Issuance of Standard
Time period to submit public input	Comment closing date	Time period to submit appeal to the council	
First draft meeting	Second draft meeting		
First draft ballot	Second draft ballot		
First draft report posted	Second draft report posted		
	Time period to submit a NITMAM		
	NITMAM received and certified		

Goodheart-Willcox Publisher

Figure 1-6. A chart showing the standards development process for the *National Electrical Code*.

Summary

- The National Fire Protection Association (NFPA) has been publishing the *National Electrical Code* since 1897.
- The *National Electrical Code* is a consensus standard.
- There are four steps in standards development process to change the *NEC*:
  - Public Input Stage
  - Public Comment Stage
  - NFPA Technical Meeting
  - NFPA Standards Council
- The public input stage gives the general public an opportunity to submit code-change proposals.
- The public comment stage gives the general public an opportunity to comment on actions taken by the code-making panels in the first draft report.
- The NFPA membership will vote on certified amending motions at the NFPA technical meeting.
- The standards council issues the next edition of the *National Electrical Code*.

Unit 1 Review

Name _____ Date _____ Class _____

Know and Understand

Answer the following questions based on information in this unit.

1. The *NEC* is published by the _____.
   A. Occupational Safety and Health Administration
   B. National Electrical Manufacturers Association
   C. International Association of Electrical Inspectors
   D. National Fire Protection Association

2. The *NEC* has _____ code-making panels.
   A. 1
   B. 9
   C. 18
   D. 19

3. _____ can submit a public input to change the *National Electrical Code*.
   A. Only NFPA members
   B. Anyone but NFPA employees
   C. Electrical inspectors only
   D. None of the above.

4. All actions other than to _____ public input requires the code-making panel to give a written explanation.
   A. accept
   B. reject
   C. table
   D. rewrite

5. When voting on a public input at the first draft meeting, a _____ vote is required for a proposal to be included in the first revision.
   A. simple majority
   B. 2/3 majority
   C. 3/4 majority
   D. unanimous

6. When the code-panel members are voting by ballot on the second revision, a _____ vote is required to have continued acceptance.
   A. simple majority
   B. 2/3 majority
   C. 3/4 majority
   D. unanimous

7. At the NFPA technical meeting, the membership can vote on any _____ that are presented.
   A. NITMAMs
   B. public input
   C. public comments
   D. certified amending motions

8. A _____ is a temporary revision to the *NEC* that is implemented between *Code* cycles.
   A. NITMAM
   B. CMP
   C. TIA
   D. TCC

9. The _____ is the final step in the standards development process.
   A. public input stage
   B. NFPA standards council
   C. NFPA annual meeting
   D. public comment stage

10. The _____ is a consensus standard concerned with worker safety while installing electrical systems.
    A. NFPA 70
    B. NFPA 70E
    C. OSHA
    D. NEMA

UNIT 2

Structure of the NEC

sockagphoto/Shutterstock.com

KEY TERMS
articles
chapter
part
section
general chapters
rough-in
trim
special chapters

LEARNING OBJECTIVES

After completing this unit, you will be able to:
- Describe the outline format of the *NEC*.
- Describe the layout of the *NEC*.
- List the chapters of the *NEC*.
- Identify the focus of each of the *NEC* chapters.
- Describe how *NEC* changes are identified.

Introduction

The key to becoming a proficient *Code* user is thoroughly understanding its structure. By understanding the layout of the *National Electrical Code*, it becomes much easier to find information. This unit will explain in detail how the *NEC* is structured and the way it is intended to be used.

NEC Outline Format

Chapter (Title in Bold Font)
    Article (Title in Bold Font)
        Part (Title in Bold Font)
            Section (Title in Bold Font)
                (A) (Subdivision 1—Uppercase Letters, Title in Bold Font)
                    (1) (Subdivision 2—Numbers, Title in Bold Font)
                        a. (*Subdivision 3—Lowercase Letters, Title in Italics*)
        1: (List of Items—Lowercase Letters, Immediately Follow a Colon)
      Exception (Numbered if there are more than one, All Italics)

Goodheart-Willcox Publisher

Figure 2-1. The *NEC* is written in an outline format. This figure shows the breakdown for *NEC Section 250.184(B)(8)a*.

Outline Format

The *National Electrical Code* is written in an outline format, **Figure 2-1**. The *NEC* is first divided into **chapters**. Chapters are divided into **articles**, and most articles are divided into **parts**. Articles and their parts are divided into **sections**. The sections are permitted to be divided into up to three levels of subdivision. The *NEC* style manual dictates the hierarchy of the *Code* and does not permit sections to have more than three levels of subdivision.

> **PRO TIP**   **Code Hierarchy**
>
> A common mistake made by new code users is to stumble across code language that sounds like it is the correct information without verifying what part of the code they are looking at. For example, a person working on a 120/240-volt single-phase system is sizing the neutral conductor. They find language in *Section 250.184(A)(2)* that appears to be what they are looking for. See **Figure 2-2**. Looking back at which part of *Article 250* the requirement is in will reveal that the answer is in *Part X*, which is for systems over 1000 volts. That section does not apply to this situation.
>
> **Part X. Grounding of Systems and Circuits of over 1000 Volts**
>
> **250.180 General.** If systems over 1000 volts are grounded, they shall comply with all applicable requirements of 250.1 through 250.178 and with 250.182 through 250.194, which supplement and modify the preceding sections.
>
> **250.182 Derived Neutral Systems.** A system neutral point derived from a grounding transformer shall be permitted to be used for grounding systems over 1 kV.
>
> *Reproduced with permission of NFPA from NFPA 70, National Electrical Code, 2023 edition. Copyright © 2022, National Fire Protection Association. For a full copy of the NFPA 70, please go to www.nfpa.org*

Figure 2-2. Care must be taken to be sure you are in the correct section or part. *Part X* of *Article 250* is only for systems over 1000V.

When there are multiple items that pertain to a requirement, they are included as a list of items. They are permitted to be incorporated into the sections as well as any of the levels of subdivision. A list of items does not have a title and will immediately follow the section or subdivision for which it applies.

Any requirement in the *NEC* applies to only the specific portion of the *Code* that contains the requirement. Requirements cannot be applied generally. For example, a specific requirement found in a subdivision only applies under the constraints of the subdivision, section, part, and article that precedes it. Be careful to not apply something found in a subdivision broadly across the *NEC* as it only applies to that specific portion of the *NEC* hierarchy in which it appears.

Finding a requirement in the wrong location of the *Code* is a common mistake, especially when using the index. The index often gives a list of several sections that could have a possible answer. Be sure to refer to the outline hierarchy to ensure the specific requirement being applied is under the correct section, part, and article.

Layout

Understanding the layout of the *NEC* is one of the most important steps to becoming a proficient code user. The *NEC* has been designed and laid out in a manner to make it easier to find information; the layout is detailed in *Section 90.3*.

The *National Electrical Code* is divided into an introduction, 9 chapters, and an annex, **Figure 2-3**.

General Chapters

Chapters 1–4 are known as the **general chapters**. They are general in nature, apply to all installations, and cover most of the day-to-day requirements an electrician will encounter.

Chapter 1, General

Chapter 1, General is the first of the of the *NEC*'s general chapters. It contains information that applies to all

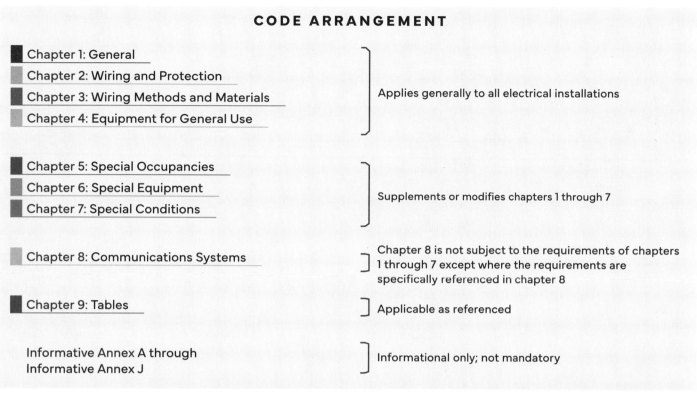

Figure 2-3. Having a thorough understanding of the layout of the *National Electrical Code* will make finding information much easier.

installations and situations. It consists of only two articles: *Article 100* and *Article 110*.

Article 100, Definitions contains definitions that must be understood to properly apply the *NEC*. *Article 110, General Requirements for Electrical Installations* consists of the basic requirements that apply to nearly everything an electrician does.

Chapter 2, Wiring and Protection

Chapter 2, Wiring and Protection is the second of the *NEC*'s general chapters. It has the information necessary to design an electrical installation as well as the requirements to protect it, **Figure 2-4**.

When the electrical project manager or estimator sits down to design the service, feeders, or branch circuits, they will find most of the relevant information in *Chapter 2*. Additionally, systems that offer protection, such as grounding, surge protection, and overcurrent protection can also be found here.

Chapter 3, Wiring Methods and Materials

Chapter 3, Wiring Methods and Materials is third of the *NEC* general chapters. *Chapter 3* has articles that address all the various wiring methods. One way to remember the general type of information contained in *Chapter 3* is it has many of the requirements followed while roughing-in an electrical installation. The electrical **rough-in** occurs after the framing crew has finished. The electrician will install boxes, raceways, cables, and conductors. All the boxes and wiring in the walls must be installed before the finished wall surface can be installed, **Figure 2-5**.

A few examples of wiring materials covered in *Chapter 3* include electrical boxes, cables, raceways, and conductors. Although not every requirement found in *Chapter 3* is related to roughing-in, this is one simple way to remember the general type of information found within the chapter.

Chapter 4, Equipment for General Use

Chapter 4, Equipment for General Use is the fourth and final of the *NEC*'s general chapters. It contains requirements for electrical equipment that control, utilize, create, or store electricity. This equipment is often installed during the trim stage of the construction process. The **trim** stage is when an electrician goes back to install the switches, outlets, lights, and other electrical components after the finished wall surface has been installed. See **Figure 2-6**.

A few examples of equipment covered in *Chapter 4* include luminaries, appliances, motors, panelboards, and batteries. Although not every requirement found in *Chapter 4* is related to trimming out a job, it a good way to remember the general type of information found within the chapter.

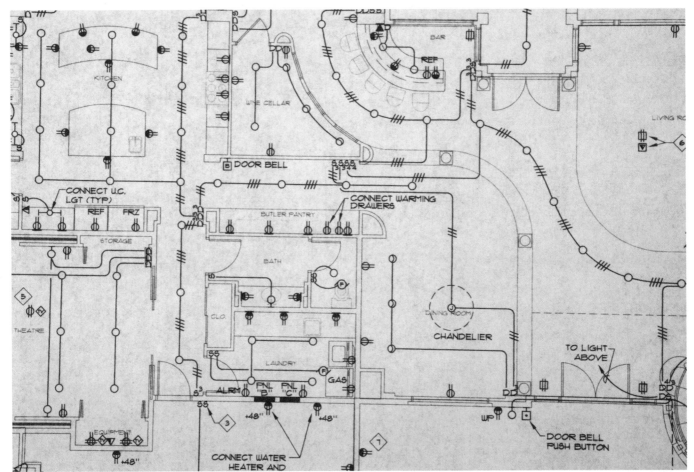

Figure 2-4. *Chapter 2* of the *NEC* has requirements for designing the electrical system.

Figure 2-5. *Chapter 3* of the *NEC* has requirements for the rough-in stage of a construction project. The *Chapter 3* wiring methods shown in this picture include nonmetallic sheathed cable, boxes, and electrical nonmetallic tubing.

Special Chapters

Chapters 5–7 are known as the **special chapters**. They supplement or modify the requirements found in other parts of the *NEC*. *Chapters 5–7* have rules that apply to special occupancies, equipment, and conditions that are not found on every jobsite. *Chapters 1–4* still pertain as they apply to all installations, but the special chapters have additional requirements that apply to those special circumstances.

Chapter 5, Special Occupancies

Chapter 5, Special Occupancies is the first of the *NEC*'s special chapters. Special occupancies have situations and hazards that are not common to every electrical installation. A few examples of special occupancies are hazardous locations, aircraft hangars, health care facilities, theaters, and motor fuel dispensing stations. See **Figure 2-7**.

Chapter 6, Special Equipment

Chapter 6, Special Equipment is the second of the *NEC*'s special chapters. The special equipment found in *Chapter 6* is not common to every electrical installation. A few examples of special equipment are cranes, elevators,

Figure 2-6. *Chapter 4* of the *NEC* has requirements for the trim stage of a construction project. The *Chapter 4* equipment shown in this picture include recessed lights, switches, and receptacles.

welders, pipe organs, X-ray equipment, and pool pumps. See **Figure 2-8**.

Chapter 7, Special Conditions

Chapter 7, Special Conditions is the third and final of the *NEC*'s special chapters. Special conditions are systems and circuits that are not common to every electrical installation. A few examples of special conditions are emergency systems, power-limited circuits, and fire alarm systems. See **Figure 2-9**.

Chapter 8, Communication Systems

Chapter 8, Communication Systems is different than the rest of the *NEC*, as its requirements typically stand alone and the rest of the *NEC* does not apply. The requirements in *Chapters 1–7* only apply to a communication system when it is specifically referenced within *Chapter 8*. The type of information in *Chapter 8* concerns communication systems, such as radio and TV equipment, antennas, and broadband systems. See **Figure 2-10**.

Figure 2-7. *Chapter 5* of the *NEC* has requirements for special occupancies. Motor fuel dispensing facilities are considered a special occupancy.

Figure 2-8. *Chapter 6* of the *NEC* has requirements for special equipment. A hydromassage tub is an example of special equipment.

Nuroon Jampaklai/Shutterstock.com

Figure 2-9. *Chapter 7* of the *NEC* has requirements for special conditions. Requirements for fire alarm systems are included in *Chapter 7*.

BCFC/Shutterstock.com

Figure 2-10. *Chapter 8* of the *NEC* has requirements for communication systems. Satellite dishes have requirements in *Chapter 8*.

NEC Chapter 9 Tables

Chapter 9, Tables is composed of reference tables. Most of the tables address conductors and raceways. The tables are only applicable when referenced by another part of the *Code*.

Informative Annexes

The *Informative Annexes* immediately follow *Chapter 9*. They contain additional information to aid the *Code* user but are not an enforceable part of the *NEC*. The *Informative Annexes* contain information such as product safety standards, sample calculations, and types of construction.

National Electrical Code Revisions

Every three years a new edition of the *NEC* is published. The *NEC* is far too vast for a person to be able to simply read it through and know what has changed. For this reason, the *NEC* gives visual cues to help identify portions of the code that are new, altered, or deleted.

Revision Symbols

The *NEC* uses several methods to help identify altered portions of the code. The revision symbols alert the *Code* user that a requirement has been altered from the previous edition and must be read carefully to ensure the new requirement is being followed. The *NEC* has a legend at the bottom of each page describing each of the revision symbols. See **Figure 2-11**.

Shaded Text

Text revisions are shaded, **Figure 2-12**. The revision may be as little as one word or up to several paragraphs. The shaded text identifies there are new or altered words within the shaded area.

Shaded Triangle Δ

Shaded triangles, **Figure 2-13**, have various meanings depending on where they appear. If the shaded triangle Δ appears before a section number, it means words within that section were deleted. A shaded triangle Δ next to a table or figure indicates the table or figure has been revised. If a shaded triangle Δ is marked throughout article, it means there was heavy revisions throughout. Using the triangle Δ in this fashion prevents the entire article from being shaded.

Shaded text = Revisions.   Δ = Text deletions and figure/table revisions.   • = Section deletions.   **N** = New material.

Reproduced with permission of NFPA from NFPA 70, National Electrical Code, 2023 edition. Copyright © 2022, National Fire Protection Association. For a full copy of the NFPA 70, please go to www.nfpa.org

Figure 2-11. The *NEC* has a legend printed at the bottom of each page showing what each of the revision symbols represents.

Part II. 1000 Volts, Nominal, or Less

110.26 Spaces About Electrical Equipment. Working space, and access to and egress from working space, shall be provided and maintained about all electrical equipment to permit ready and safe operation and maintenance of such equipment. Open equipment doors shall not impede access to and egress from the working space. Access or egress is impeded if one or more simultaneously opened equipment doors restrict working space access to be less than 610 mm (24 in.) wide and 2.0 m (6½ ft) high.

(A) Working Space. Working space for equipment operating at 1000 volts, nominal, or less to ground and likely to require examination, adjustment, servicing, or maintenance while energized shall comply with the dimensions of 110.26(A)(1), (A)(2), (A)(3), and (A)(4) or as required or permitted elsewhere in this *Code*.

Informational Note: See NFPA 70E-2021, *standard for Electrical Safety in the Workplace,* for guidance, such as, determining severity

Reproduced with permission of NFPA from NFPA 70, National Electrical Code, 2023 edition. Copyright © 2022, National Fire Protection Association. For a full copy of the NFPA 70, please go to www.nfpa.org

Figure 2-12. The shaded text indicates that there have been revisions to this section.

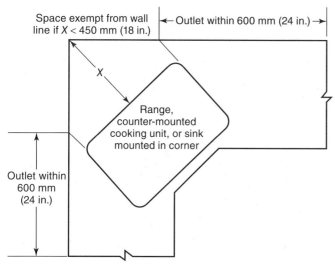

△ **Figure 210.52(C)(1)** Determination of Area Behind a Range, Counter-Mounted Cooking Unit, or Sink.

Reproduced with permission of NFPA from NFPA 70, National Electrical Code, 2023 edition. Copyright © 2022, National Fire Protection Association. For a full copy of the NFPA 70, please go to www.nfpa.org

Figure 2-13. The shaded triangle indicates that this figure has been revised.

Shaded N

A shaded letter N, **Figure 2-14**, is used to indicated that there is new language. It may be a new section, figure, table, or even chapter.

Dot •

A dot •, **Figure 2-15**, is used to identify entire portions of the *NEC* that have been removed. It may be an entire section, part, or article. When seeing a dot, it is common to review the previous edition to see exactly what language was removed.

> **PRO TIP**   **Code Changes**
>
> It may seem unimportant for new *Code* users to know past requirements while learning how to use the *NEC*. This is true to an extent, as it would be overwhelming to learn all the most current requirements, as well as all the old requirements. There is, however, value in understanding how the *Code* has evolved. Throughout an electrician's career, they are likely to come across wiring that is not legal according to the current standard. That doesn't necessarily mean it was wired incorrectly: the wiring may have been permitted by the edition of the *NEC* that was in effect when it was installed.
>
> Having knowledge of how *NEC* requirements have evolved is especially important when remodeling and troubleshooting older systems. Most of that knowledge will be absorbed throughout an electrician's career and is not something to be concerned about as an apprentice.

N **215.18 Surge Protection.**

N **(A) Surge-Protective Device.** Where a feeder supplies any of the following, a surge-protective device (SPD) shall be installed:

(1) Dwelling units
(2) Dormitory units
(3) Guest rooms and guest suites of hotels and motels
(4) Areas of nursing homes and limited-care facilities used exclusively as patient sleeping rooms

N **(B) Location.** The SPD shall be installed in or adjacent to distribution equipment, connected to the load side of the feeder, that contains branch circuit overcurrent protective device(s) that supply the locations specified in 215.18(A).

Reproduced with permission of NFPA from NFPA 70, National Electrical Code, 2023 edition. Copyright © 2022, National Fire Protection Association. For a full copy of the NFPA 70, please go to www.nfpa.org

Figure 2-14. The shaded N printed next to the section number indicates it is new material.

▲ **305.6 Protection Against Induction Heating.** Metallic raceways and associated conductors shall be arranged to avoid heating of the raceway in accordance with 300.20.

• **305.7 Covers Required.** Suitable covers shall be installed on all boxes, fittings, and similar enclosures to prevent accidental contact with energized parts or physical damage to parts or insulation.

• **305.8 Raceways in Wet Location Above Grade.** Where raceways are installed in wet locations above grade, the interior of these raceways shall be considered to be a wet location. Insulated conductors and cables installed in raceways in wet locations above grade shall be either moisture-impervious metal-sheathed or of a type listed for use in wet locations.

Reproduced with permission of NFPA from NFPA 70, National Electrical Code, 2023 edition. Copyright © 2022, National Fire Protection Association. For a full copy of the NFPA 70, please go to www.nfpa.org

Figure 2-15. The dot indicates the location where a section of the *NEC* has been deleted.

Significant Changes Class

It would be time-consuming and difficult to find all the changes in the *NEC* on your own. Most electricians will attend a class that details the significant changes in the *National Electrical Code*. In addition to discussing the changes, the classes often will give background as to why the change was necessary. Most states or jurisdictions have continuing education requirements. Taking a significant changes class is a great way to learn the new information in the *Code*, while at the same time satisfying requirements to renew a license.

Various publishers produce books that detail the major changes in the *NEC*. They will typically show the old language, the new language, a corresponding picture, and detail the substantiation for the change.

PRO TIP — **NEC Handbook**

The NFPA publishes an *NEC Handbook* to help with proper interpretation and application of the *Code*. It contains the *NEC* in its entirety as well as pictures, diagrams, explanations, and examples. One helpful feature is a *Summary of Technical Changes* that is listed in the front. It contains a list of sections that have changed, a short description of what changed, as well as reference information on where to gain additional information. See **Figure 2-16**.

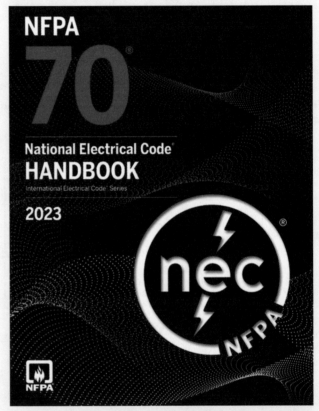

Reproduced with permission of NFPA from NFPA 70, National Electrical Code, 2023 edition. Copyright © 2022, National Fire Protection Association. For a full copy of the NFPA 70, please go to www.nfpa.org

Figure 2-16. The *NEC Handbook* contains the *National Electrical Code* as well as explanations, picture, diagrams, and much more.

Summary

- The *NEC* is written in an outline format.
- *Chapters 1–4* of the *NEC* are the general chapters. They apply to all installations.
- *Chapters 5–7* of the *NEC* are the special chapters. They supplement or modify the requirements of *Chapters 1–4*.
- *Chapter 8* of the *NEC* covers communication systems and stands alone.
- *Chapter 9* of the *NEC* contains tables that are only applicable as referenced by other parts of the *Code*.
- The *Informative Annexes* follow *Chapter 9* and are for informational purposes only and are not an enforceable part of the *Code*.
- The *NEC* uses symbols to help identify portions of the *Code* that have been revised, deleted, or added.

Unit 2 Review

Name _____ Date _____ Class _____

Know and Understand

Answer the following questions based on information in this unit.

1. The *NEC* style manual permits sections to have up to _____ levels of subdivisions.
   A. two
   B. three
   C. four
   D. five
2. The *NEC* is divided into _____ chapters.
   A. four
   B. seven
   C. eight
   D. nine
3. _____ of the *NEC* is referred to as the rough-in chapter.
   A. *Chapter 1*
   B. *Chapter 2*
   C. *Chapter 3*
   D. *Chapter 4*
4. _____ will supplement or modify the requirements found in *Chapters 1–7*.
   A. *Chapter 1*
   B. *Chapter 3*
   C. *Chapter 5*
   D. *Chapter 8*
5. What symbol is used to represent new *Code* language?
   A. Underlined text
   B. Dot (•)
   C. Shaded triangle (▲)
   D. Shaded text
6. A new edition of the *NEC* is published every _____ years.
   A. two
   B. three
   C. five
   D. six
7. The _____ stage in the construction process is where an electrician installs the switches and lights.
   A. trim
   B. rough-in
   C. service
   D. estimate
8. X-ray equipment falls under the classification of special equipment, which is found in _____.
   A. *Chapter 2*
   B. *Chapter 4*
   C. *Chapter 5*
   D. *Chapter 6*
9. Which of the following is considered one of the general chapters?
   A. *Chapter 4*
   B. *Chapter 6*
   C. *Chapter 8*
   D. *Chapter 9*
10. Which *NEC* chapter has overcurrent protection requirements?
    A. *Chapter 1*
    B. *Chapter 2*
    C. *Chapter 3*
    D. *Chapter 4*

Apply and Analyze

Answer the following questions using a copy of the National Electrical Code. *Identify the section or subsection where the answer is found.*

1. _____ are known as the general chapters which apply to all electrical installations.
   A. *Chapters 1–2*
   B. *Chapters 1–4*
   C. *Chapters 5–7*
   D. *Chapters 1–7*

   NEC _____

2. The informative annexes contain _____ relative to the use of the *NEC*.
   A. nonmandatory information
   B. supplemental information
   C. general requirements
   D. special requirements

   NEC _____

3. The tables in *Chapter 9* are _____.
   A. for informational purposes only
   B. not subject to the requirements of *Chapters 1–7*
   C. applicable as referenced
   D. cross-reference tables

   NEC _____

4. *Chapter 5* has requirements relating to _____.
   A. equipment for general use
   B. special equipment
   C. communications systems
   D. special occupancies

   NEC _____

5. _____ has requirements relating to equipment for general use.
   A. *Chapter 1*
   B. *Chapter 2*
   C. *Chapter 3*
   D. *Chapter 4*

   NEC _____

6. *Chapters 1–4* are supplemented or modified by _____.
   A. *Chapters 5–7*
   B. *Chapter 8*
   C. *Chapter 9*
   D. *Chapters 8–9*

   NEC _____

7. There are _____ informative annexes.
   A. 2
   B. 6
   C. 11
   D. 13

   NEC _____

8. _____ has requirements relating to wiring methods and materials.
   A. *Chapter 1*
   B. *Chapter 3*
   C. *Chapter 5*
   D. *Chapter 8*

   NEC _____

9. Requirements for communication systems are found in _____.
   A. *Chapter 5*
   B. *Chapter 6*
   C. *Chapter 7*
   D. *Chapter 8*

   NEC _____

10. _____ is not subject to the requirements of *Chapters 1–7* unless specifically referenced within the chapter.
    A. *Chapter 5*
    B. *Chapter 7*
    C. *Chapter 8*
    D. *Chapter 9*

    NEC _____

UNIT

3 Preparing to Use the NEC

sockagphoto/Shutterstock.com

KEY TERMS

mandatory language
permissive language
exception
informational note
keyword

LEARNING OBJECTIVES

After completing this unit, you will be able to:
- Describe the type of information in *Article 90*.
- Identify electrical work under the jurisdiction of the *NEC*.
- Describe the *NEC* arrangement.
- Explain the *NEC* enforcement.
- Interpret the language used in the *NEC*.
- Identify keywords in an *NEC* inquiry.
- Demonstrate two methods of finding information in the *NEC*.
- Describe other tools a *Code* can use to find information in the *NEC*.

Introduction

Preparing to use the *National Electrical Code* includes understanding the structure, language, and how and where it is enforced. This unit will cover *Article 90*, the introduction to the *National Electrical Code*. In addition, this unit will discuss two methods of finding information in the *NEC*.

Article 90

Article 90 is the introduction to the *National Electrical Code*. It explains the purpose of the *NEC*, how it is meant to be used, the language, and much more. *Article 90* does not have specific requirements, so it is not an article that will be looked at on a regular basis.

This unit will cover some of *Article 90*'s sections and will expand on them to stress the key information. This information is not meant to replace reading *Article 90*, it is intended to supplement it.

Section 90.2, Use and Application

Section 90.2(A) states that the purpose of the *Code* is the practical safeguarding of persons and property from the hazards arising from the use of electricity. The *NEC* aims to provide a completed installation that is free from electrical hazards.

Section 90.2(B) discusses adequacy. It says wiring according to *NEC* standards will not necessarily be convenient, the least expensive, or have future expansion in mind, it will simply ensure a safe installation. *Section 90.8* expands on the information in *Section 90.2(B)* by suggesting that leaving extra space in panelboards, boxes, and raceways will allow for future expansion. This is a statement only, not a requirement. See **Figure 3-1**.

Section 90.2(C) lists the installations that are covered by the *NEC*, while *Section 90.2(D)* lists the installations that are not covered by the *NEC*.

Most of the work electricians perform on public and private buildings is covered by the *NEC*. A few areas worth noting that are not covered are watercraft, automobiles, underground mines, railways, and electric utility generation and distribution equipment lines. See **Figure 3-2**.

Section 90.3, Code Arrangement

Section 90.3 contains information that will be repeated throughout the text and is the key to understanding where information can be found in the *NEC*. *Section 90.3* gives a written description of how the *NEC* has been structured. *Figure 90.3* gives a visual representation of the *Code* structure. See **Figure 3-3**.

> **PRO TIP** — **Article 90**
>
> One of the first things you should do as a new *Code* user is read *Article 90* in its entirety. There are not many parts of the *NEC* that should simply be read from start to finish, except for *Article 90*. Doing so will help you have a thorough understanding of how the *NEC* is set up as well as how it is meant to be used.

Goodheart-Willcox Publisher

Figure 3-1. This panel does not have any spaces left for additional circuits and has been filled with tandem breakers for additional spaces. Manufacturers often limit the number and placement of tandem breakers, likely making this installation a violation. Although it is not a requirement, *Section 90.8* states that leaving extra room allows for future expansion.

Yerv/Shutterstock.com

Figure 3-2. Utility generation equipment, substations, and power lines are not under the jurisdiction of the *NEC*. They follow standards, such as the National Electrical Safety Code, that are better suited to their installations.

Chapters 1–4

Chapters 1–4 of the *NEC* are the general chapters. They apply to all electrical installations, from a small dwelling unit to a manufacturing facility.

Chapter 1, General contains information that applies to all situations and installations. *Chapter 2, Wiring and Protection* can also be thought of as design and protection. It contains requirements that will be necessary when designing

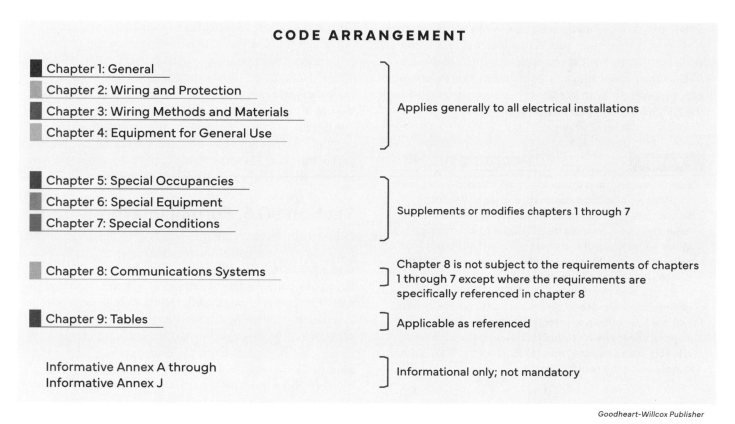

Figure 3-3. *NEC Figure 90.3* describes the structure of the *Code*.

and laying out the electrical system. It also contains requirements for electrical protection. *Chapter 3, Wiring Methods and Materials* contains many of the requirements necessary when installing wiring methods during the rough-in stage of the construction process. *Chapter 4, Equipment for General Use* discusses the equipment being installed while trimming or finishing a job.

Chapters 5-7

Chapters 5-7 of the *NEC* are the special chapters. They supplement or modify the requirements found in other parts of the *NEC*. All the requirements of *Chapters 1-4* still apply, but *Chapters 5-7* have additional information that applies to these special situations.

Chapter 8, Communications Systems

Chapter 8 of the *NEC* covers communication systems. *Chapter 8* is different than the rest of the *Code* as its requirements typically stand alone, and the rest of the *NEC* does not apply.

Chapter 9, Tables

Chapter 9 contains tables that are only applicable when referenced by another part of the *Code*. The tables that are most often referenced are used while calculating the maximum number of conductors that are permitted in a raceway.

Section 90.4, Enforcement

Section 90.4(A) addresses the enforcement of the *NEC*. It states the *NEC* is a document that is suitable to be adopted and enforced by a governing body. *Section 90.4(B)* gives the power to enforce and interpret the *NEC* to the authority having jurisdiction (AHJ). See **Figure 3-4**. In most areas, the AHJ is the electrical inspector. *Section 90.4(C)* also gives the inspector permission to waive requirements if the alternate method of completing the installation satisfies the objective and is deemed safe.

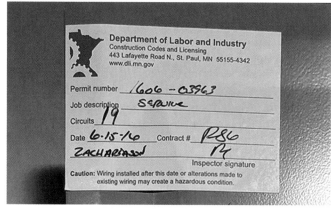

Figure 3-4. The authority having jurisdiction in most areas is the electrical inspector. The sticker on the panel indicates the date of the inspection, as well as the name of the inspector.

Situations occasionally arise where a new edition of the *NEC* requires a certain product or material that is not yet available. In this situation, *Section 90.4(D)* identifies the AHJ as the person who can grant permission to use products that comply with the most recent previous version of the *NEC* adopted in that jurisdiction.

> **PRO TIP** — **Adopting the NEC**
>
> Each state, city, and municipality will choose if they want to adopt an electrical standard, and if so, which one. Some states have a statewide adoption while others leave the decision up to the city, county, or municipality. While some parts of the country adopt the most current edition of the *NEC* shortly after it is released, others chose to use one of the previous editions.
>
> In addition to adopting the *NEC*, the adopting agency may choose to amend it. For example, North Dakota has written an electrical standard that supplements the *National Electrical Code*. Some of the requirements are more strict than the *NEC*, while others relax the requirements.

Section 90.5, Mandatory Rules, Permissive Rules, and Explanatory Material

Section 90.5 describes the language used within the *NEC*. Although *Section 90.5* describes the language used, the *NEC* style manual mandates the language used within the *Code*.

Language

Mandatory language is written using the words *shall* or *shall not*. Mandatory language is used when a specific action is required or prohibited. For example, where subject to physical damage, conductors, raceways, and cables shall be protected (*Section 300.4*).

Permissive language is written using the words *shall be permitted* or *shall not be required*. Permissive language means an action is allowed, but not required. For example, receptacles on rooftops shall not be required to be readily accessible other than from the rooftop (*Section 210.8(B) Exception No. 2*).

Exceptions are used to identify situations in which a requirement does not apply. They are placed immediately following the requirement for which the exception applies. An effort has been made to use exceptions as sparingly as possible by carefully writing the requirements so they aren't necessary. When this isn't possible, exceptions are used.

Explanatory Material and Informative Annexes

Explanatory material comes in the form of informational notes. ***Informational notes*** are scattered throughout the *Code* and are there to provide additional information. They immediately follow the section to which they apply.

Informational notes may contain references to other areas of the *NEC*, other standards, examples, or explanations. Informational notes are not an enforceable part of the *Code*, they are simply there to help the *Code* user. See **Figure 3-5**.

Like the informational notes, the *Informative Annexes* are there to help the *Code* user and are not an enforceable part of the *Code*. The annexes contain cross-reference tables, raceway fill tables, sample calculations, and more.

Section 90.6, Formal Interpretations

Occasionally there are disagreements as to the intent of the *Code*. To promote uniformity of interpretation and application, *Section 90.6* identifies a process for a formal interpretation of whether a specific situation is a *NEC* violation or not. This process is found on the NFPA website under NFPA Regulations Governing Committee Projects. The question/situation will be forwarded to the technical committee that is responsible for the part of the *Code* that applies. The question must be clearly written and have all the necessary information included so the CMP can answer with a simple yes or no.

Section 90.7 Examination of Equipment for Safety

All electrical equipment being installed must be approved. The authority having jurisdiction is responsible for that approval. *Section 90.7* gives information about the process of examining equipment for safety. It also mentions if a piece of equipment has been listed by a recognized testing agency that examination should not be necessary unless it appears to have been tampered with.

210.4 Multiwire Branch Circuits.

(A) General. Branch circuits recognized by this article shall be permitted as multiwire circuits. A multiwire circuit shall be permitted to be considered as multiple circuits. Except as permitted in 300.3(B)(4), all conductors of a multiwire branch circuit shall originate from the equipment containing the branch-circuit overcurrent protective device or protective devices.

> Informational Note No. 1: A 3-phase, 4-wire, wye-connected power system used to supply power to nonlinear loads might necessitate that the power system design allow for the possibility of high harmonic currents on the neutral conductor.
>
> Informational Note No. 2: See 300.13(B) for continuity of grounded conductors on multiwire circuits.

Reproduced with permission of NFPA from NFPA 70, National Electrical Code, 2023 edition. Copyright © 2022, National Fire Protection Association. For a full copy of the NFPA 70, please go to www.nfpa.org

Figure 3-5. Informational notes are intended to aid the *Code* user and are not an enforceable part of the *Code*. This section has an informational note that references another part of the *Code* as well as one with additional information on voltage drop.

Section 90.9 Units of Measurement

Section 90.9 covers the units of measurement used in the *National Electrical Code*. The *NEC* uses a dual system of units that contains both metric as well as inch-pound units. The metric units in the *NEC* are based on the International System of Units (SI). SI units will be listed first, with inch-pounds units immediately following. See **Figure 3-6**.

Finding Information

Finding information in the *NEC* is the primary focus of this text. Becoming proficient starts with understanding the structure of the *NEC* and learning the methods of finding information. Then it comes down to practice. To really become a proficient *Code* user, you must spend time navigating its contents.

Keywords

There are a few methods of finding information that revolve around keywords. A *keyword* is an important term related to the topic you are searching for. The process of identifying the keywords in a question is like the process used to search for something on the internet. When searching for something online, it is not necessary to type an entire sentence, just a few important words that relate to what is being searched. The same goes for identifying keywords in the *Code*. Identifying keywords is just figuring out what the question is really asking and boiling it down to its simplest form.

One of the difficulties of identifying keywords is that questions don't always have the exact same wording and language as the *National Electrical Code*. This is especially true when it comes to trade slang. Trade slang is a nickname given to electrical materials, tasks, and sometimes workers. It may describe the way something looks, the manufacturer's name, or its purpose. For example, Carflex® is a slang term used to describe liquidtight flexible nonmetallic conduit. Carflex® is the product name used by Carlon® to describe their specific brand of liquidtight flexible nonmetallic conduit.

Slang terms will vary by region, city, shop, and even crew. There are websites that are continuously updated with new slang terms used by electricians as well as other tradespeople.

Differences in terminology is one of the reasons it is so important to practice finding *NEC* requirements. After a while,

---

PROCEDURE   **Identifying Keywords**

Question: What is the maximum distance between straps when supporting nonmetallic sheathed cable? See **Figure 3-7**.

1. Determine the primary keyword. The wiring method in the question is nonmetallic sheathed cable. The primary keyword is nonmetallic sheathed cable.
2. Determine the secondary keyword. The nonmetallic sheathed cable is being supported. The second keyword is support.

Keywords: Nonmetallic sheathed cable, support

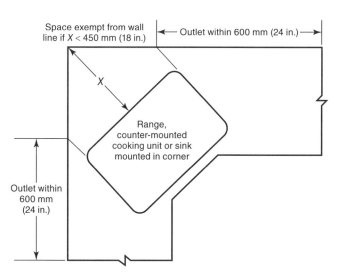

Figure 210.52(C)(1) Determination of Area Behind a Range, Counter-Mounted Cooking Unit, or Sink.

Reproduced with permission of NFPA from NFPA 70, National Electrical Code, 2023 edition. Copyright © 2022, National Fire Protection Association. For a full copy of the NFPA 70, please go to www.nfpa.org

Figure 3-6. The *NEC* contains both metric and inch-pound units of measurement. Metric units are always listed first.

Goodheart-Willcox Publisher

Figure 3-7. When roughing-in a home, nonmetallic sheathed cable must be secured and supported to prevent damage through the construction process.

the language used in the *Code* will become second nature and you will automatically know the correct keywords to use.

Code questions or inquiries will typically have multiple keywords, one being the primary subject and others being secondary. The secondary keywords will help refine the search.

Table of Contents (TOC)

One method of finding information is the table of contents. This method is often overlooked but can get you to the correct section the fastest. One of the advantages of using the table of contents to find requirements is the user must understand the structure of the *NEC*. This ensures you are following the outline hierarchy and are in the correct section, of the correct part, of the correct article. Another advantage of using the table of contents is that it is possible to get into the correct location without the correct keyword.

When spending a lot of time using the *NEC*, such as when completing worksheets or taking tests, it is helpful to have a photocopy of the table of contents. Having a separate copy of the TOC can be a time saver as it eliminates some of the time paging back and forth while looking up requirements.

PROCEDURE — Finding a Code Requirement Using the Table of Contents

The following is the process for finding information using the table of contents.

1. Identify the keywords/subject of the question.
2. Using the keywords/subject of the question, determine which chapter of the *NEC* applies.
3. Using the table of contents, search the determined chapter for the most applicable article. See **Figure 3-8**.
4. Using the table of contents, search the determined article for the most applicable part. The table of contents will have a page number of where that part begins.
5. Open the codebook to the page number and start looking through that part for a section that is applicable. See **Figure 3-9**.
6. Read that section in its entirety, including any exceptions to ensure proper application.

Question: The screw shell of a plug-type fuseholder shall be connected to the _____ side of the circuit.

1. Keyword(s): Fuseholder, Plug-type, screw shell
2. TOC: *Chapter 2, Wiring and Protection*
3. TOC: *Article 240, Overcurrent Protection*
4. TOC: *Part V, Plug Fuses, Fuseholders, and Adapters* (Page 129)
5. *Section 240.50, General*
6. *Section 240.50(E), Screw Shell*

Answer: The screw shell of a plug-type fuseholder shall be connected to the *load* side of the circuit.

Part V. Plug Fuses, Fuseholders, and Adapters

240.50 General.

(A) Maximum Voltage. Plug fuses shall be permitted to be used in the following circuits:

(1) Circuits not exceeding 125 volts between conductors
(2) Circuits supplied by a system having a grounded neutral point where the line-to-neutral voltage does not exceed 150 volts

(B) Marking. Each fuse, fuseholder, and adapter shall be marked with its ampere rating.

(C) Hexagonal Configuration. Plug fuses of 15-ampere and lower rating shall be identified by a hexagonal configuration of the window, cap, or other prominent part to distinguish them from fuses of higher ampere ratings.

(D) No Energized Parts. Plug fuses, fuseholders, and adapters shall have no exposed energized parts after fuses or fuses and adapters have been installed.

(E) Screw Shell. The screw shell of a plug-type fuseholder shall be connected to the load side of the circuit.

Chapter 2 Wiring and Protection

200	Use and Identification of Grounded Conductors	70–	76
210	Branch Circuits Not Over 1000 Volts ac, 1500 Volts dc, Nominal	70–	78
Part I.	General	70–	78
Part II.	Branch-Circuit Ratings	70–	83
Part III.	Required Outlets	70–	86
215	Feeders	70–	90
220	Branch-Circuit, Feeder, and Service Load Calculations	70–	92
Part I.	General	70–	92
Part III.	Feeders	70–	118
Part IV.	Outside Branch Circuits and Feeders	70–	119
Part V.	Services	70–	120
240	Overcurrent Protection	70–	122
Part I.	General	70–	122
Part II.	Location	70–	125
Part III.	Enclosures	70–	128
Part IV.	Disconnecting and Guarding	70–	128
Part V.	Plug Fuses, Fuseholders, and Adapters	70–	129
Part VI.	Cartridge Fuses and Fuseholders	70–	129
Part VII.	Circuit Breakers	70–	130
Part VIII.	Supervised Industrial Installations	70–	131

Reproduced with permission of NFPA from NFPA 70, National Electrical Code, 2023 edition. Copyright © 2022, National Fire Protection Association. For a full copy of the NFPA 70, please go to www.nfpa.org

Figure 3-8. A strong knowledge of the structure of the *Code* is helpful when using the table of contents to search for *NEC* requirements.

Reproduced with permission of NFPA from NFPA 70, National Electrical Code, 2023 edition. Copyright © 2022, National Fire Protection Association. For a full copy of the NFPA 70, please go to www.nfpa.org

Figure 3-9. Once the correct section has been located, read the section in its entirety to be sure all rules and exceptions are understood.

Index

Another method of finding information in the *NEC* is using the index. The index relies on having the correct keyword(s). The index will list all the sections where the keyword(s) are addressed. Having a thorough understanding of the structure of the *NEC* will allow the *Code* user to know where the requirement should be located, narrowing the number of sections to be searched. Once a specific section is found that appears to be correct, remember to back up and verify that the answer is in the correct section, of the correct part, of the correct article.

The index works well for questions that are general in nature and do not directly relate to a specific wiring method. The index does not work well when the *Code* user is not able to come up with the correct keyword.

Memorization

Becoming a proficient *Code* user is not about memorization, however some requirements will naturally be committed to memory due to repetition. It is more important to be able to find *NEC* requirements rather than simply memorizing them, as there is the potential for change or reorganization every three years with each new edition. Being comfortable finding information in the *NEC* allows you to always be able to find the correct information.

Tools to Aid the Code User

There are a few tools that can help the *Code* user find information faster and easier. Although these tools work well, caution must be taken for people who are new to the electrical industry and still need to pass an electrical exam. Most electrical licensure exams will provide *NEC* books to use during the exam to maintain its integrity. If a person has been using a book that is highlighted, has tabs, or has been using an online version, it can prove to be difficult to go back to using an unmarked codebook.

Highlighting

Highlighting a codebook is an effective way to make items stand out. When highlighting important portions of text, remember to highlight sparingly. If an entire page is highlighted, the *Code* user ends up reading the entire page anyway. When referencing the same item multiple times, it may be helpful to highlight the information so it can be seen on the page faster.

PROCEDURE — **Finding a Code Requirement Using the Index**

The following is the process for finding information using the index.
1. Identify the keywords/subject of the question.
2. Using the keywords/subject of the question, search the index for the main keyword. See **Figure 3-10**.
3. If the term that matches the primary keyword has a subset of additional terms, use the secondary keywords to narrow the focus.
4. Open the *NEC* and start reading through the part or sections listed until the correct answer is found. See **Figure 3-9**.
5. Read that section in its entirety, including any exceptions to ensure proper application.
6. Back up through the outline hierarchy to be sure the answer is in the correct section of the correct part of the correct article.

Question: The screw shell of a plug-type fuseholder shall be connected to the _____ side of the circuit.
1. Keywords: Fuseholder, Plug-type, screw shell
2. Index: *Fuseholder*
3. Index: *Fuseholder—Plug Fuse, Article 240-V*
4. Article 240 Part V, Plug Fuses, Fuseholders, and Adapters
5. Section 240.50, General
6. Section 240.50(E), Screw Shell

Answer: The screw shell of a plug-type fuseholder shall be connected to the *load* side of the circuit.

Full-load current motors
  Alternating current
    Single-phase, Table 430.248
    Three-phase, Table 430.250
    Two-phase, Table 430.249
  Direct current, Table 430.247
Furnaces, 422.12
Fuseholders
  Cartridge fuses, 240–VI
  Over 1000 volts, 245.21 (B)
  Plug fuses, 240–V
  Rating, motor controllers, 430.90
  Size, motor branch circuit
    protection, 430.57
  Type S, 240.53, 240.54
Fuses
  Cartridge, 240–VI
  Disconnection, 240.40

Reproduced with permission of NFPA from NFPA 70, National Electrical Code, 2023 edition. Copyright © 2022, National Fire Protection Association. For a full copy of the NFPA 70, please go to www.nfpa.org

Figure 3-10. The index will list the locations in the *Code* where a keyword is addressed. It is important to verify that the information found is in the correct section, of the correct part, of the correct article.

Another method of highlighting is to highlight specific portions of the *Code* to make visual stops. This helps prevent a person's eyes from mistakenly moving into the next article, part, or section which may not apply. When highlighting the *NEC* in this manner, it is wise to have a different colored highlighter for the different portions of the hierarchy. In **Figure 3-11** the articles are highlighted green, the parts are highlighted orange, and the sections are underlined with yellow.

Tabs

Tabs are a piece of paper or plastic that stick out of the codebook to allow a person to quickly reference the page the tab is attached to. They can be used to identify the location of nearly all the sections or can be used to only identify a few commonly used portions of the *Code*. Commercial tab sets are available that have been designed specifically for the *NEC* that contain a premade tab for most of the commonly used articles. See **Figure 3-12**. Another method that can be used to mark specific locations is to purchase generic tabs and install them on commonly used sections.

NEC Online Subscription

If a person has access to a computer or tablet, an online subscription to the *National Electrical Code* is available from the NFPA. The online version has a search function where keywords can be typed in and it will show you all the locations where the word appears. It does require the user to have access to the internet as it is not downloadable.

Article 210
Branch Circuits Not Over 1000 Volts ac,
1500 Volts dc, Nominal

Part I. General

210.1 Scope. This article provides the general requirements for branch circuits not over 1000 volts ac, 1500 volts dc, nominal.

Informational Note: See Part II of Article 235 for requirements for branch circuits over 1000 volts ac, 1500 volts dc, nominal.

210.2 Reconditioned Equipment. The following shall not be reconditioned:

(1) Equipment that provides ground-fault circuit-interrupter protection for personnel
(2) Equipment that provides arc-fault circuit-interrupter protection

210.3 Other Articles for Specific-Purpose Branch Circuits. Table 210.3 lists references for specific equipment and applications not located in Chapters 5, 6, and 7 that amend or supple-

Switchboards and panelboards   408.52

all ungrounded conductors at the point where the branch circuit originates.

Informational Note: See 240.15(B) for information on the use of single-pole circuit breakers as the disconnecting means.

(C) Line-to-Neutral Loads. Multiwire branch circuits shall supply only line-to-neutral loads.

Exception No. 1: A multiwire branch circuit that supplies only one utilization equipment shall be permitted to supply line-to-line loads.

Exception No. 2: A multiwire branch circuit shall be permitted to supply line-to-line loads if all ungrounded conductors of the multiwire branch circuit are opened simultaneously by the branch-circuit overcurrent device.

(D) Grouping. The ungrounded and grounded circuit conductors of each multiwire branch circuit shall be grouped in accordance with 200.4(B).

210.5 Identification for Branch Circuits.

(A) Grounded Conductor. The grounded conductor of a branch circuit shall be identified in accordance with 200.6.

Figure 3-11. Highlighting can provide visual stops that prevent *Code* users from moving into a portion of the *Code* that may not apply.

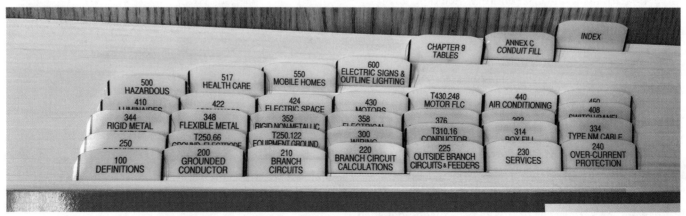

Goodheart-Willcox Publisher

Figure 3-12. Tabs can be placed to help the user navigate to the most commonly references articles.

Summary

- The purpose of the *NEC* is to ensure installations are safe from electrical hazards.
- The authority having jurisdiction (AHJ) is responsible for approving installations and interpreting the *NEC*.
- *Chapters 1–4* are the general chapters and apply to all installations.
- *Chapters 5–7* are the special chapters and will supplement or modify *Chapters 1–4*.
- Mandatory language includes *shall* or *shall not*, meaning the action is required or prohibited.
- Permissive language includes *shall be permitted* or *shall not be required*, meaning the action is allowed, but not required.
- *Code* inquiries start with identifying keywords.
- The primary methods of locating information in the *Code* are finding keywords in the table of contents or the index.
- Highlighting, pages tabs, and an online subscription are other tools that can aid in finding information in the *NEC*.

Unit 3 Review

Name _____ Date _____ Class _____

Know and Understand

Answer the following questions based on information in this unit.

1. _____ is the introduction to the *NEC*.
   A. *Chapter 1*
   B. *Chapter 2*
   C. *Chapter 3*
   D. *Article 90*

2. Which of the following is *not* under the jurisdiction of the *NEC*?
   A. Dwellings
   B. Swimming pool
   C. Underground mine
   D. Hospital

3. Which *NEC* chapter is referenced when trimming a job?
   A. *Chapter 3*
   B. *Chapter 4*
   C. *Chapter 5*
   D. *Chapter 6*

4. Which chapter has requirements for special conditions?
   A. *Chapter 5*
   B. *Chapter 6*
   C. *Chapter 7*
   D. *Chapter 8*

5. Which of the following is an example of mandatory *NEC* language?
   A. Must
   B. Shall not
   C. Shall be permitted
   D. Shall not be required

6. The *NEC* gives the power to enforce and interpret the *NEC* to the _____.
   A. authority having jurisdiction
   B. master electrician
   C. journeyman electrician
   D. insurance adjuster

7. Electrical equipment must be approved by the _____.
   A. master electrician
   B. insurance adjuster
   C. authority having jurisdiction
   D. independent testing laboratory

8. Finding _____ is the first step in finding information in the *NEC*.
   A. page numbers
   B. keywords
   C. sections
   D. None of the above.

9. Which method of finding information generally works better for requirements that are very general in nature?
   A. Table of contents
   B. Index
   C. Annex
   D. Memorization

10. Which unit is listed first in *NEC* requirements?
    A. International System of Units (SI)
    B. Inch-point units
    C. English units
    D. Imperial units

Apply and Analyze

Answer the following questions using a copy of the National Electrical Code. *Identify the section or subsection where the answer is found.*

1. The terms used in the *NEC* for mandatory rules are _____.
   A. must or must not
   B. shall be permitted or shall not be required
   C. recommended or not recommended
   D. shall or shall not

   NEC _____

2. _____ are used when the actual size of a product is not the same as the nominal size.
   A. Approximate sizes
   B. Trade sizes
   C. General requirements
   D. Special requirements

   NEC _____

3. The _____ is responsible for granting special permission that waives specific requirements of the *NEC* or permit alternative methods.
   A. insurance company
   B. building inspector
   C. authority having jurisdiction
   D. master electrician

   NEC _____

4. The purpose of the *NEC* is the practical safeguarding of _____ from hazards arising from the use of electricity.
   A. workers
   B. equipment
   C. animals
   D. persons and property

   NEC _____

5. Which of the following is not covered by the *NEC*?
   A. Underground mines
   B. Public buildings
   C. Equipment to export power from vehicles to premises wiring
   D. Electric utility office buildings

   NEC _____

6. One of the purposes of the code is to provide an installation that is efficient.
   A. True
   B. False

   NEC _____

7. _____ contains requirements for special equipment.
   A. *Chapter 4*
   B. *Chapter 5*
   C. *Chapter 6*
   D. *Chapter 7*

   NEC _____

8. Which of the following is covered by the *NEC*?
   A. Industrial substation
   B. Electric utility service drop
   C. Communication utility outdoor equipment
   D. Railway signaling and communication equipment

   NEC _____

9. _____ has tables that are applicable as referenced by other parts of the *NEC*.
   A. *Chapter 2*
   B. *Chapter 6*
   C. *Chapter 8*
   D. *Chapter 9*

   NEC _____

10. The *NEC* is not intended as a design specification or an instruction manual for _____.
    A. apprentice electricians
    B. untrained persons
    C. inspectors
    D. electrical engineers

    NEC _____

Name _____ Date _____ Class _____

Critical Thinking

Article 90

Answer the following questions using a copy of the National Electrical Code. Identify the section or subsection where the answer is found.

1. What is the purpose of the *National Electrical Code*?

   _____

   _____

   NEC _____

2. List four areas where the *NEC* does not apply.

   _____

   _____

   NEC _____

3. List the four general chapters.

   _____

   _____

   NEC _____

4. What dictates the language used by the *National Electrical Code*?

   _____

   _____

   NEC _____

5. Where can the procedures for formal interpretations be found?

   _____

   _____

   NEC _____

6. List two *Code* items that are there for informational purposes and are not an enforceable part of the *Code*.

   _____

   _____

   NEC _____

7. What chapter of the *NEC* contains requirements on equipment for general use?

   _____

   _____

   NEC _____

8. When is it permitted to use the *Chapter 9, Tables*?

   _____

   _____

   NEC _____

9. Which unit of measurement appears first within a section?

   _____

   _____

   NEC _____

10. The *NEC* is not intended as a _____ for untrained persons.

    _____

    _____

    NEC _____

UNIT 4
NEC Chapter 1

sockagphoto/Shutterstock.com

KEY TERMS

arc blast
arc flash
confined space
egress
enclosure
fault current
hazardous locations
listed
Nationally Recognized Testing Laboratory (NRTL)
nominal voltage

LEARNING OBJECTIVES

After completing this unit, you will be able to:

- Describe the content found in the *NEC Chapter 1*.
- Describe what prompts a term to be defined within the *NEC*.
- Recognize keywords that will lead you to *Article 110*.
- Locate the correct answer and section number for *NEC* questions relating to *Chapter 1*.

Introduction

This unit is an introduction to *Chapter 1* of the *NEC*. *Chapter 1* is the first of the general chapters that apply to all installations. This unit contains a description of the type of information found in *Chapter 1*, along with a few examples. At the end of the unit there are practice exercises that will require the use of the *NEC* to complete. The only way to become proficient at finding information in the *NEC* is to understand its layout and to practice looking up requirements. This unit is where we start to detail the type of information found in each *NEC* chapter and where the practice begins. Completing all the questions at the end of the unit will solidify your understanding of *Chapter 1* of the *National Electrical Code*.

Part 1 Layout and Structure: Learning to Use the NEC

CODE ARRANGEMENT

Chapter 1: General — YOU ARE HERE
Chapter 2: Wiring and Protection
Chapter 3: Wiring Methods and Materials
Chapter 4: Equipment for General Use
} Applies generally to all electrical installations

Chapter 5: Special Occupancies
Chapter 6: Special Equipment
Chapter 7: Special Conditions
} Supplements or modifies chapters 1 through 7

Chapter 8: Communications Systems
] Chapter 8 is not subject to the requirements of chapters 1 through 7 except where the requirements are specifically referenced in chapter 8

Chapter 9: Tables
] Applicable as referenced

Informative Annex A through Informative Annex J
] Informational only; not mandatory

Goodheart-Willcox Publisher

Figure 4-1. *Chapter 1* is the first of the four general chapters which contains information that applies to all installations. *Chapters 1–4* apply generally to all installations. *Chapters 5–7* will supplement or modify the other requirements of *Chapters 1–7*. *Chapter 8* covers Communications which stands alone. *Chapter 9* contains tables that are applicable as referenced by other parts of the *Code*.

Chapter 1, General

Chapter 1, General contains information and requirements that apply to all electrical installations. See **Figure 4-1**. *Chapter 1* only contains two articles: *Article 100* and *Article 110*. *Article 100* contains definitions while *Article 110* contains general requirements. Notice that all the *Chapter 1* articles start with a 1. That same format is used throughout the *Code*. All the *Chapter 2* articles start with a 2, *Chapter 3* with a 3, and so forth.

Article 100, Definitions

The *National Electrical Code* defines terms that must be understood to properly apply the *Code*. A definition found in *Article 100* is not the same definition you would find in a standard dictionary. The definitions are carefully written to allow consistent interpretation and application of the *NEC*. The definitions found in the *NEC* do not contain any requirements as they are intended only to describe the term being defined. An article number in parentheses immediately following a definition indicates that the definition only applies to that article, **Figure 4-2**.

The terms defined by the *NEC* are not necessarily the same language you will hear on jobsite. Trade slang is used throughout the country and varies by region, and sometimes even by company. When trying to identify keywords in *Code* inquiries, it takes time and practice to be able to recognize the proper *NEC* term for all the different ways questions and terminology can be written. For example, it is common to refer to a conductor with a dangerous potential voltage as a "hot" conductor. The *NEC* does not refer to that conductor as "hot," instead it is called an ungrounded conductor and is considered energized.

The term "building" is an example of how the *NEC* definition is slightly different from the general understanding or what a dictionary says. Without taking the *NEC*'s definition into consideration, you may consider the twin home in **Figure 4-3** as one building that contains two units. The *NEC* defines a "building" as a structure that stands alone or is separated from adjoining structures by fire walls. (*Article 100*)

Operator. The individual responsible for starting, stopping, and controlling an amusement ride or supervising a concession. (525) (CMP-15)

Reproduced with permission of NFPA from NFPA 70, National Electrical Code, 2023 edition. Copyright © 2022, National Fire Protection Association. For a full copy of the NFPA 70, please go to www.nfpa.org

Figure 4-2. The *NEC* definition of the term *operator* has 525 in parentheses which indicates the definition only applies within *Article 525*.

Figure 4-3. By appearance, twin homes look like one building that has two units. The *National Electrical Code* considers a twin home with a firewall that completely divides the spaces as two separate buildings. It is important to carefully read *NEC* definitions as they are often slightly different than general understanding or what would be found in a dictionary.

Thus, as far as the *Code* is concerned, a twin home with a firewall that completely separates the two units is considered two separate buildings.

Article 110, General Requirements for Electrical Installations

Article 110 contains the most general of the requirements found in the *NEC*. These are requirements that apply to the everyday activities an electrician performs. Since the requirements of *Article 110* are so basic, they are some of the most difficult items to find. It is common to search other parts of the *Code* before ending up back in *Article 110* to find the answer. See **Figure 4-4**.

Familiarizing yourself with the type of information contained within *Article 110* can help save time in the long run. Reading through all the section titles will help you recall later the type of requirements found in *Article 110*. When a question comes up relating to one of the *Article 110* topics, it may trigger your memory as to where it can be found. *Article 110* requirements are one of the situations where the index will typically lead you to the answer faster than the table of contents.

Because it is difficult to intuitively find information in *Article 110*, the remainder of this unit will introduce a few of the requirements to help give a feel for the type of information contained in the article. It is still important to spend some time getting familiar with the section titles and the types of information contained by looking through the article.

Part 1, General

Part 1 of *Article 110* is titled *General*. It has the requirements of *Article 110* that apply to all electrical installations.

Section 110.2, Approval

Section 110.2 states that the conductors and equipment that are covered by the *Code* must be approved. This is an example of why it is important to understand the definitions. In the definitions of *Article 100*, approved is defined as acceptable to the authority having jurisdiction. The authority having jurisdiction in most areas is the electrical inspector.

Section 110.3, Examination, Identification, Installation, Use, and Listing (Product Certification) of Equipment

Section 110.3 covers examination, installation, and listing of electrical equipment. It has specific requirements to look for when examining equipment to ensure it will be free from electrical hazards. Product testing is typically done by a **Nationally Recognized Testing Laboratory (NRTL)**. The Occupational Safety and Health Administration (OSHA) recognizes certain organizations as NRTLs to perform testing standards. If a product passes the minimum safety requirement, the NRTL will label the product as **listed**, meaning that the product will be safe if it is installed under the conditions of use it was tested for.

Article 110.3B states that all equipment must be installed according to the manufacturer's instructions, which is how the equipment was tested and listed. Occasionally, the instructions will provide additional requirements that are over and above what is found in the *NEC*. A good example of this is baseboard heaters. The *NEC* does not directly say that an electric baseboard heater cannot be installed under a receptacle outlet. However, the instructions included with a heater may prohibit it. See **Figure 4-6**.

Section 110.12, Mechanical Execution of Work

Section 110.12 states that electrical equipment must be installed in a professional and skillful manner. This gives the authority having jurisdiction the ability to address poor workmanship.

Section 110.12(A) requires any unused openings or knockouts in equipment to be closed. See **Figure 4-7**. This requirement applies to equipment such as boxes, panelboards, luminaires, and meter socket enclosures. Rather than have the same language repeated for each type of equipment throughout the *Code*, it is placed in *Article 110*, so it applies to all electrical equipment.

Section 110.13, Mounting and Cooling of Equipment

Section 110.13 requires equipment to be securely mounted, and if cooling is necessary, air flow is not impeded.

NEC Chapter 1
General
Article 110 Requirements for Electrical Installations

Part I General

Section 110.1 Scope
Section 110.2 Approval
Section 110.3 Examination, Identification, Installation, Use, and Listing (Product Certification) of Equipment
Section 110.4 Voltages
Section 110.5 Conductors
Section 110.6 Conductor Sizes
Section 110.7 Wiring Integrity
Section 110.8 Wiring Methods
Section 110.9 Interrupting Rating
Section 110.10 Circuit Impedance, Short-Circuit Current Ratings, and Other Characteristics
Section 110.11 Deteriorating Agents
Section 110.12 Mechanical Execution of Work
Section 110.13 Mounting and Cooling of Equipment
Section 110.14 Electrical Connections
Section 110.15 High-Leg Marking
Section 110.16 Arc-Flash Hazard Warning
Section 110.18 Arcing Parts
Section 110.19 Light and Power from Railway Conductors
Section 110.20 Reconditioned Equipment
Section 110.21 Marking
Section 110.22 Identification of Disconnecting Means
Section 110.23 Current Transformers
Section 110.24 Available Fault Current
Section 110.25 Lockable Disconnecting Means

Part II 1000 Volts, Nominal, or Less

Section 110.26 Spaces About Electrical Equipment
Section 110.27 Guarding of Live Parts
Section 110.28 Enclosure Types
Section 110.29 In Sight From (Within Sight From, Within Sight)

Part III Over 1000 Volts, Nominal

Section 110.30 General
Section 110.31 Enclosure for Electrical Installations
Section 110.32 Work Space About Equipment
Section 110.33 Entrance to Enclosures and Access to Working Space
Section 110.34 Work Space and Guarding
Section 110.36 Circuit Conductors
Section 110.40 Temperature Limitations at Terminations
Section 110.41 Inspections and Tests

Part IV Tunnel Installations over 1000 Volts, Nominal

Section 110.51 General
Section 110.52 Overcurrent Protection
Section 110.53 Conductors
Section 110.54 Bonding and Equipment Grounding Conductors
Section 110.55 Transformers, Switches, and Electrical Equipment
Section 110.56 Energized Parts
Section 110.57 Ventilation System Controls
Section 110.58 Disconnecting Means
Section 110.59 Enclosures

Part V Manholes and Other Electrical Enclosures Intended for Personnel Entry

Section 110.70 General
Section 110.71 Strength
Section 110.72 Cabling Work Space
Section 110.73 Equipment Work Space
Section 110.74 Conductor Installation
Section 110.75 Access to Manholes
Section 110.76 Access to Vaults and Tunnels
Section 110.77 Ventilation
Section 110.78 Guarding
Section 110.79 Fixed Ladders

Goodheart-Willcox Publisher

Figure 4-4. *Article 110* has general requirements that apply to all installations.

PRO TIP: Article 110 Requirements

The information found in *Article 110* applies to multiple wiring methods and situations. It is placed in *Article 110* to prevent having to repeat the same information for each situation in which it applies. Although this is logical and saves space in the *Code*, it makes the information in *Article 110* difficult to find. You will find yourself looking for a requirement specific to a certain wiring method rather than thinking about the general rule that would apply to multiple wiring methods and situations. An example of this is the prohibition of wooden plugs when mounting a box to a concrete wall, **Figure 4-5**. The keywords will lead you to look in *Chapter 3* under boxes, but the answer is found in *Section 110.13(A)*. It is placed in *Article 110* because it applies not only to boxes, but to any piece of electrical equipment. The reason the *NEC* prohibits wooden plugs as anchors in concrete walls is because the wooden plug will deteriorate if moisture is present and will become loose or fall out.

Goodheart-Willcox Publisher

Figure 4-5. When mounting equipment to a concrete surface, wooden plugs are not permitted to be used. The presence of moisture in concrete will eventually cause the wood to deteriorate, which will result in the enclosure no longer being securely attached.

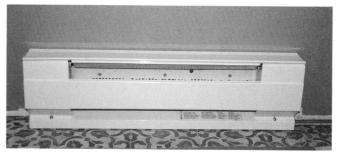

Goodheart-Willcox Publisher

Figure 4-6. The heat generated in an electric baseboard heater has the potential to deteriorate an electric cord that is draped across the heater. For this reason, the manufacturer's instructions often prohibit the heater from being installed below a receptacle outlet. It is important to read the manufacturer's instructions so you can ensure you are installing equipment according to its listing.

Goodheart-Willcox Publisher

Figure 4-7. Unused openings in boxes and other equipment must be closed to prevent fingers, rodents, and other foreign materials from entering.

Section 110.14, Electrical Connections

Section 110.14 contains information about all the various types of electrical connections and terminations. Regardless of the type of conductor or termination means, *Section 110.14* applies to them all. A common term found throughout the code is *identified*. *Section 110.14* uses this term when it states that any device used to splice or terminate must be identified for the use. Meaning that each component must be installed according to its listing and instructions and in the manner which it was designed.

Section 110.14(C) covers temperature ratings. All equipment must operate within its listed temperature range. This equipment is all of the components and tools used as part of or to install an electrical installation. This includes conductors, terminations, and devices. The *NEC* states that when

> **PRO TIP**
>
> ### Referencing the Correct NEC Section.
>
> When answering *Code* questions, it is often necessary to provide the answer as well as the specific *Code* section where the information is found. When writing down the section, be sure to write the section in its entirety. For example, the *Code* reference shown in **Figure 4-8** gives permission to have a single entrance/exit to switchgear and control panels that are rated at over 1,000 volts if certain parameters are met. The *NEC* reference is written as Section 110.33(A)(1)(a).
>
> **110.33 Entrance to Enclosures and Access to Working Space.**
>
> **(A) Entrance.** At least one entrance to enclosures for electrical installations as described in 110.31 not less than 610 mm (24 in.) wide and 2.0 m (6½ ft) high shall be provided to give access to the working space about electrical equipment.
>
> Open equipment doors shall not impede access to and egress from the working space. Access or egress is impeded if one or more simultaneously opened equipment doors restrict working space access to be less than 610 mm (24 in.) wide and 2.0 m (6½ ft) high.
>
> **(1) Large Equipment.** On switchgear and control panels exceeding 1.8 m (6 ft) in width, there shall be one entrance at each end of the equipment. A single entrance to the required working space shall be permitted where either of the conditions in 110.33(A)(1)(a) or (A)(1)(b) is met.
>
> (a) *Unobstructed Exit.* Where the location permits a continuous and unobstructed way of exit travel, a single entrance to the working space shall be permitted.
>
> *Reproduced with permission of NFPA from NFPA 70, National Electrical Code, 2023 edition. Copyright © 2022, National Fire Protection Association. For a full copy of the NFPA 70, please go to www.nfpa.org*
>
> **Figure 4-8.** When writing down the *NEC* location where a requirement is found, care must be taken to write it correctly. *NEC* exams often require the answer as well as the correct section to get credit.

Section 110.15, High-Leg Marking

Section 110.15 has requirements on how to identify a high-leg. A high-leg occurs in a three-phase, four-wire delta system where the center of one winding is grounded. The *NEC* requires the phase with the higher voltage to ground to be orange in color to draw attention to the fact that it has a higher voltage to ground than the other two phases. See **Figure 4-9**.

Section 110.16, Arc-Flash Hazard Warning

Section 110.16 lists requirements for labeling non-dwelling equipment of its arc flash hazard. An arc flash is the hidden hazard that many people are not aware of. An **arc flash** occurs when two metal components with different electrical potentials come into contact, resulting in a rapid release of energy. An arc flash has the potential to create an **arc blast**, an explosion with extremely high temperatures, blinding light, loud noise, toxic gases, and shrapnel. See **Figure 4-10**.

Section 110.24, Available Fault Current

Section 110.24 has requirements for calculating and labeling available fault current on non-dwelling service equipment. **Fault current** is the amount of current available in the event of a short circuit or ground fault. If a piece of equipment is not rated to handle the available fault current, there is the potential for an arc flash/blast.

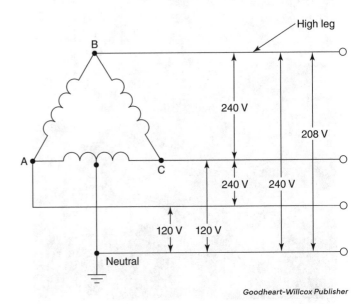

Goodheart-Willcox Publisher

Figure 4-9. A 120/240-volt, four-wire delta configuration has one phase that has a higher voltage to ground, known as the high leg and sometimes referred to as the wild leg. The high leg has 208 volts to ground rather than 120 volts like the two other phase conductors. This configuration is used in some farm applications as well as in some small industrial installations.

multiple items are used that have different temperature ratings, the temperature must not exceed the lowest rated piece of equipment. This prevents the heat from a piece of equipment with a higher temperature rating from spreading to equipment with a lower temperature rating, potentially causing it to overheat.

Section 110.14(D) covers torquing lugs and other connections. Most equipment will have torque values for its connections listed on the equipment. Torquing by an approved means is required by the *NEC*.

Figure 4-10. Article 110 requires arc flash hazard stickers where an arc flash hazard exists. The sticker alerts people working around the equipment that there is a potential hazard so they can take the necessary steps and wear the appropriate personal protective equipment (PPE) to ensure their safety.

Part II, 1000 Volts, Nominal, or Less

Part II of *Article 110* contains general requirements for systems that have a voltage of 1000 volts or less. It only has a few sections, all of which apply to equipment.

Section 110.26, Spaces About Electrical Equipment

Electrical workers need illumination and enough space to work safely around electrical equipment, and to be able escape in the event of an arcing incident. This section applies to electrical equipment that is likely to require service or maintenance while energized, such as panelboards, switchboards, and motor control centers. *Section 110.26, Spaces About Electrical Equipment* has several subsections, each addressing a different aspect of the working space. See **Figure 4-11**.

Section 110.28, Enclosure Types

Section 110.28 covers **enclosures** and the environment they are suitable for. An enclosure is the case or housing that prevents accidental personal contact with energized parts and protects the internal equipment from physical damage. *Section 110.28* gives an extensive list of examples of enclosures and refers to *Table 110.28* for a listing of enclosure types. The table contains the enclosure type numbers as well as the environmental conditions they are suitable for. For example, a Type 1 enclosure protects against incidental contact with enclosed equipment and falling dirt. See **Figure 4-12**.

Part III, Over 1000 Volts, Nominal

Part III of *Article 110* contains general requirements for systems that have a voltage over 1000 volts. Systems that operate

Figure 4-11. Electrical equipment that is likely to be worked on while energized is required to have minimum clearances. This is to ensure the worker has sufficient space to safely work and has a means of egress in the case of an emergency. The higher the potential danger, the greater the clearances as well as egress requirements.

at over 1000 volts have a greater potential and, therefore, have different requirements. The equipment is built differently, additional spacing is required, and additional requirements are in place to address the increased hazard.

Part IV, Tunnel Installations over 1000 Volts, Nominal

Part IV of *Article 110* contains general requirements for systems that have a voltage over 1000 volts that are installed in tunnels. This part of the *NEC* is not used very often by regular electricians. In the event you find yourself working on one of these systems, it is important that you familiarize yourself with this part of the *Code* as well as any applicable safety precautions.

Part V, Manholes and Other Electrical Enclosures Intended for Personnel Entry

Part V of *Article 110* contains general requirements for manholes and other enclosures intended for personnel entry.

Table 110.28 Enclosure Selection

Provides a Degree of Protection Against the Following Environmental Conditions	For Outdoor Use Enclosure Type Number									
	3	3R	3S	3X	3RX	3SX	4	4X	6	6P
Incidental contact with the enclosed equipment	X	X	X	X	X	X	X	X	X	X
Rain, snow, and sleet	X	X	X	X	X	X	X	X	X	X
Sleet*	—	—	X	—	—	X	—	—	—	—
Windblown dust	X	—	X	X	—	X	X	X	X	X
Hosedown	—	—	—	—	—	—	X	X	X	X
Corrosive agents	—	—	—	X	X	X	—	X	—	X
Temporary submersion	—	—	—	—	—	—	—	—	X	X
Prolonged submersion	—	—	—	—	—	—	—	—	—	X

Provides a Degree of Protection Against the Following Environmental Conditions	For Indoor Use Enclosure Type Number									
	1	2	4	4X	5	6	6P	12	12K	13
Incidental contact with the enclosed equipment	X	X	X	X	X	X	X	X	X	X
Falling dirt	X	X	X	X	X	X	X	X	X	X
Falling liquids and light splashing	—	X	X	X	X	X	X	X	X	X
Circulating dust, lint, fibers, and flyings	—	—	X	X	—	X	X	X	X	X
Settling airborne dust, lint, fibers, and flyings	—	—	X	X	X	X	X	X	X	X
Hosedown and splashing water	—	—	X	X	—	X	X	—	—	—
Oil and coolant seepage	—	—	—	—	—	—	—	X	X	X
Oil or coolant spraying and splashing	—	—	—	—	—	—	—	—	—	X
Corrosive agents	—	—	—	X	—	—	X	—	—	—
Temporary submersion	—	—	—	—	—	X	X	—	—	—
Prolonged submersion	—	—	—	—	—	—	X	—	—	—

Reproduced with permission of NFPA from NFPA 70, National Electrical Code, 2023 edition. Copyright © 2022, National Fire Protection Association. For a full copy of the NFPA 70, please go to www.nfpa.org

Figure 4-12. Table 110.28 has a table that lists many of the common enclosure types and the corresponding environment conditions they are listed for.

PRO TIP

Tunnels and manholes are often considered a **confined space**. Confined spaces have limited means of entry and egress. **Egress** is a term that appears in several *NEC* locations. It means to leave space. Tunnels and manholes not only have limited egress, but they may also have limited oxygen or accumulate toxic gases, which could cause a worker to lose consciousness. Follow all safety guidelines found in your employer's safety plan as well as OSHA requirements when working in confined spaces. See **Figure 4-13**.

Confined Spaces

Sakrai Sarabun/Shutterstock.com

Figure 4-13. Areas such as manholes that have limited entrance and egress are considered a confined space. The potential presence of toxic gases or limited oxygen makes them extremely dangerous. *Part V* of *Article 110* has installation requirements specific to manholes and other enclosures intended for personnel entry.

Since manholes and electrical vaults are not something an electrician runs into very frequently, this part of the *Code* is not used very often. In the event you find yourself working in or around one of these enclosures, be sure to familiarize yourself with this part of the *Code* as well as any applicable confined space safety precautions.

Summary

- *Chapter 1* is the first of the four general chapters.
- *Article 100* defines terms necessary for the proper application of the *Code*.
- *Article 110* contains general requirements that apply to all installations.
- Becoming proficient in the *NEC* requires an understanding of the layout of the *Code* as well as practicing finding requirements.

Unit 4 Review

Name _____ Date _____ Class _____

Know and Understand

Answer the following questions based on information in this unit.

1. NEC Definitions are found _____.
   A. at the beginning of the article to which it applies
   B. after the index
   C. in *Article 100*
   D. at the end of the article to which it applies
2. *Chapter 1* is the first of the _____ general chapters.
   A. two
   B. three
   C. four
   D. five
3. Which of the following statements about the definitions of *Article 100* is true?
   A. The definitions often contain requirements.
   B. The definitions are divided into three parts.
   C. The definitions are essential to properly apply the *NEC*.
   D. The definitions are the same as would be found in a standard dictionary.
4. *Article 110* contains _____ parts.
   A. two
   B. three
   C. four
   D. five
5. Definitions that only apply to a specific article _____.
   A. are defined within the article to which they apply
   B. have the article number in parentheses immediately following the definition
   C. have the article number in quotation marks before the definition
   D. are categorized by article immediately following the index
6. _____ of *Article 110* addresses manholes.
   A. *Part I*
   B. *Part III*
   C. *Part V*
   D. *Part IX*
7. *Part III* of *Article 110* is titled _____.
   A. *General*
   B. *1000 Volts, Nominal, or Less*
   C. *Over 1000 Volts, Nominal*
   D. *Hazardous Location*
8. *Table 110.28* contains information pertaining to _____.
   A. live parts
   B. enclosures
   C. working spaces
   D. electrical connections
9. *Part II* of *Article 110* contains requirements for _____.
   A. electrical systems of 1000 volts or less
   B. electrical systems that are over 1000 volts
   C. tunnel installations over 1000 volts
   D. manholes
10. Electrical conductors and equipment shall be acceptable only if _____.
    A. listed
    B. approved
    C. tested
    D. identified

Apply and Analyze

NEC Article 100

Use a copy of the National Electrical Code to identify the term corresponding to the definition provided. Identify the section or subsection where the answer is found.

1. All circuit conductors between the service equipment, the source of a separately derived system, or other power supply source and the final branch-circuit overcurrent device.
   A. Branch circuit
   B. Feeder
   C. Service
   D. Uninterruptible Power Supply

   NEC _____

2. Capable of being removed or exposed without damaging the building structure or finish, or not permanently closed in by the structure or finish of the building.
   A. Guarded
   B. Accessible (as applied to equipment)
   C. Accessible (as applied to wiring methods)
   D. Readily Accessible

   NEC _____

3. A point on the wiring system at which current is taken to supply utilization equipment.
   A. Outlet
   B. Device
   C. Power Outlet
   D. Luminaire

   NEC _____

4. Constructed so that moisture will not enter the enclosure under specified test conditions.
   A. Rainproof
   B. Raintight
   C. Watertight
   D. Weatherproof

   NEC _____

5. Type of protection where electrical parts that could ignite an explosive atmosphere by either sparking or heating are enclosed in a compound in such a way that this explosive atmosphere cannot be ignited.
   A. Flameproof
   B. Dust-Ignitionproof
   C. Hermetically Sealed
   D. Encapsulation

   NEC _____

6. The conductor connected to the neutral point of a system that is intended to carry current under normal conditions.
   A. Ungrounded Conductor
   B. Neutral Conductor
   C. Grounded Conductor
   D. Service Conductor

   NEC _____

7. One who has the skills and knowledge related to the construction and operation of the electrical equipment and installations and has received safety training to recognize and avoid the hazards involved.
   A. Qualified Person
   B. Authority Having Jurisdiction
   C. Classified Person
   D. Identified Person

   NEC _____

8. Connected to ground without inserting any resistor or impedance device.
   A. Grounded
   B. Bonded
   C. Earthed
   D. Grounded solidly

   NEC _____

9. The underground conductors between the utility electric supply system and the service point.
   A. Service Entrance Conductors, Underground System
   B. Service Conductors
   C. Service Lateral
   D. Service Drop

   NEC _____

10. A device or system that provides quality and continuity of ac power through the use of a stored energy device as a backup power source for a period of time when the normal power supply is incapable of performing acceptably.
    A. Hybrid System
    B. Uninterruptible Power Supply
    C. Interactive System
    D. Electrical Circuit Protective System

    NEC _____

Name _____ Date _____ Class _____

Critical Thinking

NEC Article 110

Answer the following questions using a copy of the National Electrical Code. Identify the section or subsection where the answer is found.

1. On a 4-wire, delta-connected system with the midpoint of one phase being grounded, the phase with the higher voltage to ground shall be marked with a finish that is _____ in color.

   _____

   _____

   NEC _____

2. The minimum width of working space in front of a 120/240-volt single-phase electrical panel shall be at least _____.

   _____

   _____

   NEC _____

3. A Type 3R enclosure will protect the enclosed equipment against _____.

   _____

   _____

   NEC _____

4. Round openings to manholes shall be at least _____ in diameter.

   _____

   _____

   NEC _____

5. The walls and roof of a vault containing over 1000-volt equipment, must have a minimum fire rating of _____ hours.

   _____

   _____

   NEC _____

6. Conductors used to carry current shall be of copper, aluminum, or _____ unless otherwise approved.

   _____

   _____

   NEC _____

7. Cable ties used to secure cables within an environmental air plenum, shall be listed as having _____ properties.

   _____

   _____

   NEC _____

8. Service equipment in _____ shall be legibly marked in the field with the available fault current.

   _____

   _____

   NEC _____

9. Enclosures for use in tunnels shall be drip-proof, weatherproof, or _____ as required by the environmental conditions.

   _____

   _____

   NEC _____

10. The temperature rating associated with the ampacity of a conductor shall be selected so as not to exceed _____.

    _____

    _____

    NEC _____

NEC Chapter 1

Answer the following questions using a copy of the National Electrical Code. Identify the section or subsection where the answer is found.

11. Unused current transformers associated with circuits that have the potential to be energized shall be _____.

    NEC _____

12. Unused openings in electrical equipment shall be _____ to afford protection equivalent to the wall of the equipment.

    NEC _____

13. Terminals for finely stranded conductors must be _____ for the specific conductor class.

    NEC _____

14. A disconnect that is *in sight from* a piece of equipment, means that it is visible and within _____ feet.

    NEC _____

15. A room containing a 480-volt, 800-ampere panelboard must have a door that opens in the direction of egress and shall be equipped with _____ or listed fire exit hardware.

    NEC _____

16. Replacement parts used for servicing and maintenance of electrical equipment shall comply with at least one of the following:
    A. Be designed by an engineer experienced in the design of replacement parts for the type of equipment being serviced or maintained.
    B. Be approved by the authority having jurisdiction.
    C. _____.

    NEC _____

17. The inside of a car wash is considered a _____ location.

    NEC _____

18. Control by _____ shall not control all illumination within the working space about service equipment, switchboards, enclosed panelboards, or motor control centers installed indoors.

    NEC _____

19. The manufacturer's equipment markings on a piece of equipment must have the durability to withstand the _____.

    NEC _____

20. A piece of equipment must have an interrupting rating at least equal to the fault current available at its _____.

    NEC _____

UNIT

5 NEC Chapter 2

sockagphoto/Shutterstock.com

LEARNING OBJECTIVES

After completing this unit, you will be able to:
- Describe the content found in *Chapter 2* of the *NEC*.
- Identify the design articles.
- Identify the protection articles.
- Recognize keywords that will lead you to *Chapter 2*.
- Find the answer and section number for *Chapter 2 Code* requirements.

KEY TERMS

bonding
branch circuit
equipment grounding conductor
feeder
ground fault
grounded conductor
grounding electrode
grounding electrode conductor
habitable room
individual branch circuit
neutral conductor
overcurrent
overload
overvoltage
service
switch leg
ungrounded

Introduction

This unit is an introduction to *Chapter 2* of the *NEC*. Each of the articles in the chapter will be introduced to help you understand the type of information they contain. The end of the unit contains practice exercises that will require the use of your *NEC* to complete. The combination of studying the type of information found in each article along with spending time looking up code sections will help to reinforce the type of information that can be found in *Chapter 2* of the *National Electrical Code*.

Part 1 Layout and Structure: Learning to Use the NEC

Chapter 2, Wiring and Protection

Chapter 2 of the *NEC* is titled *Wiring and Protection*. It is the second of the four general chapters that apply to all electrical installations. See **Figure 5-1**. The information found in *Chapter 2* can be divided into two different categories, design and protection.

Design Articles

The *Chapter 2* design articles contain requirements that must be considered while designing an installation. See **Figure 5-2**. Examples of design activities include laying out the light and receptacle locations on a set of blueprints, calculating the size and type of service, and planning the size and quantity of branch circuits and feeders. Seven articles fall under the design classification, **Figure 5-3**.

Article 200, Use and Identification of Grounded Conductors

Article 200 is dedicated to the grounded conductor, **Figure 5-4**. The *NEC* defines the grounded conductor as a system or circuit conductor that is intentionally grounded. In most situations, the grounded conductor and the neutral conductor are the same. The scope of *Article 200*, which is

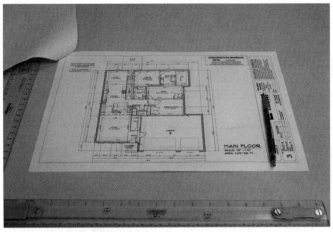

Goodheart-Willcox Publisher

Figure 5-2. One of the first steps in the construction process is to design the electrical system. This involves laying out the locations of all the receptacles, switches, luminaires, and equipment. Blueprints are created during the design stage.

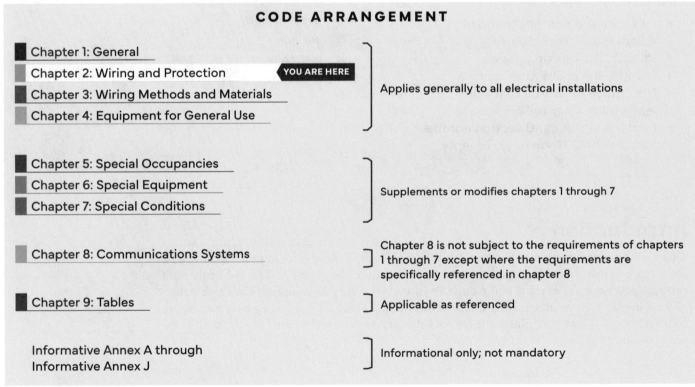

Goodheart-Willcox Publisher

Figure 5-1. The *Code* is arranged by chapter. Becoming familiar with the information in each chapter will help in finding information quickly.

NEC Chapter 2
Design Articles

Article 200 Use and Identification of Grounded Conductors
Article 210 Branch Circuits Not Over 1000 Volts ac, 1500 Volts dc, Nominal
Article 215 Feeders
Article 220 Branch-Circuit, Feeder, and Service Load Calculations
Article 225 Outside Branch Circuits and Feeders
Article 230 Services
Article 235 Branch Circuits, Feeders, and Services Over 1000 Volts ac, 1500 Volts dc, Nominal

Goodheart-Willcox Publisher

Figure 5-3. The articles of *Chapter 2* can be divided into two categories, design and protection. The design articles contain requirements to be applied when designing electrical systems.

found in *Section 200.1*, describes the type of information found within the article.
- Identification of terminals
- Grounded conductors in premises wiring systems
- Identification of grounded conductors

Article 200 is not divided into parts but is divided into nine sections. The following are a few specific examples of the type of information found in *Article 200* to help you get a feel for type of information it contains:

Section 200.6, Means of Identifying Grounded Conductors

Section 200.6 has requirements on the permitted insulation colors as well as how to identify the grounded conductor. Conductors that are 6 AWG or smaller must be white, gray, or shall have three white or gray stripes along the entire length of the conductor. Conductors 4 AWG and larger may be white or gray like the smaller conductors, but they are also permitted to be reidentified white or gray at the time of installation. Black conductors reidentified with white or gray tape is common as black conductors are readily available in almost any size. See **Figure 5-5**.

Article 200
Use and Identification of Grounded Conductors

Section 200.1 Scope
Section 200.2 General
Section 200.3 Connection to Grounded System
Section 200.4 Neutral Conductors
Section 200.6 Means of Identifying Grounded Conductors
Section 200.7 Use of Insulation of a White or Gray Color or with Three Continuous White or Gray Stripes
Section 200.9 Means of Identification of Terminals
Section 200.10 Identification of Terminals
Section 200.11 Polarity of Connections

Goodheart-Willcox Publisher

Figure 5-4. *Article 200* can be categorized as a design article and contains requirements specific to the grounded conductor.

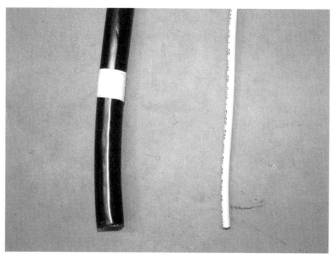

Goodheart-Willcox Publisher

Figure 5-5. Grounded and neutral conductors typically have a white or gray insulation. The *NEC* does permit conductors 4 AWG and larger to be reidentified with white or gray at the time of installation. In this picture, the larger conductor has white tape wrapped around the conductor to indicate that it is a grounded conductor.

PRO TIP: Conductor Terminology

There are four main types of conductors: ungrounded conductor, grounded conductor, neutral conductor, and equipment grounding conductor.

Ungrounded is defined as not connected to ground or to a conductive body that extends the ground connection. That makes the ungrounded conductor the conductor that is *not* connected to ground. Most people refer to the ungrounded conductor as the *hot* or *energized* conductor.

A **grounded conductor** is defined as a system or circuit conductor that is intentionally grounded.

In most systems, the grounded conductor and the neutral conductor are the same thing. An exception to that rule is the corner-grounded delta system, where one phase is grounded. In a corner-grounded delta system there is no neutral conductor.

A *neutral conductor* is defined as the conductor connected to the neutral point of a system that is intended to carry current under normal conditions.

An *equipment grounding conductor* is defined as a conductive path(s) that is part of an effective ground-fault current path and connects normally non-current-carrying metal parts of equipment together and to the system grounded conductor or to the grounding electrode conductor, or both. The equipment grounding conductor is generally thought of as an actual conductor, but sometime the raceway itself may be used as the equipment grounding conductor.

Section 200.10, Identification of Terminals

Section 200.10 has several subsections that address how the grounded terminal is identified on devices, lampholders, and other equipment. Equipment must be clearly identified so the electrician can ensure the proper polarity. *Section 200.10(B)* requires the grounded terminal to be white or silver in color or shall be marked with the word "white" or the letter "W" near the terminal.

Section 200.10(C) requires the grounded conductor to be connected to the screw shell of lampholders. The screw shell is the portion of a lampholder that a person may contact while changing a lamp. By connecting the grounded conductor to the screw shell, the ungrounded conductor gets attached to the tab in the back of the socket. This reduces the chance of a person receiving an electric shock from the ungrounded terminal, **Figure 5-6**.

The ungrounded conductor that connects to the tab in the back of a socket is often referred to as a *switch leg*.

Goodheart-Willcox Publisher

Figure 5-6. The screw shell of a lampholder has the ungrounded conductor connected to the tab in the back of the socket to prevent an electric shock.

The switch leg is the conductor that is being controlled by a switch. It will only be energized when the switch is on.

Article 210, Branch Circuits Not Over 1000 Volts ac, 1500 Volts dc, Nominal

Article 210 contains requirements for branch circuits. The *NEC* defines a **branch circuit** as the circuit conductors between the final overcurrent device protecting the circuit and the outlet(s). The requirements within *Article 210* are used when determining the size and type of branch circuit, as well as the locations of receptacles and lights (luminaires). *Article 210* is divided into three parts, **Figure 5-7**.

Part I, General

Part I of *Article 210* contains general requirements that apply to branch circuits with nominal voltages that are less than 1000 volts ac, 1500 volts dc.

Ground-fault circuit interrupter (GFCI) protection, found in *Section 210.8*, is an example of the general type of information that is found in *Part I* of *Article 210*. One example of a ground-fault circuit interrupter is a receptacle with a test button and a reset button often found near sinks, **Figure 5-8**. The purpose of a GFCI is to remove the power from the circuit in the event a person is receiving an electric shock due to current that is traveling through the person to ground, **Figure 5-9**. When working properly, the GFCI will remove the power from the circuit before the shock has the potential to become lethal. One of the most commonly known locations for GFCI receptacles is near sinks, but there are many other locations and pieces of equipment that require protection. *Section 210.8* is where most of the general GFCI requirements can be found, **Figure 5-10**.

Arc-fault circuit interrupters (AFCI) protection, found in *Section 210.12*, is another type of protection required by *Article 210*, **Figure 5-11**. An AFCI monitors the protected branch circuit for arcing faults that have the potential to start a fire. If the AFCI detects an arcing fault, it will disconnect the circuit, thereby extinguishing the fault before it can generate enough heat to start a fire. AFCI protection is required for most of the livable areas in dwelling units, hotels, and dorms. AFCI protection is typically provided by a circuit breaker as it protects the entire branch circuit.

Part II, Branch-Circuit Ratings

Part II of *Article 210* is titled *Branch-Circuit Ratings*. *Part II* has information for planning various branch circuit requirements, such as the maximum loads permitted on circuits and the size of the circuit that must be installed for a piece of equipment.

Section 210.18, Rating is often used while planning a circuit. It states that branch circuits are rated according to the maximum ampere rating of the overcurrent device, typically

Unit 5   NEC Chapter 2

Article 210
Branch Circuits

Part I General
Section 210.1 Scope
Section 210.2 Reconditioned Equipment
Section 210.3 Other Articles for Specific-Purpose Branch Circuits
Section 210.4 Multiwire Branch Circuits
Section 210.5 Identification for Branch Circuits
Section 210.6 Branch-Circuit Voltage Limitations
Section 210.7 Multiple Branch Circuits
Section 210.8 Ground-Fault Circuit-Interrupter Protection for Personnel
Section 210.9 Circuits Derived from Autotransformers
Section 210.10 Ungrounded Conductors Tapped from Grounded Systems
Section 210.11 Branch Circuits Required
Section 210.12 Arc-Fault Circuit-Interrupter Protection
Section 210.13 Ground-Fault Protection of Equipment
Section 210.17 Guest Rooms and Guest Suites

Part II Branch-Circuit Ratings
Section 210.18 Rating
Section 210.19 Conductors – Minimum Ampacity and Size
Section 210.20 Overcurrent Protection
Section 210.21 Outlet Devices
Section 210.22 Permissible Loads, Individual Branch Circuits
Section 210.23 Permissible Loads, Multiple – Outlet Branch Circuits
Section 210.24 Branch-Circuit Requirements – Summary
Section 210.25 Branch Circuits in Buildings with More Than One Occupancy

Part III Required Outlets
Section 210.50 Receptacle Outlets
Section 210.52 Dwelling Unit Receptacle Outlets
Section 210.60 Guest Rooms, Guest Suites, Dormitory Units, and Similar Occupancies
Section 210.62 Show Windows
Section 210.63 Equipment Requiring Servicing
Section 210.65 Meeting Rooms
Section 210.70 Lighting Outlets Required

Goodheart-Willcox Publisher

Figure 5-7. *Article 210* can be categorized as a design article and contains minimum branch circuit requirements that must be followed when designing an electrical system.

Goodheart-Willcox Publisher

Figure 5-8. The purpose of a ground-fault circuit interrupter is to protect people from receiving an electric shock that could be lethal. The GFCI in the picture will monitor the current passing through it. In the event of a ground fault of 4–6 milliamperes or more, it will trip and de-energize the circuit.

Effects of Electric Shock

Level of Current	Effects on Human Body
1 mA	Threshold of feeling. Slight tingling.
5 mA	Shock felt, but not painful yet. Involuntary muscle movements.
10–20 mA	Painful shock. Sustained muscle contraction. Inability to release grip.
100–300 mA	Paralysis of respiratory muscles. Can be fatal. Severe internal and external burns.
2 A	Cardiac arrest (heartbeat stops). Internal organ damage. Death is probable.

Goodheart-Willcox Publisher

Figure 5-9. Depending on the current flowing through the circuit, effects can range from minimal to lethal. Class A ground-fault circuit interrupters will open in the circuit if a ground fault in the range of 5 milliamperes exists.

General Locations Requiring GFCI Protection for Personnel

Dwelling Unit Receptacle Outlets	Non-Dwelling Unit Receptacle Outlets	Specific Appliances and Equipment
Bathrooms	Bathrooms	Crawl space lighting
Garages and accessory buildings	Kitchens/food prep	Receptacles installed for servicing equipment
Outdoors	Buffet serving areas	Automotive vacuum machines
Crawlspaces at or below grade	Rooftop	Drinking fountain/water coolers
Basements	Outdoor	High-pressure washer machines
Kitchens	Within 6' of sinks	Tire inflator machines
Areas with sinks for food prep	Indoor damp and wet locations	Vending machines
Within 6' of sinks	Locker rooms with showers	Sump pumps
Boathouses	Garages and accessory buildings	Dishwashers
Within 6' of bathtubs or shower stalls	Service bays	Ranges
Laundry areas	Crawl spaces at or below grade	Wall-mounted ovens
Indoor damp and wet locations	Unfinished basements	Counter mounted cooking units
	Aquariums, bait wells, etc.	Clothes dryers
	Laundry areas	Microwave ovens
	Within 6' of bathtubs or shower stalls	
	Outdoor receptacle outlets	

Goodheart-Willcox Publisher

Figure 5-10. *Section 210.8* has GFCI requirements that apply generally to all electrical installations.

Dwelling Unit Locations Required to Be Protected by AFCI

Kitchens
Family rooms
Dining rooms
Living rooms
Parlors
Libraries
Dens
Bedrooms
Sunrooms
Recreation rooms
Closets
Hallways
Laundry areas

Goodheart-Willcox Publisher

Figure 5-11. *Section 210.12* provides requirements for AFCI protection in dwelling units and similar occupancies where people will spend the night, other than hospitals.

Goodheart-Willcox Publisher

Figure 5-12. Circuit breakers are a type of overcurrent protective device (OCPD). The OCPD is what determines the rating of the branch circuit. These circuit breakers have the ampacity stamped on the end of the breaker handle to identify the maximum ampacity of that circuit.

a circuit breaker or fuse. See **Figure 5-12**. This means that the size of the circuit cannot be assumed by the size of the conductors. The ampacity of the overcurrent that feeds the circuit limits the current and therefore determines the circuit size.

Section 210.22 is titled *Permissible Loads, Individual Branch Circuits*. To understand branch circuits, we must know the *NEC* definition of an individual branch circuit. The *NEC* defines an ***individual branch circuit*** as a branch circuit that supplies only one utilization equipment. It is common to hear people refer to an individual branch circuit as a dedicated branch circuit, although you will not find the term "dedicated" in the *NEC*. *Section 210.22* states that the load of an individual branch circuit cannot exceed the ampere rating of the overcurrent device.

Part III, Required Outlets

When laying out the receptacle, switch, and luminaire locations, *Part III* of *Article 210* has the minimum *NEC* requirements that must be followed. Dwelling units have numerous receptacle and lighting outlet requirements, so a large portion of *Part III* is dedicated to dwelling units.

Section 210.52(A) contains the general rule for receptacle spacing in dwelling units. This requirement applies to most of the habitable rooms in a dwelling. The *NEC* defines a **habitable room** as a room in a building for living, sleeping, eating, or cooking, but excluding bathrooms, toilet rooms, closets, hallways, storage or utility spaces, and similar areas. *Section 210.52(A)* states that any point along the wall must be within 6′ of a receptacle. This ensures that floor lamps and similar equipment will reach a receptacle anyplace along the wall they may be placed without needing an extension cord. See **Figure 5-13**.

Article 215, Feeders

Article 215 contains requirements for feeders. The *NEC* defines a *feeder* as all circuit conductors between the service equipment and the source of a separately derived system, or the other power supply source and the final branch-circuit overcurrent device. Another way to look at feeders is that they are the conductors or cables that feed a subpanel. *Article 215* is short and is not divided into parts, **Figure 5-14**.

Goodheart-Willcox Publisher

Figure 5-13. *Section 210.52(A)* has receptacle outlet placement requirements. Living rooms, as shown in the picture, must have receptacle outlets placed so that any point along the wall is within 6′ of an outlet. This is to prevent the need for extension cords to plug in items such as a table lamp.

Article 215 — Feeders

Section 215.1 Scope
Section 215.2 Minimum Rating and Size
Section 215.3 Overcurrent Protection
Section 215.4 Feeders with Common Neutral Conductor
Section 215.5 Diagrams of Feeders
Section 215.6 Feeder Equipment Grounding Conductor
Section 215.7 Ungrounded Conductors Tapped from Grounded Systems
Section 215.9 Ground-Fault Circuit-Interrupter Protection for Personnel
Section 215.10 Ground-Fault Protection of Equipment
Section 215.11 Circuits Derived from Autotransformers
Section 215.12 Identification for Feeders
Section 215.15 Barriers
Section 215.18 Surge Protection

Goodheart-Willcox Publisher

Figure 5-14. *Article 215* can be categorized as a design article and contains requirements specific to feeders.

Article 220, Branch-Circuit, Feeder, and Service Load Calculations

Article 220 is the calculation article, **Figure 5-15**. It has requirements for calculating branch circuits, feeders, and the electrical service. *NEC* calculations can be a bit confusing, so *Informative Annex D* in the back of the *Code* has sample calculations to aid in understanding.

Part I, General

Part I of *Article 220* has general information that applies to all the *Article 220* calculations. An example is *Section 220.5, Calculations* which lists the voltages to be used for calculations. This section is important as it is quite common to hear people referring to equipment by a nominal voltage rather than the actual voltage available. For example, many people refer to a piece of equipment as being 110 or 220 volts when the actual voltage is 120 or 240 volts. Performing calculations using the wrong voltage will result in an incorrect current value, which may impact overcurrent protection, conductor size, or raceways size.

Part II, Branch-Circuit Load Calculations

Since *Part II* of *Article 210* contains provisions for making branch circuit calculations, much of the information found in *Part II* feeds directly into *Part III* for performing feeder and service calculations.

Article 220
Branch-Circuit, Feeder, and Service Load Calculations

Part I General
Section 220.1 Scope
Section 220.3 Other Articles for Specific-Purpose Calculations
Section 220.5 Calculations

Part II Branch-Circuit Load Calculations
Section 220.10 General
Section 220.11 Maximum Load
Section 220.14 Other Loads – All Occupancies
Section 220.16 Loads for Additions to Existing Installations

Part III Feeder and Service Load Calculations
Section 220.40 General
Section 220.41 Dwelling Units, Minimum Unit Load
Section 220.42 Lighting Load for Non-Dwelling Occupancies
Section 220.43 Office Buildings
Section 220.44 Hotel and Motel Occupancies
Section 220.45 General Lighting
Section 220.46 Show-Window and Track Lighting
Section 220.47 Receptacle Loads – Other Than Dwelling Units
Section 220.50 Motors and Air-Conditioning Equipment
Section 220.51 Fixed Electric Space Heating
Section 220.52 Small-Appliance and Laundry Loads – Dwelling Unit
Section 220.53 Appliance Load – Dwelling Unit(s)
Section 220.54 Electric Clothes Dryers – Dwelling Unit(s)
Section 220.55 Electric Cooking Appliances in Dwelling Units and Household Cooking Appliances Used in Instructional Programs
Section 220.56 Kitchen Equipment – Other Than Dwelling Units
Section 220.57 Electric Vehicle Supply Equipment (EVSE) Load
Section 220.60 Noncoincident Loads
Section 220.61 Feeder or Service Neutral Load
Section 220.70 Energy Management Systems (EMSs)

Part IV Optional Feeder and Service Load Calculations
Section 220.80 General
Section 220.82 Dwelling Unit
Section 220.83 Existing Dwelling Unit
Section 220.84 Multifamily Dwelling
Section 220.85 Two Dwelling Units
Section 220.86 Schools
Section 220.87 Determining Existing Loads
Section 220.88 New Restaurants

Part V Farm Load Calculations
Section 220.100 General
Section 220.102 Farm Loads – Buildings and Other Loads
Section 220.103 Farm Loads – Total

Part VI Health Care Facilities
Section 220.110 Receptacle Loads

Part VII Marinas, Boatyards, Floating Buildings, and Commercial and Noncommercial Docking Facilities
Section 220.120 Receptacle Loads

Goodheart-Willcox Publisher

Figure 5-15. *Article 220* can be categorized as a design article and is the calculation article. It has information on how branch circuits, feeders, and services are to be calculated. To aid the user, *Informative Annex D* in the back of the *NEC* has some sample calculations.

When laying out the branch circuits in a commercial building, one of the things to consider is receptacle outlets. The tricky part about a receptacle is that you have no idea what is going to be plugged in or how much current it will draw. *Section 220.14(I), Receptacle Outlets* addresses this issue. It states that each duplex receptacle shall be calculated at 180 volt-amperes per unit. Keep in mind that this rule does not apply to dwellings.

Part III, Feeder and Service Load Calculations

Part III of *Article 220* has minimum requirements for calculating feeders and service calculations. When performing a service calculation, all aspects of the electrical system must be included. It starts with the general lighting load and required circuits and includes all electrical equipment that will be on the premises. The calculations can be a bit confusing, so the *NEC* has provided a few sample calculations in *Informative Annex D* to help explain the process.

When performing a dwelling unit service calculation, two of the items that must be included are the small appliance and laundry branch circuits. *Section 210.11* requires a dwelling to have at least two small appliance circuits and at least one laundry circuit. *Section 220.52* states that 1500 volt-amperes must be included in the service calculation for each small appliance and laundry branch circuit.

Part IV, Optional Feeder and Service Load Calculations

Part IV of *Article 220* has an optional method of performing a service and feeder calculation. It is permitted to be used only for single and multifamily dwelling units, schools, new restaurants, and existing installations. The optional method generally provides a result that is smaller than the standard calculation.

Parts V, VI, and VII

Parts V, *VI*, and *VII* of *Article 220* provide additional demand factors for farm loads; health care facilities; marinas, boatyards, boating buildings, and docking facilities. Parts II and III of Article 220 still apply and are used to determine the service load, but Parts V, VI, and VII provide additional demand factors.

Article 225, Outside Branch Circuits and Feeders

Article 225 is dedicated to outside branch circuits and feeders, **Figure 5-16**. A common scenario for an outside feeder would be a separate building on the property that has an

Article 225
Outside Branch Circuits and Feeders

Part I General
Section 225.1 Scope
Section 225.3 Other Articles
Section 225.4 Conductor Insulation
Section 225.6 Conductor Size and Support
Section 225.10 Wiring on Buildings (or Other Structures)
Section 225.11 Feeder and Branch-Circuit Conductors Entering, Exiting, or Attached to Buildings or Structures
Section 225.12 Open-Conductor Supports
Section 225.14 Open-Conductor Spacings
Section 225.15 Supports over Buildings
Section 225.16 Attachment to Buildings
Section 225.17 Masts as Supports
Section 225.18 Clearance for Overhead Conductors and Cables
Section 225.19 Clearances from Buildings for Conductors of Not over 1000 Volts, Nominal
Section 225.20 Protection Against Physical Damage
Section 225.21 Multiconductor Cables on Exterior Surfaces of Buildings (or Other Structures)
Section 225.22 Raceways on Exterior Surfaces of Buildings or Other Structures
Section 225.24 Outdoor Lampholders
Section 225.25 Location of Outdoor Lamps
Section 225.26 Vegetation as Support
Section 225.27 Raceway Seal

Part II Buildings or Other Structures Supplied by a Feeder(s) or Branch Circuit(s)
Section 225.30 Number of Supplies
Section 225.31 Disconnecting Means
Section 225.33 Maximum Number of Disconnects
Section 225.34 Grouping of Disconnects
Section 225.35 Access to Occupants
Section 225.36 Type of Disconnecting Means
Section 225.37 Identification
Section 225.38 Disconnect Construction
Section 225.39 Rating of Disconnect
Section 225.40 Access to Overcurrent Protective Devices
Section 225.41 Emergency Disconnects
Section 225.42 Surge Protection

Goodheart-Willcox Publisher

Figure 5-16. *Article 225* can be categorized as a design article and has requirements for outside branch circuits and feeders.

electrical panel that is fed as a subpanel rather than having its own service. A detached garage or shed would be an example. Outside branch circuits may run to parking lot lights or a raceway on the exterior of a building that feeds receptacles and lights, **Figure 5-17**.

Section 225.22, Raceways on Exterior Surfaces of Buildings or Other Structures

Section 225.22 has requirements for raceways that are run on exterior surfaces of buildings. It states that they shall be listed for use in wet locations and arranged to drain. It is obvious that a raceway on the outside of a building will get wet, but we also need to consider that the inside of the raceway may also get wet. For that reason, the conductors that are installed in the raceway must be listed for wet locations, and the raceway must be installed in a manner such that water will drain from the raceway.

Article 230, Services

Article 230 is dedicated to the electrical service. *Article 100* of the *NEC* defines a **service** as the conductors and equipment connecting the serving utility to the wiring system of the premises served. All the components involved in getting power from the service point the main service disconnect are covered in *Article 230*, **Figure 5-18**.

Goodheart-Willcox Publisher

Figure 5-17. This post has a receptacle for plugging in a lake pump and a post light. Both are examples of an outside branch circuit.

Part I, General

Part I of *Article 230* is titled *General*. Like all the other general parts of the *NEC*, it has general requirements that apply to all services. A one-line diagram of the electrical service is provided in *Figure 230.1*. The diagram has descriptions next to the components and a reference to the portion of the *NEC* that applies, **Figure 5-19**.

Parts II, III, and IV

Parts II, *III*, and *IV* of *Article 230* all apply to service conductors. *Part II, Overhead Service Conductors* contains information for service conductors that are coming overhead from a pole or another building. *Part III, Underground Service Conductors* contains requirements for service conductors that are run underground to a building, typically from a pole or meter pedestal. *Part IV, Service-Entrance Conductors* contains requirements for service entrance conductors once they are at the building or structure.

Parts V, VI, and VII

Parts V, *VI*, and *VII* of *Article 230* apply to service equipment. *Part V, Service Equipment–General* contains general requirements that apply to all service equipment. For example, *Section 230.66(A)* requires all equipment to be identified as suitable for use as service equipment, and that it shall be listed, or field evaluated. See **Figure 5-20**. *Part VI, Service Equipment—Disconnecting Means* contains requirements specific to a service disconnect. *Part VII, Service Equipment—Overcurrent Protection* has requirements specific to service equipment that contain overcurrent protection.

Protection Articles

The protection articles contain information relating to protection of an electrical system. Although these articles are used while designing a project, it is easier to think of them as providing protection to the electrical system. The protection articles will help protect the electrical system from both overcurrent and overvoltage, and ensure that all metallic components are grounded and bonded together properly. There are three *Chapter 2* articles classified as protection articles, **Figure 5-21**.

Article 240, Overcurrent Protection

Article 240 is dedicated to overcurrent protection. The *NEC* defines an **overcurrent** as any current in excess of the rated current of equipment, or the ampacity of a conductor. It may result from overload, short circuit, or ground fault.

The *NEC* identifies three types of overcurrents within that definition that also must be defined. The *NEC* defines an **overload** as operation of equipment in excess of normal, full-load rating, or of a conductor in excess of its ampacity

Article 230
Services

Part I General

Section 230.1 Scope
Section 230.2 Number of Services
Section 230.3 One Building or Other Structure Not to Be Supplied Through Another
Section 230.6 Conductors Considered Outside the Building
Section 230.7 Other Conductors
Section 230.8 Raceway Seal
Section 230.9 Clearances on Buildings
Section 230.10 Vegetation as Support

Part II Overhead Service Conductors

Section 230.22 Insulation or Covering
Section 230.23 Size and Ampacity
Section 230.24 Clearances
Section 230.26 Point of Attachment
Section 230.27 Means of Attachment
Section 230.28 Service Masts as Supports
Section 230.29 Supports over Buildings

Part III Underground Service Conductors

Section 230.30 Installation
Section 230.31 Size and Ampacity
Section 230.32 Protection Against Damage
Section 230.33 Spliced Conductors

Part IV Service-Entrance Conductors

Section 230.40 Number of Service-Entrance Conductor Sets
Section 230.41 Insulation of Service-Entrance Conductors
Section 230.42 Minimum Size and Ampacity
Section 230.43 Wiring Methods for 1000 Volts, Nominal, or Less
Section 230.44 Cable Trays
Section 230.46 Spliced and Tapped Conductors
Section 230.50 Protection Against Physical Damage
Section 230.51 Mounting Supports
Section 230.52 Individual Conductors Entering Buildings or Other Structures
Section 230.53 Raceways to Drain
Section 230.54 Overhead Service Locations
Section 230.56 Service Conductor with the Higher Voltage to Ground

Part V Service Equipment – General

Section 230.62 Service Equipment – Enclosed or Guarded
Section 230.66 Marking
Section 230.67 Surge Protection

Part VI Service Equipment – Disconnecting Means

Section 230.70 General
Section 230.71 Maximum Number of Disconnects
Section 230.72 Grouping of Disconnects
Section 230.74 Simultaneous Opening of Poles
Section 230.75 Disconnection of Grounded Conductor
Section 230.76 Manually or Power Operable
Section 230.77 Indicating
Section 230.79 Rating of Service Disconnecting Means
Section 230.80 Combined Rating of Disconnects
Section 230.81 Connection to Terminals
Section 230.82 Equipment Connected to the Supply Side of the Service Disconnect
Section 230.85 Emergency Disconnects

Part VII Service Equipment – Overcurrent Protection

Section 230.90 Where Required
Section 230.91 Location
Section 230.92 Locked Service Overcurrent Devices
Section 230.93 Protection of Specific Circuits
Section 230.94 Relative Location of Overcurrent Device and Other Service Equipment
Section 230.95 Ground-Fault Protection of Equipment

Figure 5-18. *Article 230* can be categorized as a design article and contains requirements for the electric service.

Part 1 Layout and Structure: Learning to Use the NEC

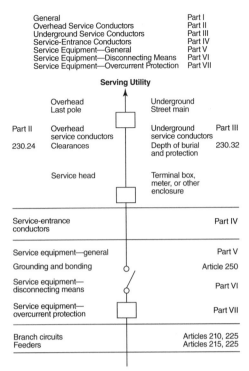

Figure 5-19. *Figure 230.1* has a diagram that shows the components of an electrical service and a reference to the applicable portion of the *NEC*.

Figure 5-20. Equipment that is being installed as part of an electrical service must be listed as suitable for use as service equipment, proving that is has been tested to be able to withstand the dangers associated with an electrical service.

NEC Chapter 2
Protection Articles

| Article 240 Overcurrent Protection |
| Article 242 Overvoltage Protection |
| Article 245 Overcurrent Protection for Systems Rated Over 1000 Volts ac, 1500 Volts dc |
| Article 250 Grounding and Bonding |

Goodheart-Willcox Publisher

Figure 5-21. There are three articles in *Chapter 2* that can be classified as protection. The requirements they contain will offer protection to the electrical equipment as well as personnel.

that, when it persists for a sufficient length of time, would cause damage or dangerous overheating. An example of an overload would be plugging two 12-ampere vacuums into one 20-ampere circuit. The two vacuums would draw a total of 24-amperes on a circuit that is only capable of safely handling 20-amperes.

A short circuit occurs when two conductive objects with a difference in potential touch, creating a low impedance path and extremely high currents. The two conductive objects could be two ungrounded conductors with a difference in potential, or an ungrounded conductor and a neutral conductor. Short circuits will result in extremely high currents that have the potential to create a dangerous arc flash and/or arc blast.

The *NEC* defines a **ground fault** as an unintentional, electrically conductive connection between an ungrounded conductor of an electrical circuit and the normally non-current-carrying conductors, metallic enclosures, metallic raceways, metallic equipment, or earth. Basically, a ground fault occurs when an energized conductor contacts something that is grounded other than the neutral conductor. A ground fault will often generate extremely high currents that have the potential to create a dangerous arc flash/blast such as a short circuit.

The primary reason for overcurrent protection is to protect the conductors. Excess current that causes heating typically results from overloads. The overcurrent protection must be sized properly to ensure that the conductors are protected from excess current to maintain the integrity of the conductor insulation. That does not mean that overcurrent devices do not offer protection to equipment and in some cases personnel, those functions are just in addition to the primary function, which is to protect the conductors.

Article 240 has nine parts to address the various scenarios, **Figure 5-22**.

Part I, General

Article 240, Part I is titled *General*. It has requirements that apply broadly to all aspects of overcurrent protection. One

Article 240
Overcurrent Protection

Part I General
Section 240.1 Scope
Section 240.2 Reconditioned Equipment
Section 240.3 Other Articles
Section 240.4 Protection of Conductors
Section 240.5 Protection of Flexible Cords, Flexible Cables, and Fixture Wires
Section 240.6 Standard Ampere Ratings
Section 240.7 Listing Requirements
Section 240.8 Fuses or Circuit Breakers in Parallel
Section 240.9 Thermal Devices
Section 240.10 Supplementary Overcurrent Protection
Section 240.11 Selective Coordination
Section 240.12 Orderly Shutdown
Section 240.13 Ground-Fault Protection of Equipment
Section 240.15 Ungrounded Conductors
Section 240.16 Interrupting Ratings

Part II Location
Section 240.21 Location in a Circuit
Section 240.22 Grounded Conductor
Section 240.24 Location in or on Premises

Part III Enclosures
Section 240.30 General
Section 240.32 Damp or Wet Locations
Section 240.24 Vertical Position

Part IV Disconnecting and Guarding
Section 240.40 Disconnecting Means for Fuses
Section 240.41 Arcing or Suddenly Moving Parts

Part V Plug Fuses, Fuseholders, and Adapters
Section 240.50 General
Section 240.51 Edison-Base Fuses
Section 240.52 Edison-Base Fuseholders
Section 240.53 Type S Fuses
Section 240.54 Type S Fuses, Adapters, and Fuseholders

Part VI Cartridge Fuses and Fuseholders
Section 240.60 General.
Section 240.61 Classification
Section 240.67 Arc Energy Reduction

Part VII Circuit Breakers
Section 240.80 Method of Operation.
Section 240.81 Indicating
Section 240.82 Nontamperable
Section 240.83 Marking
Section 240.85 Applications
Section 240.86 Series Ratings
Section 240.87 Arc Energy Reduction
Section 240.89 Replacement Trip Units

Part VIII Supervised Industrial Installations
Section 240.90 General
Section 240.91 Protection of Conductors
Section 240.92 Location in Circuit

Goodheart-Willcox Publisher

Figure 5-22. *Article 240* can be categorized as a design article and contains requirements relating to overcurrent protection. It is the first of the *Chapter 2* protection articles.

of the more commonly used sections from *Part I* is *Section 240.4, Protection of Conductors*. It describes how to choose the size of the overcurrent device.

Part II, Location

Article 240, Part II is titled *Location*. It looks at the location of the overcurrent protection two different ways: location in the circuit and physical proximity.

Section 240.21, Location in Circuit addresses the location of overcurrent protection within a circuit. This section provides a general rule that the overcurrent protection is generally installed where a conductor receives its supply.

Section 240.24, Location in or on Premises, addresses physical proximity. It provides limitations as to where overcurrent protection may be installed. For example, overcurrent devices shall not be installed over steps, in bathrooms, in the vicinity of easily ignitable material (clothes closets), or installed where they may be exposed to physical damage.

Part III and IV

Parts III and *IV* of *Article 240* are short parts that contain only a few sections. *Part III* is titled *Enclosures,* and *Part IV* is titled *Disconnecting and Guarding.*

Part V, Plug Fuses, Fuseholders, and Adapters

Part V of *Article 240* is titled *Plug Fuses, Fuseholders, and Adapters.* Plug fuses were found in early electrical systems, **Figure 5-23**. They are no longer permitted to be installed in new locations but can be replaced in existing locations.

Part VI, Cartridge Fuses and Fuseholders

Part VI of *Article 240* is titled *Cartridge Fuses and Fuseholders.* See **Figure 5-24**. Although cartridge fuses are not being installed as often as they used to be, there are some situations where they are preferable. This is particularly true in situations where specific current-limiting aspects of fuses are advantageous or when implementing selective coordination.

Part VII, Circuit Breakers

Part VII is titled *Circuit Breakers.* Circuit breakers are the most common form of overcurrent protection being used in installations today, **Figure 5-25**. In the event of an overcurrent, they do not typically need to be replaced, and they can provide more than just overcurrent protection. Circuit breakers are available with ground-fault protection, arc-fault protection, and some of the larger circuit breakers have adjustable trip settings.

Part VIII, Supervised Industrial Installations

Article 240, Part VIII has requirements for overcurrent protection that apply only to supervised industrial installations. Supervised industrial installations has a long definition found in *Section 240.2*. The overcurrent protection requirements of *Part VIII* are slightly different than other parts of

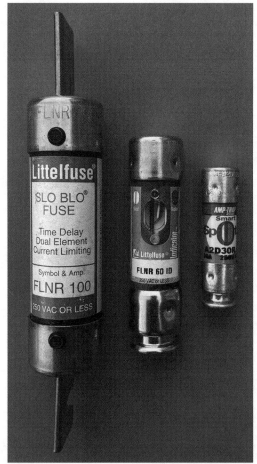

Goodheart-Willcox Publisher

Figure 5-24. Cartridge fuses come in many sizes and styles. Unlike plug fuses, cartridge fuses are still permitted to be installed as overcurrent protection in both new and existing installations.

Goodheart-Willcox Publisher

Figure 5-23. Plug fuses are screw-in fuses that can be found in some older homes and buildings. They protect the system from issues in an appliance. They are no longer permitted to be installed in new installations but can still be installed as a replacement in an existing location.

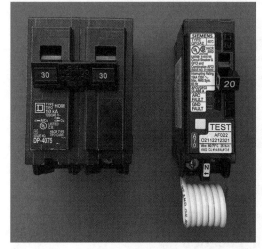

Goodheart-Willcox Publisher

Figure 5-25. Circuit breakers are the most commonly installed form of overcurrent protection in new installations. Breakers have the capability of providing additional features beyond overcurrent protection such as surge suppression, GFCI protection, AFCI protection, etc.

Article 240 as it is understood that there will be qualified people that are continually monitoring the installation.

Article 242, Overvoltage Protection

Article 242 is titled *Overvoltage Protection*. See **Figure 5-26**. An *overvoltage* is an increase in voltage beyond the maximum operating voltage of the system. It is often referred to as a voltage surge. The surge may come internally from a piece of equipment such as a motor turning on and off, from the utility company due to a failure or line switching, or it may come from nature in the form of lightning.

Part II of *Article 242* has requirements for surge-protective devices (SPDs) that are used in systems that are 1000 volts or less. See **Figure 5-27**. *Part III* of *Article 242* has requirements for surge arresters that are used in systems that are over 1000 volts.

Article 250, Grounding and Bonding

Article 250 is titled *Grounding and Bonding*. It is the last of the protection articles. There are many different aspects to grounding and bonding, making *Article 250* an extensive article consisting of ten parts. See **Figure 5-28**.

Goodheart-Willcox Publisher

Figure 5-27. Surge-protective devices (SPDs) are required in all dwelling units by *Section 230.67*. Modern electrical systems have a lot of sensitive electronic equipment, such as AFCIs, GFCIs, and smoke/CO alarms. Surge-protective devices help to protect the sensitive equipment from damage caused by voltage surges.

Article 242
Overvoltage Protection

Part I General
Section 242.1 Scope
Section 242.2 Reconditioned Equipment
Section 242.3 Other Articles

Part II Surge-Protective Devices (SPDs), 1000 Volts or Less
Section 242.6 Listing
Section 242.8 Short-Circuit Current Rating
Section 242.9 Indicating
Section 242.12 Uses Not Permitted
Section 242.13 Type 1 SPDs
Section 242.14 Type 2 SPDs
Section 242.16 Type 3 SPDs
Section 242.18 Type 4 and Other Component Type SPDs
Section 242.20 Number Required
Section 242.22 Location
Section 242.24 Routing of Conductors
Section 242.28 Conductor Size
Section 242.30 Connection Between Conductors
Section 242.32 Grounding Electrode Conductor Connections and Enclosures

Part III Surge Arresters, Over 1000 Volts
Section 242.40 Uses Not Permitted
Section 242.42 Surge Arrester Rating
Section 242.44 Number Required
Section 242.46 Location
Section 242.48 Routing of Surge Arrester Equipment Grounding Conductors
Section 242.50 Connection
Section 242.52 Surge-Arrester Conductors
Section 242.54 Interconnections
Section 242.56 Grounding Electrode Conductor Connections and Enclosures

Goodheart-Willcox Publisher

Figure 5-26. *Article 242* can be categorized as a design article and contains requirements for surge-protective devices and surge arresters.

Article 250
Grounding and Bonding

Part I General
Section 250.1 Scope
Section 250.4 General Requirements for Grounding and Bonding
Section 250.6 Objectionable Current
Section 250.8 Connection of Grounding and Bonding Equipment
Section 250.10 Protection of Ground Clamps and Fittings
Section 250.12 Clean Surfaces

Part II System Grounding
Section 250.20 Alternating-Current Systems to Be Grounded
Section 250.21 Alternating-Current Systems of 50 Volts to 1000 Volts Not Required to be Grounded
Section 250.24 Grounding of Service-Supplied Alternating-Current Systems
Section 250.25 Grounding of Systems Permitted to Be Connected on the Supply Side of the Service Disconnect
Section 250.26 Conductor to Be Grounded – Alternating-Current Systems
Section 250.28 Main Bonding Jumper and System Bonding Jumper
Section 250.30 Grounding Separately Derived Alternating-Current Systems
Section 250.32 Buildings or Structures Supplied by a Feeder(s) or Branch Circuit(s)
Section 250.34 Portable, Vehicle-Mounted, and Trailer-Mounted Generators
Section 250.35 Permanently Installed Generators
Section 250.36 Impedance Grounded Systems – 480 Volts to 1000 Volts

Part III Grounding Electrode System and Grounding Electrode Conductor
Section 250.50 Grounding Electrode System
Section 250.52 Grounding Electrodes
Section 250.53 Grounding Electrode System Installation
Section 250.54 Auxiliary Grounding Electrodes
Section 250.58 Common Grounding Electrode
Section 250.60 Use of Strike Termination Devices
Section 250.62 Grounding Electrode Conductor Material
Section 250.64 Grounding Electrode Conductor Installation
Section 250.66 Size of Alternating-Current Grounding Electrode Conductor
Section 250.68 Grounding Electrode Conductor and Bonding Jumper Connections to Grounding Electrodes
Section 250.70 Methods of Grounding and Bonding Conductors Connection to Electrodes

Part IV Enclosure, Raceway, and Service Cable Connections
Section 250.80 Service Raceways and Enclosures
Section 250.84 Underground Service Cable or Raceway
Section 250.86 Other Conductor Enclosures and Raceways

Part V Bonding
Section 250.90 General
Section 250.92 Services
Section 250.94 Bonding for Communications Systems
Section 250.96 Bonding Other Enclosures
Section 250.97 Bonding for Over 250 Volts to Ground
Section 250.98 Bonding Loosely Jointed Metal Raceways
Section 250.100 Bonding in Hazardous (Classified) Locations
Section 250.102 Grounded Conductor, Bonding Conductors, and Jumpers
Section 250.104 Bonding of Piping Systems and Exposed Structural Metal
Section 250.106 Lightning Protection Systems

(Continued)

Goodheart-Willcox Publisher

Figure 5-28. *Article 250* can be categorized as a design article and contains requirements for grounding and bonding. Electrical systems are grounded to limit the voltage imposed due to lightning or other line surges. Electrical equipment is grounded and bonded to limit the voltage to ground and to establish an effective ground-fault current path.

Article 250
Grounding and Bonding

Part VI Equipment Grounding and Equipment Grounding Conductors

Section 250.109 Metal Enclosures
Section 250.110 Equipment Fastened in Place (Fixed) or Connected by Permanent Wiring Methods
Section 250.112 Specific Equipment Fastened in Place (Fixed) or Connected by Permanent Wiring Methods
Section 250.114 Equipment Connected by Cord and Plug
Section 250.116 Nonelectrical Equipment
Section 250.118 Types of Equipment Grounding Conductors
Section 250.119 Identification of Wire-Type Equipment Grounding Conductors
Section 250.120 Equipment Grounding Conductor Installation
Section 250.122 Size of Equipment Grounding Conductors
Section 250.124 Equipment Grounding Conductor Continuity
Section 250.126 Identification of Wiring Device Terminals

Part VII Methods of Equipment Grounding Conductor Connections

Section 250.130 Equipment Grounding Conductor Connections
Section 250.132 Short Sections of Raceways or Cable Armor
Section 250.134 Equipment Fastened in Place or Connected by Permanent Wiring Methods (Fixed)
Section 250.136 Equipment Secured to a Metal Rack or Structure
Section 250.138 Cord-and-Plug-Connected Equipment
Section 250.140 Frames of Ranges and Clothes Dryers
Section 250.142 Use of Grounded Circuit Conductor for Grounding Equipment
Section 250.144 Multiple Circuit Connections
Section 250.146 Connecting Receptacle Grounding Terminal to an Equipment Grounding Conductor
Section 250.148 Continuity of Equipment Grounding Conductors and Attachment in Boxes

Part VIII Direct-Current Systems

Section 250.160 General
Section 250.162 Direct-Current Circuits and Systems to Be Grounded
Section 250.164 Point of Connection for Direct-Current Systems
Section 250.166 Size of the Direct-Current Grounding Electrode Conductor
Section 250.167 Direct-Current Ground-Fault Detection
Section 250.168 Direct-Current System Bonding Jumper
Section 250.169 Ungrounded Direct-Current Separately Derived Systems

Part IX Instruments, Meters, and Relays

Section 250.170 Instrument Transformer Circuits
Section 250.172 Instrument Transformer Cases
Section 250.174 Cases of Instruments, Meters, and Relays Operating at 1000 Volts or Less
Section 250.176 Cases of Instruments, Meters, and Relays – Operating at Over 1000 Volts
Section 250.178 Instrument Equipment Grounding Conductor

Part X Grounding of Systems and Circuits of over 1000 Volts

Section 250.180 General
Section 250.182 Derived Neutral Systems
Section 250.184 Solidly Grounded Neutral Systems
Section 250.186 Grounding Service-Supplied Alternating-Current Systems
Section 250.187 Impedance Grounded Systems
Section 250.188 Grounding of Systems Supplying Portable or Mobile Equipment
Section 250.190 Grounding of Equipment
Section 250.191 Grounding System at Alternating-Current Substations
Section 250.194 Grounding and Bonding of Fences and Other Metal Structures

Goodheart-Willcox Publisher

Figure 5-28. Continued

Part I, General

Part I of *Article 250* is titled *General* and, like all the other general portions of the *Code*, it gives general requirements and information that applies broadly. *Section 250.4* explains the purpose of grounding and bonding. Grounded electrical systems are connected to earth to limit the voltage to ground from lightning and surges and to stabilize the voltage to ground under normal circumstances. Electrical equipment must be properly grounded to ensure a low impedance path to ground, allowing the overcurrent device to quickly disconnect the circuit in the event of a short to ground. For example, if an energized conductor in a blender contacts the metal frame, the grounding path should quickly trip the fuse or breaker, so the frame does not remain energized.

Part II, System Grounding

Part II of *Article 250* is titled *System Grounding*. It describes which systems are to be grounded and how grounding is accomplished.

Part III, Grounding Electrode System and Grounding Electrode Conductor

Part III of *Article 250* is titled *Grounding Electrode System and Grounding Electrode Conductor*. The *NEC* defines a **grounding electrode** as a conducting object through which a direct connection to earth is established. Grounding electrodes are what we use to make the connection to the earth.

The **grounding electrode conductor** is defined as a conductor used to connect the system grounded conductor or the equipment to a grounding electrode or to a point on the grounding electrode system. See **Figure 5-29**. *Part III* of *Article 250* lists all the grounding electrodes that may be used, how they are to be installed, and the method of connecting them to the electrical system.

Part IV, Enclosure, Raceway, and Service Cable Connections

Part IV of *Article 250* is titled *Enclosure, Raceway, and Service Cable Connections*. It is a short part that contains three sections. It essentially tells us that all metal raceways and enclosures must be grounded. There are a few exceptions, but they are rare.

Part V, Bonding

Part V of *Article 250* is titled *Bonding*. The *NEC* defines **bonding** as connected to establish electrical continuity and

Goodheart-Willcox Publisher

Figure 5-29. A ground rod is one of the permitted grounding electrodes. If a ground rod's resistance to ground is greater than 25 ohms, the *NEC* requires an additional supplemental ground rod located at least 6′ away.

conductivity. It is connecting conductive objects together to eliminate a potential voltage between them that could lead to electric shock.

Part VI, Equipment Grounding and Equipment Grounding Conductors

Part VI of *Article 250* is titled *Equipment Grounding and Equipment Grounding Conductors*. *Part VI* lists the types of equipment that must be connected to an equipment grounding conductor. It also lists the types of equipment grounding conductors that are permitted and how to size them.

Part VII, Methods of Equipment Grounding Conductor Connections

Part VII of *Article 250* is titled *Methods of Equipment Grounding Conductor Connections*. It has requirements to ensure there is an effective equipment grounding conductor path in various scenarios.

NEC KEYWORDS

NEC Chapter 2 Keywords

Keywords that should lead you to the design articles of *Chapter 2*.

AFCI	feeder	grounded	service
branch circuit	general lighting	load	
calculation	GFCI	neutral	

Keywords that should lead you to the protection articles of *Chapter 2*.

arrester	electrode	grounding	protection
bonding	fuse	overcurrent	surge
breaker	ground	overvoltage	

Summary

- *Chapter 2* of the *NEC* is the second of the general chapters that apply to all electrical installations.
- *Chapter 2* articles can be divided into two categories: design and protection.
- The *Chapter 2* design articles have requirements necessary when designing electrical systems.
  - Article 200, Use and Identification of Grounded Conductors
  - Article 210, Branch Circuits Not Over 1000 Volts ac, 1500 Volts dc, Nominal
  - Article 215, Feeders
  - Article 220, Branch-Circuit, Feeder, and Service Load Calculations
  - Article 225, Outside Branch Circuits and Feeders
  - Article 230, Services
  - Article 235, Branch Circuits, Feeders, and Services Over 1000 Volts ac, 1500 Volts dc, Nominal
- The *Chapter 2* protection articles have requirements that address protection in electrical systems.
  - Article 240, Overcurrent Protection
  - Article 242, Overvoltage Protection
  - Article 245, Overcurrent Protection for Systems Rated Over 1000 Volts ac, 1500 Volts dc
  - Article 250, Grounding and Bonding

Unit 5 Review

Name _____ Date _____ Class _____

Know and Understand

Answer the following questions based on information in this unit.

1. Chapter 2 of the *NEC* has information necessary when _____ an electrical system.
   A. designing
   B. roughing-in
   C. trimming
   D. installing

2. *Chapter 2* contains _____ protection articles.
   A. two
   B. three
   C. four
   D. five

3. _____ has information on load calculations.
   A. *Article 200*
   B. *Article 210*
   C. *Article 220*
   D. *Article 230*

4. _____ has requirements for fuses.
   A. *Article 210*
   B. *Article 220*
   C. *Article 230*
   D. *Article 240*

5. Which part of *Article 250* has information on bonding?
   A. *Part II*
   B. *Part IV*
   C. *Part V*
   D. *Part VII*

6. _____ has requirements specific to the grounded conductor.
   A. *Article 200*
   B. *Article 220*
   C. *Article 242*
   D. *Article 250*

7. Part II of *Article 250* is titled _____.
   A. *General*
   B. *System Grounding*
   C. *Bonding*
   D. *Direct Current Systems*

8. *Part I* of all the articles that have parts contains _____ information.
   A. specific
   B. general
   C. provisional
   D. over 1000 volts

9. Which section of *Article 210* has information on arc-fault circuit interrupters?
   A. *Section 210.8*
   B. *Section 210.10*
   C. *Section 210.11*
   D. *Section 210.12*

10. Which article contains requirements that apply to outside feeders?
    A. *Article 210*
    B. *Article 215*
    C. *Article 220*
    D. *Article 225*

Apply and Analyze

NEC Chapter 2

Answer the following questions using a copy of the National Electrical Code. *Identify the section or subsection where the answer is found.*

1. All receptacles within _____ of dwelling unit sinks must be protected by a ground fault circuit interrupter.
   A. 3′
   B. 6′
   C. 8′
   D. 10′

   NEC _____

2. Overcurrent devices shall not be located in _____.
   A. bathrooms
   B. bedrooms
   C. family rooms
   D. unfinished spaces

   NEC _____

3. Which of the following is one of the required 20-ampere branch circuits required in a dwelling?
   A. Dishwasher
   B. Living room
   C. Bedroom
   D. Laundry

   NEC _____

Copyright Goodheart-Willcox Co., Inc.
May not be reproduced or posted to a publicly accessible website.

Part 1   Layout and Structure: Learning to Use the NEC

4. A raceway on the outside of a building that contains a branch circuit must be _____ and be listed for use in wet locations.
   A. made of PVC
   B. water resistant
   C. arranged to drain
   D. protected from the sun

   NEC _____

5. A ground rod type grounding electrode must be at least _____ in length.
   A. 6′
   B. 8′
   C. 9′
   D. 10′

   NEC _____

6. Service equipment rated 1000 volts or less shall be marked to identify it as being _____ for use as service equipment.
   A. designed
   B. identified
   C. intended
   D. suitable

   NEC _____

7. Terminals for the attachment of the grounded conductor shall be substantially white or _____ in color.
   A. gold
   B. silver
   C. black
   D. green

   NEC _____

8. Guest rooms of hotels must have at least one lighting outlet controlled by a listed wall mounted control device installed in every _____ and bathroom.
   A. habitable room
   B. bedroom
   C. closet
   D. kitchenette

   NEC _____

9. When performing a service load calculation, a dwelling unit small appliance load is calculated at _____.
   A. 1000 volt-amps
   B. 1200 volt-amps
   C. 1500 volt-amps
   D. 1800 volt-amps

   NEC _____

10. Plug fuses shall be permitted to be used in circuits not exceeding _____ between conductors.
    A. 120 volts
    B. 125 volts
    C. 150 volts
    D. 240 volts

    NEC _____

Name _____ Date _____ Class _____

Critical Thinking

NEC Chapter 2

Answer the following questions using a copy of the National Electrical Code. *Identify the section or subsection where the answer is found.*

1. Plug fuses that are 15 amps or lower shall be identified by a(n) _____ configuration in cap or window of the fuse.

   _____

   _____

   NEC _____

2. When calculating the general lighting load for an office building, _____ per square foot is used.

   _____

   _____

   NEC _____

3. List the four required branch circuits in a dwelling unit.

   _____

   _____

   NEC _____

4. A surge-protective device shall be installed on the _____ side of the first overcurrent device in a separately derived system.

   _____

   _____

   NEC _____

5. What is the minimum size copper equipment grounding conductor required on a 20-ampere circuit?

   _____

   _____

   NEC _____

6. Grounded conductors that are 6 AWG or smaller, shall have an outer finish that is white, _____, or has three continuous white or gray stripes along its entire length.

   _____

   _____

   NEC _____

7. The service disconnect for a one family dwelling shall be not less than _____ amps, three-wire.

   _____

   _____

   NEC _____

8. Arc-fault circuit interrupters shall be installed in a _____ location.

   _____

   _____

   NEC _____

9. A 120-volt outside branch circuit that travels overhead from a house to a garage shall have a clearance not less than _____ above a residential driveway.

   _____

   _____

   NEC _____

10. A ground rod that has a resistance of less than _____ to ground shall not be required to be supplemented by an additional electrode.

    _____

    _____

    NEC _____

Part 1   Layout and Structure: Learning to Use the NEC

NEC Chapters 1–2

Answer the following questions using a copy of the National Electrical Code. *Identify the section or subsection where the answer is found.*

11. The largest standard ampere rating for a fuse is _____.

    NEC _____

12. A ground ring type grounding electrode shall be at 20 feet of bare copper conductor not smaller than _____.

    NEC _____

13. Unless otherwise marked, equipment terminations rated 100 amperes or less shall be based on _____ °C.

    NEC _____

14. At least one receptacle must be installed within _____ feet of the bathroom sink.

    NEC _____

15. A plenum is a compartment or chamber to which one or more air ducts are connected and that forms part of the _____.

    NEC _____

16. Reconditioned equipment shall be marked with the name, trademark, or other descriptive marking by which the organization responsible for reconditioning the electrical equipment can be identified, along with _____.

    NEC _____

17. Overcurrent devices shall not be in the vicinity of easily ignitable material such as in _____.

    NEC _____

18. The voltage rating of electrical equipment shall not be less than the _____ of a circuit to which it is connected.

    NEC _____

19. Each disconnecting means shall be _____ to indicate its purpose unless located and arranged so the purpose if evident.

    NEC _____

20. Where a branch circuit supplies continuous loads or any combination of continuous and noncontinuous loads, the rating of the overcurrent device shall not be less than the noncontinuous load plus _____ percent of the continuous load.

    NEC _____

UNIT

6 NEC Chapter 3

sockagphoto/Shutterstock.com

KEY TERMS
cabinet
cable
conduit body
cutout box
enclosure
meter socket
raceway

LEARNING OBJECTIVES

After completing this unit, you will be able to:

- Describe the type of requirements in *Chapter 3* of the *NEC*.
- Identify the wiring methods found in *Chapter 3* of the *NEC*.
- Describe the common numbering format used in *Chapter 3* of the *NEC*.
- Recognize keywords that will lead you to *Chapter 3* of the *NEC*.

Introduction

This unit is an introduction to the *NEC*'s *Chapter 3, Wiring Methods and Materials*. It is the third of the general chapters that apply to all installations, **Figure 6-1**. Requirements found in some of the commonly used articles will be discussed to help you recognize the type of information found in *Chapter 3*. The end of the unit contains practice exercises that will require the use of your *Code* to complete. The combination of studying the type of information found in *Chapter 3* as well as looking up *Code* sections will help with comprehension and retention of the type of information found in *Chapter 3* of the *NEC*.

CODE ARRANGEMENT

- Chapter 1: General
- Chapter 2: Wiring and Protection
- Chapter 3: Wiring Methods and Materials **YOU ARE HERE**
- Chapter 4: Equipment for General Use

} Applies generally to all electrical installations

- Chapter 5: Special Occupancies
- Chapter 6: Special Equipment
- Chapter 7: Special Conditions

} Supplements or modifies chapters 1 through 7

- Chapter 8: Communications Systems

} Chapter 8 is not subject to the requirements of chapters 1 through 7 except where the requirements are specifically referenced in chapter 8

- Chapter 9: Tables

} Applicable as referenced

Informative Annex A through Informative Annex J

} Informational only; not mandatory

Goodheart-Willcox Publisher

Figure 6-1. *Chapter 3* is the third of the 4 general chapters that contain information applying to all installations.

Chapter 3, Wiring Methods and Materials

Chapter 3 contains requirements for all the various wiring methods installed while constructing an electrical system. The wiring methods in *Chapter 3* contain all the materials used to get the energy from the panelboard to the outlet. These materials include boxes, raceways, fittings, conductors, and cables. Another way to remember the type of information found in *Chapter 3* is that it contains many of the wiring methods installed during the rough-in stage of the construction process. See **Figure 6-2**.

Common Format

The *NEC* has a common numbering format that is used for the cable and raceway articles. The common format is used for the parts as well as the sections. Having a common format makes it easier to quickly find information.

Parts

Articles 320–396 follow a common part format. See **Figure 6-3**. These articles include all the cable and raceway

Goodheart-Willcox Publisher

Figure 6-2. The electrical rough-in is when the electrician installs the boxes, raceways, and cables so the project is ready for the wall finish to be installed. This figure shows the electrical rough-in of a wall on a commercial project.

Figure 6-3. All the cable and raceway articles in *Chapter 3* are divided into the same 3 parts.

Section	Title
Part I	General
Part II	Installation
Part III	Construction Specifications

Articles 320–396 Common Numbering Format

articles. The *Chapter 3* articles that are not included in this format are either more extensive or are too short to require separate parts.

Sections

Articles 320–362 follow a common section format. See **Figure 6-4**. The articles that follow this format are the cable and circular raceway articles. The rest of the wiring methods in *Chapter 3*, *Articles 366–396*, have a similar format with some minor terminology differences. You will not find all the sections listed in **Figure 6-4** in each article. Only the sections applicable to that article are used.

Wiring Methods and Materials

Chapter 3 contains many articles to cover the different types of wiring methods used in electrical systems, **Figure 6-5**. For the purposes of this textbook, the articles in *Chapter 3* are divided into the following classifications.

- General
- Conductors
- Enclosures
- Cables
- Circular raceways
- Noncircular raceways
- Other wiring methods

Each of these classifications is discussed in the following sections.

General Articles

There are two general articles in Chapter 3, *Article 300* and *Article 305*. *Article 300, General Requirements for Wiring Methods and Materials* applies to systems that are 1000 volts or less, **Figure 6-6**. *Article 305, General Requirements for*

Articles 320–362 Common Numbering Format

Section	Title
3XX.01	Scope
3XX.03	Reconditioned Equipment
3XX.06	Listing Requirements
3XX.10	Uses Permitted
3XX.12	Uses Not Permitted
3XX.14	Dissimilar Metals
3XX.15	Exposed Work
3XX.17	Through or Parallel to Framing Members
3XX.20	Size
3XX.22	Number of Conductors
3XX.23	In Accessible Attics
3XX.24	Bends-How Made, Bending Radius
3XX.26	Bends-Number in One Run
3XX.28	Reaming and Threading
3XX.30	Securing and Supporting
3XX.40	Boxes and Fittings
3XX.42	Couplings and Connectors
3XX.44	Expansion Fittings
3XX.46	Bushings
3XX.48	Joints
3XX.56	Splices and Taps
3XX.60	Grounding
3XX.80	Ampacity
3XX.100	Construction
3XX.104	Conductors
3XX.108	Equipment Grounding Conductor
3XX.112	Insulation
3XX.120	Marking

Figure 6-4. All the sections in the cable and circular raceway articles follow a common numbering format. Although they follow a common format, the articles only contain the sections that are applicable to that specific article.

Wiring Methods and Materials for Systems Rated Over 1000 Volts ac, 1500 Volts dc, Nominal is a new article in the 2023 edition of the *NEC* and applies systems above 1000 volts. These articles contain general requirements that apply to multiple wiring methods. Having the requirements that are common to multiple wiring methods located in these general articles saves repeating the same information for each wiring method throughout the chapter.

NEC Chapter 3
Wiring Methods and Materials

Article 300 General Requirements for Wiring Methods and Materials	Article 353 High Density Polyethylene Conduit (HDPE Conduit)
Article 305 General Requirements for Wiring Methods and Materials for Systems Rated Over 1000 Volts ac, 1500 Volts dc, Nominal	Article 354 Nonmetallic Underground Conduit with Conductors (NUCC)
	Article 355 Reinforced Thermosetting Resin Conduit (RTRC)
	Article 356 Liquidtight Flexible Nonmetallic Conduit (LFNC)
Article 310 Conductors for General Wiring	Article 358 Electrical Metallic Tubing (EMT)
Article 312 Cabinets, Cutout Boxes, and Meter Socket Enclosures	Article 360 Flexible Metallic Tubing (FMT)
	Article 362 Electrical Nonmetallic Tubing (ENT)
Article 314 Outlet, Device, Pull, and Junction Boxes; Conduit Bodies; Fittings; and Handhole Enclosures	Article 366 Auxiliary Gutters
	Article 368 Busways
Article 315 Medium Voltage Conductors, Cable, Cable Joints, and Cable Terminations	Article 369 Insulated Bus Pipe (IBP)/Tubular Covered Conductors (TCC) Systems
Article 320 Armored Cable: Type AC	Article 370 Cablebus
Article 322 Flat Cable Assemblies: Type FC	Article 371 Flexible Bus Systems
Article 324 Flat Conductor Cable: Type FCC	Article 372 Cellular Concrete Floor Raceways
Article 326 Integrated Gas Spacer Cable: Type IGS	Article 374 Cellular Metal Floor Raceways
Article 330 Metal-Clad Cable: Type MC	Article 376 Metal Wireways
Article 332 Mineral-Insulated, Metal-Sheathed Cable: Type MI	Article 378 Nonmetallic Wireways
	Article 380 Multioutlet Assemblies
Article 334 Nonmetallic-Sheathed Cable: Types NM and NMC	Article 382 Nonmetallic Extensions
	Article 384 Strut-Type Channel Raceway
Article 335 Instrumentation Tray Cable: Type ITC	Article 386 Surface Metallic Raceways
Article 336 Power and Control Tray Cable: Type TC	Article 388 Surface Nonmetallic Raceways
Article 337 Type P Cable	Article 390 Underfloor Raceways
Article 338 Service-Entrance Cable: Types SE and USE	Article 392 Cable Trays
Article 340 Underground Feeder and Branch-Circuit Cable: Type UF	Article 393 Low-Voltage Suspended Ceiling Power Distribution Systems
Article 342 Intermediate Metal Conduit (IMC)	Article 394 Concealed Knob-and-Tube Wiring
Article 344 Rigid Metal Conduit (RMC)	Article 395 Outdoor Overhead Conductors over 1000 Volts
Article 348 Flexible Metal Conduit (FMC)	Article 396 Messenger-Supporting Wiring
Article 350 Liquidtight Flexible Metal Conduit (LFMC)	Article 398 Open Wiring on Insulators
Article 352 Rigid Polyvinyl Chloride Conduit (PVC)	

Goodheart-Willcox Publisher

Figure 6-5. The *Chapter 3* articles cover the wiring methods and materials used when installing an electrical system.

Article 300
General Requirements for Wiring Methods and Materials

Section 300.1 Scope
Section 300.2 Limitations
Section 300.3 Conductors
Section 300.4 Protection Against Physical Damage
Section 300.5 Underground Installations
Section 300.6 Protection Against Corrosion and Deterioration
Section 300.7 Raceways Exposed to Different Temperatures
Section 300.8 Installation of Conductors With Other Systems
Section 300.9 Raceways in Wet Locations Above Grade
Section 300.10 Electrical Continuity of Metal Raceways, Cable Armor, and Enclosures
Section 300.11 Securing and Supporting
Section 300.12 Mechanical Continuity – Raceways and Cables
Section 300.13 Mechanical and Electrical Continuity – Conductors
Section 300.14 Length of Free Conductors at Outlets, Junctions, and Switch Points
Section 300.15 Boxes, Conduit Bodies, or Fittings – Where Required
Section 300.16 Raceway or Cable to Open or Concealed Wiring
Section 300.17 Number and Size of Conductors and Cables in Raceways
Section 300.18 Raceway Installations
Section 300.19 Supporting Conductors in Vertical Raceways
Section 300.20 Induced Currents in Ferrous Metal Enclosures or Ferrous Metal Raceways
Section 300.21 Spread of Fire or Products of Combustion
Section 300.22 Wiring in Ducts Not Used for Air Handling, Fabricated Ducts for Environmental Air, and Other Spaces for Environmental Air (Plenums)
Section 300.23 Panels Designed to Allow Access
Section 300.25 Exit Enclosures (Stair Towers).
Section 300.26 Remote-Control and Signaling Circuits Classification

Goodheart-Willcox Publisher

Figure 6-6. *Article 300* can be categorized as a general article and contains general requirements that apply to multiple wiring methods.

Section 300.4, Protection Against Physical Damage

Section 300.4, Protection Against Physical Damage is an example of one of the general requirements found in *Article 300*. It contains requirements that apply to cables and some raceways that are installed through or parallel to framing members. It requires cables and raceways to be at least 1 1/4″ away from the nearest edge of a framing member to prevent damage from fasteners such as drywall screws or trim nails, **Figure 6-7**.

Goodheart-Willcox Publisher

Figure 6-7. Nails can pop through into open space when not perfectly aligned with a stud. *Section 300.4(D)* requires cables and some raceways to be installed at least 1 1/4″ from the nearest edge of the stud to prevent damage in these situations.

PRO TIP
Article 300

Because *Article 300* is a general article, it is the one that is most difficult to quickly find information. It is common to go to a specific wiring method to find a requirement only to find that it isn't there. When this happens, the information you are looking for is likely in *Article 300*. The following are a few of the commonly used requirements that apply to multiple wiring methods that are found in *Article 300*.

- Protection of cables and raceway installed in the following areas:
  - Parallel to framing members
  - Through holes, notches, and grooves in framing members.
  - Near roof decking
- Bushings for conductors 4 AWG and larger
- Minimum burial depths for raceways and cables
- Protection against the corrosion of wiring methods
- Sealing raceways exposed to different temperatures
- Minimum length of conductors at boxes or enclosures
- Mechanical and electrical continuity of raceways, cables, and conductors
- Firestopping around wiring methods
- Wiring in air plenums

Conductor Article

Chapter 3 contains one article that is dedicated solely to conductors. *Article 310, Conductors for General Wiring* has information on all the various types of conductors used in electrical installations, **Figure 6-8**. The insulated conductors covered in *Article 310*, sometimes referred to as building wire, will be installed in one of the raceways found in *Chapter 3*. See **Figure 6-9**.

In addition to general requirements for conductors, *Article 310* contains information on conductor designations, insulations, markings, mechanical strengths, and ampacity ratings. Some of the requirements found in *Article 310* will

Article 310
Conductors for General Wiring

Part 1 General

| Section 310.1 Scope |
| Section 310.3 Conductors |

Part II Construction Specifications

| Section 310.4 Conductor Constructions and Applications |
| Section 310.6 Conductor Identification |
| Section 310.8 Marking |

Part III Installation

| Section 310.10 Uses Permitted |
| Section 310.12 Single-Phase Dwelling Services and Feeders |
| Section 310.14 Ampacities for Conductors Rated 0 Volts – 2000 Volts |
| Section 310.15 Ampacity Tables |
| Section 310.16 Ampacities of Insulated Conductors in Raceway, Cable, or Earth (Directly Buried) |
| Section 310.17 Ampacities of Single-Insulated Conductors in Free Air |
| Section 310.18 Ampacities of Insulated Conductors in Raceway or Cable |
| Section 310.19 Ampacities of Single-Insulated Conductors in Free Air |
| Section 310.20 Ampacities of Conductors Supported on a Messenger |
| Section 310.21 Ampacities of Bare or Covered Conductors in Free Air |

Goodheart-Willcox Publisher

Figure 6-8. *Article 310* can be categorized as a conductor article and has specific information about permitted applications and maximum ampacities of commonly used conductors.

Goodheart-Willcox Publisher

Figure 6-9. Conductors must be enclosed in raceways and boxes. Their insulation is not rated to be exposed or run freely in wall cavities.

also apply to the cable type wiring methods found later in *Chapter 3*. Since cables are a factory assembly of conductors, the conductor ampacities, adjustment and correction factors, and many other *Article 310* requirements are applicable.

Enclosure Articles

There are two articles in *Chapter 3* classified as enclosure articles: *Article 312, Cabinets, Cutout Boxes, and Meter Socket Enclosures,* and *Article 314, Outlet, Device, Pull, and Junction Boxes; Conduit Bodies; Fittings; and Handhole Enclosures.* The *NEC* defines an **enclosure** as the case or housing of apparatus, or the fence or walls surrounding an installation to prevent personnel from accidentally contacting energized parts or to protect the equipment from physical damage.

> **PRO TIP**   **Equipment of Article 312**
>
> It can be confusing to understand what type of enclosures *Article 312* covers and how they differ from the boxes of *Article 314*. *Article 100* has definitions that explain the differences.
> - **Cabinet**. An enclosure that is designed for either surface mounting or flush mounting and is provided with a frame, mat, or trim in which a swinging door or doors are or can be hung. An example of a cabinet is the enclosure of a panelboard. It has a cover with a hinged door, **Figure 6-10**.
> - **Cutout box**. An enclosure designed for surface mounting that has swinging doors or covers secured directly to and telescoping with the walls of the enclosure. Disconnects with a built-in hinged door are an example of a cutout box, **Figure 6-11**.
> - A **meter socket** is the enclosure that contains the utility meter that measures the amount of power consumed, **Figure 6-12**.

Goodheart-Willcox Publisher

Figure 6-10. Panelboard enclosures are an example of a cabinet. They may be installed surface mounted or flush with the finished wall surface, and they have a removable cover with a hinged door that can be removed to provide better access to their interiors.

Mike Fig Photo/Shutterstock.com

Figure 6-11. The disconnect enclosure in this figure is an example of a cutout box. They are typically surface mounted and have a hinged door.

izikMD/Shutterstock.com

Figure 6-12. Meter socket enclosures house the public utility meter that measures the amount of electrical energy used.

Article 312, Cabinets, Cutout Boxes, and Meter Socket Enclosures

The enclosures covered by *Article 312* often contain items such as relays, overcurrent protection, and meter sockets. The conductors that enter these enclosures typically terminate on lugs or terminals that are fastened to the enclosure. *Article 312* applies only to equipment rated 1000 volts or less. See **Figure 6-13**.

Article 314, Outlet, Device, Pull, and Junction Boxes; Conduit Bodies; Fittings; and Handhole Enclosures

Article 314 contains requirement for various types of boxes, **Figure 6-14**. The boxes covered by this article generally contain spliced conductors, support luminaires (light fixtures), house devices such as switches and receptacles, and may be used as a pull point when pulling conductors. See **Figure 6-15**. The following are examples of the types of requirements for boxes that can be found in *Article 314*.

- Installation requirements
- Minimum sizes
- Maximum number conductors permitted
- Support guidelines

Article 314 also has requirements for conduit bodies. The *NEC* defines a **conduit body** as a separate portion of a conduit or tubing system that provides access through a removable cover(s) to the interior of the system at a junction of two or more sections of the system or at a terminal point of the system. A common type of conduit body is the LB, which allows a conduit to transition 90° without having the sweep of a 90° bend. See **Figure 6-16**.

Cable Articles

Articles 315 and *320–340* contain requirements for the various types of cables recognized by the *NEC*, **Figure 6-17**. A ***cable***

Article 312
Cabinets, Cutout Boxes, and Meter Socket Enclosures

Part I Scope and Installation
Section 312.1 Scope
Section 312.2 Damp or Wet Locations
Section 312.3 Position in Wall
Section 312.4 Repairing Noncombustible Surfaces
Section 312.5 Cabinets, Cutout Boxes, and Meter Socket Enclosures
Section 312.6 Deflection of Conductors
Section 312.7 Space in Enclosures
Section 312.8 Switch and Overcurrent Device Enclosures
Section 312.9 Side or Back Wiring Spaces or Gutters
Section 312.10 Screws or Other Fasteners

Part II Construction Specifications
Section 312.100 Material
Section 312.101 Spacing
Section 312.102 Doors or Covers

Goodheart-Willcox Publisher

Figure 6-13. *Article 312* can be categorized as an enclosure article and contains requirements for cabinets, cutout boxes, and meter socket enclosures.

is a factory assembly of more than one conductor. Cables are permitted to be installed in their listed environment without the need to be concealed in a raceway, **Figure 6-18**.

Article 334, Nonmetallic-Sheathed Cable: Types NM and NMC

Article 334 is one of the wiring methods that can be found in *Chapter 3*. It has requirements for a nonmetallic-sheathed cable (NM cable), a wiring method that is commonly used in dwellings units. See **Figure 6-19**. *Article 334* provides requirements for items such as where NM cable is permitted, how it is to be secured and installed, and temperature ratings. See **Figure 6-20**.

PRO TIP: Nonmetallic-Sheathed Cable

It is more common to hear nonmetallic-sheathed cable referred to as Romex® than by its actual name used in the *NEC*. The term Romex originates in the early 20th century when the Rome company trademarked their nonmetallic-sheathed cable as Romex. The term is now registered by Southwire who uses it to represent their nonmetallic-sheathed cable.

Article 314
Outlet, Device, Pull, and Junction Boxes; Conduit Bodies; Fittings; and Handhole Enclosures

Part I General
Section 314.1 Scope
Section 314.2 Round Boxes
Section 314.3 Nonmetallic Boxes
Section 314.4 Metal Boxes
Section 314.5 Screws or Other Fasteners

Part II Installation
Section 314.15 Damp or Wet Locations
Section 314.16 Number of Conductors in Outlet, Device, and Junction Boxes, and Conduit Bodies
Section 314.17 Conductors and Cables Entering Boxes, Conduit Bodies, or Fittings
Section 314.19 Boxes Enclosing Flush Devices or Flush Equipment
Section 314.20 Flush-Mounted Installations
Section 314.21 Repairing Noncombustible Surfaces
Section 314.22 Surface Extensions
Section 314.23 Supports
Section 314.24 Dimensions of Boxes
Section 314.25 Covers and Canopies
Section 314.27 Outlet Boxes
Section 314.28 Pull and Junction Boxes and Conduit Bodies
Section 314.29 Boxes, Conduit Bodies, and Handhole Enclosures to Be Accessible
Section 314.30 Handhole Enclosures

Part III Pull and Junction Boxes, Conduit Bodies, and Handhole Enclosures for Use on Systems over 1000 Volts, Nominal
Section 314.70 General
Section 314.71 Size of Pull and Junction Boxes, Conduit Bodies, and Handhole Enclosures
Section 314.72 Construction and Installation Requirements

Part IV Construction Specifications
Section 314.40 Metal Boxes, Conduit Bodies, and Fittings
Section 314.41 Covers
Section 314.42 Bushings
Section 314.43 Nonmetallic Boxes
Section 314.44 Marking

Goodheart-Willcox Publisher

Figure 6-14. *Article 314* can be categorized as an enclosure article and contains requirements for various types of boxes, conduit bodies, fittings, and handhole enclosures.

Goodheart-Willcox Publisher

Figure 6-15. Electrical boxes come in various types and sizes. Nonmetallic boxes are generally used with nonmetallic cables and raceways while metal boxes are generally used with metallic cables and raceways.

Goodheart-Willcox Publisher

Figure 6-16. Conduit bodies can be used to transition the direction a raceway is going without having the sweep of a 90° bend. They have a removable cover that allows the conductors to make the transition without being damaged. The conduit body pictured is an LB which means it is L-shaped with the one of the conduits leaving the top side.

NEC Chapter 3
Cable Articles

Article 315 Medium Voltage Conductors, Cable, Cable Joints, and Cable Terminations	Article 334 Nonmetallic-Sheathed Cable: Types NM and NMC
Article 320 Armored Cable: Type AC	Article 335 Instrumentation Tray Cable: Type ITC
Article 322 Flat Cable Assemblies: Type FC	Article 336 Power and Control Tray Cable: Type TC
Article 324 Flat Conductor Cable: Type FCC	Article 337 Type P Cable
Article 326 Integrated Gas Spacer Cable: Type IGS	Article 338 Service-Entrance Cable: Types SE and USE
Article 330 Metal-Clad Cable: Type MC	Article 340 Underground Feeder and Branch-Circuit Cable: Type UF
Article 332 Mineral-Insulated, Metal-Sheathed Cable: Type MI	

Goodheart-Willcox Publisher

Figure 6-17. *Chapter 3* has eleven articles that are dedicated to cables. Cables can save time as they can be installed without the use of a raceway.

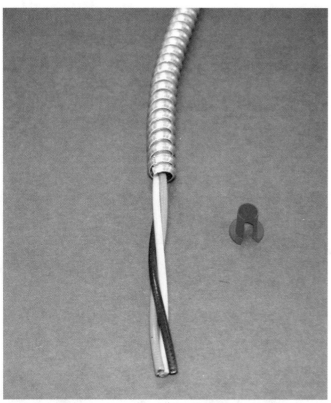

Goodheart-Willcox Publisher

Figure 6-18. *Article 330* can be categorized as a cable article and contains requirements for metal-clad cable, type MC. MC cable is a wiring method often used in commercial buildings for branch circuit and feeders.

Due to the common numbering format, the rest of the cables that are found in *Chapter 3* have their articles structured in a similar fashion to *Article 334*.

Goodheart-Willcox Publisher

Figure 6-19. Nonmetallic-sheathed cable is typically installed with nonmetallic boxes. The boxes in this picture are vapor seal boxes as they are being installed in an exterior wall. The requirement for boxes that create a vapor seal is a building code which may differ across the country.

Circular Raceways

Articles 342–362 contain requirements for the various types of circular raceways that are recognized by the *NEC*, **Figure 6-21**. The *NEC* defines a ***raceway*** as an enclosed channel designed expressly for holding wires, cables, or busbars, with additional functions as permitted in this *Code*. A circular raceway is a round tubing or conduit. Circular raceways are only permitted to contain conductors or cables. See **Figure 6-22**.

Unit 6   NEC Chapter 3

Article 334
Nonmetallic-Sheathed Cable: Types NM and NMC

Part I General

Section 334.1 Scope
Section 334.6 Listing Requirements

Part II Installation

Section 334.10 Uses Permitted
Section 334.12 Uses Not Permitted
Section 334.15 Exposed Work
Section 334.17 Through or Parallel to Framing Members
Section 334.19 Cable Entries
Section 334.23 In Accessible Attics
Section 334.24 Bending Radius
Section 334.30 Securing and Supporting
Section 334.40 Boxes and Fittings
Section 334.80 Ampacity

Part III Construction Specifications

Section 334.100 Construction
Section 334.104 Conductors
Section 334.108 Equipment Grounding Conductor
Section 334.112 Insulation
Section 334.116 Sheath

Goodheart-Willcox Publisher

Figure 6-20. *Article 334* can be categorized as a cable article and contains requirements for nonmetallic-sheathed cable. NM cable is often used with wood structures and is a popular wiring method for dwellings.

NEC Chapter 3
Circular Raceway Articles

Article 342 Intermediate Metal Conduit (IMC)
Article 344 Rigid Metal Conduit (RMC)
Article 348 Flexible Metal Conduit (FMC)
Article 350 Liquidtight Flexible Metal Conduit (LFMC)
Article 352 Rigid Polyvinyl Chloride Conduit (PVC)
Article 353 High Density Polyethylene Conduit (HDPE Conduit)
Article 354 Nonmetallic Underground Conduit with Conductors (NUCC)
Article 355 Reinforced Thermosetting Resin Conduit (RTRC)
Article 356 Liquidtight Flexible Nonmetallic Conduit (LFNC)
Article 358 Electrical Metallic Tubing (EMT)
Article 360 Flexible Metallic Tubing (FMT)
Article 362 Electrical Nonmetallic Tubing (ENT)

Goodheart-Willcox Publisher

Figure 6-21. There are twelve circular raceway articles in *Chapter 3*. They include both nonmetallic and metallic, flexible and rigid, but they are all circular.

Goodheart-Willcox Publisher

Figure 6-22. There are many different types and sizes of circular raceways. Their purpose is to provide protection to the conductors they contain.

Article 358, Electrical Metallic Tubing (EMT)

Article 358 has requirements for electrical metallic tubing (EMT), a wiring method commonly used in commercial installations. See **Figure 6-23**. Another term used to describe EMT is "thin wall" as it has a relatively thin wall compared to other types of conduit and is easy to bend, **Figure 6-24**. *Article 358* lists where EMT is permitted and requirements on how it is to be secured and installed.

The rest of the circular raceways found in *Chapter 3* have requirements that are set up in a similar fashion to *Article 358*, except they will be applicable to each respective wiring method.

Article 358
Electrical Metallic Tubing (EMT)

Part I General
- Section 358.1 Scope
- Section 358.6 Listing Requirements

Part II Installation
- Section 358.10 Uses Permitted
- Section 358.12 Uses Not Permitted
- Section 358.14 Dissimilar Metals
- Section 358.20 Size
- Section 358.22 Number of Conductors
- Section 358.24 Bends
- Section 358.28 Reaming and Threading
- Section 358.30 Securing and Supporting
- Section 358.42 Couplings and Connectors
- Section 358.56 Splices and Taps
- Section 358.60 Grounding

Part III Construction Specifications
- Section 358.100 Construction
- Section 358.120 Marking

Goodheart-Willcox Publisher

Figure 6-23. *Article 358* can be categorized as a circular raceway article and contains requirements for electrical metallic tubing (EMT). EMT is often installed in commercial installations. After the raceways have been installed, insulated conductors will be installed.

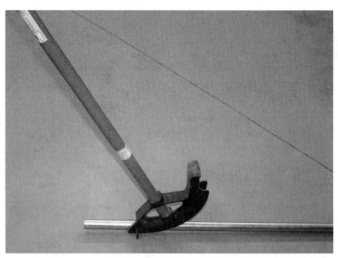

Goodheart-Willcox Publisher

Figure 6-24. Electrical metallic tubing (EMT) is a thin wall raceway that is hand bendable up to size 1 1/4″.

Noncircular Raceways

The noncircular raceways are found in *Articles 366–392*. These wiring methods come in many shapes and sizes, some of which may contain busbars or devices. See **Figure 6-25**.

Article 386, Surface Metal Raceways

An example of a noncircular raceway is *Article 386, Surface Metal Raceways.* Surface metal raceways are often referred to as Wiremold® as it is a popular brand that has been used for decades, **Figure 6-26**. Surface metal raceways are decorative raceways that can be installed on the surface in a finished room without having an industrial look. They come in many sizes and types to meet the various needs. *Article 386*

NEC Chapter 3
Noncircular Raceway Articles

Article 366 Auxiliary Gutters	Article 378 Nonmetallic Wireways
Article 368 Busways	Article 380 Multioutlet Assemblies
Article 369 Insulated Bus Pipe (IBP)/Tubular Covered Conductors (TCC) Systems	Article 382 Nonmetallic Extensions
Article 370 Cablebus	Article 384 Strut-Type Channel Raceway
Article 371 Flexible Bus Systems	Article 386 Surface Metallic Raceways
Article 372 Cellular Concrete Floor Raceways	Article 388 Surface Nonmetallic Raceways
Article 374 Cellular Metal Floor Raceways	Article 390 Underfloor Raceways
Article 376 Metal Wireways	Article 392 Cable Trays

Goodheart-Willcox Publisher

Figure 6-25. There are twelve articles that have been categorized as noncircular raceways. They vary a lot in their size and shape, but they all are meant to house and protect conductors.

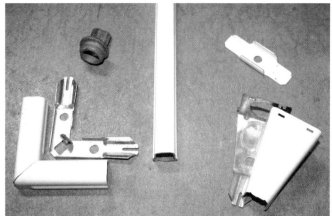

Figure 6-26. Surface metal raceways are decorative raceways that are typically surface mounted. They are available with various fittings to connect to other raceways and to transition directions and around corners.

Article 386
Surface Metal Raceways

Part I General
Section 386.1 Scope
Section 386.6 Listing Requirements

Part II Installation
Section 386.10 Uses Permitted
Section 386.12 Uses Not Permitted
Section 386.21 Size of Conductors
Section 386.22 Number of Conductors or Cables
Section 386.30 Securing and Supporting
Section 386.56 Splices and Taps
Section 386.60 Grounding
Section 386.70 Combination Raceways

Part III Construction Specifications
Section 386.100 Construction
Section 386.120 Marking

Figure 6-27. *Article 386* can be categorized as a noncircular raceway article and contains requirements for surface metal raceways.

lists where surface metal raceways are permitted and has requirements on how they are to be secured and installed, **Figure 6-27**.

Other Wiring Methods

This textbook classifies the remaining wiring methods as other wiring methods, **Figure 6-28**. These wiring methods are rarely used and do not fit well into any of the other classifications.

NEC Chapter 3
Other Wiring Method Articles

Article 393 Low-Voltage Suspended Ceiling Power Distribution Systems
Article 394 Concealed Knob-and-Tube Wiring
Article 395 Outdoor Overhead Conductors over 1000 Volts
Article 396 Messenger-Supporting Wiring
Article 398 Open Wiring on Insulators

Figure 6-28. There are five wiring methods classified as *other*.

NEC KEYWORDS — Chapter 3 Keywords

The following are examples of keywords that should lead you to *Chapter 3*:

box	conductor	installation	raceway
cabinet	conduit body	minimum burial depths	secure
cable	fittings	protection of conductors	support

Summary

- *Chapter 3* of the *NEC* is the third of the general chapters that apply to all electrical installations.
- *Chapter 3* discusses a number of different wiring method classifications used when roughing-in a job.
  - General
  - Conductors
  - Enclosures
  - Cables
  - Circular raceways
  - Noncircular raceways
  - Other wiring methods
- The sections in the cable and circular raceway articles follow a common numbering format.
- *Chapter 3* contains requirements for wiring methods that are used when constructing the electrical system.

Unit 6 Review

Name _____ Date _____ Class _____

Know and Understand

Answer the following questions based on information in this unit.

1. *Chapter 3* of the *NEC* has information necessary when _____ an electrical system.
   A. designing
   B. roughing-in
   C. trimming
   D. testing
2. The common numbering format that is used for the cable and raceway articles contains _____ parts.
   A. two
   B. three
   C. four
   D. five
3. _____ has general requirements that apply to multiple wiring methods.
   A. *Article 300*
   B. *Article 310*
   C. *Article 312*
   D. *Article 314*
4. _____ contains requirements for boxes.
   A. *Article 300*
   B. *Article 310*
   C. *Article 312*
   D. *Article 314*
5. Using the common numbering format, which section is dedicated to uses permitted?
   A. *Section 3XX.10*
   B. *Section 3XX.12*
   C. *Section 3XX.15*
   D. *Section 3XX.17*
6. _____ has requirements for liquidtight flexible nonmetallic conduit.
   A. *Article 348*
   B. *Article 350*
   C. *Article 356*
   D. *Article 360*
7. Requirements for conduit bodies are found in _____.
   A. *Article 300*
   B. *Article 312*
   C. *Article 314*
   D. *Article 320*
8. Which of the following keywords would lead you to *Chapter 3*?
   A. Grounding
   B. Luminaires
   C. Overcurrent protection
   D. Rigid metal conduit
9. Which article has requirements on cellular metal floor raceways?
   A. *Article 344*
   B. *Article 372*
   C. *Article 374*
   D. *Article 386*
10. A _____ is an enclosure designed for surface mounting that has swinging doors or covers secured directly to and telescoping with the walls of the enclosure.
    A. box
    B. cabinet
    C. handhole
    D. cutout box

Apply and Analyze

Answer the following questions using a copy of the National Electrical Code. *Identify the section or subsection where the answer is found.*

1. The minimum bending radius of nonmetallic-sheathed cable shall not be less than _____ times the diameter of the cable.
   A. 5
   B. 6
   C. 7
   D. 8

   NEC _____

2. Electrical metallic tubing shall be securely fastened within _____ of a box.
   A. 18″
   B. 2′
   C. 3′
   D. 6′

   NEC _____

3. Cable trays are not permitted to be installed in _____.
   A. schools
   B. hoistways
   C. industrial locations
   D. churches

   NEC _____

4. The maximum permitted trade size of rigid metal conduit is _____.
   A. 3
   B. 4
   C. 5
   D. 6

   NEC _____

5. Armored cable must be supported at intervals not exceeding _____.
   A. 3′
   B. 4 1/2′
   C. 6′
   D. 10′

   NEC _____

6. Floor boxes listed specially for the application shall be used for _____ located in the floor.
   A. switches
   B. outlets
   C. receptacles
   D. sensors

   NEC _____

7. The minimum size conductor permitted for general wiring is _____ copper.
   A. 16 AWG
   B. 14 AWG
   C. 12 AWG
   D. 10 AWG

   NEC _____

8. Floor mounted flat conductor cable shall be covered with carpet squared not larger than _____ square.
   A. 12.23 inches
   B. 18 inches
   C. 36 inches
   D. 39.37 inches

   NEC _____

9. Conductors installed in a cellular concrete floor raceway shall not exceed _____ of the cross-sectional area of the cell.
   A. 15%
   B. 30%
   C. 40%
   D. 50%

   NEC _____

10. The interior of an underground raceway shall be considered a _____ location.
    A. wet
    B. damp
    C. dry
    D. corrosive

    NEC _____

Name _____ Date _____ Class _____

Critical Thinking

NEC Chapter 3

Answer the following questions using a copy of the National Electrical Code. *Identify the section or subsection where the answer is found.*

1. Rigid metal conduit that is threaded in the field shall be threaded with a die that has a taper of _____ per foot.

   _____

   _____

   NEC _____

2. No conductor larger than _____ AWG shall be installed in a cellular metal floor raceway except by special permission.

   _____

   _____

   NEC _____

3. The minimum burial depth of rigid metal conduit that is not encased in concrete and not subject to vehicular traffic is _____. (480-volt system)

   _____

   _____

   NEC _____

4. What is the minimum trade size of electrical nonmetallic tubing?

   _____

   _____

   NEC _____

5. The general rule states that the minimum size conductor permitted to be installed in parallel is _____.

   _____

   _____

   NEC _____

6. Angle connectors used with flexible metallic conduit shall not be _____.

   _____

   _____

   NEC _____

7. Mineral-insulated cable shall be supported at intervals not exceeding _____.

   _____

   _____

   NEC _____

8. Where flexibility is necessary after installation, trade size 1/2″ liquidtight flexible metallic conduit shall be securely fastened within _____ of the equipment.

   _____

   _____

   NEC _____

9. A hole drilled through a wood stud for nonmetallic-sheathed cable, must be drilled so that the edge of the hole is not less than _____ from the nearest edge of the stud.

   _____

   _____

   NEC _____

10. Conductors spliced within an auxiliary gutter shall not fill the gutter to more than _____ of its area.

    _____

    _____

    NEC _____

NEC Chapters 1–3

Answer the following questions using a copy of the National Electrical Code. *Identify the section or subsection where the answer is found.*

11. The minimum length of free conductor at a box shall be at least _____ from the point where it emerges from a raceway or cable sheath.

    _____

    _____

    NEC _____

12. A circuit breaker with its interrupting rating other than _____ amperes shall have its interrupting rating shown on the breaker.

    _____

    _____

    NEC _____

13. Aluminum and copper conductors shall not be spliced directly together unless the device is _____ for the purpose and conditions of use.

    _____

    _____

    NEC _____

14. Field bends made in rigid PVC shall only be made with _____ bending equipment.

    _____

    _____

    NEC _____

15. The minimum depth of working space in front of a 120/208 panelboard shall be at least _____.

    _____

    _____

    NEC _____

16. A service disconnect shall not be installed in a _____.

    _____

    _____

    NEC _____

17. Raceways with _____ conductors or larger are required to have a fitting such as a bushing that provides a smoothly rounded edge.

    _____

    _____

    NEC _____

18. Electrical metallic tubing _____ permitted to be used as an equipment grounding conductor.

    _____

    _____

    NEC _____

19. A dwelling unit hallway _____ or longer is required to have a receptacle outlet.

    _____

    _____

    NEC _____

20. Electrical nonmetallic tubing _____ permitted to buried directly in the earth.

    _____

    _____

    NEC _____

UNIT

7 NEC Chapter 4

sockagphoto/Shutterstock.com

LEARNING OBJECTIVES

After completing this unit, you will be able to:

- Describe the type of requirements in *Chapter 4* of the *NEC*.
- Identify the equipment in *Chapter 4* of the *NEC*.
- Recognize keywords that will lead you to *Chapter 4* of the *NEC*.

KEY TERMS

adjustable speed drive
appliance
controller
equipment
general-use snap switch
general-use switch
hermetic refrigerant motor-compressor
industrial control panel
luminaire
panelboard
phase converter
receptacle
switchboard
switchgear
transformer

Introduction

This unit is an introduction to the *NEC*'s *Chapter 4, Equipment for General Use*. It is the fourth and final of the general chapters that apply to all installations. See **Figure 7-1**. The most commonly used articles will be introduced to help you recognize the type of information found in *Chapter 4*. Studying the type of information found in *Chapter 4* as well as looking up *Code* sections will help with comprehension and retention of the type of information found in *Chapter 4* of the *NEC*.

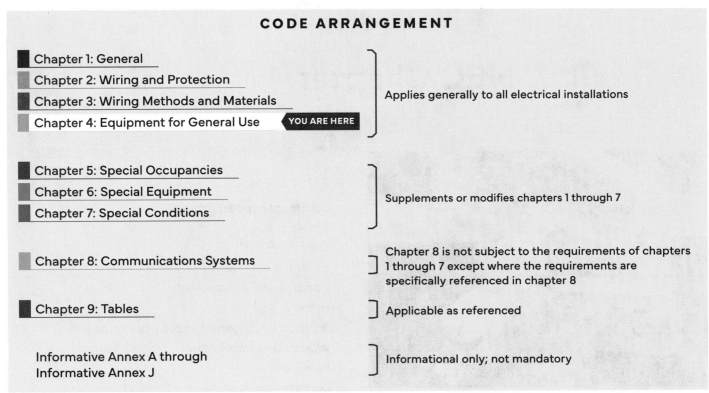

Figure 7-1. *Chapter 4* is the fourth and final of the general chapters that contain information applying to all installations.

Equipment Chapter

Chapter 4, Equipment for General Use contains requirements for all the various pieces of equipment that will be installed in an electrical installation. The *NEC* defines **equipment** as a general term, including fittings, devices, appliances, luminaires, apparatus, machinery, and the like used as a part of, or in connection with, an electrical installation. The equipment that is covered in *Chapter 4* controls, consumes, creates, or stores electrical energy. See **Figure 7-2**.

Another way to remember the type of information found in *Chapter 4* is that it contains requirements for equipment that is typically installed during the trim stage of a job, **Figure 7-3**. Trimming out a job is when the electrician comes back after the finished wall surface has been completed to install items such as devices, luminaires, appliances, and heating equipment.

This unit groups *Chapter 4* articles with similar functions together to make them easier to cover and understand.
- Equipment connection articles
- Control articles
- Consume articles
- Create, alter, and store articles

Equipment Connection Articles

The first two articles of *Chapter 4* are *Article 400, Flexible Cords and Flexible Cables,* and *Article 402, Fixture Wires*. One might think these articles belong in *Chapter 3* as they seem like a wiring method. However, they are in *Chapter 4* because they are used to connect equipment. Cords are often used to feed equipment and appliances, while fixture wires are used to connect fixtures such as luminaires.

NEC Chapter 4
Equipment for General Use

Article 400 Flexible Cords and Flexible Cables	Article 427 Fixed Electric Heating Equipment for Pipelines and Vessels
Article 402 Fixture Wires	
Article 404 Switches	Article 430 Motors, Motor Circuits, and Controllers
Article 406 Receptacles, Cord Connectors, and Attachment Plugs (Caps)	Article 440 Air-Conditioning and Refrigerating Equipment
	Article 445 Generators
Article 408 Switchboards, Switchgear, and Panelboards	Article 450 Transformers and Transformer Vaults (Including Secondary Ties)
Article 409 Industrial Control Panels	
Article 410 Luminaires, Lampholders, and Lamps	Article 455 Phase Converters
Article 411 Low-Voltage Lighting	Article 460 Capacitors
Article 422 Appliances	Article 470 Resistors and Reactors
Article 424 Fixed Electric Space-Heating Equipment	Article 480 Stationary Standby Batteries
Article 425 Fixed Resistance and Electrode Industrial Process Heating Equipment	Article 495 Equipment Over 1000 Volts ac, 1500 Volts dc, Nominal
Article 426 Fixed Outdoor Electric Deicing and Snow-Melting Equipment	

Goodheart-Willcox Publisher

Figure 7-2. The *Chapter 4* articles have general requirements for electrical equipment. Much of the equipment found in *Chapter 4* are installed during the trim stage of the construction process.

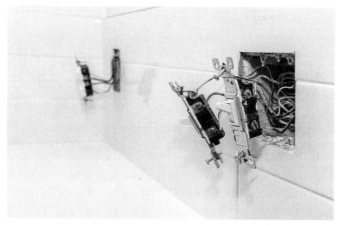

Ungvar/Shutterstock.com

Figure 7-3. This bathroom is in the process of being trimmed out. These switches and receptacle have been installed and just need the faceplates.

Article 400, Flexible Cords and Flexible Cables

Article 400 has requirements for the many different types of flexible cords and cables used in the electrical industry. See **Figure 7-4**. *Section 400.12(1)* states that a flexible cord shall not be used as a substitute for the fixed wiring of a structure. You cannot install flexible cords or cables in walls or pull them through raceways, but you can use them to feed certain pieces of equipment. *Section 400.10, Uses Permitted* allows the use of flexible cords or cables to feed equipment and appliances that move, are portable, or require flexibility.

Flexible cords and cables have finely stranded conductors to allow maximum flexibility. *Table 400.4* lists the common flexible cords and cables and has detailed information about each cord/cable. An example of a commonly used cord is a dryer cord, **Figure 7-5**.

Article 400
Flexible Cords and Flexible Cables

Part I General
Section 400.1 Scope
Section 400.2 Other Articles
Section 400.3 Suitability
Section 400.4 Types
Section 400.5 Ampacities for Flexible Cords and Flexible Cables
Section 400.6 Markings
Section 400.10 Uses Permitted
Section 400.12 Uses Not Permitted
Section 400.13 Splices
Section 400.14 Pull at Joints and Terminals
Section 400.15 In Show Windows and Showcases
Section 400.16 Overcurrent Protection
Section 400.17 Protection from Damage

Part II Construction Specifications
Section 400.20 Labels
Section 400.21 Construction
Section 400.22 Grounded-Conductor Identification
Section 400.23 Equipment Grounding Conductor Identification
Section 400.24 Attachment Plugs

Part III Portable Cables Over 600 Volts, up to 2000 Volts, Nominal
Section 400.30 General
Section 400.31 Construction
Section 400.32 Shielding
Section 400.33 Equipment Grounding Conductors
Section 400.34 Minimum Bending Radii
Section 400.35 Fittings
Section 400.36 Splices and Terminations

Part IV Portable Power Feeder Cablers Over 2000 Volts, Nominal
Section 400.40 General
Section 400.41 Portable Power Feeder Cables
Section 400.42 Uses Permitted
Section 400.43 Uses Not Permitted
Section 400.44 Construction
Section 400.45 Shielding
Section 400.46 Equipment Grounding Conductors
Section 400.47 Minimum Bending Radii
Section 400.48 Fittings
Section 400.49 Splices and Terminations
Section 400.50 Types
Section 400.51 Ampacities for Portable Power Feeder Cables Rated Greater Than 2000 Volts
Section 400.52 Markings

Goodheart-Willcox Publisher

Figure 7-4. *Article 400* can be categorized as an equipment connection article and contains requirements for flexible cords that are used to connect appliances and equipment that are portable or require flexibility.

Article 402, Fixture Wires

Article 402 has requirements for fixtures wires. Fixture wires are the conductors attached to a fixture that bring power from a lighting outlet to a fixture. This is not an article that is used very often as fixture wires generally come installed from the manufacturer on the light fixture/luminaire. Most of the article is dedicated to tables that contain information about the various fixture wires, including their ampacities. See **Figure 7-6**.

Control Articles

Chapter 4 contains four articles relating to control: *Article 404, Switches*; *Article 406, Receptacles, Cord Connectors, and Attachment Plugs (Caps)*; *Article 408, Switchboards, Switchgear, and Panelboards*; and *Article 409, Industrial Control Panels*. See **Figure 7-7**. The control articles cover equipment that will control electrical energy by opening and closing a circuit, providing a place to plug in a cord, or other means.

Article 404, Switches

Article 404 has information on all the various types of switches used in the electrical industry. See **Figure 7-8**. Switches can vary from a 15-ampere switch controlling a light in your home to a 1000-ampere disconnect controlling a large piece of equipment.

The *NEC* has six definitions to describe the different types of switches. The two most common switches used are general-use switches and general-use snap switches.

Unit 7 NEC Chapter 4

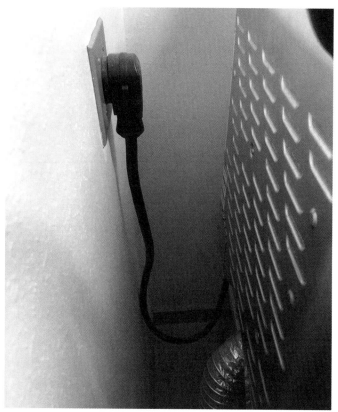

Goodheart-Willcox Publisher

Figure 7-5. Dryer cords are a commonly installed flexible cord. They are commercially available with the plug already installed. The plug serves as the disconnecting means for the dryer.

NEC Chapter 4
Control Articles

Article 404 Switches
Article 406 Receptacles, Cord Connectors, and Attachment Plugs (Caps)
Article 408 Switchboards, Switchgear, and Panelboards
Article 409 Industrial Control Panels

Goodheart-Willcox Publisher

Figure 7-7. There are four articles within *Chapter 4* that deal with equipment that control electricity.

Article 402
Fixture Wires

Section 402.1 Scope
Section 402.2 Other Articles
Section 402.3 Types
Section 402.5 Ampacities for Fixture Wires
Section 402.6 Minimum Size
Section 402.7 Number of Conductors in Conduit or Tubing
Section 402.8 Grounded Conductor Identification
Section 402.9 Marking
Section 402.10 Uses Permitted
Section 402.12 Uses Not Permitted
Section 402.14 Overcurrent Protection

Goodheart-Willcox Publisher

Figure 7-6. *Article 402* can be categorized as an equipment connection article and contains requirements for fixture wires. Fixture wires generally come preinstalled on luminaires.

Article 404
Switches

Part I General

Section 404.1 Scope
Section 404.2 Switch Connections
Section 404.3 Enclosure
Section 404.4 Damp or Wet Locations
Section 404.5 Time Switches, Flashers, and Similar Devices
Section 404.6 Position and Connection of Switches
Section 404.7 Indicating
Section 404.8 Accessibility and Grouping
Section 404.9 General-Use Snap Switches, Dimmers, and Control Switches
Section 404.10 Mounting of General-Use Snap Switches, Dimmers, and Control Switches
Section 404.11 Circuit Breakers as Switches
Section 404.12 Grounding of Enclosures
Section 404.13 Knife Switches
Section 404.14 Rating and Use of Switches
Section 404.16 Reconditioned Equipment

Part II Construction Specifications

Section 404.20 Marking
Section 404.22 Electronic Control Switches
Section 404.26 Knife Switches Rated 600 to 1000 Volts
Section 404.27 Fused Switches
Section 404.28 Wire-Bending Space
Section 404.30 Switch Enclosures with Doors

Goodheart-Willcox Publisher

Figure 7-8. *Article 404* can be categorized as a control article and contains requirements for all the various types of switches installed in electrical installations.

95

A *general-use switch* is a switch intended for use in general distribution and branch circuits. It is rated in amperes, and it is capable of interrupting its rated current at its rated voltage. A *general-use snap switch* is a form of general-use switch constructed so that it can be installed in device boxes or on box covers, or otherwise used in conjunction with wiring systems recognized by this *Code*. See **Figure 7-9**.

Article 406, Receptacles, Cord Connectors, and Attachment Plugs (Caps)

Article 406 has information and requirements for receptacles, **Figure 7-10**. The *NEC* defines a *receptacle* as a contact device installed at the outlet for the connection of an attachment plug, or for the direct connection of electrical utilization equipment designed to mate with the corresponding contact device. A single receptacle is a single-contact device with no other contact device on the same yoke or strap. A multiple receptacle is two or more contact devices on the same yoke or strap. See **Figure 7-11**. Receptacles allow current to travel from the outlet to the equipment being connected.

Article 406
Receptacles, Cord Connectors, and Attachment Plugs (Caps)

Section 406.1 Scope
Section 406.2 Reconditioned Equipment
Section 406.3 Receptacle Rating and Type
Section 406.4 General Installation Requirements
Section 406.5 Receptacle Mounting
Section 406.6 Receptacle Faceplates (Cover Plates)
Section 406.7 Attachment Plugs, Cord Connectors, and Flanged Surface Devices
Section 406.8 Noninterchangeability
Section 406.9 Receptacles in Damp or Wet Locations
Section 406.10 Grounding-Type Receptacles, Adapters, Cord Connectors, and Attachment Plugs
Section 406.11 Connecting Receptacle Grounding Terminal to Box
Section 406.12 Tamper-Resistant Receptacles
Section 406.13 Single-Pole Separable-Connector Type

Goodheart-Willcox Publisher

Figure 7-10. *Article 406* can be categorized as a control article and contains requirements for receptacles, attachment plugs, and cord connectors.

Goodheart-Willcox Publisher

Figure 7-11. A receptacle is a device that provides a place to plug in appliances. The receptacle in the picture has *TR* etched into its face, indicating that it is a tamper-resistant receptacle. *Section 406.12* requires tamper-resistant receptacles in certain locations to prevent children from sticking metal objects into the receptacle.

Goodheart-Willcox Publisher

Figure 7-9. Single-pole and double-pole general-use snap switches will have "ON" and "OFF" etched into the handle of the switch.

Unit 7 NEC Chapter 4

PRO TIP — **Receptacle Configurations**

Receptacles come in various voltages, ampacity ratings, and types. NEMA (National Electrical Manufacturers Association) has created standard configurations that receptacle and plug manufacturers follow. This ensures that the plug on the equipment will fit into the receptacle even though they may be made by different manufacturers. See **Figure 7-12**.

Article 408, Switchboards, Switchgear, and Panelboards

Article 408 has requirements for switchboards, switchgear, and panelboards. This article applies to equipment operating at 1000 volts or less, unless specifically referenced by another part of the *Code*. See **Figure 7-13**. The first step in understanding the type of information in *Article 408* is to look at the definition of the equipment it contains.

The *NEC* defines a **panelboard** as a single panel or group of panel units designed for assembly in the form of a single

NEMA Receptacle Configurations

NEMA Line No.	Volts	15 Amp	20 Amp	30 Amp	50 Amp
1	125V	White neutral			
2	250V		ϕ ϕ	ϕ ϕ	
5	125V	White neutral	White neutral	White neutral	White neutral
6	250V	ϕ ϕ	ϕ ϕ	ϕ ϕ	ϕ ϕ

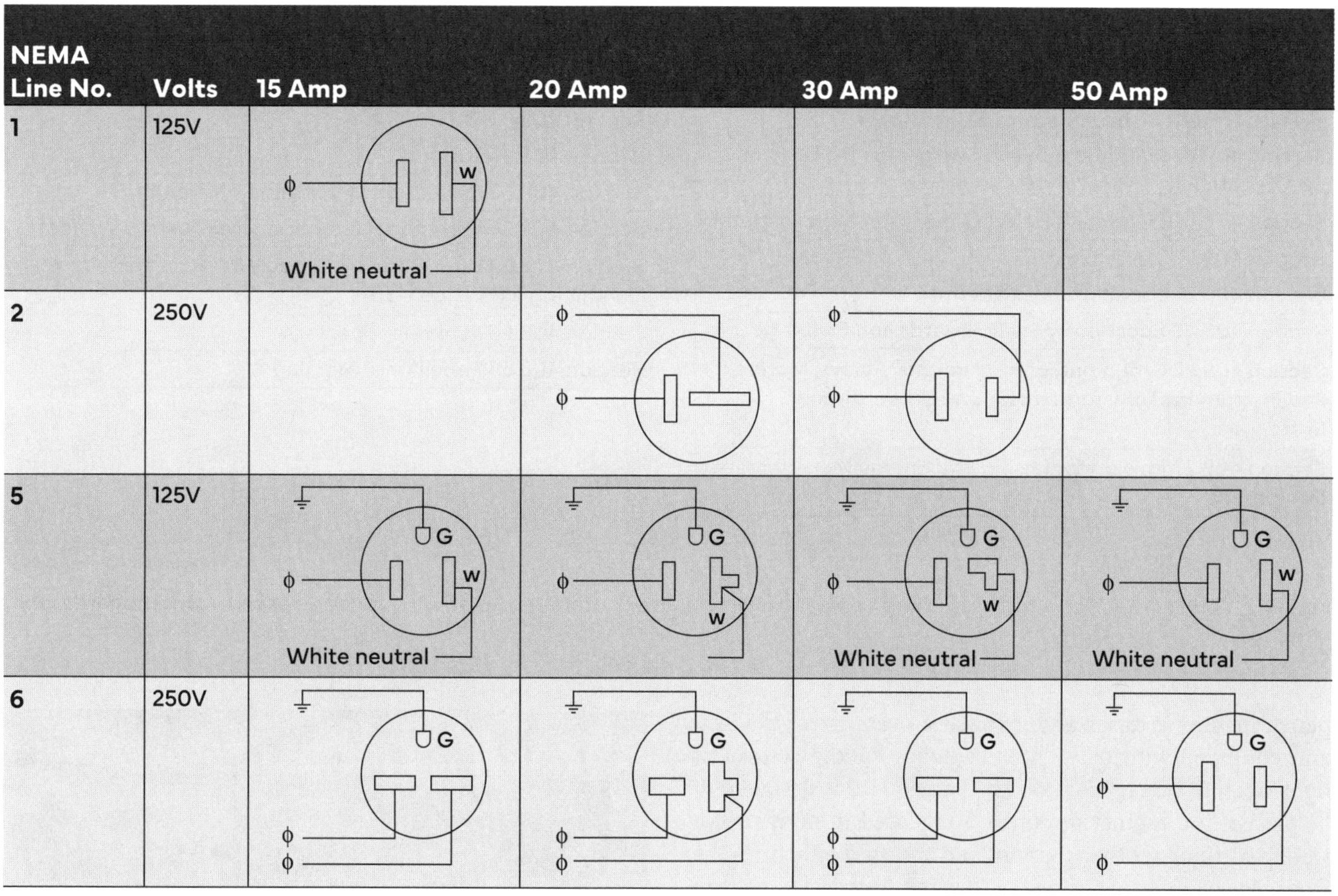

Goodheart-Willcox Publisher

Figure 7-12. This chart shows a sampling of the wide variety of NEMA receptacle configurations.

Article 408
Switchboards, Switchgear, and Panelboards

Part I General
Section 408.1 Scope
Section 408.2 Reconditioned Equipment
Section 408.3 Support and Arrangement of Busbars and Conductors
Section 408.4 Descriptions Required
Section 408.5 Clearance for Conductor Entering Bus Enclosures
Section 408.6 Short-Circuit Current Rating
Section 408.7 Unused Openings
Section 408.8 Replacement Panelboards

Part II Switchboards and Switchgear
Section 408.16 Switchboards and Switchgear in Damp or Wet Locations
Section 408.17 Location Relative to Easily Ignitable Material
Section 408.18 Clearances
Section 408.19 Conductor Insulation
Section 408.20 Location of Switchboards and Switchgear
Section 408.22 Grounding of Instruments, Relays, Meters, and Instrument Transformers on Switchboards and Switchgear
Section 408.23 Power Monitoring and Energy Management Equipment

Part III Panelboards
Section 408.30 General
Section 408.36 Overcurrent Protection
Section 408.37 Panelboards in Damp or Wet Locations
Section 408.38 Enclosure
Section 408.39 Relative Arrangement of Switches and Fuses
Section 408.40 Grounding of Panelboards
Section 408.41 Grounded Conductor Terminations
Section 408.43 Panelboard Orientation

Part IV Construction Specifications
Section 408.50 Panels
Section 408.51 Busbars
Section 408.52 Protection of Instrument Circuits
Section 408.54 Maximum Number of Overcurrent Devices
Section 408.55 Wire-Bending Space Within an Enclosure Containing a Panelboard
Section 408.56 Minimum Spacings
Section 408.58 Panelboard Marking

Goodheart-Willcox Publisher

Figure 7-13. *Article 408* can be categorized as a control article and contains requirements for panelboards, switchboards, and switchgear.

panel, including buses and automatic overcurrent devices, and equipped with or without switches for the control of light, heat, or power circuits. The panelboard is designed to be placed in a cabinet or cutout box placed in or against a wall, partition, or other support, and accessed only from the front. An example of a panelboard is the electrical panel you would find in a home, **Figure 7-14**.

The *NEC* defines a ***switchboard*** as a large single panel, frame, or assembly of panels on which are mounted on the face, back, or both, switches, overcurrent and other protective devices, buses, and usually instruments. These assemblies are generally accessible from the rear as well as from the front and are not intended to be installed in cabinets. The main difference between a panelboard and a switchboard is that switchboards may be accessed from not only the front, but also from the sides and the back. Switchboards are found in larger commercial and industrial installations and are typically rated 600 volts or less. See **Figure 7-15**.

alacatr via Getty Images

Figure 7-14. Panelboards contain overcurrent protection, typically in the form of breakers. The inside of a panelboard can be accessed by removing the cover which is on the front/face of the panel.

alacatr via Getty Images

Figure 7-15. Switchboards are similar to panelboards in the sense that they house switching and overcurrent devices. They differ in that the inside of a switchboard may be accessible from the front, sides, and back, while panelboards are only accessible from the front.

The *NEC* defines **switchgear** as an assembly completely enclosed on all sides and top with sheet metal (except ventilating openings and inspection windows) and containing primary power circuit switching, interrupting devices, or both, with buses and connections. The assembly may include control and auxiliary devices. Access to the interior of the enclosure is provided by doors, removable covers, or both. Switchgear is a heavy-duty piece of equipment and is constructed to a different standard that panelboards and switchboards, **Figure 7-16**.

One common area of confusion when looking up *Code* requirements is clearances in front of equipment such as panelboards. *Section 110.26* provides requirements such as minimum clearances and minimum illumination, rather

Nutthapat Matphongtavorn/Shutterstock.com

Figure 7-16. Switchgear is built to a higher standard than switchboards and is typically surrounded by sheet metal except for ventilation openings and inspection windows. Switchgear is typically rated for higher voltages than switchboards are.

than *Article 408*. The clearances found in *Article 110* apply not only to panelboards, but to other equipment as well, such as industrial control panels. Rather than repeating the same information in multiple articles, it is located in *NEC Chapter 1* so that it applies to all equipment and installations.

Article 409, Industrial Control Panels

Article 409 has requirements for industrial control panels, **Figure 7-17**. Industrial control panels may be factory or field assembled. The requirements in *Article 409* are for control panels operating at 1000 volts or less.

The *NEC* defines an **industrial control panel** as an assembly of two or more components consisting of one of the following: 1) power circuit components only, such as motor controllers, overload relays, fused disconnect switches, and circuit breakers; 2) control circuit components only, such as push buttons, pilot lights, selector switches, timers, switches, and control relays; 3) a combination of power and control circuit components. These components, with associated wiring and terminals, are

Article 409
Industrial Control Panels

Part I General

Section 409.1 Scope
Section 409.3 Other Articles

Part II Installation

Section 409.20 Conductor-Minimum Size and Ampacity
Section 409.21 Overcurrent Protection
Section 409.22 Short-Circuit Current Rating
Section 409.30 Disconnecting Means
Section 409.60 Bonding
Section 409.60 Surge Protection

Part III Construction Specifications

Section 409.100 Enclosures
Section 409.102 Busbars
Section 409.104 Wiring Space
Section 409.106 Spacings
Section 409.108 Service Equipment
Section 409.110 Marking

Goodheart-Willcox Publisher

Figure 7-17. *Article 409* can be categorized as a control article and contains requirements for industrial control panels.

mounted on, or contained within, an enclosure or mounted on a subpanel. See **Figure 7-18**.

Electricians rarely design and build their own industrial control panel. Instead, it is far more common for an electrician to make connections to a factory-produced industrial control panel. However, there are some electrical contractors that specialize in this type of work and will design, build, and install industrial control panels.

Consume Articles

There are nine articles in *Chapter 4* that are dedicated to equipment that consume or use electrical energy to perform a task., **Figure 7-19**. The consume articles apply to the very end of the electrical system where electricity does the work and is converted into a useful form of energy, such as light, heat, or mechanical rotation.

Article 410, Luminaires, Lampholders, and Lamps

Article 410 contains requirements for luminaires. The *NEC* defines a **luminaire** as a complete lighting unit consisting of a light source, such as a lamp or lamps, together with the parts designed to position the light source and connect it to the power supply. It may also include parts to protect the light source or the ballast or to distribute the light. A lampholder itself is not a luminaire. See **Figure 7-20**. The term *luminaire* may seem like a strange term to new code users, as *light fixture* is the standard language used by the general public. The *NEC* changed the term from *light fixture* to *luminaire* in 2002 to describe the lighting unit more accurately and to match the language used by other standards.

Luminaires are a portion of the electrical system that people interact with frequently. They may be installed in

NEC Chapter 4
Use and Consume Articles

Article 410 Luminaires, Lampholders, and Lamps
Article 411 Low-Voltage Lighting
Article 422 Appliances
Article 424 Fixed Electric Space-Heating Equipment
Article 425 Fixed Resistance and Electrode Industrial Process Heating Equipment
Article 426 Fixed Outdoor Electric Deicing and Snow-Melting Equipment
Article 427 Fixed Electric Heating Equipment for Pipelines and Vessels
Article 430 Motors, Motor Circuits, and Controllers
Article 440 Air-Conditioning and Refrigerating Equipment

Goodheart-Willcox Publisher

Figure 7-19. There are nine articles in *Chapter 4* discussing equipment that will use/consume electric energy to perform a task.

Anton Belo/Ahutterstock.com

Figure 7-20. Luminaires come in many styles. The luminaire pictured is installed outdoors and therefore must be listed for wet locations.

areas where they could create a shock hazard. They have the potential to produce a lot of heat. They often require maintenance, such as cleaning and changing lamps. There are also many different types and styles of luminaires. *Article 410* is a large article containing seventeen parts to address all the various scenarios and types of luminaires. See **Figure 7-21**.

NavinTar/Shutterstock.com

Figure 7-18. Industrial control panels may be built on site or purchased to control a piece of equipment or industrial process.

Article 410
Luminaires, Lampholders, and Lamps

Part I. General
Section 410.1 Scope
Section 410.2 Reconditioned Equipment
Section 410.5 Live Parts
Section 410.6 Listing Required
Section 410.8 Inspection

Part II. Luminaire Locations
Section 410.10 Luminaires in Specific Locations
Section 410.11 Luminaires Near Combustible Material
Section 410.12 Luminaires over Combustible Material
Section 410.14 Luminaires in Show Windows
Section 410.16 Luminaires in Clothes Closets
Section 410.18 Space for Cove Lighting

Part III Luminaire Outlet Boxes, Canopies, and Pans
Section 410.20 Space for Conductors
Section 410.21 Temperature Limit of Conductors in Outlet Boxes
Section 410.22 Outlet Boxes to Be Covered
Section 410.23 Covering of Combustible Material at Outlet Boxes
Section 410.24 Connection of Electric-Discharge and LED Luminaires

Part IV Luminaire Supports
Section 410.30 Supports
Section 410.36 Means of Support

Part V Grounding
Section 410.40 Equipment Grounding Conductor
Section 410.42 Luminaire(s) with Exposed Conductive Surfaces
Section 410.44 Connection to the Equipment Grounding Conductor

Part VI Wiring of Luminaires
Section 410.48 Luminaire Wiring - General
Section 410.50 Polarization of Luminaires
Section 410.52 Conductor Insulation
Section 410.54 Pendant Conductors for Incandescent Filament Lamps
Section 410.56 Protection of Conductors and Insulation
Section 410.59 Cord-Connected Showcases
Section 410.62 Cord-Connected Lampholders and Luminaires
Section 410.64 Luminaires as Raceways
Section 410.68 Feeder and Branch-Circuit Conductors and Ballasts
Section 410.69 Identification of Control Conductor Insulation
Section 410.70 Combustible Shades and Enclosures
Section 410.71 Disconnecting Means for Fluorescent or LED Luminaires that Utilize Double-Ended Lamps

Part VII Construction of Luminaires
Section 410.74 Luminaire Rating
Section 410.82 Portable Luminaire
Section 410.84 Cord Bushings

Part VIII Installation of Lampholders
Section 410.90 Screw Shell Type
Section 410.93 Double-Pole Switched Lampholders
Section 410.96 Lampholders in Wet or Damp Locations
Section 410.97 Lampholders Near Combustible Material

Part IX Lamps and Auxiliary Equipment
Section 410.103 Bases, Incandescent Lamps
Section 410.104 Electric-Discharge Lamp Auxiliary Equipment

Part X Special Provisions for Flush and Recessed Luminaires
Section 410.110 General
Section 410.115 Temperature
Section 410.116 Clearance and Installation
Section 410.117 Wiring
Section 410.118 Access to Other Boxes

(Continued)

Goodheart-Willcox Publisher

Figure 7-21. *Article 410* can be categorized as a consume article and contains all the various types of luminaires and all the locations they may be installed.

Article 410
Luminaires, Lampholders, and Lamps

Part XI Construction of Flush and Recessed Luminaires
- Section 410.119 Temperature
- Section 410.120 Lamp Wattage Marking
- Section 410.121 Solder Prohibited
- Section 410.122 Lampholders

Part XII Special Provisions for Electric-Discharge Lighting Systems of 1000 Volts or Less
- Section 410.130 General
- Section 410.134 Direct-Current Equipment
- Section 410.135 Open-Circuit Voltage Exceeding 300 Volts
- Section 410.136 Luminaire Mounting
- Section 410.137 Equipment Not Integral with Luminaire
- Section 410.138 Autotransformers
- Section 410.139 Switches

Part XIII Special Provisions for Electric-Discharge Lighting Systems of More Than 1000 Volts
- Section 410.140 General
- Section 410.141 Control
- Section 410.142 Lamp Terminals and Lampholders
- Section 410.143 Transformers
- Section 410.144 Transformer Locations
- Section 410.145 Exposure to Damage
- Section 410.146 Marking

Part XIV Lighting Track
- Section 410.151 Installation
- Section 410.153 Heavy-Duty Lighting Track
- Section 410.154 Fastening
- Section 410.155 Construction Requirements

Part XV Decorative Lighting and Similar Accessories.
- Section 410.160 Listing of Decorative Lighting

Part XVI Special Provisions for Horticultural Lighting Equipment
- Section 410.170 General
- Section 410.172 Listing
- Section 410.174 Installation and Use
- Section 410.176 Locations Not Permitted
- Section 410.178 Flexible Cord
- Section 410.180 Fittings and Connectors
- Section 410.182 Equipment Grounding Conductor
- Section 410.184 Ground-Fault Circuit-Interrupter (GFCI) Protection and Special Purpose Ground-Fault Circuit-Interrupter (SPGFCI) Protection
- Section 410.186 Support

Part XVII Special Provisions for Germicidal Irradiation Luminaires
- Section 410.190 General
- Section 410.191 Listing
- Section 410.193 Installation
- Section 410.195 Locations Not Permitted
- Section 410.197 Germicidal Irradiation Systems

Goodheart-Willcox Publisher

Figure 7-21. Contined

With the advancements of LED lighting, some of the dangers involved with luminaires are beginning to diminish. Luminaires with integrated LEDs do not have lamps to change and they operate at lower temperatures than some other types of fixtures.

Article 411, Low-Voltage Lighting

Article 411 contains requirements for low-voltage lighting, which operates at no more than 30 volts ac or 60 volts dc. Low-voltage lighting has become more popular in recent years with the advancement of LED technology, **Figure 7-22.**

Article 422, Appliances

Article 422 contains requirements for appliances. The *NEC* defines an ***appliance*** as utilization equipment, generally other than industrial, that is normally built in standardized sizes or types and is installed or connected as a unit to perform one or more functions, such as clothes washing, air-conditioning, food mixing, or deep frying. See **Figure 7-23.**

Dwelling units are not the only places you find appliances, but they are the ones that are the most identifiable. Common dwelling unit appliances include water heaters, gas furnaces, dishwashers, garbage disposals, trash compactors, ranges, dryers, and microwaves. *Article 422* does have some requirements that are specific to certain types of appliances, but most of the requirements are general and apply to all. See **Figure 7-24**.

Article 424, Fixed Electric Space-Heating Equipment

Article 424 contains requirements for fixed electric space-heating equipment, **Figure 7-25**. The requirements in *Article 424* apply to space heating equipment that uses electricity

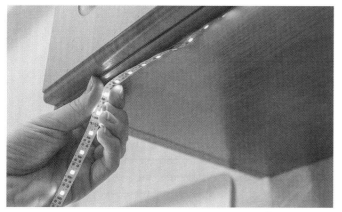

Kapustin Igor/Shutterstock.com

Figure 7-22. LED tape light is a popular type of low voltage lighting used for undercabinet lighting in kitchens.

Article 422
Appliances

■ Part I General
Section 422.1 Scope
Section 422.5 GFCI Protection
Section 422.6 Listing Required

■ Part II Installation
Section 422.10 Branch Circuits
Section 422.11 Overcurrent Protection
Section 422.12 Central Heating Equipment
Section 422.13 Storage-Type Water Heaters
Section 422.16 Flexible Cords
Section 422.17 Protection of Combustible Material
Section 422.18 Ceiling-Suspended (Paddle) Fans
Section 422.19 Space for Conductors
Section 422.20 Outlets Boxes to Be Covered
Section 422.21 Covering of Combustible Material at Outlet Boxes
Section 422.22 Utilizing Separable Attachment Fitting

■ Part III Disconnecting Means
Section 422.30 General
Section 422.31 Disconnection of Permanently Connected Appliances
Section 422.33 Disconnection of Cord-and-Plug-Connected or Attachment Fitting-Connected Appliances
Section 422.34 Unit Switch(es) as Disconnecting Means
Section 422.35 Switch and Circuit Breaker to Be Indicating

■ Part IV Construction
Section 422.40 Polarity in Cord-and-Plug-Connected Appliances
Section 422.41 Cord-and-Plug-Connected Appliances Subject to Immersion
Section 422.42 Signals for Heated Appliances
Section 422.44 Cord-and-Plug-Connected Immersion Heaters
Section 422.45 Stands for Cord-and-Plug-Connected Appliances
Section 422.47 Water Heater Controls
Section 422.48 Infrared Lamp Industrial Heating Appliances

■ Part V Marking
Section 422.60 Nameplate
Section 422.61 Marking of Heating Elements
Section 422.62 Appliances Consisting of Motors and Other Loads

Goodheart-Willcox Publisher

Figure 7-23. *Article 422* can be categorized as a consume article and contains requirements for appliances.

Goodheart-Willcox Publisher

Figure 7-24. Ranges and microwaves are examples of appliances that are commonly installed in dwelling units.

to produce heat. Heating equipment that use other methods, such as gas or oil, to produce heat are covered in *Article 422*. See **Figure 7-26**.

Article 430, Motors, Motor Circuits, and Controllers

Article 430 contains requirements for motors, motor circuits, and motor controllers, **Figure 7-27**. Electric motors are typically found in larger commercial and industrial jobs. Small commercial and residential installations will have motors, but they are typically a part of an appliance and not a stand-alone motor.

In addition to covering motors, *Article 430* has requirements for motor controllers and adjustable speed drives. The *NEC* defines a **controller** as a device or group of devices that serves to govern in some predetermined manner the electric

Article 424
Fixed Electric Space-Heating Equipment

Part I General
Section 424.1 Scope
Section 424.3 Other Articles
Section 424.4 Branch Circuits
Section 424.6 Listed Equipment

Part II. Installation
Section 424.10 General
Section 424.11 Supply Conductors
Section 424.12 Locations
Section 424.13 Spacing from Combustible Materials

Part III Control and Protection of Fixed Electric Space-Heating Equipment
Section 424.19 Disconnecting Means
Section 424.20 Thermostatically Controlled Switching Devices
Section 424.21 Switch and Circuit Breaker to Be Indicating
Section 424.22 Overcurrent Protection

Part IV Marking of Heating Equipment
Section 424.28 Nameplate
Section 424.29 Marking of Heating Elements

(*Continued*)

Goodheart-Willcox Publisher

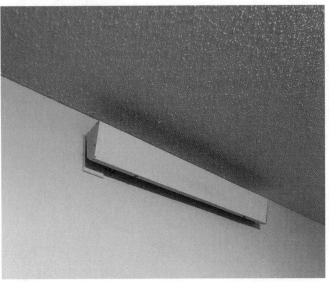

Goodheart-Willcox Publisher

Figure 7-25. A cove heater is pictured. Cove heaters are an electric radiant heater that is installed on the wall near the ceiling.

Figure 7-26. *Article 424* can be categorized as a consume article and contains requirements for the various types of electric heaters.

Article 424
Fixed Electric Space-Heating Equipment

Part V Electric Space-Heating Cables

Section 424.34 Heating Cable Construction
Section 424.35 Marking of Heating Cables
Section 424.36 Clearances of Wiring in Ceilings
Section 424.38 Area Restrictions
Section 424.39 Clearance from Other Objects and Openings
Section 424.40 Splices
Section 424.41 Ceiling Installation of Heating Cables on Dry Board, in Plaster, and on Concrete
Section 424.42 Finished Ceilings
Section 424.43 Installation of Nonheating Leads of Cables
Section 424.44 Installation of Cables in Concrete or Poured Masonry Floors
Section 424.45 Installation of Cables Under Floor Coverings
Section 424.46 Inspection
Section 424.47 Label Provided by Manufacturer
Section 424.48 Installation of Cables in Walls

Part VI. Duct Heaters

Section 424.57 General
Section 424.58 Identification
Section 424.59 Airflow
Section 424.60 Elevated Inlet Temperature
Section 424.61 Installation of Duct Heaters with Heat Pumps and Air Conditioners
Section 424.62 Condensation
Section 424.63 Fan Circuit Interlock
Section 424.64 Limit Controls
Section 424.65 Location of Disconnecting Means
Section 424.66 Installation

Part VII Resistance-Type Boilers

Section 424.70 Scope
Section 424.71 Identification
Section 424.72 Overcurrent Protection
Section 424.73 Overtemperature Limit Control
Section 424.74 Overpressure Limit Control

Part VIII Electrode-Type Boilers

Section 424.80 Scope
Section 424.81 Identification
Section 424.82 Branch-Circuit Requirements
Section 424.83 Overtemperature Limit Control
Section 424.84 Overpressure Limit Control
Section 424.85 Grounding
Section 424.86 Markings

Part IX Electric Radiant Heating Panels and Heating Panel Sets

Section 424.90 Scope
Section 424.92 Markings
Section 424.93 Installation
Section 424.94 Clearances of Wiring in Ceilings
Section 424.95 Location of Branch-Circuit and Feeder Wiring in Walls
Section 429.96 Connection to Branch-Circuit Conductors
Section 424.97 Nonheating Leads
Section 424.98 Installation in Concrete or Poured Masonry
Section 424.99 Installation Under Floor Covering

Part X Low-Voltage Fixed Electric Space-Heating Equipment

Section 424.100 Scope
Section 424.101 Energy Source
Section 424.102 Listed Equipment
Section 424.103 Installation
Section 424.104 Branch Circuit

Goodheart-Willcox Publisher

Figure 7-26. Continued

Wisarut pumipak/Shutterstock.com

Figure 7-27. Electric motors that are stand-alone and not a part of an appliance are typically found on large commercial and industrial jobs.

power delivered to the apparatus to which it is connected. Motor controllers are used to start and stop motors and may offer overload protection that will protect the motor from damage due to excess current. *Adjustable-speed drives* are defined as power conversion equipment that provide a means of adjusting the speed of an electric motor.

The end of *Article 430 (Part XIV)* has tables that give full load current values for various types of motors, **Figure 7-28**.

Article 440, Air-Conditioning and Refrigerating Equipment

Article 440 contains requirements for air-conditioning and refrigeration equipment, **Figure 7-29**. A common component in air-conditioning and refrigeration equipment is a hermetic refrigerant motor-compressor. The *NEC* defines a *hermetic refrigerant motor-compressor* as a combination consisting of a compressor and motor, both of which are enclosed in the same housing, with no external shaft or shaft seals, with the motor operating in the refrigerant.

Article 440 has requirements for all different types of air-conditioning equipment. This includes everything from small window air conditioners to convenience store coolers to large rooftop units used to cool commercial buildings. See **Figure 7-30**.

Create, Alter, and Store Articles

Articles 445–480 have requirements for equipment that will create, alter, or store electricity. See **Figure 7-31**.

Article 445, Generators

Article 445 has requirements specific to generators, **Figure 7-32**. Generators are used to create electricity. The generators installed by electricians are typically used as a backup to the utility power. Most generators use a fossil fuel such as gasoline, diesel, natural gas, or propane to spin the generator to create electricity, **Figure 7-33**. The requirements found in *Article 445* apply directly to the generators themselves, while some of the articles in *Chapter 7* address the different systems that may contain a generator.

Article 450, Transformers and Transformer Vaults (Including Secondary Ties)

Article 450 has requirements for transformers, **Figure 7-34**. A *transformer* changes the voltage of an ac system either up or down. It is more common to see transformers that step the voltage down in building. Supplying a building with a higher voltage reduces the amount of current necessary, saving money on equipment and material costs. The voltage can then be stepped down with a transformer to feed the loads that require a lower voltage. See **Figure 7-35**.

Article 455, Phase Converters

Article 455 has requirements for phase converters, **Figure 7-37**. The *NEC* defines a *phase converter* as an electrical device that converts single-phase power to 3-phase electric power. Phase converters are used when a piece of equipment requires 3-phase power, but the building only has a single-phase electric service. It is expensive to upgrade the electric service to 3-phase, so installing a phase converter to create 3-phase power for that one piece of equipment can be more economical.

> **PRO TIP** — **Doorbell Transformers**
>
> Nearly all buildings have transformers of one type or another, even dwelling units. A common transformer found in dwellings is for the doorbell. These doorbell transformers typically have a 120-volt input and a 16-volt output, which is the voltage at which the doorbell operates. With video doorbells becoming more common, some transformers will need to be upgraded to a higher volt-ampere rating to operate the camera and the doorbell simultaneously. See **Figure 7-36**.

Article 430
Motors, Motor Circuits, and Controllers

Part I. General
Section 430.1 Scope
Section 430.2 Reconditioned Motors
Section 430.4 Part-Winding Motors
Section 430.5 Other Articles
Section 430.6 Conductor Ampacity and Motor Rating Determination
Section 430.7 Marking on Motors and Multimotor Equipment
Section 430.8 Marking on Motor Controllers
Section 430.9 Terminals
Section 430.10 Wiring Space in Enclosures
Section 430.11 Protection Against Liquids
Section 430.12 Motor Terminal Housings
Section 430.13 Bushings
Section 430.14 Location of Motors
Section 430.16 Exposure to Dust Accumulations
Section 430.17 Highest Rated or Smallest Rated Motor
Section 430.18 Nominal Voltage or Rectifier Systems

Part II Motor Circuit Conductors.
Section 430.21 General
Section 430.22 Single Motor
Section 430.23 Wound-Rotor Secondary
Section 430.24 Several Motors or a Motor(s) and Other Load(s)
Section 430.25 Multimotor and Combination-Load Equipment
Section 430.26 Feeder Demand Factor
Section 430.27 Capacitors with Motors
Section 430.28 Feeder Taps
Section 430.29 Constant Voltage Direct-Current Motors-Power Resistors

Part III Motor and Motor Branch-Circuit Overload Protection
Section 430.31 General
Section 430.32 Continuous-Duty Motors
Section 430.33 Intermittent and Similar Duty
Section 430.35 Shunting During Starting Period
Section 430.36 Fuses - In Which Conductor
Section 430.37 Devices Other Than Fuses - In Which Conductors
Section 430.38 Number of Conductors Opened by Overload Device
Section 430.39 Motor Controller as Overload Protection
Section 430.40 Overload Relays
Section 430.42 Motors on General-Purpose Branch Circuits
Section 430.43 Automatic Restarting
Section 430.44 Orderly Shutdown

Part IV Motor Branch-Circuit Short-Circuit and Ground-Fault Protection
Section 430.51 General
Section 430.52 Rating or Setting for Individual Motor Circuit
Section 430.53 Several Motors or Loads on One Branch Circuit
Section 430.54 Multimotor and Combination-Load Equipment
Section 430.55 Combined Overcurrent Protection
Section 430.56 Branch-Circuit Protective Devices - In Which Conductor
Section 430.57 Size of Fuseholder
Section 430.58 Rating of Circuit Breaker

Part V Motor Feeder Short-Circuit and Ground-Fault Protection
Section 430.61 General
Section 430.62 Rating or Setting-Motor Load
Section 430.63 Rating or Setting-Motor Load and Other Load(s)

(Continued)

Goodheart-Willcox Publisher

Figure 7-28. *Article 430* can be categorized as a consume article and contains requirements for electric motors.

Article 430
Motors, Motor Circuits, and Controllers

Part VI Motor Control Circuits
- Section 430.71 General
- Section 430.72 Overcurrent Protection
- Section 430.73 Protection of Conductors from Physical Damage
- Section 430.74 Electrical Arrangement of Control Circuits
- Section 430.75 Disconnection

Part VII Motor Controllers
- Section 430.81 General
- Section 430.82 Motor Controller Design
- Section 430.83 Ratings
- Section 430.84 Need Not Open All Conductors
- Section 430.85 In Grounded Conductors
- Section 430.87 Number of Motors Served by Each Controller
- Section 430.88 Adjustable-Speed Motors
- Section 430.89 Speed Limitation
- Section 430.90 Combination Fuseholder and Switch as Controller

Part VIII Motor Control Centers
- Section 430.92 General
- Section 430.94 Overcurrent Protection
- Section 430.95 Service Equipment
- Section 430.96 Grounding
- Section 430.97 Busbars and Conductors
- Section 430.98 Marking
- Section 430.99 Available Fault Current

Part IX. Disconnecting Means
- Section 430.101 General
- Section 430.102 Location
- Section 430.103 Operation
- Section 430.104 To Be Indicating
- Section 430.105 Grounded Conductors
- Section 430.107 Readily Accessible
- Section 430.108 Every Disconnecting Means
- Section 430.109 Type
- Section 430.110 Current Rating and Interrupting Capacity
- Section 430.111 Switch or Circuit Breaker as Both Controller and Disconnecting Means
- Section 430.112 Motors Served by Single Disconnecting Means
- Section 430.113 Energy from More than One Source

Part X Adjustable-Speed Drive Systems
- Section 430.120 General
- Section 430.122 Conductors-Minimum Size and Ampacity
- Section 430.124 Overload Protection
- Section 430.126 Motor Overtemperature Protection
- Section 430.128 Disconnecting Means
- Section 430.130 Branch-Circuit Short-Circuit and Ground-Fault Protection for Single Motor Circuits Containing Power Conversion Equipment
- Section 430.131 Several Motors or Loads on One Branch Circuit Including Power Conversion Equipment

Part XI Over 1000 Volts, Nominal
- Section 430.201 General
- Section 430.202 Marking on Motor Controllers
- Section 430.203 Raceway Connection to Motors
- Section 430.204 Wire-Bending Space in Enclosures
- Section 430.205 Size of Conductors
- Section 430.206 Motor-Circuit Overcurrent Protection
- Section 430.207 Rating of Motor Control Apparatus
- Section 430.208 Disconnecting Means

Part XII Protection of Live Parts – All Voltages
- Section 430.231 General
- Section 430.232 Where Required
- Section 430.233 Guards for Attendants

Part XIII Grounding-All Voltages
- Section 430.241 General
- Section 430.242 Stationary Motors
- Section 430.243 Portable Motors
- Section 430.244 Controllers
- Section 430.245 Methods of Grounding

Part XIV Tables

Goodheart-Willcox Publisher

Figure 7-28. Continued

Unit 7 NEC Chapter 4

galinast/Getty Images

Figure 7-29. This central air-conditioning unit will cool the air blown through a ductwork by an air handler or furnace blower. *Article 440.14* requires a disconnect, which is the gray enclosure on the wall, to be within sight from, and readily accessible from the air-conditioning equipment.

Article 440
Air-Conditioning and Refrigerating Equipment

Part I General
Section 440.1 Scope
Section 440.4 Marking on Hermetic Refrigerant Motor-Compressors and Equipment
Section 440.5 Marking on Controllers
Section 440.6 Ampacity and Rating
Section 440.7 Highest Rated (Largest) Motor
Section 440.8 Single Machine and Location
Section 440.9 Grounding and Bonding
Section 440.10 Short-Circuit Current Rating

Part II Disconnecting Means
Section 440.11 General
Section 440.12 Rating and Interrupting Capacity
Section 440.13 Cord-Connected Equipment
Section 440.14 Location

Part III Branch-Circuit Short-Circuit and Ground-Fault Protection.
Section 440.21 General
Section 440.22 Application and Selection

(*Continued*)

Goodheart-Willcox Publisher

Figure 7-30. *Article 440* can be categorized as a consume article and contains requirements for air-conditioning and refrigerating equipment.

Article 440
Air-Conditioning and Refrigerating Equipment

Part IV Circuit Conductors
Section 440.31 General
Section 440.32 Single Motor-Compressor
Section 440.33 Motor-Compressor(s) With or Without Additional Motor Loads
Section 440.34 Combination Load
Section 440.35 Multimotor and Combination-Load Equipment

Part V Controllers for Motor-Compressors
Section 440.41 Rating

Part VI. Motor-Compressor and Branch-Circuit Overload Protection
Section 440.51 General
Section 440.52 Application and Selection
Section 440.53 Overload Relays
Section 440.54 Motor-Compressors and Equipment on 15- or 20-Ampere Branch Circuits - Not Cord- and Attachment-Plug-Connected
Section 440.55 Cord- and Attachment-Plug-Connected Motor-Compressors and Equipment on 15- or 20-Ampere Branch Circuits

Part VII Room Air Conditioners
Section 440.60 General
Section 440.61 Grounding
Section 440.62 Branch-Circuit Requirements
Section 440.63 Disconnecting Means
Section 440.64 Supply Cords
Section 440.65 Protection Devices

Goodheart-Willcox Publisher

Figure 7-30. Continued

Article 480, Stationary Standby Batteries

Article 480 has requirements for storage batteries, **Figure 7-38**. It is becoming more common to see homes and businesses install renewable sources of energy such as solar and wind on their premises. If they want to be able to store the energy for times of the day when production is down, they will need batteries. Battery technology has been rapidly changing over the past few years, making batteries safer and able to store more energy. See **Figure 7-39**.

NEC Chapter 4
Create, Alter, and Store Articles

Article 445 Generators
Article 450 Transformers and Transformer Vaults (Including Secondary Ties)
Article 455 Phase Converters
Article 460 Capacitors
Article 470 Resistors and Reactors
Article 480 Stationary Standby Batteries

Goodheart-Willcox Publisher

Figure 7-31. There are six articles that address equipment that create, alter, or store electricity.

Goodheart-Willcox Publisher

Figure 7-33. The generator pictured is fed with natural gas and has an automatic transfer switch located near the electrical panel. In the case of a power outage, the generator will come on automatically and provide power to select loads in the house until the utility power returns.

Article 445
Generators

Section 445.1 Scope
Section 445.6 Listing
Section 445.10 Location
Section 445.11 Marking
Section 445.12 Overcurrent Protection
Section 445.13 Ampacity of Conductors
Section 445.14 Protection of Live Parts
Section 445.15 Guards for Attendants
Section 445.16 Bushings
Section 445.17 Generator Terminal Housings
Section 445.18 Disconnecting Means
Section 445.19 Emergency Shutdown of Prime Mover
Section 445.20 Ground-Fault Circuit-Interrupter Protection for Receptacles on 15-kW or Smaller Portable Generators

Goodheart-Willcox Publisher

Figure 7-32. *Article 445* can be categorized as a create, alter, and store article and contains requirements for generators.

Article 450
Transformers and Transformer Vaults (Including Secondary Ties)

Part 1 General
Section 450.1 Scope
Section 450.2 Interconnection of Transformers
Section 450.3 Overcurrent Protection
Section 450.4 Autotransformers 1000 Volts, Nominal, or Less
Section 450.5 Grounding Autotransformers
Section 450.6 Secondary Ties
Section 450.7 Parallel Operation
Section 450.8 Guarding
Section 450.9 Ventilation
Section 450.10 Grounding and Bonding
Section 450.11 Marking
Section 450.12 Terminal Wiring Space
Section 450.13 Accessibility
Section 450.14 Disconnecting Means

(*Continued*)

Goodheart-Willcox Publisher

Figure 7-34. *Article 450* can be categorized as a create, alter, and store article and contains requirements for transformers.

Article 450
Transformers and Transformer Vaults (Including Secondary Ties)

Part II Installation

Section 450.21 Dry-Type Transformers Installed Indoors
Section 450.22 Dry-Type Transformers Installed Outdoors
Section 450.23 Less-Flammable Liquid-Insulated Transformers
Section 450.24 Nonflammable Fluid-Insulated Transformers
Section 450.25 Askarel-Insulated Transformers Installed Indoors
Section 450.26 Oil-Insulated Transformers Installed Indoors
Section 450.27 Oil-Insulated Transformers Installed Outdoors
Section 450.28 Modification of Transformers

Part III Transformer Vaults

Section 450.41 Location
Section 450.41 Walls, Roofs, and Floors
Section 450.43 Doorways
Section 450.45 Ventilation Openings
Section 450.46 Drainage
Section 450.47 Water Pipes and Accessories
Section 450.48 Storage in Vaults

Goodheart-Willcox Publisher

Figure 7-34. Continued

Goodheart-Willcox Publisher

Figure 7-35. The box in between the two panelboards is a transformer. It is stepping the voltage from the 480-volt panelboard on the right down to 120/208 volts to feed the panelboard on the left.

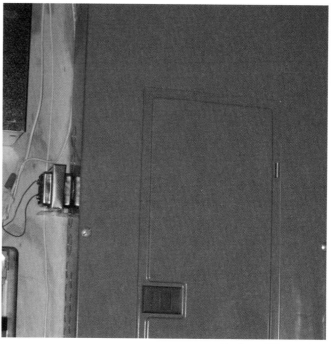

Goodheart-Willcox Publisher

Figure 7-36. The small transformer mounted on the side of the panel steps the voltage down from 120 volts to 16 volts to feed the doorbell.

Article 455
Phase Converters

Part 1 General

Section 455.1 Scope
Section 455.3 Other Articles
Section 455.4 Marking
Section 455.5 Equipment Grounding Connection
Section 455.6 Conductors
Section 455.7 Overcurrent Protection
Section 455.8 Disconnecting Means
Section 455.9 Connection of Single-Phase Loads
Section 455.10 Terminal Housings

Part II Specific Provisions Applicable to Different Types of Phase Converters

Section 455.20 Disconnecting Means
Section 455.21 Start-Up
Section 455.22 Power Interruption
Section 455.23 Capacitors

Goodheart-Willcox Publisher

Figure 7-37. *Article 455* can be categorized as a create, alter, and store article and contains requirements for phase converters.

Article 480
Stationary Standby Batteries

Section 480.1 Scope
Section 480.3 Equipment
Section 480.4 Battery and Cell Terminations
Section 480.5 Wiring and Equipment Supplied from Batteries
Section 480.6 Overcurrent Protection for Prime Movers
Section 480.7 DC Disconnect Methods
Section 480.8 Insulation of Batteries
Section 480.9 Battery Support Systems
Section 480.10 Battery Locations
Section 480.11 Vents
Section 480.12 Battery Interconnections
Section 480.13 Ground-Fault Detection

Goodheart-Willcox Publisher

Figure 7-38. *Article 480* can be categorized as a create, alter, and store article and contains requirements for storage batteries.

Goodheart-Willcox Publisher

Figure 7-39. New battery technologies are increasing storage capacity, lengthening life expectancy, and making batteries safer. The battery in the image is a 48-volt 300-amphour lithium iron phosphate battery.

NEC KEYWORDS

The following are examples of keywords that should lead you to *Chapter 4*:

			Chapter 3 Keywords
air conditioning	device	lighting	switch
appliance	equipment	luminaire	
battery	generator	motor	
controller	heat	receptacle	

Summary

- *Chapter 4* of the *NEC* is the fourth and last of the general chapters that apply to all electrical installations.
- *Chapter 4* contains general requirements for equipment installed in an electrical system.
- Requirements related to the equipment installed while trimming a job are found in *Chapter 4*
- The articles in *Chapter 4* can be grouped into the following categories.
  - Equipment connection articles
  - Control articles
  - Consume articles
  - Create, alter, and store articles

Unit 7 Review

Name _____ Date _____ Class _____

Know and Understand

Answer the following questions based on information in this unit.

1. *Chapter 4* of the *NEC* has information necessary when _____ an electrical system.
   A. designing
   B. roughing-in
   C. trimming
   D. testing

2. *Chapter 4* contains _____ requirements.
   A. wiring method and material
   B. equipment for general use
   C. wiring and protection
   D. special equipment

3. _____ has general requirements on appliances.
   A. *Article 400*
   B. *Article 409*
   C. *Article 410*
   D. *Article 422*

4. _____ contains requirements on air-conditioning and refrigeration equipment.
   A. *Article 424*
   B. *Article 430*
   C. *Article 440*
   D. *Article 450*

5. A(n) _____ switch is intended for use in general distribution and branch circuits. It is rated in amperes and is capable of interrupting its rated current at its rated voltage.
   A. bypass isolation
   B. general use
   C. isolating
   D. transfer

6. _____ has requirements on equipment rated over 1000 volts, nominal.
   A. *Article 460*
   B. *Article 470*
   C. *Article 480*
   D. *Article 495*

7. Requirements for motor controllers are found in _____.
   A. *Article 410*
   B. *Article 422*
   C. *Article 430*
   D. *Article 445*

8. Which of the following keywords would lead you to Chapter 4?
   A. Grounding
   B. Luminaires
   C. Overcurrent protection
   D. Rigid metallic conduit

9. _____ has requirements on flexible cords and cables.
   A. *Article 400*
   B. *Article 402*
   C. *Article 410*
   D. *Article 450*

10. _____ has requirements that apply to appliance disconnects.
    A. *Article 410–Part V*
    B. *Article 422–Part III*
    C. *Article 430–Part IX*
    D. *Article 440–Part II*

Apply and Analyze

Answer the following questions using a copy of the National Electrical Code. *Identify the section or subsection where the answer is found.*

1. A surface-mounted incandescent luminaire must be at least _____ from the storage space in a clothes closet.
   A. 6″
   B. 8″
   C. 12″
   D. 24″

   NEC _____

2. Switches shall not disconnect the _____ conductor of a circuit.
   A. hot
   B. ungrounded
   C. grounded
   D. switch leg

   NEC _____

3. The maximum length cord permitted to be installed on a dishwasher is _____, measured from the attachment plug to the plane of the rear of the appliance.
   A. 3′   C. 4.5′
   B. 4′   D. 6.5′

   NEC _____

4. When installing heating cables in a dry board ceiling, the ceiling below the heating cable shall be covered with a gypsum board not exceeding _____ in thickness.
   A. 1/2"
   B. 5/8"
   C. 3/4"
   D. 1"

   NEC _____

5. Which of the following is *not* one of the protection devices that a manufacturer may install on the cord of a room air conditioner?
   A. Leakage-current detector-interrupter
   B. Arc-fault circuit interrupter
   C. Ground-fault circuit interrupter
   D. Heat detecting circuit interrupter

   NEC _____

6. A 16 AWG fixture wire has a maximum ampacity of _____ amperes.
   A. 6
   B. 8
   C. 10
   D. 13

   NEC _____

7. An attachment plug and receptacle or cord connector is permitted to serve as the motor controller for a portable motor of _____ or less.
   A. 1/3 hp
   B. 1/2 hp
   C. 3/4 hp
   D. 1 hp

   NEC _____

8. Receptacles located more than _____ above the floor are not required to be tamper-resistant.
   A. 4′
   B. 4 1/2′
   C. 5 1/2′
   D. 6′

   NEC _____

9. _____ piping is not permitted in dedicated battery rooms.
   A. Water
   B. Gas
   C. Sewage
   D. All of the above.

   NEC _____

10. Where transformers are protected with an automatic sprinkler, water spray, carbon dioxide, or halon, the transformer vault is permitted to have a _____ fire-resistance rating.
    A. 30-minute
    B. 1-hour
    C. 2-hour
    D. 3-hour

    NEC _____

Name _____ Date _____ Class _____

Critical Thinking

NEC Chapter 4

Answer the following questions using a copy of the National Electrical Code. *Identify the section or subsection where the answer is found.*

1. The nonheating leads of electric space heating cables shall be at least _____ in length.

   _____

   _____

   NEC _____

2. Low-voltage lighting systems shall be supplied by maximum _____ branch circuit.

   _____

   _____

   NEC _____

3. All 15/20-ampere, 125/250-volt nonlocking receptacles installed in damp locations shall be a listed _____ type.

   _____

   _____

   NEC _____

4. Heavy-duty lighting track is identified for use exceeding _____ amperes.

   _____

   _____

   NEC _____

5. Resistors and reactors shall not be placed where exposed to _____.

   _____

   _____

   NEC _____

6. The branch circuit overcurrent devices and conductors for a 50-gallon storage-type water heater shall be sized not smaller than _____ of the ampere rating of the water heater.

   _____

   _____

   NEC _____

7. When installing a duct heater, means shall be provided to ensure _____ airflow across the face of the heater in accordance with the manufacturer's instructions.

   _____

   _____

   NEC _____

8. Conductors supplying a single motor in a continuous-duty application shall have an ampacity of not less than _____ of the motor full-load current rating.

   _____

   _____

   NEC _____

9. Flexible cords and their fittings shall be suitable for the conditions of use and _____.

   _____

   _____

   NEC _____

10. Where single-phase loads are connected on the load side of a phase converter, they shall not be connected to the _____.

    _____

    _____

    NEC _____

Part 1   Layout and Structure: Learning to Use the NEC

NEC Chapters 1–4

Answer the following questions using a copy of the National Electrical Code. *Identify the section or subsection where the answer is found.*

11. A receptacle outlet within _____ of a dwelling unit wet bar sink shall be GFCI protected.

    _____

    _____

    NEC _____

12. Recessed luminaires that are non-type IC shall be at least _____ from thermal insulation.

    _____

    _____

    NEC _____

13. Reinforced thermosetting resin conduit (type RTRC) is not permitted in ambient temperatures exceeding _____ unless listed otherwise.

    _____

    _____

    NEC _____

14. A garbage disposer is permitted to be connected by a cord that does not exceed _____ in length.

    _____

    _____

    NEC _____

15. Conductors and equipment recognized by the *NEC* shall be _____ only if approved.

    _____

    _____

    NEC _____

16. A building is a structure that stands alone or that is separated from adjoining structures by _____.

    _____

    _____

    NEC _____

17. Receptacles shall not be installed _____ in the area below a sink.

    _____

    _____

    NEC _____

18. Type AC cable shall be supported and secured at intervals not exceeding _____.

    _____

    _____

    NEC _____

19. An overhead service raceway equipped with a service head shall have conductors of different potential brought out through _____ openings.

    _____

    _____

    NEC _____

20. Direct buried conductors emerging from grade shall be protected by an enclosure or raceway to at least _____ above finished grade.

    _____

    _____

    NEC _____

UNIT

8 NEC Chapter 5

sockagphoto/Shutterstock.com

KEY TERMS

assembly occupancy
Class I, Division 1 location
Class I, Division 2 location
Class II, Division 1 location
Class II, Division 2 location
Class III, Division 1 location
Class III, Division 2 location
combustible dust
intrinsically safe circuit
manufactured building
park trailer
relocatable structure
shore power
temporary installation
Zone 0 location
Zone 1 location
Zone 2 location
Zone 20 location
Zone 21 location
Zone 22 location

LEARNING OBJECTIVES

After completing this unit, you will be able to:
- Describe the type of requirements in *Chapter 5* of the *NEC*.
- Identify the occupancies covered by *Chapter 5* of the *NEC*.
- Recognize keywords that will lead you to *Chapter 5* of the *NEC*.

Introduction

This unit is an introduction to *Chapter 5* of the *NEC*, titled *Special Occupancies*. It is the first of the special chapters that supplement or modify the information found in *Chapters 1–7*, **Figure 8-1**. This unit will introduce the type of occupancies covered by *Chapter 5*. The end of the unit contains practice exercises that will require the use of your *National Electrical Code* to complete.

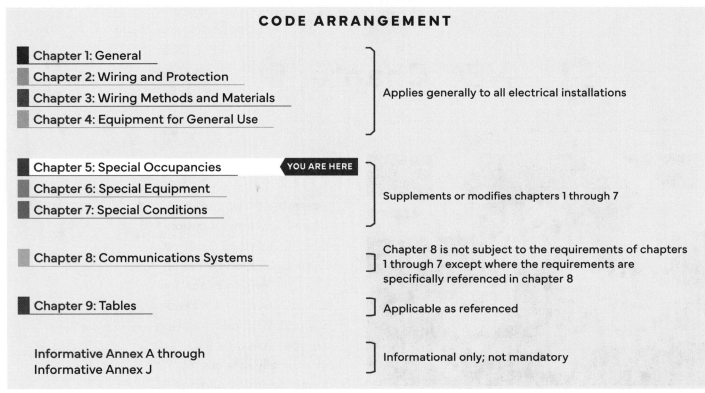

Figure 8-1. Chapter 5 is the first of the special chapters. It contains requirements that will supplement or modify the information found in Chapters 1–7.

Special Occupancies

Chapter 5 of the *NEC* contains requirements to address unique hazards that are not common to every occupancy, **Figure 8**-2. Additional requirements are necessary to ensure that these occupancies remain safe from hazards that may arise from the use of electricity. For example, the occupancies may contain combustible atmospheres, large groups of people, corrosive environments, or bodies of water.

Hazardous Locations

Articles 500 through *517* cover occupancies and locations that can be classified as hazardous or a portion of which may be considered hazardous. A location classified as hazardous may have flammable liquids, gases, or vapors, or the accumulation of particles and dusts that may be readily ignitable. The *NEC* has special requirements to ensure that the electrical system remains safe despite the additional hazard.

PRO TIP

There are many types of installations and different aspects of the electrical industry that electricians will encounter. It is impossible for any one person to know everything there is to know about the electrical industry. "You don't know what you don't know," is a quote that sums this up well. It is possible that you could end up working on a jobsite that falls under the *NEC* classification of special occupancy and not realize it. This can result in wasted time and money if the special requirements found in *Chapter 5* were not followed and therefore work needs to be corrected.

Table of Contents–Chapter 5

One way to help prevent this from happening is to become familiar with the types of occupancies covered by *Chapter 5*. Spending some time simply looking at the titles of the *Chapter 5* articles in the table of contents may be enough to create a mental note. Then, when a situation or job arises that falls under the classification of special occupancy, you will recognize it as such and check your codebook to see if there are any applicable special requirements.

Articles 500 through 506 have general requirements for hazardous locations. Articles 511 through 517 have requirements for specific occupancies that are likely to have one of these hazards present.

Article 500, Hazardous (Classified) Locations, Classes I, II, and III, Divisions 1 and 2

Chapter 5 starts off with *Article 500, Hazardous (Classified) Locations, Class I, II, and III, Divisions 1 and 2*, **Figure 8**-3. *Article 500* contains a lot of general information on hazardous locations and dedicates a large portion of the article to defining the three different types of locations. It also describes the different classifications of materials that may be flammable or combustible.

Article 501, Class I Locations

Article 501 covers Class I locations. Class I locations may have flammable gases or vapors present, **Figure 8**-4. Class I locations are divided into two divisions. *Class I, Division 1 locations* are areas where flammable gases or vapors may exist under normal operating conditions. *Class I, Division 2 locations* are areas where flammable gases or vapors are stored or handled but are confined within containers and therefore not present under normal conditions. Oil wells and gasification plants are examples of occupancies that will have areas that are classified as Class I locations, **Figure 8**-5.

Article 502, Class II Locations

Article 502 covers Class II locations. Class II locations may have combustible dusts present, **Figure 8**-6. *Combustible*

NEC Chapter 5
Special Occupancies

Article 500 Hazardous (Classified) Locations, Classes I, II, and III, Divisions 1 and 2
Article 501 Class I Locations
Article 502 Class II Locations
Article 503 Class III Locations
Article 504 Intrinsically Safe Systems
Article 505 Zone 0, 1, and 2 Locations
Article 506 Zone 20, 21, and 22 Locations for Combustible Dusts or Ignitible Fibers/Flyings
Article 511 Commercial Garages, Repair and Storage
Article 512 Cannabis Oil Equipment and Cannabis Oil Systems Using Flammable Materials
Article 513 Aircraft Hangars
Article 514 Motor Fuel Dispensing Facilities
Article 515 Bulk Storage Plants
Article 516 Spray Application, Dipping, Coating, and Printing Processes Using Flammable or Combustible Materials
Article 517 Health Care Facilities
Article 518 Assembly Occupancies
Article 520 Theaters, Audience Areas of Motion Picture and Television Studios, Performance Areas, and Similar Locations
Article 522 Control Systems for Permanent Amusement Attractions
Article 525 Carnivals, Circuses, Fairs, and Similar Events
Article 530 Motion Picture and Television Studios and Remote Locations
Article 540 Motion Picture Projection Rooms
Article 545 Manufactured Buildings and Relocatable Structures
Article 547 Agricultural Buildings
Article 550 Mobile Homes, Manufactured Homes, and Mobile Home Parks
Article 551 Recreational Vehicles and Recreational Vehicle Parks
Article 552 Park Trailers
Article 555 Marinas, Boatyards, Floating Buildings, and Commercial and Noncommercial Docking Facilities
Article 590 Temporary Installations

Goodheart-Willcox Publisher

Figure 8-2. *Chapter 5* contains information on special occupancies. It is important to become familiar with these types of occupancies so when you encounter them on the job you can recognize that they may have special requirements.

Article 500
Hazardous (Classified) Locations, Classes I, II, and III, Divisions 1 and 2

Section 500.1 Scope
Section 500.4 Documentation
Section 500.5 Classification of Locations
Section 500.6 Materials
Section 500.7 Protection Techniques
Section 500.8 Equipment

Goodheart-Willcox Publisher

Figure 8-3. *Article 500* defines and describes Class I, Class II, and Class III hazardous locations. It contains some requirements, but much of the article is dedicated to explaining the classifications.

Part 1   Layout and Structure: Learning to Use the NEC

Article 501
Class I Locations

Part I General
Section 501.1 Scope
Section 501.5 Zone Equipment

Part II Wiring
Section 501.10 Wiring Methods
Section 501.15 Sealing and Drainage
Section 501.17 Process Sealing
Section 501.20 Conductor Insulation, Class I, Divisions 1 and 2
Section 501.25 Uninsulated Exposed Parts, Class I, Divisions 1 and 2
Section 501.30 Grounding and Bonding
Section 501.35 Surge Protection

Part III Equipment
Section 501.100 Transformers and Capacitors
Section 501.105 Meters, Instruments, and Relays
Section 501.115 Switches, Circuit Breakers, Motor Controllers, and Fuses
Section 501.120 Control Transformers and Resistors
Section 501.125 Motors and Generators
Section 501.130 Luminaires
Section 501.135 Utilization Equipment
Section 501.140 Flexible Cords, Class I, Division 1 and 2
Section 501.141 Flexible Cables, Class I, Division 2
Section 501.145 Receptacles and Attachment Plugs, Class I, Division 1 and Division 2
Section 501.150 Signaling, Alarm, Remote-Control, and Communications Systems

Goodheart-Willcox Publisher

Figure 8-4. *Article 501* has requirements for Class I locations. Class I locations are areas that have flammable gases or vapors.

GE_4530/Shutterstock.com

Figure 8-5. Oil refineries and gasification plants are examples of Class I locations because flammable gases may be present.

Unit 8 NEC Chapter 5

Article 502
Class II Locations

Part I General
Section 502.1 Scope
Section 502.5 Explosionproof Equipment
Section 502.6 Zone Equipment

Part II Wiring
Section 502.10 Wiring Methods
Section 502.15 Sealing, Class II, Divisions 1 and 2
Section 502.25 Uninsulated Exposed Parts, Class II, Division 1 and 2
Section 502.30 Grounding and Bonding
Section 502.35 Surge Protection — Class II, Divisions 1 and 2

Part III Equipment
Section 502.100 Transformers and Capacitors
Section 502.115 Switches, Circuit Breakers, Motor Controllers, and Fuses
Section 502.120 Control Transformers and Resistors
Section 501.125 Motors and Generators
Section 502.128 Ventilating Piping
Section 502.130 Luminaires
Section 502.135 Utilization Equipment
Section 502.140 Flexible Cords — Class II, Division 1 and 2
Section 502.145 Receptacles and Attachment Plugs
Section 502.150 Signaling, Alarm, Remote-Control, and Communication Systems; and Meters, Instruments, and Relays

Goodheart-Willcox Publisher

Figure 8-6. *Article 502* has requirements for locations where combustible dusts may be present.

dusts are dust particles that are 500 microns or smaller and present a fire or explosion hazard when dispersed and ignited in the air. Like Class I locations, Class II locations are divided into two divisions. ***Class II, Division 1 locations*** are areas where combustible dusts may exist in quantities sufficient to produce ignition under normal operating conditions. ***Class II, Division 2 locations*** are areas where combustible dusts may be present but not in quantities sufficient to cause ignition under normal conditions. Grain bins and sawmills are examples of occupancies that will have areas that are classified as Class II locations, **Figure 8-7**.

jantsarik/Shutterstock.com

Figure 8-7. Grain elevators are an example of a Class II location where combustible dusts may be present.

Article 503, Class III Locations

Article 503 covers Class III locations. Class III locations may have ignitable fibers or flyings present, **Figure 8-8**. The size of the particles is what determines if it is a Class II dust or Class III fiber. Examples of Class III fibers are the particles present in a textile mill where fabric and clothing are manufactured. Like Class I and Class II locations, Class III locations are divided into two divisions. ***Class III, Division 1 locations*** are areas where ignitable fibers or flyings are handled or manufactured and may exist under normal operating conditions. ***Class III, Division 2 locations*** are areas where ignitable fibers are stored or handled and should not be present under normal conditions.

Article 504, Intrinsically Safe Systems

Article 504 covers intrinsically safe systems, **Figure 8-9**. Intrinsically safe systems may be installed in hazardous/classified areas with relaxed wiring method requirements. This is because intrinsically safe systems are not capable of producing a spark that could ignite the gas, dust, or fiber that may be present. The *NEC* defines an ***intrinsically safe circuit*** as a circuit in which any spark or thermal effect is incapable of causing ignition of a mixture of flammable or combustible material in air under prescribed test conditions.

Article 503 — Class III Locations

Part I General
- Section 503.1 Scope
- Section 503.5 General
- Section 503.6 Zone Equipment

Part II Wiring
- Section 503.10 Wiring Methods
- Section 503.25 Uninsulated Exposed Parts, Class III, Division 1 and 2
- Section 503.30 Grounding and Bonding

Part III Equipment
- Section 503.100 Transformers and Capacitors — Class III, Divisions 1 and 2
- Section 503.115 Switches, Circuit Breakers, Motor Controllers, and Fuses — Class III, Divisions 1 and 2
- Section 503.120 Control Transformers and Resistors — Class III, Divisions 1 and 2
- Section 503.125 Motors and Generators — Class III, Division 1 and Division 2
- Section 503.128 Ventilating Piping — Class III, Divisions 1 and 2
- Section 503.130 Luminaires — Class III, Divisions 1 and 2
- Section 503.135 Utilization Equipment — Class III, Divisions 1 and 2
- Section 503.140 Flexible Cords — Class III, Divisions 1 and 2
- Section 503.145 Receptacles and Attachment Plugs — Class III, Division 1 and Division 2
- Section 503.150 Signaling, Alarm, Remote-Control, and Local Loudspeaker Intercommunications Systems — Class III, Division 1 and Division 2
- Section 503.155 Electric Cranes, Hoists, and Similar Equipment — Class III, Divisions 1 and 2
- Section 503.160 Storage Battery Charging Equipment — Class III, Divisions 1 and 2

Goodheart-Willcox Publisher

Figure 8-8. *Article 503* has requirements for areas with ignitable fibers or flyings.

Article 504 — Intrinsically Safe Systems

- Section 504.1 Scope
- Section 504.3 Application of Other Articles
- Section 504.4 Equipment
- Section 504.10 Equipment Installation
- Section 504.20 Wiring Methods
- Section 504.30 Separation of Intrinsically Safe Conductors
- Section 504.50 Grounding
- Section 504.60 Bonding
- Section 504.70 Sealing
- Section 504.80 Identification

Goodheart-Willcox Publisher

Figure 8-9. *Article 504* has requirements for intrinsically safe systems. Intrinsically safe systems are designed and controlled so they cannot produce a spark that is capable of igniting gases or vapors.

Article 505, Zone 0, 1, and 2 Locations

Article 505 has an alternate way of classifying locations that contain flammable gases, **Figure 8-10**. *Article 505* provides requirements that fall in line with other standards as well as the international community.

Zone 0 and 1 locations are similar to Class I, Division 1 locations, except they are split into two different classifications. *Zone 0 locations* will have flammable gases present continuously. *Zone 1 locations* are likely to have flammable gases present under normal conditions. *Zone 2 locations* are similar to Class I, Division II locations in that gases will not be present unless there is a malfunction.

> **PRO TIP** — **Hazardous Locations—Class vs Zone**
>
> North America uses the class/division system to classify hazardous locations. Outside of the United States, a zone system has predominately been used to classify hazardous locations. Since the world isn't in agreement with one way of classifying hazardous locations, the *NEC* has included both. This, in addition to having metric equivalents, allows the *NEC* to be an international standard that is easier to adopt outside of the United States. It also clears up confusion about design specifications from foreign companies working within the United States.

Article 505 Zone 0, 1, and 2 Locations

Section 505.1 Scope
Section 505.4 Documentation
Section 505.5 Classification of Locations
Section 505.6 Material Groups
Section 505.7 Special Precaution
Section 505.8 Protection Techniques
Section 505.9 Equipment
Section 505.15 Wiring Methods
Section 505.16 Sealing and Drainage
Section 505.17 Flexible Cables, Cords, and Connections
Section 505.18 Conductors and Conductor Insulation
Section 505.19 Uninsulated Exposed Parts
Section 505.20 Equipment Requirements
Section 505.22 Increased Safety "e" Motors and Generators
Section 505.26 Process Sealing
Section 505.30 Grounding and Bonding

Goodheart-Willcox Publisher

Figure 8-10. *Article 505* describes and has requirements for Zone 0, 1, and 2 locations. The zone system is an alternate way of classifying hazardous locations.

Article 506 Zone 20, 21, and 22 Locations for Combustible Dusts or Ignitible Fibers/Flyings

Section 506.1 Scope
Section 506.4 Documentation
Section 506.5 Classification of Locations
Section 506.6 Material Groups
Section 506.7 Special Precaution
Section 506.8 Protection Techniques
Section 506.9 Equipment Requirements
Section 506.15 Wiring Methods
Section 506.16 Sealing
Section 506.17 Flexible Cords
Section 506.20 Equipment Installation
Section 506.30 Grounding and Bonding

Goodheart-Willcox Publisher

Figure 8-11. *Article 506* describes and has requirements for Zone 20, 21, and 22 locations. The Zone system is an alternate way of classifying hazardous locations.

Article 506, Zone 20, 21, and 22 Locations for Combustible Dusts or Ignitible Fibers/Flyings

Article 506 has an alternate way of classifying Class II and Class III locations, **Figure 8-11**. *Article 506* provides requirements that align with other standards as well as other parts of the international community.

Dusts and fibers are combined in *Article 506*. The locations are divided by the likelihood of the hazard being present. ***Zone 20 locations*** will have combustible dusts and ignitable fibers present continuously. ***Zone 21 locations*** will have combustible dusts and ignitable fibers present under normal conditions. ***Zone 22 locations*** are where dusts and fibers are stored or handled but are not likely to be present unless there is a malfunction.

Article 511, Commercial Garages, Repair and Storage

Article 511 covers occupancies where vehicles such as automobiles, buses, trucks, and tractors, are serviced or repaired, **Figure 8-12**. The key to considering these occupancies "special" is when the vehicles use flammable liquids or gases as fuel.

Article 511 Commercial Garages, Repair and Storage

Section 511.1 Scope
Section 511.2 Other Articles
Section 511.3 Area Classification, General
Section 511.4 Wiring and Equipment in Class I Locations
Section 511.7 Wiring and Equipment Installed Above Hazardous (Classified) Locations
Section 511.8 Underground Wiring Below Hazardous (Classified) Locations
Section 511.9 Sealing
Section 511.10 Special Equipment
Section 511.12 Ground-Fault Circuit-Interrupter Protection for Personnel
Section 511.16 Grounding and Bonding Requirements

Goodheart-Willcox Publisher

Figure 8-12. *Article 511* has requirements for commercial garages, repair, and storage.

The entire structure does not necessarily have special requirements, only certain areas where there is an increased hazard due to the liquids, vapors, or gases that may be present, **Figure 8-13**.

Article 512, Cannabis Oil Equipment and Cannabis Oil Systems Using Flammable Materials

Article 512 was added in the 2023 edition of the *NEC*. It covers areas in commercial and industrial facilities where flammable gases, liquid-produced vapors, or combustible liquid-produced vapors are a part of preparing, processing, or extracting cannabis oil. **Figure 8-14**.

Article 513, Aircraft Hangars

Article 513 covers occupancies where fueled aircrafts are stored or serviced, **Figure 8-15**. The classification of these occupancies also depends on the type of fuel present and whether the temperature is above or below its flash point. Like commercial garages, the entire structure does not necessarily have special requirements, only the areas where there is an increased hazard due to the liquids, vapors, or gases that may be present, **Figure 8-16**.

Goodheart-Willcox Publisher

Figure 8-13. Automobile repair shops are an example of a commercial garage where the requirements of *Article 511* apply due to the likely existence of gasoline and other flammable vapors.

Article 512
Cannabis Oil Equipment and Cannabis Oil Systems Using Flammable Materials

Part I General
- Section 512.1 Scope
- Section 512.2 Other Articles
- Section 512.3 Classified Locations

Part II Wiring
- Section 512.10 Wiring Installation and Operation
- Section 512.13 Wiring Installed Above Hazardous (Classified) Locations

Part III Equipment
- Section 512.20 Equipment and Systems
- Section 512.22 Equipment Installed in Hazardous (Classified) Locations
- Section 512.30 Equipment Installed Above Hazardous (Classified) Locations
- Section 512.32 Marking

Goodheart-Willcox Publisher

Figure 8-14. *Article 512* has requirements for cannabis oil equipment and systems that use flammable materials.

Article 513
Aircraft Hangars

- Section 513.1 Scope
- Section 513.2 Other Articles
- Section 513.3 Classification of Locations
- Section 513.4 Wiring and Equipment in Class I Locations
- Section 513.7 Wiring and Equipment Not Installed in Class I Locations
- Section 513.8 Underground Wiring
- Section 513.9 Sealing
- Section 513.10 Special Equipment
- Section 513.12 Ground-Fault Circuit-Interrupter Protection for Personnel
- Section 513.16 Grounding and Bonding Requirements

Goodheart-Willcox Publisher

Figure 8-15. *Article 513* applies to aircraft hangars.

Werner Gillmer/Shutterstock.com

Figure 8-16. *Article 511* applies to aircraft hangars where fueled airplanes will be repaired or serviced due to the likely existence of flammable vapors.

Article 514, Motor Fuel Dispensing Facilities

Article 514 has requirements for wiring dispensing equipment as well as the areas around motor fuel dispensers, **Figure 8-17**. The hazards around fuel dispensers are evident, but there are also considerations that must be given to the ground beneath, as flammable vapors are heavier than air and tend to accumulate in low-lying areas. The *NEC* has included a few diagrams within the article to help identify the classified areas, **Figure 8-18**.

Article 514
Motor Fuel Dispensing Facilities

Section 514.1 Scope
Section 514.2 Other Articles
Section 514.3 Classification of Locations
Section 514.4 Wiring and Equipment Installed in Hazardous (Classified) Locations
Section 514.7 Wiring and Equipment Above Hazardous (Classified) Locations
Section 514.8 Underground Wiring
Section 514.9 Sealing
Section 514.11 Circuit Disconnects
Section 514.13 Provisions for Maintenance and Service of Dispensing Equipment
Section 514.16 Grounding and Bonding

Goodheart-Willcox Publisher

Figure 8-17. *Article 514* applies to motor fuel dispensing facilities.

Carolyn Franks/Shutterstock.com

Figure 8-18. The requirements of *Article 514* classify the areas around gas pumps, as well as underneath them, as hazardous locations due to the presence of gasoline vapors.

Article 515, Bulk Storage Plants

Article 515 has requirements for areas where flammable liquids are stored or blended in bulk, **Figure 8-19**. Like fuel dispensers, the areas below and around the storage vessels as well as the apparatus used to move the liquids are hazardous and have special requirements, **Figure 8-20**.

Article 515
Bulk Storage Plants

Section 515.1 Scope
Section 515.2 Other Articles
Section 515.3 Classified Locations
Section 515.4 Wiring and Equipment Located in Hazardous (Classified) Locations
Section 515.7 Wiring and Equipment Above Hazardous (Classified) Locations
Section 515.8 Underground Wiring
Section 515.9 Sealing
Section 515.10 Special Equipment — Motor Fuel Dispensers
Section 515.16 Grounding and Bonding

Goodheart-Willcox Publisher

Figure 8-19. *Article 515* applies to bulk storage plants.

ymgerman/Shutterstock.com

Figure 8-20. Bulk storage plants are locations where gasoline or other flammable liquids are stored or transferred.

Article 516
Spray Application, Dipping, Coating, and Printing Processes Using Flammable or Combustible Materials

Part I General
- Section 516.1 Scope
- Section 516.2 Other Articles
- Section 516.3 Class I Locations

Part II Open Containers
- Section 516.4 Area Classification

Part III Spray Application Processes
- Section 516.5 Area Classification
- Section 516.6 Wiring and Equipment in Class I Locations
- Section 516.7 Wiring and Equipment Not Within Hazardous (Classified) Locations
- Section 516.10 Special Equipment
- Section 516.16 Grounding

Part IV Spray Application Operations in Membrane Enclosures
- Section 516.18 Area Classification for Temporary Membrane Enclosures
- Section 516.23 Electrical and Other Sources of Ignition

Part V Printing, Dipping, and Coating Processes
- Section 516.29 Classification of Locations
- Section 516.35 Areas Adjacent to Enclosed Dipping and Coating Processes
- Section 516.36 Equipment and Containers in Ventilated Areas
- Section 516.37 Luminaires
- Section 516.38 Wiring and Equipment Not Within Hazardous (Classified) Locations
- Section 516.40 Static Electric Discharges

Goodheart-Willcox Publisher

Figure 8-21. *Article 516* applies to areas where materials have a finish coating that is flammable or combustible.

Article 516, Spray Application, Dipping, Coating, and Printing Processes Using Flammable or Combustible Materials

Article 516 has special requirements for areas where items receive a finish coat such as paint or varnish, **Figure 8-21**. This may be done by spraying, powder coating, printing, dipping, and more. Some of the products used in the process may be flammable or combustible, while others are not. The type of product used as well as its flash point will determine how the space is classified, which in turn will determine the necessary wiring methods, **Figure 8-22**.

Article 517, Health Care Facilities

Article 517, **Figure 8-23**, is often lumped in with the hazardous occupancies by *Article 510* due to the storage and use of flammable anesthetics. However, very little of the article is dedicated to these hazards. The majority of the article addresses the vast number of life-safety systems, as well as specialized equipment that necessitate special electrical requirements. It includes topics such as flammable anesthetics, special power requirements in patient care areas, essential life safety systems, and X-ray equipment, **Figure 8-24**.

Shevel Artur/Shutterstock.com

Figure 8-22. Paint booths often spray paint or varnish that is flammable or combustible, making them a hazardous area. If the booth only uses water-based materials, it is not a hazardous location, and the requirements of *Article 516* will not apply.

Article 517
Health Care Facilities

Part I General
Section 517.1 Scope
Section 517.6 Patient Care-Related Electrical Equipment

Part II Wiring and Protection
Section 517.10 Applicability
Section 517.12 Wiring Methods
Section 517.13 Equipment Grounding Conductor for Receptacles and Fixed Electrical Equipment in Patient Care Spaces
Section 517.14 Panelboard Bonding
Section 517.16 Use of Isolated Ground Receptacles
Section 517.17 Ground-Fault Protection of Equipment
Section 517.18 Category 2 Spaces
Section 517.19 Category 1 Spaces
Section 517.20 Wet Procedure Locations
Section 517.21 Ground-Fault Circuit-Interrupter Protection for Personnel in Category 2 and Category 1 Spaces
Section 517.22 Demand Factors

Part III Essential Electrical System (EES)
Section 517.25 Essential Electrical Systems for Health Care Facilities
Section 517.26 Application of Other Articles
Section 517.29 Type 1 Essential Electrical Systems
Section 517.30 Sources of Power
Section 517.31 Requirements for the Essential Electrical System
Section 517.32 Branches Requiring Automatic Connection
Section 517.33 Life Safety Branch
Section 517.34 Critical Branch
Section 517.35 Equipment Branch Connection to Alternate Power Source
Section 517.40 Type 2 Essential Electrical Systems
Section 517.41 Required Power Sources
Section 517.42 Essential Electrical Systems for Nursing Homes and Limited Care Facilities
Section 517.43 Automatic Connection to Life Safety and Equipment Branch
Section 517.44 Connection to Equipment Branch
Section 517.45 Essential Electrical Systems for Other Health Care Facilities

Part IV Inhalation Anesthetizing Locations
Section 517.60 Anesthetizing Location Classification
Section 517.61 Wiring and Equipment
Section 517.62 Grounding
Section 517.63 Grounded Power Systems in Anesthetizing Locations
Section 517.64 Low-Voltage Equipment and Instruments

Part V Diagnostic Imaging and Treatment Equipment
Section 517.70 Applicability
Section 517.71 Connection to Supply Circuit
Section 517.72 Disconnecting Means
Section 517.73 Rating of Supply Conductors and Overcurrent Protection
Section 517.74 Control Circuit Conductors
Section 517.76 Transformers and Capacitors
Section 517.77 Installation of Cables with Grounded Shields
Section 517.78 Guarding and Grounding

Part VI Communications, Signaling Systems, Data Systems, Fire Alarm Systems, and Systems Less than 120 Volts, Nominal
Section 517.80 Patient Care Spaces
Section 517.81 Other-Than-Patient-Care Spaces
Section 517.82 Signal Transmission Between Appliances

Part VII Isolated Power Systems
Section 517.160 Isolated Power Systems

Figure 8-23. *Article 517* applies to health care facilities.

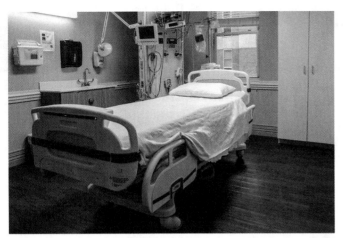

InkaOne/Shutterstock.com

Figure 8-24. In addition to having special requirements for areas containing flammable anesthetics, health care facilities must follow requirements for back-up power in case of emergencies.

Gatherings and Entertainment

Articles 518 through *540* have special requirements for spaces where people gather and entertainment venues. When large numbers of people gather, there is the potential for a dangerous situation. If people panic and/or try to quickly exit due to fire or loss of power, people can be trampled or seriously hurt. The *NEC* has special requirements to ensure the electrical system is robust enough to reduce the likelihood of a fire and to extend the time it operates, providing light and power during an evacuation.

Article 518
Assembly Occupancies

Section 518.1 Scope
Section 518.2 General Classification
Section 518.3 Temporary Wiring
Section 518.4 Wiring Methods
Section 518.5 Supply
Section 518.6 Illumination

Goodheart-Willcox Publisher

Figure 8-25. *Article 518* has requirements for assembly occupancies.

Article 518, Assembly Occupancies

Article 518 has requirements for assembly occupancies, **Figure 8-25**. *Assembly occupancies* are areas that are designed or intended for 100 or more persons to gather. *Section 518.2(A)* has examples of types of occupancies that are considered assembly occupancies. The reason for the special requirements is in the case of a fire or power outage, people may panic and get hurt in the process of exiting or moving to a safe location. The types of wiring methods that are permitted in a place of assembly will help ensure that the electrical system is not going to cause a fire. The entire structure is not considered an assembly if the gathering of people is only in a certain area. In that case, only the area where people gather has the special requirements, **Figure 8-26**.

Goodheart-Willcox Publisher

Figure 8-26. Churches are an example of an assembly occupancy, as it is likely for 100 or more people to be gathered.

Article 520, Theaters, Audience Areas of Motion Picture and Television Studios, Performance Areas, and Similar Locations

Article 520 is titled *Theaters, Audience Areas of Motion Picture and Television Studios, Performance Areas, and Similar Locations,* **Figure 8-27**. In addition to the fact that these areas are intended for people to assemble and be entertained, they contain stage equipment that has special requirements to ensure safety.

Article 522, Control Systems for Permanent Amusement Attractions

Article 522 is titled *Control Systems for Permanent Amusement Attractions,* **Figure 8-28**. This article is meant for amusement attractions, like Universal Studios and Six Flags. These are installations where the amusement attractions are permanently in place and are not moved around.

Article 520
Theaters, Audience Areas of Motion Picture and Television Studios, Performance Areas, and Similar Locations

Part I. General
- Section 520.1 Scope
- Section 520.5 Wiring Methods
- Section 520.6 Number of Conductors in Raceway
- Section 520.9 Branch Circuits
- Section 520.10 Portable Equipment Used Outdoors

Part II. Fixed Stage Switchboards
- Section 520.21 General
- Section 520.25 Dimmers
- Section 520.26 Type of Switchboard
- Section 520.27 Stage Switchboard Feeders

Part III. Fixed Stage Equipment Other Than Switchboards
- Section 520.40 Stage Lighting Hoists
- Section 520.41 Circuit Loads
- Section 520.42 Conductor Insulation
- Section 520.43 Footlights
- Section 520.44 Borders, Proscenium Sidelights, Drop Boxes, and Connector Strips
- Section 520.45 Receptacles
- Section 520.46 Connector Strips, Drop Boxes, Floor Pockets, and Other Outlet Enclosures
- Section 520.47 Backstage Lamps (Bare Bulbs)
- Section 520.48 Curtain Machines
- Section 520.49 Smoke Ventilator Control

Part IV. Portable Switchboards on Stage
- Section 520.50 Road Show Connection Panel (A Type of Patch Panel)
- Section 520.51 Supply
- Section 520.52 Overcurrent Protection for Branch Circuits
- Section 520.53 Construction
- Section 520.54 Supply Conductors

Part V. Portable Stage Equipment Other Than Switchboards
- Section 520.61 Arc Lamps
- Section 520.62 Portable Power Distribution Units
- Section 520.63 Bracket Fixture Wiring
- Section 520.64 Portable Strips
- Section 520.65 Festoons
- Section 520.66 Special Effects
- Section 520.67 Multiple Branch-Circuit Cable Connectors
- Section 520.68 Conductors for Portables
- Section 520.69 Adapters

Part VI. Dressing Rooms, Dressing Areas, and Makeup Areas
- Section 520.71 Pendant Lampholders
- Section 520.72 Lamp Guards
- Section 520.73 Switches Required
- Section 520.74 Pilot Lights Required

Part VII. Equipment Grounding Conductor
- Section 520.81 Equipment Grounding Conductor

Goodheart-Willcox Publisher

Figure 8-27. *Article 520* applies to theaters and audience areas of motion picture and television studios.

Article 522
Control Systems for Permanent Amusement Attractions

Part I. General
- Section 522.1 Scope
- Section 522.5 Voltage Limitations
- Section 522.7 Maintenance

Part II. Control Circuits
- Section 522.10 Power Sources for Control Circuits

Part III Control Circuit Wiring Methods
- Section 522.20 Conductors, Busbars, and Slip Rings
- Section 522.21 Conductor Sizing
- Section 522.22 Conductor Ampacity
- Section 522.23 Overcurrent Protection for Conductors
- Section 522.24 Conductors of Different Circuits in the Same Cable, Cable Tray, Enclosure, or Raceway
- Section 522.25 Ungrounded Control Circuits
- Section 522.28 Control Circuits in Wet Locations

Goodheart-Willcox Publisher

Figure 8-28. *Article 522* applies to control systems for permanent amusement attractions.

Article 525 Carnivals, Circuses, Fairs, and Similar Events

Article 525 is titled *Carnivals, Circuses, Fairs, and Similar Events*, **Figure 8-29**. These events are set up for a short amount of time and then taken down. They contain a lot of electrical equipment with temporary electrical connections that people are interacting with or riding. Special precautions must be taken to ensure safety because much of the wiring is done with portable cables that are lying on the ground, **Figure 8-30**.

Article 530, Motion Picture and Television Studios and Remote Locations

Article 530 is titled *Motion Picture and Television Studios and Remote Locations*, **Figure 8-31**. These locations are where television programs or motion pictures are created. This includes the permanent structures where filming occurs, as well as remote locations where equipment is set up temporarily.

Article 525
Carnivals, Circuses, Fairs, and Similar Events

Part I General
- Section 525.1 Scope
- Section 525.3 Other Articles
- Section 525.5 Overhead Conductor Clearances
- Section 525.6 Protection of Electrical Equipment

Part II Power Sources
- Section 525.10 Services
- Section 525.11 Multiple Sources of Supply

Part III Wiring Methods
- Section 525.20 Wiring Methods
- Section 525.21 Rides, Tents, and Concessions
- Section 525.22 Portable Distribution or Termination Boxes
- Section 525.23 Ground-Fault Circuit-Interrupter (GFCI) Protection

Part IV Equipment Grounding and Bonding
- Section 525.30 Equipment Bonding
- Section 525.31 Equipment Grounding
- Section 525.32 Equipment Grounding Conductor Continuity Assurance

Goodheart-Willcox Publisher

Figure 8-29. *Article 525* applies to carnivals, fairs, and similar events.

Unit 8 NEC Chapter 5

Figure 8-30. Carnivals that are set up for short periods of time will have temporary wiring methods that are installed to power all the equipment and rides.

Article 540, Motion Picture Projection Rooms

Article 540 is titled *Motion Picture Projection Rooms*, **Figure 8-32**. In a movie theater, the projection room is where the projector that projects the movie onto the screen is housed. Professional movie projectors have light sources that create extremely high temperatures and may develop hazardous gases, dust, or radiation, **Figure 8-33**.

Manufactured Structures and Agricultural Buildings

Articles 545 through *555* address agricultural buildings and structures that are manufactured elsewhere and moved onto the site. Examples of manufactured structures include modular homes, mobile homes, recreational vehicles, and

---

Article 530
Motion Picture and Television Studios and Remote Locations

Part I General
- Section 530.1 Scope
- Section 530.3 Restricted Public Access
- Section 520.4 Supervision by Qualified Personnel
- Section 520.5 Wiring Methods
- Section 520.7 Sizing of Feeder Conductors for Motion Picture and/or Television Studio Sets
- Section 520.8 Equipment Grounding Conductor
- Section 520.9 Plugs and Receptacles
- Section 520.10 Single-Pole Separable Connectors
- Section 520.11 Branch Circuits
- Section 520.12 Enclosing and Guarding Live Parts

Part II Portable Equipment in Production Areas of Studios and Remote Locations
- Section 520.21 Portable Equipment
- Section 530.22 Portable Wiring
- Section 530.23 Overcurrent Protection
- Section 530.24 Purpose-Built Luminaires, Lighting, and Effects Equipment
- Section 530.26 Portable Luminaires

Part III Portable Equipment in Support Areas
- Section 530.41 Restricted Public Access
- Section 530.42 Overcurrent Protection for Portable Cable
- Section 530.43 Portable Generators
- Section 530.44 Ground-Fault Circuit-Interrupter (GFCI) Protection
- Section 530.45 Production Vehicles and Trailers
- Section 530.46 Protection

Part IV Dressing Rooms
- Section 530.61 Fixed Wiring in Dressing Rooms

Part V Portable Substations
- Section 530.71 General
- Section 530.72 Over 1000 Volts, Nominal

Figure 8-31. *Article 530* has requirements for motion picture and television studios.

Article 540
Motion Picture Projection Rooms

Part I General
Section 540.1 Scope

Part II Equipment and Projectors of the Professional Type
Section 540.10 Motion Picture Projection Room Required
Section 540.11 Location of Associated Electrical Equipment
Section 540.12 Work Space
Section 540.13 Conductor Size
Section 540.14 Conductors on Lamps and Hot Equipment
Section 540.15 Flexible Cords
Section 540.20 Listing Requirements
Section 540.21 Marking

Part III Nonprofessional Projectors
Section 540.31 Motion Picture Projection Room Not Required
Section 540.32 Listing Requirements

Goodheart-Willcox Publisher

Figure 8-32. *Article 540* applies to motion picture projection rooms.

Goodheart-Willcox Publisher

Figure 8-33. Motion picture projector rooms have projectors with light sources that reach high temperatures, some of which may develop hazardous gases, dust, or radiation.

floating buildings. There are also requirements for locations where the structures will be parked, such as marinas or trailer parks.

Article 545, Manufactured Buildings and Relocatable Structures

Article 545 has requirements for manufactured buildings and relocatable structures, **Figure 8-34**. A ***manufactured building*** is defined as any building (other than manufactured homes, mobile homes, park trailers, or recreational vehicles) that is of closed construction and made or assembled in a manufacturing facility for installation, or for assembly and installation, on the building site.

Article 545
Manufactured Buildings and Relocatable Structures

Part I General
Section 545.1 Scope
Section 545.4 Wiring Methods
Section 545.5 Supply Conductors
Section 545.6 Installation of Service-Entrance Conductors
Section 545.7 Service Equipment
Section 545.8 Protection of Conductors and Equipment
Section 545.9 Boxes
Section 545.10 Receptacle or Switch with Integral Enclosure
Section 545.11 Bonding and Grounding
Section 545.12 Grounding Electrode Conductor
Section 545.13 Component Interconnections

Part II Relocatable Structures
Section 545.20 Application Provisions
Section 545.22 Power Supply
Section 545.24 Disconnecting Means and Branch-Circuit Overcurrent Protection
Section 545.26 Bonding of Exposed Non-Current-Carrying Metal Parts
Section 545.27 Intersystem Bonding
Section 545.28 Ground-Fault Circuit-Interrupters (GFCI)

Goodheart-Willcox Publisher

Figure 8-34. *Article 545* applies to manufactured buildings and relocatable structures.

A *relocatable structure* is a factory-assembled structure, or structures transportable in one or more sections, built on a permanent chassis and designed to be used as other than a dwelling unit without a permanent foundation. Examples include temporary job trailers, studio dressing rooms, and display or demonstration of merchandise, **Figure 8-35**.

Article 547, Agricultural Buildings

Article 547 is titled *Agricultural Buildings*, **Figure 8-36**. Agricultural buildings may have excessive dust accumulations and/or a corrosive atmosphere related to the presence of animals or fish. The wiring in agricultural buildings is not a one-size-fits-all method and will vary based on the type of animal being housed and the hazard they present, **Figure 8-37**.

Michael O'Keene/Shutterstock.com

Figure 8-35. Construction job trailers are carried to a job site for use while the project is under construction. The trailer will then be transported to the next job site.

Article 547
Agricultural Buildings

Part I General
Section 547.1 Scope
Section 547.3 Other Articles
Section 547.4 Surface Temperatures

Part II Installations
Section 547.20 Wiring Methods
Section 547.21 Mounting
Section 547.22 Equipment Enclosures, Boxes, Conduit Bodies, and Fittings
Section 547.23 Damp or Wet Locations
Section 547.24 Corrosive Atmosphere
Section 547.25 Flexible Connections
Section 547.26 Physical Protection
Section 547.27 Separate Equipment Grounding Conductor
Section 547.28 Ground-Fault Circuit-Interrupter Protection
Section 547.29 Switches, Receptacles, Circuit Breakers, Controllers, and Fuses
Section 547.30 Motors
Section 547.31 Luminaires

Part III Distribution
Section 547.40 Electrical Supply to Building(s) or Structure(s) from a Distribution Point
Section 547.41 Overhead Service
Section 547.42 Service Disconnecting Means and Overcurrent Protection at the Distribution Point
Section 547.43 Identification
Section 547.44 Equipotential Plates and Bonding of Equipotential Planes

Goodheart-Willcox Publisher

Figure 8-36. *Article 547* applies to agricultural buildings.

Pressmaster/Shutterstock.com

Figure 8-37. Areas where livestock are present will have corrosive atmospheres due to the animal waste, as well as excessive dust. The wiring methods used must be able to withstand the environment.

Article 550, Mobile Homes, Manufactured Homes, and Mobile Home Parks

Article 550 is titled *Mobile Homes, Manufactured Homes, and Mobile Home Parks,* **Figure 8-38**. Mobile and manufactured

Michael O'Keene/Shutterstock.com

Figure 8-39. Mobile homes are delivered as a completed unit and are connected to the power system available at the mobile home park.

homes are built in a manufacturing facility and moved to the destination. The interior wiring of the structures is completed by the company manufacturing the unit, but the connection to the electric service is done by an electrician, **Figure 8-39**.

Article 550
Mobile Homes, Manufactured Homes, and Mobile Home Parks

Part I General
Section 550.1 Scope
Section 550.4 General Requirements

Part II Mobile and Manufactured Homes
Section 550.10 Power Supply
Section 550.11 Disconnecting Means and Branch-Circuit Protective Equipment
Section 550.12 Branch Circuits
Section 550.13 Receptacle Outlets
Section 550.14 Luminaires and Appliances
Section 550.15 Wiring Methods and Materials
Section 550.16 Grounding
Section 550.17 Testing
Section 550.18 Calculations
Section 550.19 Interconnection of Multiple-Section Mobile or Manufactured Home Units
Section 550.20 Outdoor Outlets, Luminaires, Air-Cooling Equipment, and So Forth
Section 550.25 Arc-Fault Circuit-Interrupter Protection

Part III Services and Feeders
Section 550.30 Distribution System
Section 550.31 Allowable Demand Factors
Section 550.32 Service Equipment
Section 550.33 Feeder

Goodheart-Willcox Publisher

Figure 8-38. *Article 550* applies to mobile homes, manufactured homes, and mobile home parks.

Article 551, Recreational Vehicles and Recreational Vehicle Parks

Article 551 is titled *Recreational Vehicles and Recreational Vehicle Parks,* **Figure 8-40**. These vehicles often have dual-voltage systems so they can run on direct current received from the vehicle or alternating current from a generator or RV park power. Most of the requirements for recreational vehicles relate to the connection to power at the RV park or to a generator, but there are some requirements that must be followed by the manufacturer when assembling the vehicle.

Article 551
Recreational Vehicles and Recreational Vehicle Parks

Part I General
Section 551.1 Scope
Section 551.3 Electrical Datum Plane Distances
Section 551.4 General Requirements

Part II Combination Electrical Systems
Section 551.20 Combination Electrical Systems

Part III Other Power Sources
Section 551.30 Generator Installations
Section 551.31 Multiple Supply Source
Section 551.32 Other Sources
Section 551.33 Alternate Source Restrictions

Part IV Nominal 120-Volt or 120/240-Volt Systems
Section 551.40 120-Volt or 120/240-Volt, Nominal, Systems
Section 551.41 Receptacle Outlets Required
Section 551.42 Branch Circuits Required
Section 551.43 Branch-Circuit Protection
Section 551.44 Feeder Assembly
Section 551.45 Panelboard
Section 551.46 Means for Connecting to Power Supply
Section 551.47 Wiring Methods
Section 551.48 Conductors and Boxes
Section 551.49 Grounded Conductors
Section 551.50 Connection of Terminals and Splices
Section 551.51 Switches
Section 551.52 Receptacles
Section 551.53 Luminaires and Other Equipment
Section 551.54 Grounding
Section 551.55 Interior Equipment Grounding
Section 551.56 Bonding of Non-Current-Carrying Metal Parts
Section 551.57 Appliance Accessibility and Fastening

Part V Factory Tests
Section 551.60 Factory Tests (Electrical)

Part VI Recreational Vehicle Parks
Section 551.71 Type Receptacles Provided
Section 551.72 Distribution System
Section 551.73 Calculated Load
Section 551.74 Overcurrent Protection
Section 551.76 Grounding — Recreational Vehicle Site Supply Equipment
Section 551.77 Recreational Vehicle Site Supply Equipment
Section 551.78 Protection of Outdoor Equipment
Section 551.79 Clearance for Overhead Conductors
Section 551.80 Underground Service, Feeder, Branch-Circuit, and Recreational Vehicle Site Feeder-Circuit Conductors
Section 551.81 Receptacles

Goodheart-Willcox Publisher

Figure 8-40. *Article 551* has requirements for recreational vehicles and recreational vehicle parks.

Article 552, Park Trailers

Article 552 has requirements for park trailers. A ***park trailer*** is defined as a unit that is built on a single chassis mounted on wheels and has a gross trailer area not exceeding 37 m^2 (400 ft^2) in the set-up mode, **Figure 8-41**. Park trailers are intended for seasonal use and are *not* meant to be a permanent dwelling unit or to be used for commercial uses, such as banks, clinics, or offices, **Figure 8-42**.

Article 555, Marinas, Boatyards, Floating Buildings, and Commercial and Noncommercial Docking Facilities

Article 555 is titled *Marinas, Boatyards, Floating Buildings, and Commercial and Noncommercial Docking Facilities*, **Figure 8-43**. Boats and floating buildings will often be

Cegli/Shutterstock.com

Figure 8-42. Park trailers are often parked on seasonal lots in locations such as campgrounds, resorts, and RV parks.

Article 552 — Park Trailers

Part I General
- Section 552.1 Scope
- Section 552.4 General Requirements
- Section 552.5 Labels

Part II Low-Voltage Systems
- Section 552.10 Low-Voltage Systems

Part III Combination Electrical Systems
- Section 552.20 Combination Electrical Systems

Part IV Nominal 120-Volt or 120/240-Volt Systems
- Section 552.40 120-Volt or 120/240-Volt, Nominal, Systems
- Section 552.41 Receptacle Outlets Required
- Section 552.42 Branch-Circuit Protection
- Section 552.43 Power Supply
- Section 552.44 Cord
- Section 552.45 Panelboard
- Section 552.46 Branch Circuits
- Section 552.47 Calculations
- Section 552.48 Wiring Methods
- Section 552.49 Maximum Number of Conductors in Boxes
- Section 552.50 Grounded Conductors
- Section 552.51 Connection of Terminals and Splices
- Section 552.52 Switches
- Section 552.53 Receptacles
- Section 552.54 Luminaires
- Section 552.55 Grounding
- Section 552.56 Interior Equipment Grounding
- Section 552.57 Bonding of Non-Current-Carrying Metal Parts
- Section 552.58 Appliance Accessibility and Fastening
- Section 552.59 Outdoor Outlets, Fixtures, Including Luminaires, Air-Cooling Equipment, and So On

Part V Factory Tests
- Section 552.60 Factory Tests (Electrical)

Goodheart-Willcox Publisher

Figure 8-41. *Article 552* applies to park trailers.

Article 555
Marinas, Boatyards, Floating Buildings, and Commercial and Noncommercial Docking Facilities

Part I General
Section 555.1 Scope
Section 555.3 Electrical Datum Plane Distances
Section 555.4 Location of Service Equipment
Section 555.5 Maximum Voltage
Section 555.6 Load Calculations for Service and Feeder Conductors
Section 555.7 Transformers
Section 555.8 Marine Hoists, Railways, Cranes, and Monorails
Section 555.10 Signage
Section 555.11 Motor Fuel Dispending Stations — Hazardous (Classified) Locations
Section 555.12 Repair Facilities — Hazardous (Classified) Locations
Section 555.13 Bonding of Non-Current-Carrying Metal Parts
Section 555.14 Equipotential Planes and Bonding of Equipotential Planes
Section 555.15 Replacement of Equipment

Part II Marinas, Boatyards, and Docking Facilities
Section 555.30 Electrical Equipment and Connections
Section 555.31 Electrical Equipment Enclosures
Section 555.32 Circuit Breakers, Switches, Panelboards, and Marina Power Outlets
Section 555.33 Receptacles
Section 555.34 Wiring Methods and Installation
Section 555.35 Ground-Fault Protection of Equipment (GFPE) and Ground-Fault Circuit Interrupter
Section 555.36 Disconnecting Means for Shore Power Connection(s)
Section 555.37 Equipment Grounding Conductor
Section 555.38 Luminaires

Part III Floating Buildings
Section 555.50 Service Conductors
Section 555.51 Feeder Conductors
Section 555.52 Installation of Services and Feeders
Section 555.53 Ground-Fault Protection
Section 555.54 Grounding
Section 555.55 Insulated Neutral
Section 555.56 Equipment Grounding

Goodheart-Willcox Publisher

Figure 8-43. *Article 555* covers marinas, boatyards, floating buildings, and commercial and noncommercial docking facilities.

connected to shore power. ***Shore power*** is defined as the electrical equipment required to power a floating vessel including, but not limited to, the receptacle and cords. Having boats and structures that have electrical connections while in the water creates a unique electrical hazard which necessitates a special article to ensure safety, **Figure 8-44**.

Article 590, Temporary Installations

Article 590 is titled *Temporary Installations*, **Figure 8-45**. Temporary Installations are installations with time constraints where the wiring is installed for a limited amount of time, after which it will be removed. Temporary wiring is less restrictive and does not follow all the wiring requirements found in other parts of the *NEC*. A common example of temporary wiring is a construction site. Nearly every construction site has temporary power distributed throughout

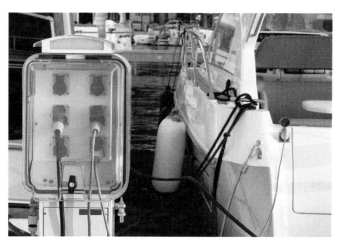

Ga_Na/Shutterstock.com

Figure 8-44. Some marinas will have shore power available for the connections of boats. Having electricity in close proximities to bodies of water has the potential to be extremely dangerous. *Article 555* has requirements to ensure the electrical installation is safe.

Part 1 Layout and Structure: Learning to Use the NEC

Article 590
Temporary Installations

Section 590.1 Scope
Section 590.2 All Wiring Installations
Section 590.3 Time Constraints
Section 590.4 General
Section 590.5 Listing of Decorative Lighting
Section 590.6 Ground-Fault Protection for Personnel
Section 590.7 Guarding
Section 590.8 Overcurrent Protective Devices

Goodheart-Willcox Publisher

Figure 8-45. *Article 590* has requirements for temporary installations.

the site so the tradespeople have lights and access to electrical energy, **Figure 8-46**. A couple of other examples of temporary wiring are holiday lighting, which is permitted to be installed for up to 90 days, and emergency response wiring, which can provide power during an emergency.

SB7/Shutterstock.com

Figure 8-46. Temporary power is installed on most job sites to provide power and lighting for the construction of the building. Jobsite temporary power is only in place for the duration of the project and is often moveable. *Article 590* has special rules that apply to temporary installations to ensure that they remain safe.

NEC KEYWORDS **Chapter 5 Keywords**

The following are examples of keywords that should lead you to *Chapter 5*:

agricultural buildings	dance hall	hospital	performance
aircraft hangars	dental office	houseboat	pier
amusement attractions	dip tank	inhalation	projection room
anesthetizing location	docks	intrinsically safe	recreational vehicle
arenas	dressing room	lint	shore power
armories	fair	manufactured buildings	site-isolation device
auditorium	fibers	manufactured homes	spray application
bulk storage	flammable	marinas	stage
carnival	floating building	mobile home	television
circus	flyings	motion picture	theater
church	garage	nightclub	vapors
combustible	gasoline	operating room	zone
commercial garage	hazardous	park trailer	
critical care	health care	patient	

Summary

- *Chapter 5* of the *NEC* is the first of the special chapters that supplement or modify *Chapters 1–7*.
- *Chapter 5* contains requirements for various special occupancies that may contain hazards or situations not found in a typical occupancy such as:
  - Gatherings and entertainment
  - Manufactured structures and agricultural buildings
  - Temporary installations
- *Chapter 5* contains requirements for occupancies that have flammable gases, combustible dusts, or ignitable fibers.
- Hazardous locations can be categorized by a class/division or a zone system.

Unit 8 Review

Name _____ Date _____ Class _____

Know and Understand

Answer the following questions based on information in this unit.

1. *Chapter 5* of the *NEC* is titled *Special* _____.
   A. *Conditions*
   B. *Installations*
   C. *Occupancies*
   D. *Equipment*

2. The requirements in *Chapter 5* will supplement or modify the information found in _____.
   A. *Chapter 1*
   B. *Chapters 1–4*
   C. *Chapters 1–7*
   D. *Chapters 1–8*

3. _____ has requirements that apply to temporary installations.
   A. *Article 500*
   B. *Article 520*
   C. *Article 550*
   D. *Article 590*

4. An area where flammable gases are confined within closed containers from which they can escape only in the case of a rupture or breakdown of the container is classified as a _____ location.
   A. Class I, Division 1
   B. Class I, Division 2
   C. Class II, Division 1
   D. Class II, Division 2

5. Places of assembly are locations where _____ or more people are gathered.
   A. 50
   B. 75
   C. 100
   D. 300

6. _____ has requirements for manufactured buildings.
   A. *Article 545*
   B. *Article 550*
   C. *Article 551*
   D. *Article 552*

7. A Class III location contains _____.
   A. flammable gases
   B. combustible dusts
   C. combustible vapors
   D. ignitable fibers

8. Which of the following keywords would lead you to *Chapter 5*?
   A. Capacitor
   B. Mobile homes
   C. Hot tub
   D. Communication cable

9. Which article has requirements on motion picture projection rooms?
   A. *Article 520*
   B. *Article 522*
   C. *Article 530*
   D. *Article 540*

10. *Article 517* applies to electrical installations that provide health care services to _____.
    A. farm animals
    B. domestic animals
    C. human beings
    D. All of the above.

Apply and Analyze

NEC Chapter 5

Use a copy of the National Electrical Code *to identify the term corresponding to the definition provided. Identify the section or subsection where the answer is found.*

1. All boxes and fittings in a Class II, Division 2 location shall be _____.
   A. dust-tight
   B. dust-ignition proof
   C. explosion proof
   D. threaded

   NEC _____

2. A power-limited control circuit used with a permanent amusement attraction shall be supplied from a source that has a rated output of not more than 30 volts and _____ volt-amperes.
   A. 300
   B. 500
   C. 600
   D. 1000

   NEC _____

3. A patient bed location in a Category 2 space shall have at least _____ receptacles.
   A. 4
   B. 6
   C. 8
   D. 12

   NEC _____

4. The minimum thickness of the sealing compound installed in a conduit seal in a Class I location is _____.
   A. 1/8″
   B. 1/4″
   C. 3/8″
   D. 5/8″

   NEC _____

5. Intrinsically safe apparatus shall be permitted to be installed in any hazardous location for which it has been _____.
   A. identified
   B. listed
   C. tested
   D. approved

   NEC _____

6. The main overcurrent protective device that feeds a floating building shall have ground-fault protection not exceeding _____ mA.
   A. 4-6
   B. 10
   C. 50
   D. 100

   NEC _____

7. A major repair garage without ventilation is considered a Class I, Division 2 location up to _____ above the floor (Heavier-than-air fuel).
   A. 6″
   B. 12″
   C. 18″
   D. 24″

   NEC _____

8. A motion picture projection room shall have a clear working space of not less than _____ wide on each side and at the rear of the equipment.
   A. 24″
   B. 30″
   C. 36″
   D. 42″

   NEC _____

9. A park trailer panelboard shall have a clearance of not less than _____ wide and 30″ deep.
   A. 24″
   B. 30″
   C. 36″
   D. 42″

   NEC _____

10. An autotransformer-type dimmer installed in a fixed stage switchboard of a theater shall not exceed _____ volts between conductors.
    A. 120
    B. 125
    C. 150
    D. 300

    NEC _____

Name _____ Date _____ Class _____

Critical Thinking

NEC Chapter 5

Answer the following questions using a copy of the National Electrical Code. *Identify the section or subsection where the answer is found.*

1. Carnival service equipment shall not be installed in a location that is accessible to unqualified persons, unless the equipment is _____.

   _____

   _____

   NEC _____

2. In a health care location where flammable anesthetics are employed, the entire area shall be considered a Class I, Division 1 location up to a level _____ above the floor.

   _____

   _____

   NEC _____

3. A Zone _____ location is a location where flammable gases or vapors are present continuously.

   _____

   _____

   NEC _____

4. The area within 3′ of any opening in an enclosed spray booth shall be classified as a _____ location.

   _____

   _____

   NEC _____

5. A park trailer is intended for _____ use and is not intended as a permanent dwelling unit or for commercial uses.

   _____

   _____

   NEC _____

6. A receptacle on a fixed pier shall be at least 12″ above the deck and not below _____.

   _____

   _____

   NEC _____

7. A generator installed on a recreational vehicle shall be mounted in such a manner as to be effectively bonded to _____.

   _____

   _____

   NEC _____

8. A fixed luminaire installed in Class III locations shall clearly show the maximum wattage of the lamps that shall be permitted without exceeding an exposed surface temperature of _____ under normal conditions of use.

   _____

   _____

   NEC _____

9. In an aircraft painting hangar, the area _____ horizontally from and _____ above the aircraft shall be classified as a Class I, Division 1 location.

   _____

   _____

   NEC _____

10. An agricultural building site isolation device shall simultaneously disconnect _____.

    _____

    _____

    NEC _____

Part 1   Layout and Structure: Learning to Use the NEC

NEC Chapters 1–5

Answer the following questions using a copy of the National Electrical Code. *Identify the section or subsection where the answer is found.*

11. The general provision for required receptacle outlets in a dwelling unit living room states that no space measured horizontally along the floor line shall be more than _____.

    _____

    _____

    _____

    _____

    NEC _____

12. Panelboards shall not be orientated so they are in the _____ position.

    _____

    _____

    NEC _____

13. A listed seal shall be provided in each conduit run _____ a fuel dispenser.

    _____

    _____

    NEC _____

14. Switchboards, switchgear, panelboards, and motor control centers shall be located in dedicated spaces and _____.

    _____

    _____

    NEC _____

15. Meter sockets shall have an air space of at least _____ between the enclosure and the wall or other supporting surface.

    _____

    _____

    NEC _____

16. The voltage of a circuit is the greatest _____ difference of potential between any two conductors of the circuit concerned.

    _____

    _____

    NEC _____

17. Motors used in agricultural building shall be _____ or designed so as to minimize the entrance of dust, moisture, or corrosive particles.

    _____

    _____

    NEC _____

18. Embedded deicing cables shall be spaced apart according to the rating of the cable but shall not be less than _____ on centers.

    _____

    _____

    NEC _____

19. High density polyethylene conduit shall not be installed exposed, _____, in hazardous locations, or where subjected to temperatures in excess of 50°C (122°F).

    _____

    _____

    NEC _____

20. Line and grounding conductors to a surge-protective device shall be at least _____.

    _____

    _____

    _____

    _____

    NEC _____

UNIT

9 NEC Chapter 6

sockagphoto/Shutterstock.com

LEARNING OBJECTIVES

After completing this unit, you will be able to:
- Describe the type of requirements in *Chapter 6* of the *NEC*.
- Identify the occupancies covered by *Chapter 6* of the *NEC*.
- Recognize keywords that will lead you to *Chapter 6* of the *NEC*.

KEY TERMS

artificially made body of water
audio signal processing equipment
electric sign
electric welder
electrified truck parking space
electrolytic cell
electroplating
fire pump
fuel cell system
induction heating, melting, and welding
industrial machinery
information technology equipment (ITE)
irrigation machine
manufactured wiring system
modular data center
natural body of water
office furnishing
pipe organ
solar photovoltaic (PV) system
wind turbine

Introduction

This unit is an introduction to *Chapter 6, Special Equipment*. It is the second of the special chapters that supplement or modify the information found in *Chapters 1–7*, **Figure 9-1**. This unit will introduce the type of equipment covered in *Chapter 6*. The end of the unit contains practice exercises that will require the use of your *National Electrical Code* to complete. The combination of studying the type of information found in *Chapter 6* as well as looking up *Code* sections will help with comprehension and retention of the types of equipment found in *Chapter 6* of the *NEC*.

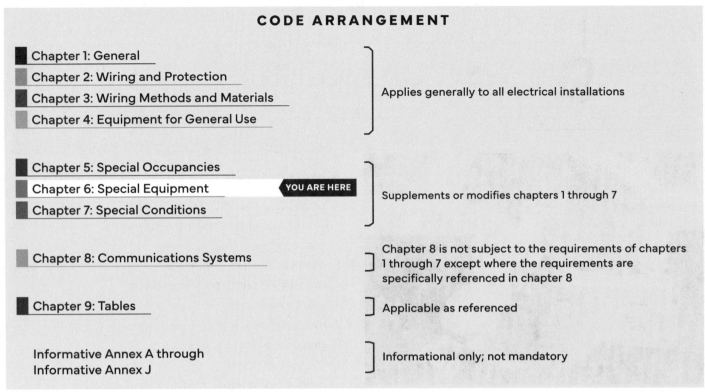

Figure 9-1. *Chapter 6* is the second of the special chapters. It has requirements that supplement or modify the information found in *Chapters 1–7*.

Special Equipment

Chapter 4 contains requirements for equipment that is considered general in nature and common to many electrical installations. The equipment found in *Chapter 6* is considered special, as it is not common to every installation and may have unique characteristics that make special requirements necessary. One of the reasons that some of the equipment found in *Chapter 6* is considered special is that if there is an installation error or a malfunction, it carries an additional risk of becoming dangerous. The general requirements found in *Chapters 1–4* still apply, but *Chapter 6* contains information and requirements that will supplement or modify the other portions of the *Code*, **Figure 9-2**.

Article 600, Electric Signs and Outline Lighting

Chapter 6 starts off with *Article 600*, *Electric Signs and Outline Lighting*, **Figure 9-3**. Most commercial and industrial buildings will have an electric sign. The *NEC* defines an ***electric sign*** as a fixed, stationary, or portable self-contained electrically operated and/or electrically illuminated utilization equipment with words or symbols designed to convey information or attract attention. Signs are typically prefabricated by a sign manufacturer and installed or assembled on-site. The electrician is responsible for connecting the branch circuit to the sign. There are many methods to light signs, some incorporating high voltages, such as neon. *Article 600* has special requirements to ensure they are electrically safe, **Figure 9-4**.

> **PRO TIP** **Table of Contents–Chapter 6**
>
> *Chapter 6* of the *National Electrical Code* is not used by most electricians very often. For this reason, it is important to become familiar with the equipment that is covered by *Chapter 6*. Simply going through the table of contents and spending time looking at the titles of the *Chapter 6* articles is all it takes to get you acquainted with the type of equipment that it includes. With this preparation, when a special piece of equipment is encountered on the job, you will recognize it as such and check the *NEC* for any applicable requirements.

NEC Chapter 6
Special Equipment

Article 600 Electric Signs and Outline Lighting
Article 604 Manufactured Wiring Systems
Article 605 Office Furnishings
Article 610 Cranes and Hoists
Article 620 Elevators, Dumbwaiters, Escalators, Moving Walks, Platform Lifts, and Stairway Chairlifts
Article 625 Electric Vehicle Power Transfer System
Article 626 Electrified Truck Parking Spaces
Article 630 Electric Welder
Article 640 Audio Signal Processing, Amplification, and Reproduction Equipment
Article 645 Information Technology Equipment
Article 646 Modular Data Centers
Article 647 Sensitive Electronic Equipment
Article 650 Pipe Organs
Article 660 X-Ray Equipment
Article 665 Induction and Dielectric Heating Equipment
Article 668 Electrolytic Cells
Article 669 Electroplating
Article 670 Industrial Machinery
Article 675 Electrically Driven or Controlled Irrigation Machines
Article 680 Swimming Pools, Fountains, and Similar Installations
Article 682 Natural and Artificially Made Bodies of Water
Article 685 Integrated Electrical Systems
Article 690 Solar Photovoltaic (PV) Systems
Article 691 Large-Scale Photovoltaic (PV) Electric Supply Stations
Article 692 Fuel Cell Systems
Article 694 Wind Electric Systems
Article 695 Fire Pumps

Goodheart-Willcox Publisher

Figure 9-2. *Chapter 6* has requirements for special equipment not found on every job with special requirements to ensure it remains electrically safe.

Article 604, Manufactured Wiring Systems

Article 604 has requirements for manufactured wiring systems, **Figure 9-5**. The *NEC* defines a ***manufactured wiring system*** as a system containing component parts that are assembled in the process of manufacturing and cannot be

Article 600
Electric Signs and Outline Lighting

Part I General

Section 600.1 Scope
Section 600.3 Listing
Section 600.4 Markings
Section 600.5 Branch Circuits
Section 600.6 Disconnects
Section 600.7 Grounding and Bonding
Section 600.8 Enclosures
Section 600.9 Location
Section 600.10 Portable or Mobile Signs
Section 600.12 Field-Installed Secondary Wiring
Section 600.21 Ballasts, Transformers, Electronic Power Supplies, and Class 2 Power Sources
Section 600.22 Ballasts
Section 600.23 Transformers and Electronic Power Supplies
Section 600.24 Class 2 Power Sources

Part II Field-Installed Skeleton Tubing, Outline Lighting, and Secondary Wiring

Section 600.30 Applicability
Section 600.31 Neon Secondary-Circuit Wiring, 1000 Volts or Less, Nominal
Section 600.32 Neon Secondary-Circuit Wiring, over 1000 Volts, Nominal
Section 600.33 Class 2 Sign Illumination Systems, Secondary Wiring
Section 600.34 Photovoltaic (PV) Powered Sign
Section 600.35 Retrofit Kits
Section 600.41 Neon Tubing
Section 600.42 Electrode Connections

Goodheart-Willcox Publisher

Figure 9-3. *Article 600* has requirements for electric signs. Although they are found on most commercial and industrial jobs, they are not standardized, are typically custom made, and in some cases, they operate at high voltages. For those reasons, electric signs have their own special article.

Goodheart-Willcox Publisher

Figure 9-4. Electric signs are often custom made to fit the building and match the design or logo of the business.

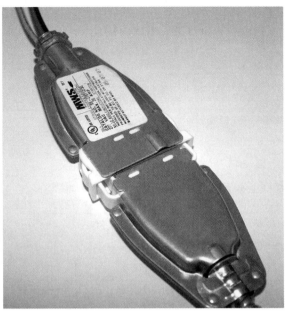

Cooper Lighting Solutions

Figure 9-6. Lighting whips are an example of a manufactured wiring system. They come prewired from the manufacturer and snap together to make the connection, saving time during installation.

Article 604 Manufactured Wiring Systems

Section 604.1 Scope
Section 604.6 Listing Requirements
Section 604.7 Installation
Section 604.10 Uses Permitted
Section 604.12 Uses Not Permitted
Section 604.100 Construction

Goodheart-Willcox Publisher

Figure 9-5. *Article 604* has requirements for manufactured wiring systems.

Article 605 Office Furnishings

Section 605.1 Scope
Section 605.3 General
Section 605.4 Wireways
Section 605.5 Office Furnishing Interconnections
Section 605.6 Lighting Accessories
Section 605.7 Fixed-Type Office Furnishings
Section 605.8 Freestanding-Type Office Furnishings
Section 605.9 Freestanding-Type Office Furnishings, Cord- and Plug-Connected

Goodheart-Willcox Publisher

Figure 9-7. *Article 605* has requirements for office furnishings.

inspected at the building site without damage or destruction to the assembly and used for the connection of luminaires, utilization equipment, continuous plug-in type busways, and other devices. Manufactured wiring systems are becoming more popular because they can save labor during installation, **Figure 9-6**. They are often used in conjunction with prefabrication of buildings where portions of the building are built/assembled off-site in a controlled environment. The finished product is then delivered to the jobsite, saving time and money.

Article 605, Office Furnishings

Article 605 has requirements for office furnishings, **Figure 9-7**. The *NEC* defines **office furnishings** as cubicle panels, partitions, study carrels, workstations, desks, shelving systems, and storage units that may be mechanically and electrically interconnected to form an office furnishing system. It is common to see office cubicles with power integrated within the walls. Powered cubicle walls have wiring harnesses that clip together with a feed point that either plugs into the wall or is directly wired, **Figure 9-8**.

Article 610, Cranes and Hoists

Article 610 has requirements for cranes and hoists, **Figure 9-9**. Cranes and hoists are used to lift heavy objects into place. They must be able to move around and possibly extend out, which requires the electrical connections to provide flexibility. These are commonly found on jobsites to help with the construction process, but they are also found in many industrial sites and warehouses to lift heavy objects and move them around, **Figure 9-10**.

Gill Thompson/Shutterstock.com

Figure 9-8. Office cubicles often have electrical run through the units for receptacles and desk lights. The cubicles are prewired from the manufacturer and have feed points where the building wiring is attached. The power is distributed through the cubicles by internal connectors that plug into one another.

Goodheart-Willcox Publisher

Figure 9-10. The hoist pictured is in a diesel repair shop. It is on a track system that enables it to move heavy objects up and down as well as back and forth and side to side. The wiring methods used must be flexible enough to accommodate the movement.

Article 610
Cranes and Hoists

Part I General
- Section 610.1 Scope
- Section 610.3 Special Requirements for Particular Locations

Part II Wiring
- Section 610.11 Wiring Method
- Section 610.12 Raceway or Cable Terminal Fittings
- Section 610.13 Types of Conductors
- Section 610.14 Rating and Size of Conductors
- Section 610.15 Common Return

Part III Contact Conductors
- Section 610.21 Installation of Contact Conductors
- Section 610.22 Collectors

Part IV Disconnecting Means
- Section 610.31 Runway Conductor Disconnecting Means
- Section 610.32 Disconnecting Means for Cranes and Monorail Hoists
- Section 610.33 Rating of Disconnecting Means

Part V Overcurrent Protection
- Section 610.41 Feeders, Runway Conductors
- Section 610.42 Branch-Circuit Short-Circuit and Ground-Fault Protection
- Section 610.43 Overload Protection

Part VI Control
- Section 610.51 Separate Controllers
- Section 610.53 Overcurrent Protection
- Section 610.57 Clearance

Part VII Grounding and Bonding
- Section 610.61 Grounding and Bonding

Goodheart-Willcox Publisher

Figure 9-9. *Article 610* has requirements for cranes and hoists. This content is broken up into 7 different parts.

Part 1   Layout and Structure: Learning to Use the NEC

Article 620, Elevators, Dumbwaiters, Escalators, Moving Walks, Platform Lifts, and Stairway Chairlifts

Article 620 has requirements for elevators, dumbwaiters, escalators, moving walks, platform lifts, and stairway chairlifts, **Figure 9-11**. Since these pieces of equipment are used to move people, special requirements are necessary to ensure they provide safe passage that is free from electrical hazards, **Figure 9-12**.

Article 620
Elevators, Dumbwaiters, Escalators, Moving Walks, Platform Lifts, and Stairway Chairlifts

Part I General
- Section 620.1 Scope
- Section 620.3 Voltage Limitations
- Section 620.4 Live Parts Enclosed
- Section 620.5 Working Clearances
- Section 620.6 Ground-Fault Circuit-Interrupter Protection for Personnel

Part II Conductors
- Section 620.11 Insulation of Conductors
- Section 620.12 Minimum Size of Conductors
- Section 620.13 Feeder and Branch-Circuit Conductors
- Section 620.14 Feeder Demand Factor
- Section 620.15 Motor Controller Rating
- Section 620.16 Short-Circuit Current Rating

Part III Wiring
- Section 620.21 Wiring Methods
- Section 620.22 Branch Circuits for Car Lighting, Receptacle(s), Ventilation, Heating, and Air-Conditioning
- Section 620.23 Branch Circuits for Machine Room, Control Room/Machinery Space, Control Space, or Truss Interior Lighting and Receptacle(s)
- Section 620.24 Branch Circuit for Hoistway Pit Lighting and Receptacles
- Section 620.25 Branch Circuits for Other Utilization Equipment

Part IV Installation of Conductors
- Section 620.32 Metal Wireways and Nonmetallic Wireways
- Section 620.33 Number of Conductors in Raceways
- Section 620.34 Supports
- Section 620.35 Auxiliary Gutters
- Section 620.36 Different Systems in One Raceway or Traveling Cable
- Section 620.37 Wiring in Hoistways, Machine Rooms, Control Rooms, Machinery Spaces, and Control Spaces
- Section 620.38 Electrical Equipment in Garages and Similar Occupancies

Part V Traveling Cables
- Section 620.41 Suspension of Traveling Cables
- Section 620.42 Hazardous (Classified) Locations
- Section 620.43 Location of and Protection for Cables
- Section 620.44 Installation of Traveling Cables

Part VI Disconnecting Means and Control
- Section 620.51 Disconnecting Means
- Section 620.52 Power from More Than One Source
- Section 620.53 Car Light, Receptacle(s), and Ventilation Disconnecting Means
- Section 620.54 Heating and Air-Conditioning Disconnecting Means
- Section 620.55 Utilization Equipment Disconnecting Means

Part VII Overcurrent Protection
- Section 620.61 Overcurrent Protection
- Section 620.62 Selective Coordination
- Section 620.65 Signage

Part VIII Machine Rooms, Control Rooms, Machinery Spaces, and Control Spaces
- Section 620.71 Guarding Equipment

Part IX Grounding and Bonding
- Section 620.81 Metal Raceways Attached to Cars
- Section 620.82 Electric Elevators
- Section 620.83 Nonelectric Elevators
- Section 620.84 Escalators, Moving Walks, Platform Lifts, and Stairway Chairlifts

Part X Emergency and Standby Power Systems
- Section 620.91 Emergency and Standby Power Systems

Goodheart-Willcox Publisher

Figure 9-11. *Article 620* has requirements for elevators, dumbwaiters, escalators, moving walks, platform lifts, and stairway lifts.

Unit 9 NEC Chapter 6

Figure 9-12. Elevators have special requirements to ensure its occupants remain safe under normal operation as well as in the event of an emergency or power outage.

Article 625, Electric Vehicle Power Transfer System

Article 625 has requirements for electric vehicle power transfer systems, **Figure 9-13**. It is becoming more common to see electric vehicle charging incorporated into houses, commercial buildings, and parking spaces, **Figure 9-14**. In addition to charging electric vehicles, the transfer equipment may be capable of bidirectional current flow so the vehicle may be used to export power back to the premises.

Article 626, Electrified Truck Parking Spaces

Article 626 has requirements for electrified truck parking spaces, **Figure 9-15**. An *electrified truck parking space* is defined as a truck parking space that has been provided with an electrical system that allows truck operators to connect their vehicles and use off-board power sources to operate onboard systems, such as air conditioning, heating, and appliances, without any engine idling.

Truck drivers are only allowed to drive a certain number of hours before they are required to pull over and rest. If they are hauling a load that requires refrigeration, or if they are traveling through an area with temperature extremes, the truck would have to idle the entire time they are stopped if an electrified truck parking space isn't available. Since many states have adopted regulations that are meant to reduce the amount of time that a truck idles, electrified truck parking spaces are becoming more common, **Figure 9-16**.

Article 630, Electric Welders

Article 630, Electric Welders contains electrical requirements for electric welders, **Figure 9-17**. *Electric welders*

Article 625
Electric Vehicle Power Transfer System

Part I General
| Section 625.1 Scope |
| Section 625.4 Voltages |
| Section 625.5 Listed |

Part II Equipment Construction
| Section 625.17 Cords and Cables |
| Section 625.22 Personnel Protection System |

Part III Installation
| Section 625.40 Electric Vehicle Branch Circuit |
| Section 625.41 Overcurrent Protection |
| Section 625.42 Rating |
| Section 625.43 Disconnecting Means |
| Section 625.44 Equipment Connection |
| Section 625.46 Loss of Primary Source |
| Section 625.47 Multiple Feeder or Branch Circuits |
| Section 625.48 Interactive Equipment |
| Section 625.49 Island Mode |
| Section 625.50 Location |
| Section 625.52 Ventilation |
| Section 625.54 Ground-Fault Circuit-Interrupter Protection for Personnel |
| Section 625.56 Receptacle Enclosures |
| Section 625.60 AC Receptacle Outlets Used for EVPE |

Part IV Wireless Power Transfer Equipment
| Section 625.101 Grounding |
| Section 625.102 Installation |

Figure 9-13. *Article 625* has requirements for electrical vehicle power transfer systems.

use electricity to provide the heat necessary to join metals together. There are many different types of welders, but the *NEC* groups them into two types, arc welders and resistance welders. See **Figure 9-18**.

Article 640, Audio Signal Processing, Amplification, and Reproduction Equipment

Article 640 has requirements for audio signal processing, amplification, and reproduction equipment, **Figure 9-19**. The *NEC* defines *audio signal processing equipment* as

Figure 9-14. Electrical car chargers are showing up in parking lots for charging vehicles.

Figure 9-16. Electrified truck parking spaces provide a place to plug a truck and trailer in while parking to eliminate idling.

Article 626
Electrified Truck Parking Spaces

Part I General
- Section 626.1 Scope
- Section 626.3 Other Articles
- Section 626.4 General Requirements

Part II Electrified Truck Parking Space Electrical Wiring Systems
- Section 626.10 Branch Circuits
- Section 626.11 Feeder and Service Load Calculations

Part III Electrified Truck Parking Space Supply Equipment
- Section 626.22 Wiring Methods and Materials
- Section 626.23 Overhead Gantry or Cable Management System
- Section 626.24 Electrified Truck Parking Space Supply Equipment Connection Means
- Section 626.25 Separable Power-Supply Cable Assembly
- Section 626.26 Loss of Primary Power
- Section 626.27 Interactive Systems

Part IV Transport Refrigerated Units (TRUs)
- Section 626.30 Transport Refrigerated Units
- Section 626.31 Disconnecting Means and Receptacles
- Section 626.32 Separable Power Supply Cable Assembly

Figure 9-15. *Article 626* has requirements for electrified truck parking spaces.

Article 630
Electric Welders

Part I General
- Section 630.1 Scope
- Section 630.6 Listing
- Section 630.8 Ground-Fault Circuit-Interrupter Protection for Personnel

Part II Arc Welders
- Section 630.11 Ampacity of Supply Conductors
- Section 630.12 Overcurrent Protection
- Section 630.13 Disconnecting Means
- Section 630.14 Marking
- Section 630.15 Grounding of Welder Secondary Circuit

Part III Resistance Welders
- Section 630.31 Ampacity of Supply Conductors
- Section 630.32 Overcurrent Protection
- Section 630.33 Disconnecting Means
- Section 630.34 Marking

Part IV Welding Cable
- Section 630.41 Conductors
- Section 630.42 Installation

Figure 9-17. *Chapter 6* has requirements for electric welders.

Goodheart-Willcox Publisher

Figure 9-18. Electrical welders are available in many different types and have varying electrical requirements. This metal inert gas (MIG) welding setup requires a power supply for operation.

electrically operated equipment that produces and/or processes electronic signals that, when appropriately amplified and reproduced by a loudspeaker, produce an acoustic signal within the range of normal human hearing, typically 20 Hz to 20 kHz. Audio signal processing equipment can be found at a concert, theater, or church to control the sound system. It is common to see equipment like this in a place of assembly, **Figure 9-20**.

Article 645, Information Technology Equipment

Article 645 has requirements for information technology equipment, **Figure 9-21**. The *NEC* defines *information technology equipment (ITE)* as equipment and systems rated 1000 volts or less, normally found in offices or other business establishments and similar environments classified as ordinary locations, that are used for creation and manipulation of data, voice, video, and similar signals that

Article 640
Audio Signal Processing, Amplification, and Reproduction Equipment

Part I General
Section 640.1 Scope
Section 640.3 Locations and Other Articles
Section 640.4 Protection of Electrical Equipment
Section 640.5 Access to Electrical Equipment Behind Panels Designed to Allow Access
Section 640.6 Mechanical Execution of Work
Section 640.7 Grounding
Section 640.8 Grouping of Conductors
Section 640.9 Wiring Methods
Section 640.10 Audio Systems Near Bodies of Water

Part II Permanent Audio System Installations
Section 640.21 Use of Flexible Cords and Cables
Section 640.22 Wiring of Equipment Racks and Enclosures
Section 640.23 Conduit or Tubing
Section 640.24 Wireways, Gutters, and Auxiliary Gutters
Section 640.25 Loudspeaker Installation in Fire Resistance-Rated Partitions, Walls, and Ceilings

Part III Portable and Temporary Audio System Installations
Section 640.41 Multipole Branch-Circuit Cable Connectors
Section 640.42 Use of Flexible Cords and Cables
Section 640.43 Wiring of Equipment Racks
Section 640.44 Environmental Protection of Equipment
Section 640.45 Protection of Wiring
Section 640.46 Equipment Access

Goodheart-Willcox Publisher

Figure 9-19. *Article 640* has requirements for signal processing, amplification, and reproduction equipment.

Marko Poplasen/Shutterstock.com

Figure 9-20. Recording studios have equipment that falls under the jurisdiction of *Article 640*.

are not communications equipment. This article refers to the equipment that is installed in an IT room, such as servers, equipment racks, and patch panels, **Figure 9-22**.

Article 645
Information Technology Equipment

Section 645.1 Scope
Section 645.3 Other Articles
Section 645.4 Special Requirements for Information Technology Equipment Room
Section 645.5 Supply Circuits and Interconnecting Cables
Section 645.10 Disconnecting Means
Section 645.11 Uninterruptible Power Supply (UPS)
Section 645.14 System Grounding
Section 645.15 Equipment Grounding and Bonding
Section 645.16 Marking
Section 645.17 Power Distribution Units
Section 645.18 Surge Protection for Critical Operations Data Systems
Section 645.25 Engineering Supervision
Section 645.27 Selective Coordination

Goodheart-Willcox Publisher

Figure 9-21. *Article 645* has requirements for information technology equipment.

Article 646, Modular Data Centers

Article 646 has requirements for modular data centers, **Figure 9-23**. The *NEC* defines a ***modular data center*** as prefabricated units, rated 1000 volts or less, consisting of an outer enclosure housing multiple racks or cabinets of information technology equipment (ITE) and various support equipment, such as electrical service and distribution equipment, HVAC systems, and the like. Modular data centers

Article 646
Modular Data Centers

Part I General
Section 646.1 Scope
Section 646.3 Other Articles
Section 646.4 Applicable Requirements
Section 646.5 Nameplate Data
Section 646.6 Supply Conductors and Overcurrent Protection
Section 646.7 Short-Circuit Current Rating
Section 646.8 Field-Wiring Compartments
Section 646.9 Flexible Power Cords and Cables for Connecting Equipment Enclosures of an MDC System

Part II Equipment
Section 646.10 Electrical Supply and Distribution
Section 646.11 Distribution Transformers
Section 646.12 Receptacles
Section 646.13 Other Electrical Equipment
Section 646.14 Installation and Use

Part III Lighting
Section 646.15 General Illumination
Section 646.16 Emergency Lighting
Section 646.17 Emergency Lighting Circuits

Part IV Workspace
Section 646.18 General
Section 646.19 Entrance to and Egress from Working Space
Section 646.20 Working Space for ITE
Section 646.21 Work Areas and Working Space About Batteries
Section 646.22 Workspace for Routine Service and Maintenance

Goodheart-Willcox Publisher

Figure 9-23. *Article 646* has requirements for modular data centers.

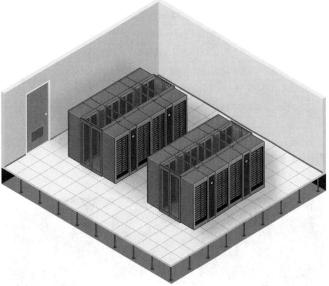

Stock image/Shutterstock.com

Figure 9-22. Larger information technology (IT) systems will often be installed in a room with a raised floor. This allows for ventilation from the floor cavity to be drawn through the equipment to help with cooling. The *NEC* has special requirements to address the raised floor and ensure the electrical system does not play a role in causing damage to or a shutdown of the IT equipment.

are similar to information technology rooms in the type of equipment they will contain but are different in their construction process. IT rooms are built in place along with the rest of the building. Modular data centers are built in a manufacturing facility and transported to the jobsite/building.

Article 647, Sensitive Electronic Equipment

Article 647 has requirements for sensitive electronic equipment, **Figure 9-24**. The systems described in this article are intended to reduce electronic noise in audio and video equipment in commercial and industrial applications. The system that is specified in this article is a separately derived system with 120V line to line and 60 volts to ground.

Article 650, Pipe Organs

Article 650 has requirements for pipe organs, **Figure 9-25**. Pipe organs are found in occupancies that are typically

Article 647
Sensitive Electronic Equipment

Section 647.1 Scope
Section 647.3 General
Section 647.4 Wiring Methods
Section 647.5 Three-Phase Systems
Section 647.6 Grounding
Section 647.7 Receptacles
Section 647.8 Lighting Equipment

Goodheart-Willcox Publisher

Figure 9-24. *Article 647* has requirements for sensitive electronic equipment.

Article 650
Pipe Organs

Section 650.1 Scope
Section 650.3 Other Articles
Section 650.4 Source of Energy
Section 650.5 Grounding or Double Insulation of the DC Power Supply
Section 650.6 Conductors
Section 650.7 Installation of Conductors
Section 650.8 Overcurrent Protection
Section 650.9 Protection from Accidental Contact

Goodheart-Willcox Publisher

Figure 9-25. *Article 650* has requirements for pipe organs.

Goodheart-Willcox Publisher

Figure 9-26. Pipe organs are found in churches, theaters, and similar installations.

classified as a place of assembly, such as churches and theaters. A *pipe organ* is a musical instrument that produces sound by driving pressurized air (called wind) through pipes selected via a keyboard. See **Figure 9-26**.

Article 660, X-Ray Equipment

Article 660 has requirements for X-ray equipment. **Figure 9-27** The X-ray equipment covered in *Article 660* is not the type used in medical facilities. Instead, it is the X-ray equipment used in industrial and scientific applications. Using X-rays is a nondestructive way of inspecting a product for structural defects, **Figure 9-28**.

Article 665, Induction and Dielectric Heating

Article 665 has requirements for inducting and dielectric heating equipment, **Figure 9-29**. The *NEC* defines *induction heating, melting, and welding* as the heating, melting, or welding of a nominally conductive material due to its own I^2R losses when the material is placed in a varying electromagnetic field. The equipment covered by this article is found in industrial establishments only and does not apply to medical applications or pipeline and vessel heating. The heating equipment covered by this article uses electromagnetic fields to quickly heat objects, **Figure 9-30**.

Article 668, Electrolytic Cells

Article 668 has requirements for electrolytic cells, **Figure 9-31**. The *NEC* defines an *electrolytic cell* as a tank or vat in which electrochemical reactions are caused by applying electric energy for the purpose of refining or producing usable materials. Electrolytic cells are used in industrial establishments to produce materials such as aluminum, cadmium, chlorine, copper, fluorine, hydrogen peroxide, magnesium, sodium, sodium chlorate, and zinc.

Article 660
X-Ray Equipment

Part I General
- Section 660.1 Scope
- Section 660.3 Hazardous (Classified) Locations
- Section 660.4 Connection to Supply Circuit
- Section 660.5 Disconnecting Means
- Section 660.6 Rating of Supply Conductors and Overcurrent Protection
- Section 660.7 Wiring Terminals
- Section 660.9 Minimum Size of Conductors
- Section 660.10 Equipment Installations

Part II Control
- Section 660.20 Fixed and Stationary Equipment
- Section 660.21 Portable and Mobile Equipment
- Section 660.23 Industrial and Commercial Laboratory Equipment
- Section 660.24 Independent Control

Part III Transformers and Capacitors
- Section 660.35 General
- Section 660.36 Capacitors

Part IV Guarding and Grounding
- Section 660.47 General
- Section 660.48 Grounding

Goodheart-Willcox Publisher

Figure 9-27. *Article 660* has requirements for X-ray equipment used in industrial processes.

shinobi/Shutterstock.com

Figure 9-28. X-ray equipment is a nondestructive way of testing the integrity of components and connections. In this image, the X-ray machine is checking where two pipes have been welded together.

Article 665
Induction and Dielectric Heating Equipment

Part I General
- Section 665.1 Scope
- Section 665.5 Output Circuit
- Section 665.7 Remote Control
- Section 665.10 Ampacity of Supply Conductors
- Section 665.12 Disconnecting Means

Part II Guarding, Grounding, and Labeling
- Section 665.19 Component Interconnection
- Section 665.20 Enclosures
- Section 665.21 Control Panels
- Section 665.22 Access to Internal Equipment
- Section 665.23 Hazard Labels or Signs
- Section 665.24 Capacitors
- Section 665.25 Dielectric Heating Applicator Shielding
- Section 665.26 Grounding and Bonding
- Section 665.27 Marking

Goodheart-Willcox Publisher

Figure 9-29. *Article 665* has requirements for induction and dielectric heating equipment.

Surasak_Photo/Shutterstock.com

Figure 9-30. Dielectric heating equipment uses magnetic fields and high frequencies to quickly heat up conductive objects, such as steel.

Article 669, Electroplating

Article 669 is titled electroplating, **Figure 9-32**. *Electroplating* is the process of coating of a metal object with another metal by the process of electrolytic deposition. In addition to electroplating, *Article 669* has requirements for anodizing, electropolishing, and electrostripping. These processes are found in industrial production facilities.

Article 670, Industrial Machinery

Article 670 has requirements for industrial machinery, **Figure 9-33**. The *NEC* defines *industrial machinery* as a power-driven machine or system, not portable by hand while working, that is used to process materials by cutting; forming; pressure; electrical, thermal, or optical techniques; lamination; or a combination of these processes. It can include associated equipment used to transfer material or tooling, including fixtures, to assemble/disassemble, to inspect or test, or to package. Industrial machinery is often designed and built specifically for an individual manufacturer's process and does not fit into any other article very well, **Figure 9-34**. *Article 670* has requirements that can be applied to these situations.

Article 668 — Electrolytic Cells

Section 668.1 Scope
Section 668.3 Other Articles
Section 668.10 Cell Line Working Zone
Section 668.11 Direct-Current Cell Line Process Power Supply
Section 668.12 Cell Line Conductors
Section 668.13 Disconnecting Means
Section 668.14 Shunting Means
Section 668.15 Grounding
Section 668.20 Portable Electrical Equipment
Section 668.21 Power-Supply Circuits and Receptacles for Portable Electrical Equipment
Section 668.30 Fixed and Portable Electrical Equipment
Section 668.31 Auxiliary Nonelectrical Connections
Section 668.32 Cranes and Hoists
Section 668.40 Enclosures

Goodheart-Willcox Publisher

Figure 9-31. *Article 668* has requirements for electrolytic cells.

Article 669 — Electroplating

Section 669.1 Scope
Section 669.3 General
Section 669.5 Branch-Circuit Conductors
Section 669.6 Wiring Methods
Section 669.7 Warning Signs
Section 669.8 Disconnecting Means
Section 669.9 Overcurrent Protection

Goodheart-Willcox Publisher

Figure 9-32. *Article 669* has requirements for electroplating.

Article 670 — Industrial Machinery

Section 670.1 Scope
Section 670.3 Machine Nameplate Data
Section 670.4 Supply Conductors and Overcurrent Protection
Section 670.5 Short-Circuit Current Rating
Section 670.6 Overvoltage Protection

Goodheart-Willcox Publisher

Figure 9-33. *Article 670* has requirements for industrial machinery.

Article 675, Electrically Driven or Controlled Irrigation Machines

Article 675 has requirements for electrically driven or controlled irrigation machines, **Figure 9-35**. The *NEC* defines an *irrigation machine* as an electrically driven or controlled

Nordroden/Shutterstock.com

Figure 9-34. This sheet metal roller was specifically designed to perform a single task.

Article 675
Electrically Driven or Controlled Irrigation Machines

Part I General
- Section 675.1 Scope
- Section 675.4 Irrigation Cable
- Section 675.5 More Than Three Conductors in a Raceway or Cable
- Section 675.6 Marking on Main Control Panel
- Section 675.7 Equivalent Current Ratings
- Section 675.8 Disconnecting Means
- Section 675.9 Branch-Circuit Conductors
- Section 675.10 Several Motors on One Branch Circuit
- Section 675.11 Collector Rings
- Section 675.12 Grounding
- Section 675.13 Methods of Grounding
- Section 675.14 Bonding
- Section 675.15 Lighting Protection
- Section 675.16 Energy from More Than One Source
- Section 675.17 Connectors

Part II Center Pivot Irrigation Machines
- Section 675.21 General
- Section 6755.22 Equivalent Current Ratings

Goodheart-Willcox Publisher

Figure 9-35. *Article 675* has requirements for electrically driven or controlled irrigation machines.

Kenneth Keifer/Shutterstock.com

Figure 9-36. Center pivot irrigation machines are commonly found in farms across the United States. Electric power is used to draw water from a well and also run the motors to move the unit.

machine, with one or more motors, not hand-portable, and used primarily to transport and distribute water for agricultural purposes. Irrigation machines are found in agricultural fields and will typically draw water from a well to water the crops, **Figure 9-36**.

Article 680, Swimming Pools, Fountains, and Similar Installations

Article 680 has requirements for swimming pools, fountains, and similar installations, **Figure 9-37**. In addition to swimming pools and fountains, *Article 680* covers wading, therapeutic, and decorative pools; fountains; hot tubs; spas; hydromassage bathtubs; and all the associated equipment. The electrical requirements associated with *Article 680* ensure people's safety as they are in and around water, **Figure 9-38**.

Article 682, Natural and Artificially Made Bodies of Water

Article 682 has requirements for natural and artificially made bodies of water, **Figure 9-39**. The *NEC* defines **natural bodies of water** as bodies of water such as lakes, streams, ponds, rivers, and other naturally occurring bodies of water, which may vary in depth throughout the year. People who have property with access to a river or lake will often have docks, boat lifts, or irrigation equipment in and around the water that have an electrical connection.

The *NEC* defines **artificially made bodies of water** as bodies of water that have been constructed or modified to fit some decorative or commercial purpose, such as, but not limited to, aeration ponds, fish farm ponds, storm retention basins, treatment ponds, and irrigation (channel) facilities. Water depths may vary seasonally or be controlled. Many residential neighborhoods will create an artificial pond or lake to give the residents the feeling of living along a shoreline. It is common for there to be a fountain and circulation equipment with an electrical connection. This article applies to electrical equipment or structures that are in or on the shore of natural or artificial body of water, **Figure 9-40**.

Article 685, Integrated Electrical Systems

Article 685 is a short article that covers integrated electrical systems, **Figure 9-41**. This article applies to industrial establishments where orderly power shutdown is necessary to minimize personnel hazard and equipment damage.

Article 680
Swimming Pools, Fountains, and Similar Installations

Part I General

Section 680.1 Scope
Section 680.4 Inspections After Installation
Section 680.5 Ground-Fault Circuit-Interrupter (GFCI) and Special Purpose Ground-Fault Circuit-Interrupter (SPGFCI) Protection
Section 680.6 Listing Requirements
Section 680.7 Grounding and Bonding
Section 680.8 Cord-and-Plug-Connected Equipment
Section 680.9 Overhead Conductor Clearances
Section 680.10 Electric Pool Water Heaters Incorporating Resistive Heating Elements and Electrically Powered Swimming Pool Heat Pumps and Chillers
Section 680.11 Underground Wiring
Section 680.12 Equipment Rooms, Vaults, and Pits
Section 680.13 Maintenance Disconnecting Means
Section 680.14 Corrosive Environments

Part II Permanently Installed Pools

Section 680.20 General
Section 680.21 Motors
Section 680.22 Lighting, Receptacles, and Equipment
Section 680.23 Underwater Luminaires
Section 680.24 Junction Boxes and Electrical Enclosures for Transformers or Ground-Fault Circuit Interrupters
Section 680.26 Equipotential Bonding
Section 680.27 Specialized Pool Equipment
Section 680.28 Gas-Fired Water Heater

Part III Storable Pools, Storable Spas, Storable Hot Tubs, and Storage Immersion Pools

Section 680.30 General
Section 680.31 Pumps
Section 680.32 Ground-Fault Circuit-Interrupter (GFCI) and Special Purpose Ground-Fault Circuit-Interrupter (SPGFCI) Protection
Section 680.33 Luminaires
Section 680.34 Receptacle Locations
Section 680.35 Storable and Portable Immersion Pools

Part IV Permanently Installed and Self-Contained Spas and Hot Tubs and Permanently Installed Immersion Pools

Section 680.40 General
Section 680.41 Location of Other Equipment
Section 680.42 Outdoor Installations
Section 680.43 Indoor Installations
Section 680.44 Ground-Fault Circuit-Interrupters (GFCI) and Special Purpose Ground-Fault Circuit-Interrupter (SPGFCI) Protection
Section 680.45 Permanently Installed Immersion Pools

Part V Fountains

Section 680.50 General
Section 680.51 Luminaires, Submersible Pumps, and Other Submersible Equipment
Section 680.52 Junction Boxes and Other Enclosures
Section 680.54 Grounding and Bonding
Section 680.55 Methods of Grounding
Section 680.56 Cord-and-Plug-Connected Equipment
Section 680.57 Signs
Section 680.58 Ground-Fault Circuit-Interrupter (GFCI) and Special Purpose Ground-Fault Circuit-Interrupter (SPGFCI) Protection for Adjacent Receptacle Outlets
Section 680.59 Ground-Fault Circuit-Interrupter (GFCI) and Special Purpose Ground-Fault Circuit-Interrupter (SPGFCI) Protection for Permanently Installed Nonsubmersible Pumps

Part VI Pools and Tubs for Therapeutic Use

Section 680.60 General
Section 680.61 Permanently Installed Therapeutic Pools
Section 680.62 Therapeutic Tubs (Hydrotherapeutic Tanks)

Part VII Hydromassage Bathtubs

Section 680.70 General
Section 680.71 Protection
Section 680.72 Other Electrical Equipment
Section 680.73 Accessibility
Section 680.74 Bonding

Part VIII Electrically Powered Pool Lifts

Section 680.80 General
Section 680.81 Equipment Approval
Section 680.82 Protection
Section 680.83 Equipotential Bonding
Section 680.84 Switching Devices and Receptacles
Section 680.85 Nameplate Marking

Goodheart-Willcox Publisher

Figure 9-37. *Article 680* has requirements for swimming pools, fountains, and similar installations.

Part 1   Layout and Structure: Learning to Use the NEC

Goodheart-Willcox Publisher

Figure 9-38. Swimming pools typically have electrical connections to their circulation equipment, heaters, and lights. The *NEC* has extensive requirements to ensure there are not any electrical hazards in and around the pool.

Nancy Salmon/Shutterstock.com

Figure 9-40. Some housing developments have artificially made bodies of water, so the residents have the feeling of living near a lake. Any power requirements in and around the water falls under the jurisdiction of *Article 682*.

Article 682
Natural and Artificially Made Bodies of Water

Part I General
- Section 682.1 Scope
- Section 682.3 Other Articles
- Section 682.4 Industrial Application
- Section 682.5 Electrical Datum Plane Distances

Part II Installation
- Section 682.10 Electrical Equipment and Transformers
- Section 682.11 Location of Electrical Distribution Equipment
- Section 682.12 Electrical Connections
- Section 682.13 Wiring Methods and Installation
- Section 682.14 Disconnecting Means
- Section 682.15 Ground-Fault Protection

Part III Grounding and Bonding
- Section 682.30 Grounding
- Section 682.31 Equipment Grounding Conductors
- Section 682.32 Bonding of Non-Current-Carrying Metal Parts
- Section 682.33 Equipotential Planes and Bonding of Equipotential Planes

Goodheart-Willcox Publisher

Figure 9-39. *Article 462* has requirements for natural and artificially made bodies of water.

Article 685
Integrated Electrical Systems

Part I General
- Section 685.1 Scope
- Section 685.3 Application of Other Articles

Part II Orderly Shutdown
- Section 685.10 Location of Overcurrent Devices in or on Premises
- Section 685.12 Direct-Current System Grounding
- Section 685.14 Ungrounded Control Circuits

Goodheart-Willcox Publisher

Figure 9-41. *Article 685* has requirements for integrated electrical systems.

Article 690, Solar Photovoltaic (PV) Systems

Article 690 has requirements for solar photovoltaic (PV) systems, **Figure 9-42**. The *NEC* defines a *solar photovoltaic (PV) system* as the total components, circuits, and equipment up to and including the PV system disconnecting means that, in combination, convert solar energy into electric energy. This article covers stand-alone, grid tied, and interactive systems. Solar cells use energy from the sun (photons) to knock electrons free, which creates a current flow. As technologies have improved and solar power has become more affordable, solar installations are gaining popularity for homes and businesses, **Figure 9-43**.

Article 691, Large-Scale Photovoltaic (PV) Electric Supply Stations

Article 691 has requirements for large-scale photovoltaic electric supply stations, **Figure 9-44**. *Article 691* applies to supply stations fed from large solar installations that are 5 MW or greater, but not under the exclusive control of the utility, **Figure 9-45**.

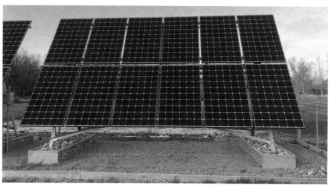

Goodheart-Willcox Publisher

Figure 9-43. Solar installations are becoming more popular for commercial as well as residential customers because the cost of solar has decreased while the reliability has increased.

Article 690 — Solar Photovoltaic (PV) Systems

Part I General
- Section 690.1 Scope
- Section 690.4 General Requirements
- Section 690.6 Alternating-Current (ac) Modules and Systems

Part II Circuit Requirements
- Section 690.7 Maximum Voltage
- Section 690.8 Circuit Sizing and Current
- Section 690.9 Overcurrent Protection
- Section 690.11 Arc-Fault Circuit Protection (dc)
- Section 690.12 Rapid Shutdown of PV Systems on Buildings

Part III Disconnecting Means
- Section 690.13 Photovoltaic System Disconnecting Means
- Section 690.15 Disconnecting Means for Isolating Photovoltaic Equipment

Part IV Wiring Methods and Materials
- Section 690.31 Wiring Methods
- Section 690.32 Component Interconnections
- Section 690.33 Mating Connectors
- Section 690.34 Access to Boxes

Part V Grounding and Bonding
- Section 690.41 PV System DC Circuit Grounding and Protection
- Section 690.42 Point of PV System DC Circuit Grounding Connection
- Section 690.43 Equipment Grounding and Bonding
- Section 690.45 Size of Equipment Grounding Conductors
- Section 690.47 Grounding Electrode System

Part VI Source Connections
- Section 690.56 Identification of Power Sources
- Section 690.59 Connection to Other Sources
- Section 690.72 Self-Regulated PV Charge Control

Goodheart-Willcox Publisher

Figure 9-42. *Article 690* has requirements for solar photovoltaic (PV) systems.

Article 691 — Large-Scale Photovoltaic (PV) Electric Supply Stations

- Section 691.1 Scope
- Section 691.4 Special Requirements for Large-Scale PV Electric Supply Stations
- Section 691.5 Equipment
- Section 691.6 Engineered Design
- Section 691.7 Conformance of Construction to Engineered Design
- Section 691.8 Direct Current Operating Voltage
- Section 691.9 Disconnecting Means for Isolating Photovoltaic Equipment
- Section 691.10 Fire Mitigation
- Section 691.11 Fence Bonding and Grounding

Goodheart-Willcox Publisher

Figure 9-44. *Article 691* has requirements for large-scale photovoltaic (PV) electric supply stations.

Jenson/Shutterstock.com

Figure 9-45. The equipment used in this large-scale photovoltaic electric supply station falls under the jurisdiction of *Article 691*.

Article 692, Fuel Cell Systems

Article 692 has requirements for fuel systems, **Figure 9-46**. The *NEC* defines a ***fuel cell system*** as the complete aggregate of equipment used to convert chemical fuel into usable electricity and typically consisting of a reformer, stack, power inverter, and auxiliary equipment. Fuel cells use a fuel, such as hydrogen, and an oxidizing agent, such as oxygen, to create a chemical reaction, resulting in a current flow. The exhaust/byproduct of most fuel cells is water or carbon dioxide, **Figure 9-47**.

Article 692
Fuel Cell Systems

Part I General
- Section 692.1 Scope
- Section 692.4 Installation
- Section 692.6 Listing Requirement

Part II Circuit Requirements
- Section 692.8 Circuit Sizing and Current
- Section 692.9 Overcurrent Protection

Part III Disconnecting Means
- Section 692.13 All Conductors
- Section 692.17 Switch or Circuit Breaker

Part IV Wiring Methods
- Section 692.31 Wiring Systems

Part V Marking
- Section 692.50 Fuel Cell Power Sources
- Section 692.51 Fuel Shut-Off
- Section 692.52 Stored Energy

Part VI Connection to Other Circuits
- Section 692.60 Connection to Other Systems
- Section 692.61 Transfer Switch

Goodheart-Willcox Publisher

Figure 9-46. *Article 692* has requirements for fuel cell systems.

Bloom Energy

Figure 9-47. Fuel cells can be used as a backup or in conjunction with other power supplies or can be stand-alone.

Article 694, Wind Electric Systems

Article 694 has requirements for wind electrical systems, **Figure 9-48**. The *NEC* defines a ***wind turbine*** as a mechanical device that converts wind energy to electrical energy. Wind turbines are available in a wide variety of sizes and outputs. The smaller ones are often installed at dwellings or small commercial installations, while the larger ones are incorporated into wind farms. **Figure 9-49**.

Article 695, Fire Pumps

Article 695, Fire Pumps has requirements for fire pumps, **Figure 9-50**. ***Fire pumps*** are electric pumps used to increase the water pressure to the fire suppression system in certain buildings. This is often necessary in high-rise buildings and buildings with fire suppression systems that require higher pressures than what is available from the water supply. Fire pumps have different requirements from regular motors as they must operate in emergency situations, **Figure 9-51**.

Unit 9 NEC Chapter 6

Article 694
Wind Electric Systems

Part I General
- Section 694.1 Scope
- Section 694.7 Construction and Maintenance

Part II Circuit Requirements
- Section 694.10 Maximum Voltage
- Section 694.12 Circuit Sizing and Current
- Section 694.15 Overcurrent Protection

Part III Disconnecting Means
- Section 694.20 All Conductors
- Section 694.22 Additional Provisions
- Section 694.23 Turbine Shutdown
- Section 694.24 Disconnection of Wind Electric System Equipment
- Section 694.26 Fuses
- Section 694.28 Installation and Service of a Wind Turbine

Part IV Wiring Methods
- Section 694.30 Permitted Methods

Part V Grounding and Bonding
- Section 694.40 Equipment Grounding and Bonding

Part VI Marking
- Section 694.52 Power Systems Employing Energy Storage
- Section 694.54 Identification of Power Sources
- Section 694.56 Instructions for Disabling Turbine

Part VII Connection to Other Sources
- Section 694.60 Identified Interactive Equipment
- Section 694.62 Installation
- Section 694.66 Operating Voltage Range

Goodheart-Willcox Publisher

Figure 9-48. *Article 694* has requirements for wind electrical systems.

Utsa Barua/Shutterstock.com

Figure 9-49. Wind generators use the wind to spin a generator that creates electrical energy. Their size varies from large units used by the electric utility to smaller units that can be installed on a dwelling.

Article 695
Fire Pumps

- Section 695.1 Scope
- Section 695.2 Reconditioned Equipment
- Section 695.3 Power Source(s) for Electric Motor-Driven Fire Pumps
- Section 695.4 Continuity of Power
- Section 695.5 Transformers
- Section 695.6 Power Wiring
- Section 695.7 Voltage Drop
- Section 695.10 Listed Equipment
- Section 695.12 Equipment Location
- Section 695.14 Control Wiring
- Section 695.15 Surge Protection

Goodheart-Willcox Publisher

Figure 9-50. *Article 695* has requirements for fire pumps.

ABCDstock/Shutterstock.com

Figure 9-51. In emergency situations, fire pumps can be used to boost the water pressure to sprinklers and hydrants in large buildings.

NEC KEYWORDS

Chapter 6 Keywords

The following are examples of keywords that should lead you to *Chapter 6*:

- array
- artificially made bodies of water
- audio equipment
- cranes
- dielectric heating
- electric signs
- electric vehicle
- electrical noise
- electrified truck parking space
- electrolytic cell
- electronic power converter
- electroplating
- elevator
- escalator
- festoon cable
- fire pump
- forming shell (pool lighting)
- fountain
- fuel cell
- hoists
- hot tub
- hydromassage tub
- immersion pool
- induction heating
- industrial machinery
- industrial manufacturing systems
- industrial process heating equipment
- information technology (IT)
- integrated electrical systems
- interactive inverters
- interactive systems
- inverter
- irrigation machines
- lighting assembly
- machine rooms
- machinery space
- manufactured wiring system
- modular data center
- modules (solar)
- multimode inverter
- nacelle
- natural bodies of water
- neon tubing
- office furnishing
- organ
- outline lighting
- output circuits
- overhead gantry
- photovoltaic
- pipe organ
- power production
- radiograph
- recording system
- sensitive electronic equipment
- sign
- skeleton tubing
- solar cell
- solar photovoltaic
- splash pads
- stairway chairlift
- submersible equipment
- swimming pool
- technical power system
- therapeutic pool/tub
- tower
- truck
- wading pool
- water
- welder
- wind electric system
- wind turbine
- wireless power transfer
- X-ray

Summary

- *Chapter 6* of the *NEC* is the second of the special chapters that supplement or modify *Chapters 1–7*.
- *Chapter 6* contains requirements for various types of special equipment, such as:
  - Prefabricated equipment
  - Lifting equipment
  - Vehicle equipment
  - Welding equipment
  - Audio, video, and data equipment
  - Industrial equipment
  - Equipment in or around water
  - Electrical power production equipment
  - Fire pump equipment
- The special equipment found in *Chapter 6* has unique characteristics that necessitate special requirements to ensure that an installation is free from electrical hazards.

Unit 9 Review

Name _____ Date _____ Class _____

Know and Understand

Answer the following questions based on information in this unit.

1. Chapter 6 of the *NEC* is titled _____.
   A. *Special Conditions*
   B. *Special Installations*
   C. *Special Occupancies*
   D. *Special Equipment*

2. The requirements in *Chapter 6* will supplement or modify the information found in _____.
   A. *Chapter 1*
   B. *Chapters 1–4*
   C. *Chapters 1–7*
   D. *Chapters 1–8*

3. _____ has requirements that apply to swimming pools.
   A. *Article 600*
   B. *Article 620*
   C. *Article 680*
   D. *Article 695*

4. A pipe organ is a musical instrument that produces sound by driving _____ through pipes selected via a keyboard.
   A. pressurized air
   B. compressed gas
   C. compressed liquid
   D. None of the above.

5. *Article 660* applies to X-ray equipment used for _____ use.
   A. medical
   B. dental
   C. industrial
   D. All of the above.

6. _____ has requirements for cranes and hoists.
   A. *Article 600*
   B. *Article 610*
   C. *Article 620*
   D. *Article 630*

7. Which of the following is not covered by *Article 630*?
   A. Acetylene torches
   B. Resistance welders
   C. Arc welders
   D. Welding cable

8. Which of the following keywords would lead you to *Chapter 6*?
   A. Capacitor
   B. Mobile homes
   C. Overcurrent protection
   D. Hot tub

9. Which article has requirements for fuel cell systems?
   A. *Article 690*
   B. *Article 692*
   C. *Article 694*
   D. *Article 695*

10. Induction heaters heat conductive objects by placing them in a _____.
    A. varying electromagnetic field
    B. industrial furnace
    C. path of concentrated photons
    D. chemical heater

Apply and Analyze

NEC Chapter 6

Answer the following questions using a copy of the National Electrical Code. *Identify the section or subsection where the answer is found.*

1. The branch circuit that feeds a neon sign shall not exceed _____ amperes.
   A. 15
   B. 20
   C. 30
   D. 40

   NEC _____

2. Electroplating equipment shall have warning signs posted to indicate the presence of _____.
   A. bare conductors
   B. high voltages
   C. arc-flash hazard
   D. corrosive materials

   NEC _____

3. Submersible pumps installed in a fountain shall operate at _____ volts or less between conductors.
   A. 120
   B. 150
   C. 300
   D. 600

   NEC _____

4. Which of the following is not a permitted wiring method that can be used on an elevator car assembly?
   A. Flexible metal conduit
   B. Electrical nonmetallic tubing
   C. Liquidtight flexible metal conduit
   D. Liquidtight flexible nonmetallic conduit

   NEC _____

5. Conductors that supply an arc welder shall be protected by an overcurrent protective device rated or set at no more than _____% of the conductor ampacity.
   A. 100
   B. 125
   C. 150
   D. 200

   NEC _____

6. A listed surge protective device shall be installed _____ the fire pump controller.
   A. in or on
   B. ahead of
   C. on the circuit feeding
   D. after

   NEC _____

7. Photovoltaic circuits installed on a one-family dwelling shall have a maximum voltage no greater than _____ volts.
   A. 300
   B. 600
   C. 1000
   D. 1500

   NEC _____

8. Each outlet installed for the purpose of supplying EVSE greater than 16 amperes or 120 volts shall be supplied by a(n) _____.
   A. 20-ampere branch circuit
   B. AFCI protected circuit
   C. multi-wire branch circuit
   D. individual branch circuit

   NEC _____

9. Only _____ power transformers shall be permitted to be installed in the modular data center equipment enclosure.
   A. oil-filled
   B. dry-type
   C. step-up
   D. laminated

   NEC _____

10. Conductors used with pipe organs shall have thermoplastic or _____ insulation.
    A. thermosetting
    B. rubber
    C. polyethylene
    D. silicone

    NEC _____

Name _____ Date _____ Class _____

Critical Thinking

NEC Chapter 6

Answer the following questions using a copy of the National Electrical Code. *Identify the section or subsection where the answer is found.*

1. Unless _____, X-ray and related equipment shall not be installed or operated in hazardous (classified) locations.

   _____

   _____

   NEC _____

2. Irrigation machine collector rings shall be protected from the expected environment and from accidental contact by means of a _____.

   _____

   _____

   NEC _____

3. When a pump is replaced on a permanently installed pool, the new pump motor shall be provided with _____.

   _____

   _____

   NEC _____

4. Information technology equipment interconnecting cables that may be exposed to physical damage shall be _____.

   _____

   _____

   NEC _____

5. What are three requirements for the required electric sign branch circuit in a commercial building?

   _____

   _____

   _____

   NEC _____

6. A hoistway pit light switch shall be located so as to be _____ from the pit access door.

   _____

   _____

   NEC _____

7. The maximum voltage drop on a branch circuit feeding sensitive electronic equipment shall not exceed _____.

   _____

   NEC _____

8. A surge-protective device shall be installed between a wind electric system and any loads served by the _____.

   _____

   _____

   NEC _____

9. Electric motor-driven fire pumps shall have a _____ source of power.

   _____

   NEC _____

10. Freestanding-type office furnishings that are not cord and plug connected shall be permitted to be connected to the building electrical system by _____.

    _____

    _____

    NEC _____

Part 1   Layout and Structure: Learning to Use the NEC

NEC Chapters 1–6

Answer the following questions using a copy of the National Electrical Code. *Identify the section or subsection where the answer is found.*

11. Operating rooms in health care facilities shall be provided with a minimum of 36 receptacles divided between at least _____ circuits.

    _____

    _____

    NEC _____

12. A hot tub in a hotel must have an emergency switch that is readily accessible to the users, at least five feet away from the edge of the tub, _____, and within sight of the hot tub.

    _____

    _____

    NEC _____

13. Oil immersion is immersing _____ in a protective liquid in such a way that an explosive atmosphere that may be above the liquid or outside the enclosure cannot be ignited.

    _____

    _____

    NEC _____

14. List three locations where nonmetallic underground conduit with conductors (Type NUCC) is not permitted.

    _____

    _____

    _____

    NEC _____

15. Each multiwire branch circuit shall be provided with a means that will simultaneously disconnect _____ at the point where the branch circuit originates.

    _____

    _____

    NEC _____

16. When flexibility is required for crane wiring, a listed flexible cord or cable or _____ shall be permitted to be used, and where necessary, cable reels or take-up devices shall be used.

    _____

    _____

    NEC _____

17. The ampacity of capacitor circuit conductors shall not be less than _____ of the rated current of the capacitor.

    _____

    _____

    NEC _____

18. A Zone 20 location is a location in which ignitable concentrations of combustible dust or ignitable fibers/flyings are present continuously or _____.

    _____

    _____

    NEC _____

19. The minimum mounting height for an electrified truck parking space pedestal shall be not less than _____ above ground.

    _____

    _____

    NEC _____

20. When installing liquidtight flexible nonmetallic conduit (LFNC) in lengths exceeding 6′, the conduit shall be securely fastened at intervals not exceeding _____.

    _____

    _____

    NEC _____

UNIT

10 NEC Chapter 7

sockagphoto/Shutterstock.com

LEARNING OBJECTIVES

After completing this unit, you will be able to:

- Describe the type of requirements in *Chapter 7* of the *NEC*.
- Identify the special conditions covered by *Chapter 7* of the *NEC*.
- Recognize keywords that will lead you to *Chapter 7* of the *NEC*.

KEY TERMS

Class 1 circuit
Class 2 circuit
Class 3 circuit
critical operations power system (COPS)
emergency system
energy management system
energy storage system
fault-managed power system
fire alarm system
fire-resistive cable system
interconnected electric power production sources
legally required standby system
optional standby system
remote-control circuit
stand-alone system

Introduction

This unit is an introduction to *Chapter 7* of the *NEC*, called *Special Conditions*. It is the third and final of the special chapters that supplement or modify the information found in *Chapters 1–7*, **Figure 10-1**. The end of the unit contains practice exercises that will require the use of your *National Electrical Code* to complete. The combination of studying the type of information found in *Chapter 7*, as well as looking up *Code* sections, will help with comprehension and retention of the special conditions found in the *NEC*.

Part 1   Layout and Structure: Learning to Use the NEC

CODE ARRANGEMENT

- Chapter 1: General
- Chapter 2: Wiring and Protection
- Chapter 3: Wiring Methods and Materials
- Chapter 4: Equipment for General Use

} Applies generally to all electrical installations

- Chapter 5: Special Occupancies
- Chapter 6: Special Equipment
- Chapter 7: Special Conditions — **YOU ARE HERE**

} Supplements or modifies chapters 1 through 7

- Chapter 8: Communications Systems

} Chapter 8 is not subject to the requirements of chapters 1 through 7 except where the requirements are specifically referenced in chapter 8

- Chapter 9: Tables

} Applicable as referenced

Informative Annex A through Informative Annex J

} Informational only; not mandatory

Goodheart-Willcox Publisher

Figure 10-1. *Chapter 7* is the third and final of the special chapters. It has requirements that supplement or modify the information found in *Chapters 1–7*.

Special Conditions

Special conditions is a broad description that includes several systems not found on every job, such as backup systems, energy storage and production systems, power-limited circuits, and optical fiber cables. These requirements for special conditions will supplement or modify *Chapters 1–7* of the *NEC*, **Figure 10-2**.

PRO TIP — Backup Systems

One of the important considerations when installing a backup system is the prevention of backfeeding the utility. Consumer backup generators are becoming more common as they can keep heat and appliances running during a utility power outage. If a consumer generator imposes a voltage back onto the utility wires, it has the potential to electrocute a lineman who is working to repair the line. Listed transfer equipment is designed to prevent this possibility, but incorrect equipment or improper connections can create this hazard. The *NEC* has requirements throughout *Chapter 7* to address this issue, **Figure 10-3**.

NEC Chapter 7
Special Conditions

Article 700 Emergency Systems
Article 701 Legally Required Standby Systems
Article 702 Optional Standby Systems
Article 705 Interconnected Electric Power Production
Article 706 Energy Storage Systems
Article 708 Critical Operations Power Systems (COPS)
Article 710 Stand-Alone Systems
Article 722 Cables for Power-Limited Circuits and Fault-Managed Power Circuits
Article 724 Class 1 Power-Limited Circuits and Class 1 Power-Limited Remote-Control and Signaling Circuits
Article 725 Class 2 and Class 3 Power-Limited Circuits
Article 726 Class 4 Fault-Managed Power Systems
Article 728 Fire-Resistive Cable Systems
Article 750 Energy Management Systems
Article 760 Fire Alarm Systems
Article 770 Optical Fiber Cables

Goodheart-Willcox Publisher

Figure 10-2. *Chapter 7* has requirements for special conditions. Most of the articles in *Chapter 7* relate to systems that have special requirements and are not common to every jobsite.

Cheryl Ann Quigley/Shutterstock.com

Figure 10-3. Due to working on high-voltage wires at great heights and in the elements, linemen have a very dangerous profession. Backup systems must be configured so there is no chance that it can back feed and put the life of a lineman at risk.

Energy Backup and Storage Systems

There are several articles that address backup energy systems. Some of these systems are legally required, while others are optional. The two main types of backup systems are manual and automatic. Manual systems must be manually started in the event of a power outage, while automatic systems have transfer equipment that will start them automatically. One way to differentiate the types of systems is the amount of time it takes for the backup system to be operational. A backup system doesn't necessarily supply power to the entire premises. In many cases, it only provides backup power to specific equipment or portions of a building.

Article 700, Emergency Systems

Article 700 has requirements for emergency systems, **Figure 10-4**. The *NEC* defines an ***emergency system*** as systems legally required and classed as emergency by municipal, state, federal, or other codes or by any governmental agency having jurisdiction. These systems are intended to automatically supply illumination, power, or both to designated areas and equipment in the event of a loss of power. This loss can be of the normal supply or an accident to elements of a system intended to supply, distribute,

Article 700 Emergency Systems

Part I General
- Section 700.1 Scope
- Section 700.2 Reconditioned Equipment
- Section 700.3 Tests and Maintenance
- Section 700.4 Capacity and Rating
- Section 700.5 Transfer Equipment
- Section 700.6 Signals
- Section 700.7 Signs
- Section 700.8 Surge Protection

Part II Circuit Wiring
- Section 700.10 Wiring, Emergency System
- Section 700.11 Wiring, Class-2-Powered Emergency Lighting Systems

Part III Sources of Power
- Section 700.12 General Requirements

Part IV Emergency System Circuits for Lighting and Power
- Section 700.15 Loads on Emergency Branch Circuits
- Section 700.16 Emergency Illumination
- Section 700.17 Branch Circuits for Emergency Lighting
- Section 700.18 Circuits for Emergency Power
- Section 700.19 Multiwire Branch Circuits

Part V Control — Emergency Lighting Circuits
- Section 700.20 Switch Requirements
- Section 700.21 Switch Location
- Section 700.22 Exterior Lights
- Section 700.23 Dimmer and Relay Systems
- Section 700.24 Directly Controlled Emergency Luminaires
- Section 700.25 Branch Circuit Emergency Lighting Transfer Switch
- Section 700.26 Automatic Load Control Relay
- Section 700.27 Class 2 Powered Emergency Lighting Systems

Part VI Overcurrent Protection
- Section 700.30 Accessibility
- Section 700.31 Ground-Fault Protection of Equipment
- Section 700.32 Selective Coordination

Goodheart-Willcox Publisher

Figure 10-4. Emergency systems provide power to emergency circuits to allow occupants to move about and exit safely.

and control power and illumination essential for safety to human life.

Emergency systems are found in many buildings, including schools, places of assembly, and hospitals. The purpose of the emergency system is to provide enough illumination and building systems to allow the general public to safely move around and exit the building if the normal power supply is lost. Emergency power may be supplied by batteries, as are often found in emergency and exit lights, or by a generator that supplies critical loads, such as elevators, emergency lighting circuits, fire pumps, and communication circuits, **Figure 10-5**. The *NEC* requires an emergency system to restore power within 10 seconds. When the backup is provided by batteries, they must be able to maintain the emergency load for at least 1 1/2 hours.

Article 701, Legally Required Standby Systems

Article 701 has requirements for legally required standby systems, **Figure 10-6**. The *NEC* defines ***legally required standby systems*** as systems required and so classed as legally required standby by municipal, state, federal, or other codes or by any governmental agency having jurisdiction. These systems are intended to automatically supply power to selected loads (other than those classed as emergency systems) in the event of a failure of the normal source.

Legally required standby systems focus on powering equipment that has the potential to create a hazard that could hinder rescue and firefighting operations. Examples include ventilation and smoke removal systems, heating and cooling systems, and some industrial processes, **Figure 10-7**. The *NEC* requires a legally required standby system to restore power within 60 seconds. When the backup is provided by batteries, they must be able to maintain the emergency load for at least 1 1/2 hours.

Article 701
Legally Required Standby Systems

Part I General
Section 701.1 Scope
Section 701.2 Reconditioned Equipment
Section 701.3 Commissioning and Maintenance
Section 701.4 Capacity and Rating
Section 701.5 Transfer Equipment
Section 701.6 Signals
Section 701.7 Signs

Part II Circuit Wiring
Section 701.10 Wiring Legally Required Standby Systems

Part III Sources of Power
Section 701.12 General Requirements

Part IV Overcurrent Protection
Section 701.30 Accessibility
Section 701.31 Ground-Fault Protection of Equipment
Section 701.32 Selective Coordination

Goodheart-Willcox Publisher

Figure 10-6. Legally required standby systems provide power to circuits that are necessary to ensure the safety of emergency and rescue personnel.

Goodheart-Willcox Publisher

Figure 10-5. An emergency light often receives its backup power from an internal battery that turns the light on if power is lost to the branch circuit that feeds it.

John_T/Shutterstock.com

Figure 10-7. Local, federal, and other codes will determine if a building must have a legally required standby system. If required, ventilation and smoke removal equipment are fed from these systems to ensure the safety of first responders and firefighters.

Article 702, Optional Standby Systems

Article 702 has requirements for optional standby systems, **Figure 10-8**. The *NEC* defines ***optional standby systems*** as systems intended to supply power to public or private facilities or property where life safety does not depend on the performance of the system. These systems are intended to supply on-site generated or stored power to selected loads either automatically or manually.

Residential backup generators are an example of an optional standby system. They are typically a small generator and therefore not capable of running all the loads in a home. Instead, they will feed specific branch circuits that have a greater need for power during an outage. Examples include heating systems, sump pumps, refrigerators, freezers, and wells, **Figure 10-9**.

Because these systems are optional, the *NEC* does not have a requirement for the time it takes for them to restore power, nor the amount of time they will maintain power.

Article 705, Interconnected Electric Power Production Sources

Article 705 has requirements for interconnected electric power and production sources, **Figure 10-10**. An

Article 702 Optional Standby Systems

Part I General
- Section 702.1 Scope
- Section 702.2 Reconditioned Equipment
- Section 702.4 Capacity and Rating
- Section 702.5 Interconnection or Transfer Equipment
- Section 702.6 Signals
- Section 702.7 Signs

Part II Wiring
- Section 702.10 Wiring Optional Standby Systems
- Section 702.11 Portable Generator Grounding
- Section 702.12 Outdoor Generator Sets

Goodheart-Willcox Publisher

Figure 10-8. Article *702* has requirements for optional standby systems.

Article 705 Interconnected Electric Power Production Sources

Part I General
- Section 705.1 Scope
- Section 705.5 Parallel Operation
- Section 705.6 Equipment Approval
- Section 705.8 System Installation
- Section 705.10 Identification of Power Sources
- Section 705.11 Source Connections to a Service
- Section 705.12 Load-Side Source Connections
- Section 705.13 Energy Management Systems (EMS)
- Section 705.20 Source Disconnecting Means
- Section 705.25 Wiring Methods
- Section 705.28 Circuit Sizing and Current
- Section 705.30 Overcurrent Protection
- Section 705.32 Ground-Fault Protection
- Section 705.40 Loss of Primary Source
- Section 705.45 Unbalanced Interconnections

Part II Microgrid Systems
- Section 705.50 System Operation
- Section 705.60 Primary Power Source Connection
- Section 705.65 Reconnection to Primary Power Source
- Section 705.70 Microgrid Interconnect Devices (MID)
- Section 705.76 Microgrid Control System (MCS)

Part III Interconnected Systems Operating in Island Mode
- Section 705.80 Power Source Capacity
- Section 705.81 Voltage and Frequency Control
- Section 705.82 Single 120-Volt Supply

Goodheart-Willcox Publisher

Goodheart-Willcox Publisher

Figure 10-9. A residential generator is an example of an optional standby system. This generator has transfer equipment that turns it on automatically if there is a power outage.

Figure 10-10. Article *705* may apply along with other backup articles in *Chapter 7*, depending on the configuration of the system. If the backup supply is in parallel with the primary source of electricity, then *Article 705* applies.

interconnected electric power production source is a power supply that is installed in parallel with the primary source of electricity. *Article 705* often applies in conjunction with other *Chapter 7* articles if their backup source is installed in parallel with the primary source of electricity, **Figure 10-11**.

Article 706, Energy Storage Systems

Article 706 has requirements for energy storage systems, **Figure 10-12**. The *NEC* defines an *energy storage system* as one or more components assembled capable of storing energy and providing electrical energy into the premises wiring system or an electric power production and distribution network. Batteries alone do not count as an energy storage system. The batteries, inverters, etc. must be listed as a unit for *Article 706* to apply. Batteries that are not listed as part of an energy storage system are covered by *Article 480*.

Article 708, Critical Operations Power Systems (COPS)

Article 708 has requirements for critical operations power systems (COPS), **Figure 10-13**. The *NEC* defines *critical operations power systems (COPS)* as power systems for facilities or parts of facilities that require continuous operation for the reasons of public safety, emergency management, national security, or business continuity.

Article 708 ensures that power remains intact to these facilities or at least specific portions of them during the emergency. Examples of facilities that would have critical operations power systems are police stations, fire stations,

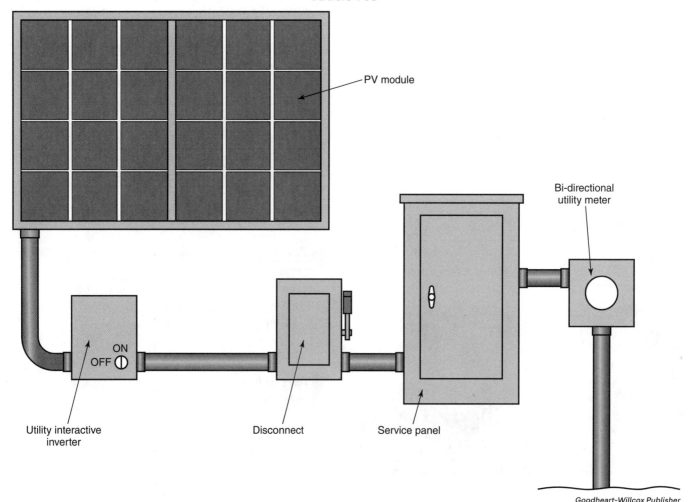

Figure 10-11. An interconnected electric power production source is connected in parallel with the utility power. In the event of a primary power outage, the backup source must disconnect power from the primary source to prevent a backfeed scenario where a lineman's life could be put at risk.

Article 706
Energy Storage Systems

Part I General
Section 706.1 Scope
Section 706.3 Qualified Personnel
Section 706.4 System Requirements
Section 706.5 Listing
Section 706.6 Multiple Systems
Section 706.7 Commissioning and Maintenance
Section 706.9 Maximum Voltage

Part II Disconnecting Means
Section 706.15 Disconnecting Means
Section 706.16 Connection to Energy Sources

Part III Installation Requirements
Section 706.20 General
Section 706.21 Directory (Identification of Power Sources)

Part IV Circuit Requirements
Section 706.30 Circuit Sizing and Current
Section 706.31 Overcurrent Protection
Section 706.33 Charge Control

Part V Flow Battery EESs
Section 706.40 General
Section 706.41 Electrolyte Classification
Section 706.42 Electrolyte Containment
Section 706.43 Flow Controls
Section 706.44 Pumps and Other Fluid Handling Equipment

Part VI Other Energy Storage Technologies
Section 706.50 General
Section 706-51 Flywheel ESS (FESS)

Goodheart-Willcox Publisher

Figure 10-12. Energy storage systems are more than just batteries. They are a listed unit that contains the backup source, inverter, and control equipment.

Article 708
Critical Operations Power Systems (COPS)

Part I General
Section 708.1 Scope
Section 708.2 Reconditioned Equipment
Section 708.4 Risk Assessment
Section 708.5 Physical Security
Section 708.6 Testing and Maintenance
Section 708.7 Cybersecurity
Section 708.8 Commissioning

Part II Circuit Wiring and Equipment
Section 708.10 Feeder and Branch Circuit Wiring
Section 708.11 Branch Circuit and Feeder Distribution Equipment
Section 708.12 Feeders and Branch Circuits Supplied by COPS
Section 708.14 Wiring of HVAC, Fire Alarm, Security, Emergency Communications, and Signaling Systems

Part III Power Sources and Connection
Section 708.20 Sources of Power
Section 708.21 Ventilation
Section 708.22 Capacity of Power Sources
Section 708.24 Transfer Equipment
Section 708.30 Branch Circuits Supplied by COPS

Part IV Overcurrent Protection
Section 708.50 Accessibility
Section 708.52 Ground-Fault Protection of Equipment
Section 708.54 Selective Coordination

Part V System Performance and Analysis
Section 708.64 Emergency Operations Plan

Goodheart-Willcox Publisher

Figure 10-13. Critical operations power systems are intended to maintain power to essential buildings in case of a natural disaster or other emergency that disrupts the normal power supply. Federal, state, and other municipal codes make the determination for when *Article 708* applies.

wastewater treatment facilities, health care facilities, and air traffic control. In the event of a natural disaster or other emergency, these facilities must remain operational. This classification is determined by Homeland Security or by another government authority, **Figure 10-14**. The *NEC* requires a criterial operations power system to have power restored in the time required for the application. The requirement is nonspecific and puts the burden on the engineering of the system.

Figure 10-14. Air traffic control towers are an example of a structure that has *Article 708* requirements. They must maintain power to help navigate planes in the air.

Autonomous Power Production Systems

Autonomous systems are capable of working independently from the utility power grid. It is possible for some of the systems to have a utility grid connection but also be capable of isolating and working independently.

Article 710, Stand-Alone Systems

Article 710 has requirements for stand-alone systems, **Figure 10-15**. The *NEC* defines a ***stand-alone system*** as a system that is capable of supplying power independent of an electric power production and distribution network. Stand-alone

Article 710
Stand-Alone Systems
Section 710.1 Scope
Section 710.6 Equipment Approval
Section 710.10 Identification of Power Sources
Section 710.12 Stand-Alone Inverter Input Circuit Current
Section 710.15 General

Figure 10-15. *Article 710* has requirements for stand-alone systems.

Figure 10-16. Road signs without an electrical supply are an example of a stand-alone system. They can operate autonomously as they have solar panels, batteries, lights, and a control system.

systems do not receive power from a public utility. Instead, they produce their own power using a power source, such as a generator, solar, wind, fuel cells, or batteries, **Figure 10-16**.

Power-Limited Circuits, Control Circuits, and Managed Power Systems

Power-limited circuits have a power supply that limits the amount of available power. They are generally associated with a system or piece of equipment. Instead of wiring being contained within an enclosure or piece of equipment, it leaves for the purpose of control. Since the control wiring is installed on-site and is not contained within a listed piece of equipment, the *NEC* has requirements to ensure a safe installation.

> **PRO TIP**
>
> **Class I, II, III Hazardous locations vs Class 1, 2, 3 Remote-Control and Power-Limited Circuits**
>
> Be careful not to confuse Class I, II, and III hazardous locations with Class 1, 2, and 3 remote-control and power-limited circuits. These classifications can be confusing at first since they use similar terminology, but keeping the context in mind will help to differentiate. Class I, II, and III hazardous locations use roman numerals to denote the class and are associated with locations with combustible gases, dust, or fibers. Class 1, 2, and 3 power-limited circuits use numerical digits and are associated with control wiring.

Article 722, Cables for Power-Limited Circuits and Fault-Managed Power Circuits

Article 722 has requirements for Class 2 and Class 3 power-limited circuits, power-limited fire alarm circuits, Class 4 fault-managed power circuits, and optical fiber cables, **Figure 10-17**. This is a new article for the 2023 edition of the *National Electrical Code*. *Article 722* focuses on the requirements related to the installation of the cables and wiring methods that are associated with these systems. *Article 722* prevents repeating similar language in multiple articles as the wiring methods and cables used are similar for the covered systems.

Article 724, Class 1 Power-Limited Circuits and Class 1 Power-Limited Remote-Control and Signaling Circuits

Article 724 has requirements for Class 1 power-limited, remote-control, and signaling circuits that are not integral to a piece of equipment, **Figure 10-18**. The *NEC* defines a Class 1 circuit as the portion of the wiring system between the load side of the overcurrent device or power-limited supply and the connected equipment. *Section 724.40* limits Class 1 circuits to no more than 1000 volt-amperes at 30 volts or less. Remote-control, signaling, and power-limited circuits

Article 722
Cables for Power-Limited Circuits and Fault-Managed Power Circuits

Part I General
Section 720.1 Scope
Section 720.3 Other Articles
Section 720.10 Hazardous (Classified) Locations
Section 720.12 Uses Not Permitted
Section 720.21 Access to Electrical Equipment Behind Panels Designed to Allow Access
Section 720.24 Mechanical Execution of Work
Section 720.25 Abandoned Cables
Section 720.31 Safety-Control Equipment
Section 720.135 Installation of Cables
Part II Listing Requirements
Section 722.179 Listing and Making of Cables

Goodheart-Willcox Publisher

Figure 10-17. *Article 722* has requirements for cables used in power-limited circuits, fault-managed circuits, and optical fiber cables.

Article 724
Class 1 Power-Limited Circuits and Class 1 Power-Limited Remote-Control and Signaling Circuits

Section 724.1 Scope
Section 724.3 Other Articles
Section 724.21 Access to Electrical Equipment Behind Panels Designed to Allow Access
Section 724.24 Mechanical Execution of Work
Section 724.30 Class 1 Circuit Identification
Section 724.31 Safety-Control Equipment
Section 724.40 Class 1 Circuits
Section 724.43 Class 1 Circuit Overcurrent Protection
Section 724.45 Class 1 Circuit Overcurrent Device Location
Section 724.46 Class 1 Circuit Wiring Methods
Section 724.48 Conductors of Different Circuits in the Same Cable, Cable Tray, Enclosure, or Raceway
Section 724.49 Class 1 Circuit Conductors
Section 724.51 Number of Conductors in Cable Trays and Raceway, and Ampacity Adjustment
Section 724.52 Circuits Extending Beyond One Building

Goodheart-Willcox Publisher

Figure 10-18. *Article 724* has requirements for Class 1 Power-Limited Circuits and Class 1 Power-Limited Remote Control and Signaling Circuits.

will have a power supply that alters the source voltage to the desired voltage of the control circuit. A *remote-control circuit* is defined as any electrical circuit that controls another circuit through a relay or an equivalent device.

Article 725, Class 2 and Class 3 Power-Limited Circuits

Article 725 has requirements for Class 2 and Class 3 remote-control, signaling, and power-limited circuits, **Figure 10-19**. The first step in understanding the type of information in this article is to understand how the *NEC* defines a remote-control circuit. A *remote-control circuit* is defined as any electrical circuit that controls another circuit through a relay or an equivalent device. Remote-control, signaling, and power-limited circuits will have a power supply that alters the source voltage to the desired voltage of the control circuit. The *NEC* provides the following definitions for Class 2 and Class 3 circuits.

- *Class 2 circuit*: The portion of the wiring system between the load side of a Class 2 power source and the connected equipment. Due to its power limitations, a Class 2 circuit considers safety from a fire initiation standpoint and provides acceptable protection from electric shock. Residential doorbell wiring is an example of a Class 2 circuit, **Figure 10-20**.
- *Class 3 circuit*: The portion of the wiring system between the load side of a Class 3 power source and the connected equipment. Due to its power limitations, a Class 3 circuit considers safety from a fire initiation standpoint. Since a Class 3 circuit permits higher levels of voltage than a Class 2 circuit, additional safeguards are specified to provide protection from an electric shock hazard that could be encountered.

Article 725
Class 2 and Class 3 Power-Limited Circuits

Part I General
- Section 725.1 Scope
- Section 725.3 Other Articles
- Section 725.10 Hazardous (Classified) Locations
- Section 725.21 Access to Electrical Equipment Behind Panels Designed to Allow Access
- Section 725.24 Mechanical Execution of Work
- Section 725.30 Class 2 and Class 3 Circuit Identification
- Section 725.31 Safety-Control Equipment

Part II Class 2 and Class 3 Circuits
- Section 725.60 Power Sources for Class 2 and Class 3 Circuits
- Section 725.127 Wiring Methods on Supply Side of the Class 2 or Class 3 Power Source
- Section 725.130 Wiring Methods and Materials on Load Side of the Class 2 or Class 3 Power Source
- Section 725.136 Separation from Electric Light, Power, Class 1, Non-Power-Limited Fire Alarm Circuit Conductors, and Medium-Power Network-Powered Broadband Communications Cables
- Section 725.139 Installation of Conductors of Different Circuits in the Same Cable, Enclosure, Cable Tray, Raceway, or Cable Routing Assembly
- Section 725.144 Bundling of Cables Transmitting Power and Data

Part III Listing Requirements
- Section 725.160 Listening and Marking of Equipment for Power and Data Transmission

Goodheart-Willcox Publisher

Figure 10-19. *Article 725* has requirements for Class 2, and Class 3 power-limited circuits.

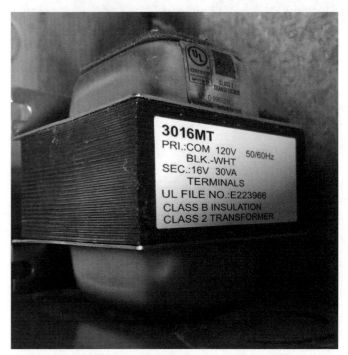

Goodheart-Willcox Publisher

Figure 10-20. A doorbell transformer is typically a Class 2 circuit. The power supply of a power-limited and signaling circuit is what determines the classification of the circuit. *Section 725.60(C)* requires the manufacturer to clearly and durably mark Class 2 and 3 power supplies.

Article 726, Class 4 Fault-Managed Power Systems

Article 726 has requirements for Class 4 fault-managed power systems, **Figure 10-21**. *Article 726* is a new article for the 2023 edition of the *NEC*. ***Fault-managed power systems*** are defined as powering systems that monitor for faults and control current delivered to ensure fault energy is limited. They consist of a Class 4 power transmitter and receiver that are connected by a Class 4 cable. By monitoring the system, the transmitter can limit the amount of current delivered to a fault.

Article 728, Fire-Resistive Cable Systems

Article 728 has requirements for fire-resistive cable systems, **Figure 10-22**. The *NEC* defines a ***fire-resistive cable system*** as a cable and components used to ensure survivability of critical circuits for a specified time under fire conditions. Fire-resistive cables are used to ensure circuit integrity for items such as fire pumps and other equipment that must remain operational for an extended period during a fire.

Article 750, Energy Management Systems

Article 750 has requirements for energy management systems, **Figure 10-23**. The *NEC* defines an ***energy management system*** as a system consisting of any of the following: monitor(s), communications equipment, controller(s), timer(s), or other device(s) that monitors and/or controls an electrical load or a power production of a storage source. Energy management systems are used to shed loads and/or switch to alternate sources of power. Energy management systems can reduce the overall energy the building consumes and allow it to qualify for utility savings programs.

Article 760, Fire Alarm Systems

Article 760 has requirements for fire alarm systems, **Figure 10-24**. *Fire alarm systems* are often found in commercial and industrial installations. They have pull stations,

Article 726
Class 4 Fault-Managed Power Systems

Part I General
- Section 726.1 Scope
- Section 726.3 Other Articles
- Section 726.10 Hazardous (Classified) Locations
- Section 726.12 Uses Not Permitted
- Section 726.24 Mechanical Execution of Work

Part II Class 4 Circuits
- Section 726.121 Power Sources for Class 4 Circuits
- Section 726.122 Class 4 Loads
- Section 726.124 Class 4 Marking
- Section 726.130 Terminals and Connectors
- Section 726.136 Separation from Electric Light, Power, Class 1, Non-Power-Limited Fire Alarm Circuit, and Medium-Power Network-Powered Broadband Communications Cables
- Section 726.139 Installation of Conductors of Different Circuits in the Same Cable, Enclosure, Cable Tray, Raceway, or Cable Routing Assembly
- Section 726.144 Ampacity

Part III Listing Requirements
- Section 726.170 Listing of Equipment for Class 4 Systems

Goodheart-Willcox Publisher

Figure 10-21. *Article 726* requirements for Class 4 fault-managed power systems.

Article 728
Fire-Resistive Cable Systems

- Section 728.1 Scope
- Section 728.3 Other Articles
- Section 728.4 General
- Section 728.5 Installations
- Section 728.60 Equipment Grounding Conductor
- Section 728.120 Marking

Goodheart-Willcox Publisher

Figure 10-22. *Article 728* has requirements for fire-resistive cable systems.

Article 750
Energy Management Systems

- Section 750.1 Scope
- Section 750.6 Listing
- Section 750.20 Alternate Power Sources
- Section 750.30 Load Management
- Section 750.50 Directory

Goodheart-Willcox Publisher

Figure 10-23. *Article 750* has requirements for energy management systems.

Article 760
Fire Alarm Systems

Part I General
Section 760.1 Scope
Section 760.3 Other Articles
Section 760.10 Hazardous (Classified) Locations
Section 760.21 Access to Electrical Equipment Behind Panels Designed to Allow Access
Section 760.24 Mechanical Execution of Work
Section 760.25 Abandoned Cables
Section 760.30 Fire Alarm Circuit Identification
Section 760.32 Fire Alarm Circuits Extending Beyond One Building
Section 760.33 Supply-Side Overvoltage Protection
Section 760.35 Fire Alarm Circuit Requirements

Part II Non-Power-Limited Fire Alarm (NPLFA) Circuits
Section 760.41 NPLFA Circuit Power Source Requirements
Section 760.43 NPLFA Circuit Overcurrent Protection
Section 760.45 NPLFA Circuit Overcurrent Device Location
Section 760.46 NPLFA Circuit Wiring
Section 760.48 Conductors of Different Circuits in Same Cable, Enclosure, or Raceway
Section 760.49 NPLFA Circuit Conductors
Section 760.51 Number of Conductors in Cable Trays and Raceways, and Ampacity Adjustment Factors
Section 760.53 Multiconductor NPLFA Cables

Part III Power-Limited Fire Alarm (PLFA) Circuits
Section 760.121 Power Sources for PLFA Circuits
Section 760.124 Circuit Marking
Section 760.127 Wiring Methods on Supply Side of the PLFA Power Source
Section 760.130 Wiring Methods and Materials on Load Side of the PLFA Power Source
Section 760.133 Installation of Conductors and Equipment in Cables, Compartments, Cable Trays, Enclosures, Manholes, Outlet Boxes, Device Boxes, Raceways, and Cable Routing Assemblies for Power-Limited Fire Alarm Circuits
Section 760.136 Separation from Electric Light, Power, Class 1, NPLFA, and Medium-Power Network-Powered Broadband Communications Circuit Conductors
Section 760.139 Installation of Conductors of Different PLFA Circuits, Class 2, Class 3, and Communications Circuits in the Same Cable, Enclosure, Cable Tray, Raceway, or Cable Routing Assembly
Section 760.142 Conductor Size
Section 760.143 Support of Conductors
Section 760.145 Current-Carrying Continuous Line-Type Fire Detectors
Section 760.154 Applications of Listed PLFA Cables

Part IV Listing Requirements
Section 760.176 Listing and Marking of NPLFA Cables
Section 760.179 Listing and Marking of Insulated Continuous Line-Type Fire Detectors

Goodheart-Willcox Publisher

Figure 10-24. *Article 760* has requirements for fire alarm systems.

smoke and heat detectors, horns, speakers, and strobe lights. Fire alarm systems have a fire alarm control panel that monitors the building, connects to other building systems (such as elevators and HVAC), will signal the alarms in the case of a fire, and can notify the fire department. It is rare to have fire alarm systems installed in dwellings. The smoke, heat, and carbon monoxide alarms found in most homes are self-contained units and are not a fire alarm system, **Figure 10-25**.

The information on fire alarm systems is limited in the *NEC*. Installation method requirements are provided, but nothing about the design or the protection the system must provide. The National Fire Alarm and Signaling Code (NFPA 72) is a standard published by the NFPA that contains rules and requirements specific to fire alarm systems.

Article 770, Optical Fiber Cables

Article 770 has requirements for optical fiber cables, often referred to as fiber optic cables, **Figure 10-26**. Optical fiber cables contain a strand or strands of glass that can carry

Figure 10-25. This fire alarm panel is found in a commercial building. It monitors inputs, such as pull stations or smoke detectors, for indication of a fire and signals outputs, such as a horn or strobe, when necessary. Fire alarm panels are capable of communicating with a monitoring service or municipal authorities in the case of a fire, drastically reducing response time.

pulses of light, **Figure 10-27**. The ends of the cable will have a light source and receiver that create and read pulses of light. Optical fiber cables use the light pulse to transmit communication signals. They can pass information at a much higher rate than traditional copper communication cables. Because glass is nonconductive, it will not be affected by the magnetic fields of nearby conductors. This allows power or communication conductors to be bundled with optical fibers in the same cable. Since cables are available with optical fiber as well as power conductors, they are covered in *Chapter 7* rather than *Chapter 8, Communications Systems.*

Article 770
Optical Fiber Cables

Part I General
| Section 770.1 Scope |
| Section 770.3 Other Articles |
| Section 770.21 Access to Electrical Equipment Behind Panels Designed to Allow Access |
| Section 770.24 Mechanical Execution of Work |
| Section 770.25 Abandoned Cables |
| Section 770.26 Spread of Fire or Products of Combustion |
| Section 770.27 Temperature Limitation of Optical Fiber Cables |

Part II Cables Outside and Entering Buildings
| Section 770.44 Overhead (Aerial) Optical Fiber Cables |
| Section 770.47 Underground Optical Fiber Cables Entering Buildings |
| Section 770.48 Unlisted Cables Entering Buildings |
| Section 770.49 Metal Entrance Conduit Grounding |

Part III Protection
| Section 770.93 Grounding, Bonding, or Interruption of Non-Current-Carrying Metallic Members of Optical Fiber Cables |

Part IV Grounding Methods
| Section 770.100 Entrance Cable Bonding and Grounding |
| Section 770.106 Grounding and Bonding of Entrance Cables at Mobile Homes |

Part V Installation Methods Within Buildings
| Section 770.110 Raceways, Cable Routing Assemblies, and Cable Trays for Optical Fiber Cables |
| Section 770.111 Innerduct for Optical Fiber Cables |
| Section 770.113 Installation of Optical Fiber Cables |
| Section 770.114 Grounding |
| Section 770.133 Installation of Optical Fibers and Electrical Conductors |
| Section 770.154 Applications of Listed Optical Fiber Cables |

Part VI Listing Requirements
| Section 770.179 Optical Fiber Cables |
| Section 770.180 Grounding Devices |

Figure 10-26. *Article 770* has requirements for optical fiber cables.

Figure 10-27. Each strand of glass in an optical fiber cable can transmit a different signal.

NEC KEYWORDS

Chapter 7 Keywords

The following are examples of keywords that should lead you to *Chapter 7*:

- alarm system
- alternate power source
- battery system
- burglar alarm system
- circuit integrity cable
- Class 1 circuit
- Class 2 circuit
- Class 3 circuit
- critical operations power system
- direct current microgrid
- emergency system
- energy management system
- energy storage system
- fire alarm system
- fire detector
- fire-resistive cable
- generator
- ground fault detection equipment
- interactive inverters
- interactive system
- interconnected electric power production sources
- inverter
- island mode
- legally required standby system
- microgrid
- optical fiber cable
- optional standby system
- paralleled source
- power-limited
- remote-control
- signaling circuit
- stand-alone system
- standby system
- uninterruptible power supply (UPS)
- utility interactive inverters

Summary

- *Chapter 7* of the *NEC* is the third and final of the special chapters.
- *Chapter 7* will supplement or modify *Chapters 1–7*.
- *Chapter 7* contains requirements for various types of special conditions.
  - Backup energy and storage
  - Autonomous power production systems
  - Power-limited and control circuits
  - Fiber optics
- The special conditions found in *Chapter 7* have unique characteristics that necessitate special requirements.

Unit 10 Review

Name _____ Date _____ Class _____

Know and Understand

Answer the following questions based on information in this unit.

1. *Chapter 7* of the *NEC* is titled _____.
   A. *Special Conditions*
   B. *Special Installations*
   C. *Special Occupancies*
   D. *Special Equipment*
2. The requirements in *Chapter 7* will supplement or modify the information found in _____.
   A. *Chapter 1*
   B. *Chapters 1–4*
   C. *Chapters 1–7*
   D. *Chapters 1–8*
3. _____ has requirements that apply to legally required standby systems.
   A. *Article 700*
   B. *Article 701*
   C. *Article 702*
   D. *Article 705*
4. *Part IV* of *Article 708* (COPS) covers _____.
   A. power sources and connections
   B. circuit requirements
   C. system performance and analysis
   D. overcurrent protection
5. A _____ circuit is the portion of the system between the load side of the Class 1 power source and the connected equipment.
   A. Class 1
   B. Class 2
   C. Class 3
   D. None of the above.
6. _____ has requirements for optical fiber cables.
   A. *Article 706*
   B. *Article 710*
   C. *Article 760*
   D. *Article 770*
7. Which of the following keywords will lead you to *Chapter 7*?
   A. Fire alarm
   B. Photovoltaic
   C. Communication circuits
   D. Welding cable
8. Which article has information on flow batteries?
   A. *Article 700*
   B. *Article 702*
   C. *Article 706*
   D. *Article 710*
9. Which part of *Article 760* covers power-limited fire alarm (PFLA) circuits?
   A. *Part I*
   B. *Part II*
   C. *Part III*
   D. *Part IV*
10. Residential generators are an example of a(n) _____.
    A. emergency system
    B. optional standby system
    C. legally required standby system
    D. critical operations power system

Apply and Analyze

Answer the following questions using a copy of the National Electrical Code. *Identify the section or subsection where the answer is found.*

1. A fault-managed power system with a Class 4 receiver shall be marked with its maximum _____ and current for each connection point.
   A. volt-ampere
   B. output voltage
   C. input voltage
   D. power

   NEC _____

2. Where critical operations power system feeders are below the level of the _____ year flood plain, the insulated circuit conductors shall be listed for use in a wet location.
   A. 50
   B. 100
   C. 250
   D. 500

   NEC _____

3. Optional standby systems with a power inlet _____ amperes or greater that is not rated as a disconnecting means and is used to connect a portable generator shall be listed for the intended use and shall be equipped with an interlocked disconnecting means.
   A. 30
   B. 50
   C. 60
   D. 100

   NEC _____

4. Class 1 circuits shall not exceed _____ volts.
   A. 30
   B. 120
   C. 300
   D. 600

   NEC _____

5. The power source of a non-power-limited fire alarm (NPFLA) circuit shall not have an output voltage that exceeds _____ volts.
   A. 120
   B. 300
   C. 600
   D. 1,000

   NEC _____

6. Which of the following is a plenum-rated optical fiber cable?
   A. OFNP
   B. OFNG
   C. OFPR
   D. OFCR

   NEC _____

7. A _____ shall be kept of all test and maintenance performed on an emergency system.
   A. documentation file
   B. dated certification
   C. detailed list
   D. written record

   NEC _____

8. Flow battery energy storage systems shall have flow controls that will safely shut down the system in the event of a(n) _____.
   A. increase in flow
   B. electrolyte blockage
   C. increase in temperature
   D. decrease in voltage

   NEC _____

9. Storage batteries that supply standby illumination in a legally required standby system shall be of a suitable rating and capacity to supply and maintain the total load for a minimum period of _____ hours without the voltage applied to the load falling below 87 1/2% of normal.
   A. 1 1/2
   B. 2
   C. 2 1/2
   D. 3

   NEC _____

10. An interconnected electric power production source connected to the service disconnecting means shall have conductors not smaller than _____ AWG copper.
    A. 8
    B. 6
    C. 4
    D. 2

    NEC _____

Name _____ Date _____ Class _____

Critical Thinking

NEC Chapter 7

Answer the following questions using a copy of the National Electrical Code. *Identify the section or subsection where the answer is found.*

1. An energy storage system for a one-family dwelling shall not exceed _____ between conductors or to ground.

   NEC _____

2. A stand-alone system power supply shall be controlled so that the voltage and frequency remain within _____ for the connected loads.

   NEC _____

3. Circuit integrity cable used in a power limited circuit shall be supported at distances not exceeding _____.

   NEC _____

4. Optional standby system wiring _____ to occupy the same raceways, cables, boxes, and cabinets with other general wiring.

   NEC _____

5. An emergency power source shall be of suitable rating and capacity to supply and maintain the total load for not less than _____ of system operation.

   NEC _____

6. Class 2 and Class 3 power-limited power sources shall not have their output circuits connected in _____ unless they are listed for such an interconnection.

   NEC _____

7. Cable supports and fasteners for fire alarm circuit integrity (CI) cable shall be made of _____.

   NEC _____

8. Optical fiber cable shall be installed in a(n) _____.

   NEC _____

9. The legally required standby system wiring shall be permitted to occupy the same _____ with other general wiring.

   NEC _____

10. The _____ shall conduct or witness a test of the complete critical operations power system upon installation and periodically afterward.

    NEC _____

Part 1 Layout and Structure: Learning to Use the NEC

NEC Chapters 1–7

Answer the following questions using a copy of the National Electrical Code. *Identify the section or subsection where the answer is found.*

11. Which three wiring methods are permitted in escalator and moving walk wellways?

    _____

    _____

    _____

    NEC _____

12. Pumps and other fluid handling equipment used in conjunction with a flow battery energy storage system are to be rated/specified suitable for _____.

    _____

    NEC _____

13. Transformers with ventilating openings shall be installed so that the openings are not blocked by _____.

    _____

    NEC _____

14. A habitable room is a room in a building for living, sleeping, eating, or _____, but excluding bathrooms, toilet rooms, closets, hallways, storage or utility spaces, and similar areas.

    _____

    NEC _____

15. When installed in raceways, conductors _____ AWG and larger shall be stranded, unless specifically permitted elsewhere in the code.

    _____

    NEC _____

16. Connectors used with Class 4 circuits shall be designed so they are not interchangeable with _____ located on the same premises.

    _____

    NEC _____

17. What percentages of recreational vehicle park sites with an electrical supply are required to have a 30-ampere, 125-volt receptacle?

    _____

    NEC _____

18. Grounding and bonding connection devices that depend solely on _____ shall not be used.

    _____

    NEC _____

19. Round boxes shall not have conduits that require the use of locknuts connected to the _____.

    _____

    NEC _____

20. Switches shall not be installed within tub or shower spaces unless installed as part of _____.

    _____

    NEC _____

UNIT 11 NEC Chapter 8

sockagphoto/Shutterstock.com

KEY TERMS
communication circuit
network interface unit
network terminal

LEARNING OBJECTIVES

After completing this unit, you will be able to:
- Describe the type of requirements found in *Chapter 8* of the *NEC*.
- Identify the types of requirements found in *Chapter 8* of the *NEC*.
- Recognize keywords that will lead you to *Chapter 8* of the *NEC*.

Introduction

Unit 11 is an introduction to *Chapter 8, Communications Systems* of the *NEC*. The types of systems covered include telephone, television, and broadband communication systems. The end of the unit contains practice exercises that will require the use of your *National Electrical Code* to complete. The combination of studying the type of information found in *Chapter 8* as well as looking up *Code* sections will help with comprehension and retention of the type of communication systems found in *Chapter 8*.

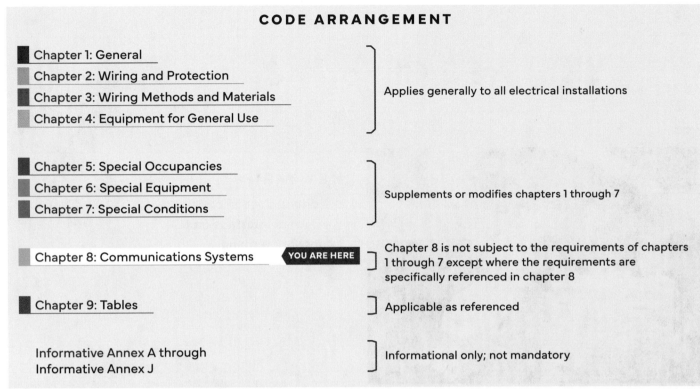

Figure 11-1. *Chapter 8* is a stand-alone chapter. The rest of the *NEC* does not apply unless specifically referenced by a *Chapter 8* article.

Communications Systems

Chapter 8 is a stand-alone chapter, **Figure 11-1**. Communication systems are not subject to the requirements of *Chapters 1–7* unless a specific reference is made in one of the *Chapter 8* articles. The communication requirements found in the *NEC* apply mostly to grounding, cable types and ratings, and restrictions with respect to other wiring methods. The *NEC* looks at communication systems cabling from a safety perspective. To achieve maximum speeds and performance from communication networks, construction specifications often require other voluntary standards to be followed, such as ANSI/TIA and ISO/IEC standards, **Figure 11-2**.

Article 800, General Requirements for Communications Systems

Article 800 contains general requirements for communication systems, **Figure 11-3**. The requirements found in *Article 800* apply to all the *Chapter 8* articles. This article shares requirements that were previously found in multiple articles. These requirements were combined into one article to reduce redundant information and lower the page count of the *NEC*.

Figure 11-2. *Chapter 8* has requirements for communication systems.

Article 805, Communications Circuits

Article 805 has requirements for communication circuits, **Figure 11-4**. The *NEC* defines a **communication circuit** as a metallic, fiber, or wireless circuit that provides voice/

Article 800
General Requirements for Communications Systems

Part I General
- Section 800.1 Scope
- Section 800.2 Reconditioned Equipment
- Section 800.3 Other Articles
- Section 800.21 Access to Electrical Equipment Behind Panels Designed to Allow Access
- Section 800.24 Mechanical Execution of Work
- Section 800.25 Abandoned Cables
- Section 800.26 Spread of Fire or Products of Combustion
- Section 800.27 Temperature Limitation of Wires and Cables

Part II Wires and Cables Outside and Entering Buildings
- Section 800.44 Overhead (Aerial) Wires and Cables
- Section 800.47 Underground Systems Entering Buildings
- Section 800.48 Unlisted Cables Entering Buildings
- Section 800.49 Metal Entrance Conduit Grounding
- Section 800.53 Separation from Lightning Conductors

Part III Grounding Methods
- Section 800.100 Cable and Primary Protector Bonding and Grounding
- Section 800.106 Primary Protector Grounding and Bonding at Mobile Homes

Part IV Installation Methods Within Buildings
- Section 800.110 Raceways, Cable Routing Assemblies, and Cable Trays
- Section 800.113 Installation of Cables Used for Communications Circuits, Communications Wires, Cable Routing Assemblies, and Communication Raceways
- Section 800.133 Installation of Communications Wires and Cables and CATV-Type Coaxial Cables
- Section 800.154 Applications of Listed Communications Wires, Cables, and Raceways, and Listed Cable Routing Assemblies

Part V Listing Requirements
- Section 800.170 Plenum Cable Ties
- Section 800.171 Communications Equipment
- Section 800.179 Wires and Cables
- Section 800.180 Grounding Devices
- Section 800.182 Cable Routing Assemblies and Communications Raceways

Goodheart-Willcox Publisher

Figure 11-3. *Article 800* has general requirements that apply to multiple *Chapter 8* articles.

Article 805
Communications Circuits

Part I General
- Section 805.1 Scope
- Section 805.18 Installation of Equipment

Part II Wires and Cables Outside and Entering Buildings
- Section 805.47 Underground Communications Wires and Cables Entering Buildings - Underground Block Distribution
- Section 805.50 Circuits Requiring Primary Protectors

Part III Protection
- Section 805.90 Protective Devices
- Section 805.93 Grounding, Bonding, or Interruption of Non-Current-Carrying Metallic Sheath Members of Communications Cables

Part IV Installation Methods Within Buildings
- Section 805.154 Substitutions for Listed Communications Cables
- Section 805.156 Dwelling Unit Communications Outlet

Part V Listing Requirements
- Section 805.170 Protectors
- Section 805.173 Drop Wire and Cable

Goodheart-Willcox Publisher

Figure 11-4. *Article 805* has requirements for communications systems.

data (and associated power) for communications-related services between communication equipment. Although the use of telephone lines in homes is no longer common, *Section 805.156* requires at least one communication outlet to be installed in new homes. This provides access to a signal in the event a homeowner wants a telephone line, or if a dial-up service is necessary for a security or fire alarm system. The *NEC* is not specific about where the outlet is located, only that it must be accessible and connected to the service provider demarcation point. See **Figure 11-5**.

Article 810, Antenna Systems

Article 810 has requirements for antenna systems, **Figure 11-6**. This article applies primarily to the equipment that provides radio and television signals. Examples of the type of equipment covered in *Article 810* includes television antennas, satellite dishes, and HAM radio antennas. See **Figure 11-7**.

Part 1   Layout and Structure: Learning to Use the NEC

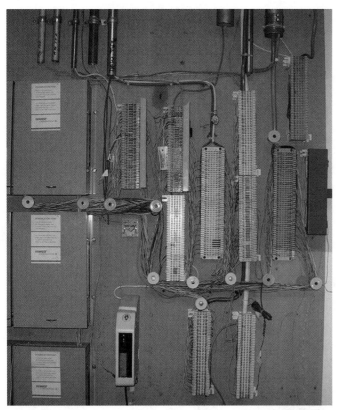

Goodheart-Willcox Publisher

Figure 11-5. Pictured are terminal blocks where conductors leave the demarcation point and connect to the building's communication circuits.

Article 810
Antenna Systems

Part I General
Section 810.1 Scope
Section 810.3 Other Articles
Section 810.4 Community Television Antenna
Section 810.5 Radio Noise Suppression
Section 810.6 Antenna Lead-In Protectors
Section 810.7 Grounding Devices

Part II Receiving Equipment - Antenna Systems
Section 810.11 Material
Section 810.12 Supports
Section 810.13 Avoidance of Contacts with Conductors of Other Systems
Section 810.14 Splices
Section 810.15 Grounding or Bonding
Section 810.16 Size of Wire-Strung Antenna - Receiving Station
Section 810.17 Size of Lead-in - Receiving Station
Section 810.18 Clearances - Receiving Stations
Section 810.19 Electrical Supply Circuits Used in Lieu of Antenna - Receiving Stations
Section 810.20 Antenna Discharge Units - Receiving Stations
Section 810.21 Bonding Conductors and Grounding Electrode Conductors - Receiving Stations

Part III Amateur and Citizen Band Transmitting and Receiving Stations - Antenna Systems
Section 810.51 Other Sections
Section 810.52 Size of Conductors
Section 810.53 Size of Lead-in Conductors
Section 810.54 Clearance on Building
Section 810.55 Entrance to Building
Section 810.56 Protection Against Accidental Contact
Section 810.57 Antenna Discharge Units - Transmitting Stations
Section 810.58 Bonding Conductors and Grounding Electrode Conductors - Amateur and Citizen Band Transmitting and Receiving Stations

Part IV Interior Installation - Transmitting Stations
Section 810.70 Separation from Other Conductors
Section 810.71 General

Goodheart-Willcox Publisher

Figure 11-6. *Article 810* has requirements that apply to radio and television equipment.

yoshi0511/Shutterstock.com

Figure 11-7. Residential antennas and satellite dishes are examples of the types of communications equipment that are covered by *Article 810*.

188

Copyright Goodheart-Willcox Co., Inc.
May not be reproduced or posted to a publicly accessible website.

Article 820, Community Antenna Television and Radio Distribution Systems

Article 820 has requirements for community antenna television and radio distribution systems, **Figure 11-8**. The primary focus of *Article 820* is the installation of coaxial cables for television signals and security cameras. The requirements in *Article 820* refer to proper grounding of the cable shield, clearances and restrictions as applied to other wiring methods, and permitted environments of the cables based on their listings, **Figure 11-9**.

Article 830, Network-Powered Broadband Communications Systems

Article 830 has requirements for network-powered broadband systems, **Figure 11-10**. In a network-powered broadband system, the network interface unit receives its power as well as the signal from the supplying utility. A *network interface unit* is defined as a device that converts a broadband signal into component voice, audio, video, data, and interactive services signals and provides isolation between

The Toidi/Shutterstock.com

Figure 11-9. Coaxial cables installed in dwellings are covered by *Article 820*.

the network power and the premises signal circuits. These devices often contain primary and secondary protectors that offer some protection against overvoltages due to lightning or accidental contact with another electrical source, such as a power line. Most residential landline telephone systems are an example of a network-powered system. All the power needed to ring, dial, and carry the voice signal comes from the telephone company. In the event of a neighborhood power outage, the telephones will still work because the telephone provider maintains power at all its sites. Although most people have switched from landline phones to cell phones, some keep their landline for emergencies, **Figure 11-11**.

Article 840, Premises-Powered Broadband Communications Systems

Article 840 has requirements for premises-powered broadband communication systems, **Figure 11-12**. In a premises-powered broadband system, the network terminal receives the broadband signal from the supplying utility, but its power supply comes from the premises. The *NEC* defines a **network terminal** as a device that converts network-provided signals (optical, electrical, or wireless) into component signals, including voice, audio, video, data, wireless, optical, and interactive services, and is considered a network device on the premises that is connected to a communications service

---

Article 820
Community Antenna Television and Radio Distribution Systems

Part I General
| Section 820.1 Scope |
| Section 820.3 Other Articles |
| Section 820.15 Power Limitations |

Part III Protection
| Section 820.93 Grounding of the Outer Conductive Shield of Coaxial Cables |

Part IV Grounding Methods
| Section 820.100 Cable Bonding and Grounding |
| Section 820.103 Equipment Grounding |

Part V Installation Methods Within Buildings
| Section 820.154 Substitutions of Listed CATV Cables |

Goodheart-Willcox Publisher

Figure 11-8. *Article 820* has requirements that apply to cables installed for items like televisions and security cameras.

provider and is powered at the premises. Signals that are received from the serving utility over a fiber optic cable is an example of a premises-powered broadband communication system. Building power is used to power the equipment that transmits and receives the signals passed through the fiber-optic cable. See **Figure 11-13**.

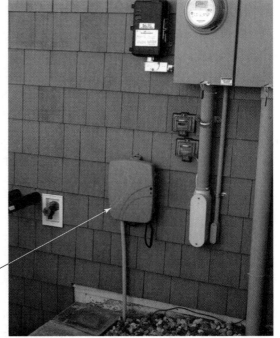

Goodheart-Willcox Publisher

Figure 11-11. The network interface is typically installed outside for dwellings.

Article 830
Network-Powered Broadband Communications Systems

Part I General
- Section 830.1 Scope
- Section 830.15 Power Limitations

Part II Cables Outside and Entering Buildings
- Section 830.40 Entrance Cables
- Section 830.44 Overhead (Aerial) Cables
- Section 830.47 Underground Network-Powered Broadband Communications Cables Entering Buildings

Part III Protection
- Section 830.90 Primary Electrical Protection

Part IV Grounding Methods
- Section 830.93 Grounding or Interruption of Metallic Members of Network-Powered Broadband Communications Cables

Part V Installation Methods Within Buildings
- Section 830.133 Installation of Network-Powered Broadband Communications Cables and Equipment
- Section 830.154 Substitutions of Network-Powered Broadband Communications System Cables
- Section 830.160 Bends

Part VI Listing Requirements
- Section 830.179 Network-Powered Broadband Communications Equipment and Cables

Goodheart-Willcox Publisher

Figure 11-10. *Article 830* has requirements for network-powered broadband communications systems.

Article 840
Premises-Powered Broadband Communications Systems

Part I General
- Section 840.1 Scope

Part II Cables Outside and Entering Buildings
- Section 840.47 Underground Wires and Cables Entering Buildings

Part III Protection
- Section 840.90 Protective Devices
- Section 840.93 Grounding or Interruption
- Section 840.94 Premises Circuits Leaving the Building

Part IV Grounding Methods
- Section 840.101 Premises Circuits Not Leaving the Building
- Section 840.102 Premises Circuits Leaving the Building

Part VI Premises Powering of Communications Equipment over Communications Cables
- Section 840.160 Powering Circuits

Part VII Listing Requirements
- Section 840.170 Equipment and Cables

Goodheart-Willcox Publisher

Figure 11-12. *Article 840* has requirements for premises-powered broadband communications equipment.

Unit 11 NEC Chapter 8

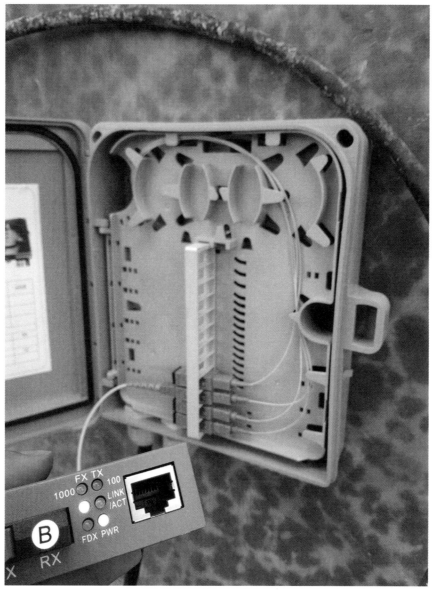

Mohamad Rifki Insani/Shutterstock.com

Figure 11-13. Fiber optic cables are an example of a premises-powered broadband communications system.

NEC KEYWORDS

Chapter 8 Keywords

The following are examples of keywords that should lead you to *Chapter 8*:

- antenna
- cable (coaxial)
- CATV
- circuit protector
- communications
- community antenna television and radio distribution
- grounding (communications)
- grounding electrode conductor (radio and TV equipment)
- network interface
- network-powered broadband communications system
- network terminal
- optical network terminal
- plenum (coaxial and communication cable ratings)
- poles (communication)
- premises (communication and CATV circuits)
- premises-powered broadband communication system
- radio equipment
- suppressor (radio noise)
- surge arresters (radio and tv equipment)
- telecommunications
- telephone
- television
- transmitting
- TV
- underground communication circuit

Summary

- *Chapter 8* is a stand-alone chapter that is not subject to the requirements of *Chapters 1–7* unless they are specifically referenced in a *Chapter 8* article.
- *Chapter 8* has requirements for various communication systems, such as:
  - Communication circuits
  - Radio and television equipment
  - Community antenna television and radio distribution systems
  - Network-powered broadband communications systems
  - Premises-powered broadband communications systems

Unit 11 Review

Name _____ Date _____ Class _____

Know and Understand

Answer the following questions based on information in this unit.

1. Chapter 8 of the *NEC* is titled _____.
   A. *Special Conditions*
   B. *Special Installations*
   C. *Special Occupancies*
   D. *Communications Systems*

2. Which of the following is true of requirements found in *Chapter 8* of the *NEC*?
   A. They will supplement or modify the requirements of *Chapters 1–7*.
   B. They are for informational purposes only.
   C. They are stand-alone and are not subject to the requirements of *Chapters 1–7* unless specifically referenced from within *Chapter 8*.
   D. All of the above.

3. _____ has general requirements for communication systems.
   A. *Article 800*   C. *Article 810*
   B. *Article 805*   D. *Article 820*

4. *Part III* of *Article 805, Communications Circuits* covers _____.
   A. listing requirements
   B. installation methods
   C. protection
   D. overcurrent protection

5. A _____ is a metallic, fiber, or wireless circuit that provides voice/data (and associated power) for communications-related services between communications equipment.
   A. service drop
   B. communication circuit
   C. network interface
   D. demarcation circuit

6. _____ has requirements for antenna systems.
   A. *Article 800*   C. *Article 810*
   B. *Article 805*   D. *Article 830*

7. Which of the following keywords will lead you to *Chapter 8*?
   A. Radio and television equipment
   B. Television studio
   C. Churches
   D. Cell

8. _____ contains requirements that apply to multiple *Chapter 8* Articles.
   A. *Article 800*   C. *Article 820*
   B. *Article 810*   D. *Article 830*

9. Which part of *Article 810* covers receiving equipment for antenna systems?
   A. *Part I*   C. *Part III*
   B. *Part II*   D. *Part IV*

10. A _____-powered broadband system receives its power from the utility company.
    A. premises   C. distribution
    B. utility   D. network

Apply and Analyze

Answer the following questions using a copy of the National Electrical Code. *Identify the section or subsection where the answer is found.*

1. The accessible portion of abandoned communication cables shall be _____.
   A. labeled   C. removed
   B. tagged   D. grounded

   NEC _____

2. Fused primary protectors installed on communication circuits shall be located in, on, or immediately adjacent to the structure or building served and as close as practicable to the _____.
   A. point of entrance
   B. communication closet
   C. network interface
   D. patch panel

   NEC _____

3. Communication system wires, cables, cable routing assemblies, and communications raceways installed in buildings shall be _____.
   A. labeled   C. secured every 6′
   B. orange in color   D. listed

   NEC _____

4. Broadband is defined as wide bandwidth data transmission that transports multiple signals, protocols, and _____ over various media types.
   A. traffic types
   B. currents
   C. data
   D. frequencies

   NEC _____

5. Direct buried coaxial cable entering buildings shall be separated at least _____ from conductors of any light or power, non-power-limited fire alarm circuit conductors, or Class 1 circuits.
   A. 4″
   B. 6″
   C. 12″
   D. 24″

   NEC _____

6. The power limitation for a network-powered broadband communication system is _____ volt-amperes.
   A. 100
   B. 150
   C. 250
   D. 1000

   NEC _____

7. Outdoor television antennas and lead in conductors shall not be attached to the _____.
   A. side of a dwelling
   B. electric service mast
   C. roof sheathing
   D. roof overhang

   NEC _____

8. A communication circuit cable with a CMX rating is permitted to be installed in which of the following?
   A. Risers
   B. Dwellings
   C. Plenums
   D. Places of assembly

   NEC _____

9. Interior television transmitters shall have interlocks on access doors that disconnect all voltages of over _____ volts between conductors when any access door is opened.
   A. 50
   B. 150
   C. 200
   D. 350

   NEC _____

10. Coaxial cable shall be permitted to deliver power to equipment that is directly associated with the radio frequency distribution system if the voltage is not over _____ volts and if the current is supplied by a transformer or other device that has power-limiting characteristics.
    A. 24
    B. 30
    C. 60
    D. 150

    NEC _____

Name _____ Date _____ Class _____

Critical Thinking

NEC Chapter 8

Answer the following questions using a copy of the National Electrical Code. *Identify the section or subsection where the answer is found.*

1. Coaxial cable installed within a building shall be separated from conductors of any electric light, power, Class 1, non-power-limited fire alarm, or medium-power network-powered broadband communication circuits by at least _____.

   _____

   NEC _____

2. Where practicable, a separation of at least _____ shall be maintained between lightning conductors and all communication cables entering a building.

   _____

   NEC _____

3. Coaxial cables shall not be _____ to the exterior of any conduit or raceway as a means of support.

   _____
   _____

   NEC _____

4. The general requirements found in *Article 800* states that a communication system bonding or grounding electrode conductor shall not be smaller than _____ and shall not be required to be larger than _____.

   _____
   _____

   NEC _____

5. A network-powered broadband communication cable installed in a nonmetallic raceway under a residential driveway shall be buried at least _____ deep.

   _____

   NEC _____

6. Premises-powered broadband communication circuits _____ required to be grounded.

   _____

   NEC _____

7. Dwelling units shall have a minimum of _____ communication outlet(s) in a readily accessible area and cabled to the service provider demarcation point.

   _____

   NEC _____

8. Plenum-rated cables up to _____ long are permitted to be installed in ducts specifically fabricated for environmental air.

   _____

   NEC _____

9. Radio noise suppressors shall be of the listed type and not exposed to _____.

   _____

   NEC _____

10. Overhead (aerial) network-powered broadband communication cables shall be at least _____ above a residential driveway.

    _____

    NEC _____

Part 1   Layout and Structure: Learning to Use the NEC

NEC Chapters 1–8

Answer the following questions using a copy of the National Electrical Code. *Identify the section or subsection where the answer is found.*

11. List the four occupancies required to have surge-protective devices installed on the service.

    _____

    _____

    _____

    _____

    NEC _____

12. What is the difference between a Type 3 and Type 3S enclosure?

    _____

    _____

    NEC _____

13. Luminaires installed under roof decking shall be installed so that there is at least _____ from the lowest surface of the roof decking to the top of the luminaire.

    _____

    NEC _____

14. The accessible portion of communication system cables shall be removed unless _____.

    _____

    _____

    NEC _____

15. Flywheel energy storage systems (FESS) shall not be used in _____.

    _____

    NEC _____

16. Vertical metal wireways shall be securely supported at intervals not exceeding _____ and shall not have more than one joint between supports.

    _____

    NEC _____

17. Each _____ from an outdoor antenna shall be provided with a listed antenna discharge unit.

    _____

    NEC _____

18. Electrically powered pool lifts shall be bonded using a solid copper conductor that is not smaller than _____.

    _____

    NEC _____

19. Nonprofessional motion picture projectors are permitted to be operated without a _____.

    _____

    NEC _____

20. Raceways shall be installed complete between outlet, junction, or splicing points prior to _____.

    _____

    NEC _____

UNIT 12

NEC Chapter 9 and Informative Annexes

sockagphoto/Shutterstock.com

KEY TERMS

circular mil
compact conductor
impedance
informative annex
knockout
mil
nipple
one-shot bender
reactance
stranded conductor
torque

LEARNING OBJECTIVES

After completing this unit, you will be able to:
- Recognize the tables found in *Chapter 9* of the *NEC*.
- Explain when *Chapter 9* tables apply.
- Describe the type of information found in the *Informative Annexes*.

Introduction

Unit 12 is an introduction to tables and informative annexes of the *NEC*, **Figure 12-1**. The *Chapter 9* tables relating to circular raceways and conductors will be covered in detail in Unit 14. The **informative annexes** have additional information to help with understanding and application but are not an enforceable part of the *NEC*. This unit will help you become familiar with the type of information found in both the *Chapter 9* tables as well as the *Informative Annexes*. The end of the unit has practice problems where you will look up information to help with understanding and comprehension.

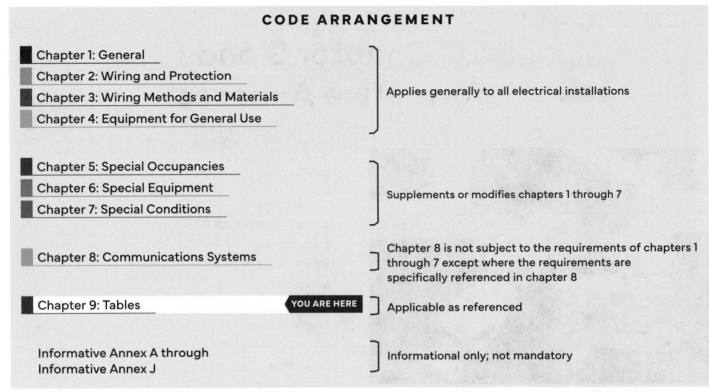

Figure 12-1. *Chapter 9* has tables that are applicable as referenced by other parts of the *NEC*. The *Informative Annexes* are for informational purposes only and are not a mandatory portion of the *Code*.

Chapter 9, Tables

Chapter 9 has several tables that provide detailed additional information on circular raceways, conductors, and power limited sources. According to *Section 90.3*, these tables are only applicable when referenced by another part of the *Code*. Instead of going right to the table to find information, the *NEC* intends for you to first go to the applicable wiring method, which will send you back to *Chapter 9* if it is appropriate. For example, if you are looking to determine the maximum number of conductors in electrical metallic tubing, you would look in *Section 358.22, Number of Conductors*. Its language reads: "The number of conductors shall not exceed that permitted by the percentage fill specified in Table 1, Chapter 9." (*NEC* Section 348.22)

The reason for going to the wiring method first is that there may be important information that must be understood before applying any of the *Chapter 9* tables. This helps prevent misapplication and ensure all the requirements have been considered.

Table 1—Percent of Cross Section of Conduit and Tubing for Conductors and Cables

Table 1 gives the maximum percentage of cross-sectional area of a conduit or tubing that is permitted to be filled by conductors, **Figure 12-2**. The table has three categories: one

Table 1 Percent of Cross Section of Conduit and Tubing for Conductors and Cables

Number of Conductors and/or Cables	Cross-Sectional Area (%)
1	53
2	31
Over 2	40

Reproduced with permission of NFPA from NFPA 70, National Electrical Code, 2023 edition. Copyright © 2022, National Fire Protection Association. For a full copy of the NFPA 70, please go to www.nfpa.org

Figure 12-2. *Table 1* allows the user to calculate the cross-sectional area of conduit that is allowed to be filled by conductors.

conductor, two conductors, and more than two conductors. The amount of space that conductors are permitted to occupy in a conduit varies by the number of conductors being installed. This is to prevent damage to conductors during installation due to excessive pulling force and/or binding while going around bends.

Notes to Tables

The *Notes to Tables* include information on how to properly apply the conduit fill tables. There are endless scenarios where conductors and cables are installed in a conduit or tubing, so it is impossible to create a single table that applies to all situations. The notes give guidance on how to address some of these various scenarios, **Figure 12-3**.

Table 2—Radius of Conduit and Tubing Bends

Table 2 has a minimum bending radius for the various sizes of conduit and tubing used in the electrical industry. See **Figure 12-4**. It has one column for one-shot and full shoe benders and another column for other bends. A *one-shot bender* makes the complete bend in one step rather in multiple steps, **Figure 12-5**. Other bends refers to bends made by other means, such as heating PVC so it can be formed into the desired bend or using a hickey bender to bend rigid metal conduit, **Figure 12-6**.

The *NEC* provides minimum bending radius requirements to prevent raceways from collapsing or kinking while making a bend, as well as to prevent making a bend that is so tight that it causes excess stress on the conductors during installation. The commercially available benders are designed and manufactured to ensure that the bender meets the minimum *NEC* requirements.

Goodheart-Willcox Publisher

Figure 12-3. *Chapter 9* defines a conduit nipple as a conduit or tubing that is 24″ or less in length. Due to its short length, the *NEC* allows a nipple to be filled to 60% of its cross-sectional area, and ampacity derating based on the number of conductors is not required.

Table 2 Radius of Conduit and Tubing Bends

Conduit or Tubing Size		One Shot and Full Shoe Benders		Other Bends	
Metric Designator	Trade Size	mm	in.	mm	in.
16	½	101.6	4	101.6	4
21	¾	114.3	4½	127	5
27	1	146.05	5¾	152.4	6
35	1¼	184.15	7¼	203.2	8

Reproduced with permission of NFPA from NFPA 70, National Electrical Code, 2023 edition. Copyright © 2022, National Fire Protection Association. For a full copy of the NFPA 70, please go to www.nfpa.org

Figure 12-4. *Table 2* contains radii of conduit and tubing bends.

Goodheart-Willcox Publisher

Figure 12-5. Hand benders that are used to bend EMT are an example of a one-shot bender that is identified in *Table 2*. It makes a complete bend in one movement (shot).

Goodheart-Willcox Publisher

Figure 12-6. In order to bend rigid PVC, it must be heated in a PVC heater such as this one. The minimum bending radius when heating PVC would fall under the other bends column of *Table 2*.

Table 4—Dimensions and Percent Area of Conduit and Tubing

There are many types of raceways used in the electrical industry to protect and house the conductors. They all have different wall thicknesses, internal diameters, and attributes. Although the various types of raceways are made to fit into standard trade-size knockouts, each has a slightly different internal area, **Figure 12-7**. *Knockouts* are openings in boxes and enclosures for the installation cables and raceways.

A 1/2″ electrical metallic tubing has an internal cross-sectional area of 0.304 in^2, while 1/2″ rigid metallic conduit has an internal cross-sectional area of 0.314 in^2. Due to their different sizes, they have varying capacities for conductors. See **Figure 12-8**. Regardless of the type of raceway, its connector will allow it to be installed in a standard size knockout. For example, all 1/2″ trade sizes raceways will be accepted by a 1/2″ trade size knockout.

Table 5—Dimensions of Insulated Conductors and Fixture Wires

Just as we have different raceways to handle all the various applications, there are also different types of insulated

Goodheart-Willcox Publisher

Figure 12-7. All the raceways in this figure are 1/2″ trade size. Although they will all connect to a 1/2″ knockout, the internal diameter of each raceway is slightly different.

Table 4 Dimensions and Percent Area of Conduit and Tubing (Areas of Conduit or Tubing for the Combinations of Wires Permitted in Table 1, Chapter 9)

		Article 358 — Electrical Metallic Tubing (EMT)											
Metric Designator	Trade Size	Over 2 Wires 40%		60%		1 Wire 53%		2 Wires 31%		Nominal Internal Diameter		Total Area 100%	
		mm^2	in.^{2}	mm^2	in.^{2}	mm^2	in.^{2}	mm^2	in.^{2}	mm	in.	mm^2	in.^{2}
16	½	78	0.122	118	0.182	104	0.161	61	0.094	15.8	0.622	196	0.304
21	¾	137	0.213	206	0.320	182	0.283	106	0.165	20.9	0.824	343	0.533
27	1	222	0.346	333	0.519	295	0.458	172	0.268	26.6	1.049	556	0.864
35	1¼	387	0.598	581	0.897	513	0.793	300	0.464	35.1	1.380	968	1.496

Reproduced with permission of NFPA from NFPA 70, National Electrical Code, 2023 edition. Copyright © 2022, National Fire Protection Association. For a full copy of the NFPA 70, please go to www.nfpa.org

Figure 12-8. *Table 4* lists the total diameter and volume for raceways as well as the amount of available space for conductors.

Figure 12-9. Conductors come in a variety of different sizes and types of insulation. Both the size of the conductor as well as the type of insulation will factor into the number of conductors that will fit in a raceway.

allows for a conductor with a smaller overall diameter while maintaining the same circular mil area and current carrying capacity, **Figure 12-11**.

Table 8—Conductor Properties

Table 8 has detailed information on conductors. It considers only the conductive portion of a conductor and does not include any insulation the conductor may have into the values provided. It lists the size of the conductors in AWG as well as the equivalent circular mil, stranded vs solid, the area and diameter of uninsulated conductors, and the direct-current resistance of the various sizes of conductors covered by the *NEC*. See **Figure 12-12**.

conductors. The varying types of insulation have different thicknesses, causing the amount of space the conductor takes up to differ, **Figure 12-9**.

Table 5 has information on the diameter and area of the commonly used types of conductor insulation, **Figure 12-10**. This information can be used in conjunction with *Table 4* to determine how many conductors can safely be installed in a raceway.

Table 5A—Compact Copper and Aluminum Building Wire Nominal Dimensions and Areas

Table 5A has similar information to *Table 5* but for compact conductors. A traditional **stranded conductor** has round strands that leave small air spaces between the strands. A **compact conductor** has strands that are shaped in a manner that eliminates the air spaces. Eliminating the air spaces

Figure 12-11. The conductor on the left is a copper conductor with circular strands. The conductor on the right is an aluminum compact conductor. Compact conductors have the strands formed in a way to fit together and eliminate air gaps.

Table 5 Dimensions of Insulated Conductors and Fixture Wires

Type	Size (AWG or kcmil)	Approximate Area		Approximate Diameter	
		mm²	in.²	mm	in.
Type: FFH-2, RFH-1, RFH-2, RFHH-2, RHH*, RHW*, RHW-2*, RHH, RHW, RHW-2, SF-1, SF-2, SFF-1, SFF-2, TF, TFF, THHW, THW, THW-2, TW, XF, XFF					
RFH-2, FFH-2, RFHH-2	18	9.355	0.0145	3.454	0.136
	16	11.10	0.0172	3.759	0.148
RHH, RHW, RHW-2	14	18.90	0.0293	4.902	0.193
	12	22.77	0.0353	5.385	0.212
	10	28.19	0.0437	5.994	0.236
	8	53.87	0.0835	8.280	0.326
	6	67.16	0.1041	9.246	0.364

Reproduced with permission of NFPA from NFPA 70, National Electrical Code, 2023 edition. Copyright © 2022, National Fire Protection Association. For a full copy of the NFPA 70, please go to www.nfpa.org

Figure 12-10. *Table 5* lists the amount of space a conductor will take up in a raceway. The approximate area of each conductor is based on its size as well as the type of insulation.

Table 8 Conductor Properties

Size (AWG or kcmil)	Area		Conductors						Direct-Current Resistance at 75°C (167°F)					
			Stranding			Overall			Copper				Aluminum	
				Diameter		Diameter		Area	Uncoated		Coated			
	mm²	Circular mils	Quantity	mm	in.	mm	in.	mm²	in.²	ohm/km	ohm/kFT	ohm/km	ohm/kFT	ohm/km	ohm/kFT
18	0.823	1620	1	—	—	1.02	0.040	0.823	0.001	25.5	7.77	26.5	8.08	42.0	12.8
18	0.823	1620	7	0.39	0.015	1.16	0.046	1.06	0.002	26.1	7.95	27.7	8.45	42.8	13.1
16	1.31	2580	1	—	—	1.29	0.051	1.31	0.002	16.0	4.89	16.7	5.08	26.4	8.05
16	1.31	2580	7	0.49	0.019	1.46	0.058	1.68	0.003	16.4	4.99	17.3	5.29	26.9	8.21
14	2.08	4110	1	—	—	1.63	0.064	2.08	0.003	10.1	3.07	10.4	3.19	16.6	5.06
14	2.08	4110	7	0.62	0.024	1.85	0.073	2.68	0.004	10.3	3.14	10.7	3.26	16.9	5.17
12	3.31	6530	1	—	—	2.05	0.081	3.31	0.005	6.34	1.93	6.57	2.01	10.45	3.18
12	3.31	6530	7	0.78	0.030	2.32	0.092	4.25	0.006	6.50	1.98	6.73	2.05	10.69	3.25

Reproduced with permission of NFPA from NFPA 70, National Electrical Code, 2023 edition. Copyright © 2022, National Fire Protection Association. For a full copy of the NFPA 70, please go to www.nfpa.org

Figure 12-12. *Table 8* lists the properties of commonly used conductors. These values are for the bare conductor alone and do not including any insulation.

PRO TIP — Circular Mils

A *circular mil* is the unit of area of a circle with a diameter of one mil, **Figure 12-13**. One *mil* is a unit of measurement that is equal to 1/1000 of an inch. This measurement is used to find the area of circles. The *NEC* uses both the American Wire Gauge (AWG) system as well as circular mils for sizing conductors. Conductors 4/0 and smaller use AWG with conductors larger than 4/0 being expressed in kcmils (one thousand circular mils). For example, a 250 kcmil conductor has an area of 250,000 circular mils.

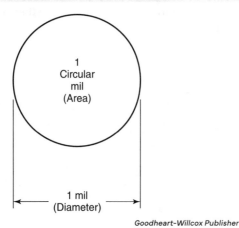

Goodheart-Willcox Publisher

Figure 12-13. A circular mil is a circle with the area of one mil. Larger conductors that are used in the electrical industry are identified by their size in circular mils.

Table 9—Alternating-Current Resistance and Reactance for 600-Volt Cables, 3-Phase, 60 Hz, 75°C (167°F)—Three Single Conductors in Conduit

Table 9 has information on the inductive reactance, ac resistance, and the overall impedance of conductors when passing alternating-current (ac) through them. AC circuits will have reactance in addition to resistance. **Reactance** is the nonresistive opposition to current flow that is present in ac circuits. It may be capacitive, inductive, or a combination of the two. **Impedance** is the total opposition to current flow in an ac circuit. Impedance is the combination of reactance and resistance. Since the type of raceway housing the conductors has an effect on the reactance and impedance, the table has information for PVC, aluminum, and steel conduits, **Figure 12-14**.

Table 10—Conductor Stranding

Table 10 has information on the stranding of conductors. It lists the number of strands based on the class of conductor. Conductors with greater numbers of strands allow for less transmission of vibration, which can loosen connections on equipment that moves or vibrates, and greater bendability, making them easier to install and terminate, **Figure 12-15**.

Table 9 Alternating-Current Resistance and Reactance for 600-Volt Cables, 3-Phase, 60Hz, 75°C (167°F)—Three Single Conductors in Conduit

	\multicolumn{13}{c}{Ohms to Neutral per Kilometer / Ohms to Neutral per 1000 Feet}	
	X_L (Reactance) for All Wires		Alternating-Current Resistance for Uncoated Copper Wires			Alternating-Current Resistance for Aluminum Wires			Effective Z at 0.85 PF for Uncoated Copper Wires			Effective Z at 0.85 PF for Aluminum Wires		
Size (AWG or kcmil)	PVC, Aluminum Conduits	Steel Conduit	PVC Conduit	Aluminum Conduit	Steel Conduit	PVC Conduit	Aluminum Conduit	Steel Conduit	PVC Conduit	Aluminum Conduit	Steel Conduit	PVC Conduit	Aluminum Conduit	Steel Conduit	Size (AWG or kcmil)
14	0.190 / 0.058	0.240 / 0.073	10.2 / 3.1	10.2 / 3.1	10.2 / 3.1	—	—	—	8.9 / 2.7	8.9 / 2.7	8.9 / 2.7	—	—	—	14
12	0.177 / 0.054	0.223 / 0.068	6.6 / 2.0	6.6 / 2.0	6.6 / 2.0	10.5 / 3.2	10.5 / 3.2	10.5 / 3.2	5.6 / 1.7	5.6 / 1.7	5.6 / 1.7	9.2 / 2.8	9.2 / 2.8	9.2 / 2.8	12
10	0.164 / 0.050	0.207 / 0.063	3.9 / 1.2	3.9 / 1.2	3.9 / 1.2	6.6 / 2.0	6.6 / 2.0	6.6 / 2.0	3.6 / 1.1	3.6 / 1.1	3.6 / 1.1	5.9 / 1.8	5.9 / 1.8	5.9 / 1.8	10
8	0.171 / 0.052	0.213 / 0.065	2.56 / 0.78	2.56 / 0.78	2.56 / 0.78	4.3 / 1.3	4.3 / 1.3	4.3 / 1.3	2.26 / 0.69	2.26 / 0.69	2.30 / 0.70	3.6 / 1.1	3.6 / 1.1	3.6 / 1.1	8

Reproduced with permission of NFPA from NFPA 70, National Electrical Code, 2023 edition. Copyright © 2022, National Fire Protection Association. For a full copy of the NFPA 70, please go to www.nfpa.org

Figure 12-14. *Table 9* lists inductive reactance, ac resistance, and the overall impedance of conductors.

Table 10 Conductor Stranding

Conductor Size		Number of Strands		
		Copper		Aluminum
AWG or kcmil	mm²	Class B[a]	Class C	Class B[a]
24–30	0.20–0.05	b	—	—
22	0.32	7	—	—
20	0.52	10	—	—
18	0.82	16	—	—
16	1.3	26	—	—
14–2	2.1–33.6	7	19	7[c]
1–4/0	42.4–107	19	37	19

Reproduced with permission of NFPA from NFPA 70, National Electrical Code, 2023 edition. Copyright © 2022, National Fire Protection Association. For a full copy of the NFPA 70, please go to www.nfpa.org

Figure 12-15. *Table 10* lists the number of strands in the commonly used classes of stranded conductors.

Table 11(A)—Class 2 and Class 3 Alternating-Current Power Source Limitations and Table 11(B)—Class 2 and Class 3 Direct-Current Power Source Limitations

Tables 11(A) and *11(B)* detail the limitations of power limited power supplies. *Table 11(A)* covers ac power limited supplies, and *Table 11(B)* covers dc power limited supplies. They contain information such as maximum voltages, currents, and volt/amp ratings for Class 2 and Class 3 power sources and identify when overcurrent protection is required, **Figure 12-16.**

Table 12(A)—PLFA Alternating-Current Power Source Limitations and Table 12(B)—PLFA Direct-Current Power Source Limitations

Tables 12(A) and *12(B)* contain similar information to that found in *Tables 11(A)* and *(B)*, except it pertains to power limited fire alarm power supplies. See **Figure 12-17.**

Table 13—Equipment Suitable for Hazardous (Classified) Locations

Tables 13 is new for the 2023 *NEC*. It lists the various classifications of hazardous locations and details the types of

Table 11(A) Class 2 and Class 3 Alternating-Current Power Source Limitations

Power Source		Inherently Limited Power Source (Overcurrent Protection Not Required)			Not Inherently Limited Power Source (Overcurrent Protection Required)			
		Class 2		Class 3	Class 2		Class 3	
Source voltage V_{max} (volts) (see Note 1)		0 through 20*	Over 20 and through 30*	Over 30 and through 150	Over 30 and through 100	0 through 20*	Over 20 and through 30*	Over 30 and through 100	Over 100 and through 150
Power limitations VA_{max} (volt-amperes) (see Note 1)		—	—	—	—	250 (see Note 3)	250	250	N.A.
Current limitations I_{max} (amperes) (see Note 1)		8.0	8.0	0.005	$150/V_{max}$	$1000/V_{max}$	$1000/V_{max}$	$1000/V_{max}$	1.0
Maximum overcurrent protection (amperes)		—	—	—	—	5.0	$100/V_{max}$	$100/V_{max}$	1.0
Power source maximum nameplate rating	VA (volt-amperes)	$5.0 \times V_{max}$	100	$0.005 \times V_{max}$	100	$5.0 \times V_{max}$	100	100	100
	Current (amperes)	5.0	$100/V_{max}$	0.005	$100/V_{max}$	5.0	$100/V_{max}$	$100/V_{max}$	$100/V_{max}$

Note: Notes for this table can be found following Table 11 (B).
*Voltage ranges shown are for sinusoidal ac in indoor locations or where wet contact is not likely to occur.
For nonsinusoidal or wet contact conditions, see Note 2.

Reproduced with permission of NFPA from NFPA 70, National Electrical Code, 2023 edition. Copyright © 2022, National Fire Protection Association. For a full copy of the NFPA 70, please go to www.nfpa.org

Figure 12-16. *Table 11(A)* has information on ac power limited power sources. Values for dc power limited power sources can be found in *Table 11(B)*.

Table 12(A) PLFA Alternating-Current Power Source Limitations

Power Source		Inherently Limited Power Source (Overcurrent Protection Not Required)			Not Inherently Limited Power Source (Overcurrent Protection Required)		
Circuit voltage V_{max} (volts) (see Note 1)		0 through 20	Over 20 and through 30	Over 30 and through 100	0 through 20	Over 20 and through 100	Over 100 and through 150
Power limitations VA_{max} (volt-amperes) (see Note 1)		—	—	—	250 (see Note 2)	250	N.A.
Current limitations I_{max} (amperes) (see Note 1)		8.0	8.0	$150/V_{max}$	$1000/V_{max}$	$1000/V_{max}$	1.0
Maximum overcurrent protection (amperes)		—	—	—	5.0	$100/V_{max}$	1.0
Power source maximum nameplate ratings	VA (volt-amperes)	$5.0 \times V_{max}$	100	100	$5.0 \times V_{max}$	100	100
	Current (amperes)	5.0	$100/V_{max}$	$100/V_{max}$	5.0	$100/V_{max}$	$100/V_{max}$

Note: Notes for this table can be found following Table 12(B).

Reproduced with permission of NFPA from NFPA 70, National Electrical Code, 2023 edition. Copyright © 2022, National Fire Protection Association. For a full copy of the NFPA 70, please go to www.nfpa.org

Figure 12-17. *Table 12(A)* has information on ac power limited fire alarm power sources. Values for dc power limited fire alarm power sources can be found in *Table 12(B)*.

equipment suitable to be installed within these locations. The table contains zone and group classifications as well as class and division classifications. See **Figure 12-18**.

Informative Annexes

The *Informative Annexes* are for informational purposes and are not an enforceable portion of the *National Electrical Code*. They are a resource that can help the *Code* user understand *NEC* requirements, save time performing calculations, and acquire additional information.

Informative Annex A— Product Safety Standards

Informative Annex A has a listing of UL standards that are cross-referenced to their related *NEC* articles. The *NEC* makes note that this information changes frequently and that it is only accurate as of when the *NEC* was published, **Figure 12-19**.

Table 13 Equipment Suitable for Hazardous (Classified) Locations

Area Classification		Type (Level) of Protection
Zone 0	Intrinsically safe	Intrinsically safe for Class I, Division 1
	Intrinsic safety (Group II)	ia
	Encapsulation (Group II)	ma
	Flameproof (Group II)	da[1]
	Inherently safe optical radiation	op is, with EPL Ga[2]
	Optical system with interlock	op sh, with EPL Ga[2]
	Special protection (Group II)	sa
	EPL[3]	Ga
Zone 1	Equipment suitable for use in Zone 0	
	Equipment suitable for use in Class I, Division 1	
	Flameproof (Group II)	d, db
	Intrinsic safety (Group II)	ib
	Increased safety (Group II)	e, eb
	Pressurized enclosure (Group II)	p, px, pxb, py, pyb
	Encapsulation (Group II)	m, mb
	Pressurized room (Group II)	pb
	Powder filling (Group II)	q, qb
	Liquid immersion (Group II)	o, ob
	Electrical resistance trace heating	60079-30-1, with EPL Gb[2]
	Skin effect trace heating	IEEE 844.1, with EPL Gb[2]

Reproduced with permission of NFPA from NFPA 70, National Electrical Code, 2023 edition. Copyright © 2022, National Fire Protection Association. For a full copy of the NFPA 70, please go to www.nfpa.org

Figure 12-18. *Table 13* has information on the equipment permitted to be installed in hazardous locations.

Informative Annex A—Product Safety Standards

Informative Annex A is not a part of the requirements of this NFPA document but is included for informational purposes only.

This informative annex provides a list of product safety standards used for product listing where that listing is required by this *Code*. It is recognized that this list is current at the time of publication but that new standards or modifications to existing standards can occur at any time while this edition of the *Code* is in effect.

This informative annex does not form a mandatory part of the requirements of this *Code* but is intended to identify for the *Code* users the standards upon which *Code* requirements have been based.

Table A.1(a) Product Safety Standards for Conductors and Equipment That Have an Associated Listing Requirement

Article	Standard Number	Standard Title
110	UL 10C	Positive Pressure Fire Tests of Door Assemblies
	UL 305	Panic Hardware
	UL 486D	Sealed Wire Connector Systems
	UL 2043	Fire Test for Heat and Visible Smoke Release for Discrete Products and Their Accessories Installed in Air-Handling Spaces
	UL 62275	Cable Management Systems — Cable Ties for Electrical Installation
210	UL 498	Attachment Plugs and Receptacles
	UL 935	Fluorescent-Lamp Ballasts
	UL 943	Ground Fault Circuit Interrupters
	UL 1029	High-Intensity-Discharge Lamp Ballast
	UL 1699	Arc-Fault Circuit-Interrupters
	UL 1699A	Outlet Branch Circuit AFCIs
225	UL 6	Electrical Rigid Metal Conduit — Steel
	UL 6A	Electrical Rigid Metal Conduit — Aluminum, Red Brass and Stainless Steel
	UL 360	Liquid-Tight Flexible Metal Conduit

Reproduced with permission of NFPA from NFPA 70, National Electrical Code, 2023 edition. Copyright © 2022, National Fire Protection Association. For a full copy of the NFPA 70, please go to www.nfpa.org

Figure 12-19. *Informative Annex A* is a cross-reference from the *NEC* to *UL Standards*.

Informative Annex B—Application Information for Ampacity Calculation

Informative Annex B has additional ampacity information beyond what is found in *Article 310*. It has a few tables that address conductors installed in free air, but most of it is focused on underground conductors either directly buried or in electrical ducts. The information is meant to be applied with engineering supervision. See **Figure 12-20**.

Informative Annex C—Conduit, Tubing, and Cable Tray Fill Tables for Conductors and Fixture Wires of the Same Size

Informative Annex C contains tables that list the maximum number of conductors permitted in each size and type of raceway. These tables only apply when the conductors are all the same size and have the same type of insulation. *Informative Annex C* tables save a little time performing calculations in situations where all the conductors are the same size, **Figure 12-21**.

Informative Annex D—Examples

Informative Annex D contains several sample electrical service and feeder calculations. The calculations that are included show step by step how the calculation is to be performed and references the *NEC* Section where the requirements are found. See **Figure 12-22**.

Informative Annex E—Types of Construction

Informative Annex E describes some of the most common construction types and their required fire rating. It also has a table with fire resistance ratings for walls and columns for various types of construction. The information is taken from the *NFPA 5000*, the *Building and Construction Safety Code*. This information is helpful in determining which wiring method is permitted based on the type of construction, **Figure 12-23**.

Detail 1
290 mm × 290mm
(11.5 in. × 11.5 in.)
Electrical duct bank
One electrical duct

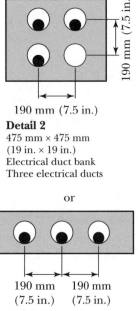

Detail 2
475 mm × 475 mm
(19 in. × 19 in.)
Electrical duct bank
Three electrical ducts

or

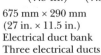

675 mm × 290 mm
(27 in. × 11.5 in.)
Electrical duct bank
Three electrical ducts

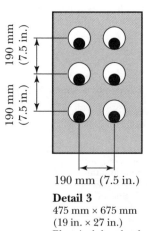

Detail 3
475 mm × 675 mm
(19 in. × 27 in.)
Electrical duct bank
Six electrical ducts

or

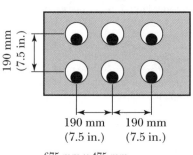

675 mm × 475 mm
(27 in. × 19 in.)
Electrical duct bank

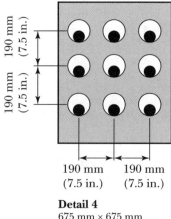

Detail 4
675 mm × 675 mm
(27 in. × 27 in.)
Electrical duct bank
Nine electrical ducts

Reproduced with permission of NFPA from NFPA 70, National Electrical Code, 2023 edition. Copyright © 2022, National Fire Protection Association. For a full copy of the NFPA 70, please go to www.nfpa.org

Figure 12-20. *Informative Annex B* has additional ampacity calculations that can be done under engineering supervision. Some of the calculations address underground cables in duct banks. *Informative Annex B* has several figures to help explain the scenarios.

Table C.1 Maximum Number of Conductors or Fixture Wires in Electrical Metallic Tubing (EMT) (Based on Chapter 9: Table 1, Table 4, and Table 5)

Type	Conductor Size (AWG/kcmil)	Trade Size (Metric Designator)											
		⅜ (12)	½ (16)	¾ (21)	1 (27)	1¼ (35)	1½ (41)	2 (53)	2½ (63)	3 (78)	3½ (91)	4 (103)	5 (129)	6 (155)
		CONDUCTORS											
RHH, RHW, RHW-2	14	—	4	7	11	20	27	46	80	120	157	201	302	427
	12	—	3	6	9	17	23	38	66	100	131	167	251	354
	10	—	2	5	8	13	18	30	53	81	105	135	203	286
	8	—	1	2	4	7	9	16	28	42	55	70	106	150
	6	—	1	1	3	5	8	13	22	34	44	56	85	120
	4	—	1	1	2	4	6	10	17	26	34	44	66	94
	3	—	1	1	1	4	5	9	15	23	30	38	58	82
	2	—	1	1	1	3	4	7	13	20	26	33	50	71

Reproduced with permission of NFPA from NFPA 70, National Electrical Code, 2023 edition. Copyright © 2022, National Fire Protection Association. For a full copy of the NFPA 70, please go to www.nfpa.org

Figure 12-21. *Informative Annex Table C* is a quick reference guide to the number of conductors that will fit in a raceway. For these tables to apply, all of the conductors must be the same size and have the same type of insulation.

Example D1 (a) One-Family Dwelling

The dwelling has a floor area of 1500 ft², exclusive of an unfinished cellar not adaptable for future use, unfinished attic, and open porches. Appliances are a 12-kW range and a 5.5-kW, 240-V dryer. Assume range and dryer kW ratings equivalent to kVA ratings in accordance with 220.54 and 220.55.

Calculated Load *(see 220.40)*

General Lighting Load 1500 ft² at 3 VA/ft² = 4500 VA

Minimum Number of Branch Circuits Required *[see 210.11(A)]*

General Lighting Load: 4500 VA ÷ 120 V = 38 A

This requires three 15-A, 2-wire or two 20-A, 2-wire circuits.

Small-Appliance Load: Two 2-wire, 20-A circuits *[see 210.11(C)(1)]*

Laundry Load: One 2-wire, 20-A circuit *[see 210.11(C)(2)]*

Bathroom Branch Circuit: One 2-wire, 20-A circuit (no additional load calculation is required for this circuit) *[see 210.11(C)(3)]*

Reproduced with permission of NFPA from NFPA 70, National Electrical Code, 2023 edition. Copyright © 2022, National Fire Protection Association. For a full copy of the NFPA 70, please go to www.nfpa.org

Figure 12-22. *Informative Annex D* has a few sample service calculations. The sample calculations are intended to show the step-by-step process of performing a service calculation.

Informative Annex F— Availability and Reliability for Critical Operations Power Systems; and Development and Implementation of Functional Performance Tests (FPTs) for Critical Operations Power Systems

Informative Annex F has information on reliability and testing of critical operations power systems. This is important information to have when designing a system and creating job specifications.

Informative Annex G— Supervisory Control and Data Acquisition (SCADA)

Supervisory control and data acquisition (SCADA) systems are used for monitory and control processes associated with a building. They are not typically installed on dwellings and small commercial buildings. Instead, they are found on large commercial and industrial jobs. *Informative Annex G* has suggestions that can help ensure proper operation that can be written into the job specifications and implemented.

Table E.1 Fire Resistance Ratings for Type I Through Type V Construction (hr)

	Type I		Type II			Type III		Type IV	Type V	
	442	332	222	111	000	211	200	2HH	111	000
Exterior Bearing Walls[a]										
Supporting more than one floor, columns, or other bearing walls	4	3	2	1	0[b]	2	2	2	1	0[b]
Supporting one floor only	4	3	2	1	0[b]	2	2	2	1	0[b]
Supporting a roof only	4	3	1	1	0[b]	2	2	2	1	0[b]
Interior Bearing Walls										
Supporting more than one floor, columns, or other bearing walls	4	3	2	1	0	1	0	2	1	0
Supporting one floor only	3	2	2	1	0	1	0	1	1	0
Supporting roofs only	3	2	1	1	0	1	0	1	1	0

Reproduced with permission of NFPA from NFPA 70, National Electrical Code, 2023 edition. Copyright © 2022, National Fire Protection Association. For a full copy of the NFPA 70, please go to www.nfpa.org

Figure 12-23. *Informative Annex E* has details on types of construction.

Informative Annex H— Administration and Enforcement

Informative Annex H is a template that can be used to set up and administer an electrical inspection program. It is written in such a way that it can be adopted as is, or the enforcing entity can pick and choose the portions of it they would like to use. It is structured and numbered to be *Article 80*, but since it is for informational purposes only and not an enforceable part of the *NEC*, it is in the Annexes, **Figure 12-24**.

Informative Annex I— Recommended Tightening Torque Tables from UL Standard 486A-486B

When tightening screw type connections, it is important to tighten them to the proper torque, **Figure 12-25**. *Torque* is the rotational force being applied while tightening a fastener or termination. A lot of equipment will have the torque specifications written on the equipment or available in the product specification sheets. In the case that the torque specifications are not available, *Informative Annex I* provides recommended torque specifications, **Figure 12-26**.

Informative Annex J—ADA Standards for Accessible Design

Informative Annex J contains information taken from Americans with Disabilities Act (ADA) standards that may be incorporated into the design of the electrical system. An example is minimum and maximum height of receptacles to be reachable from a person in a wheelchair. It contains several figures to help with understanding the requirements. See **Figure 12-27**.

Informative Annex H is not a part of the requirements of this NFPA document and is included for informational purposes only. Informative Annex H is intended to provide a template and sample language for local jurisdictions adopting the National Electrical Code®.

80.1 Scope. The following functions are covered:

(1) The inspection of electrical installations as covered by 90.2
(2) The investigation of fires caused by electrical installations
(3) The review of construction plans, drawings, and specifications for electrical systems
(4) The design, alteration, modification, construction, maintenance, and testing of electrical systems and equipment
(5) The regulation and control of electrical installations at special events including but not limited to exhibits, trade shows, amusement parks, and other similar special occupancies

80.2 Definitions.
Authority Having Jurisdiction. The organization, office, or individual responsible for enforcing the requirements of a code or standard, or for approving equipment, materials, an installation, or a procedure.

Reproduced with permission of NFPA from NFPA 70, National Electrical Code, 2023 edition. Copyright © 2022, National Fire Protection Association. For a full copy of the NFPA 70, please go to www.nfpa.org

Figure 12-24. Pictured is the scope of *Informative Annex H*. Although it is for informational purposes only, it is written in way where it could be adopted and implement.

Informative Annex K—Use of Medical Electrical Equipment in Dwellings and Residential Board-and-Care Occupancies

Informative Annex K contains general information about medical equipment that is connected to the electrical systems of a dwelling. It is projected that the use of this equipment will increase in the near future. It is important to consider the circuitry and power supplies to help prevent circuits from tripping so that power is maintained to the medical equipment.

Denis Faraktinov/Shutterstock.com

Figure 12-25. A torque wrench is used to ensure that lugs are tightened to the appropriate setting.

Table I.1 Tightening Torque for Screws

Test Conductor Installed in Connector		Tightening Torque, N-m (lbf-in.)							
		Slotted head No. 10 and larger[a]							
AWG or kcmil	mm^2	Slot width 1.2 mm (0.047 in.) or less and slot length 6.4 mm (¼ in.) or less		Slot width over 1.2 mm (0.047 in.) or slot length over 6.4 mm (¼ in.)		Split-bolt connectors		Other connectors	
30–10	0.05–5.3	2.3	(20)	4.0	(35)	9.0	(80)	8.5	(75)
8	8.4	2.8	(25)	4.5	(40)	9.0	(80)	8.5	(75)
6–4	13.2–21.2	4.0	(35)	5.1	(45)	18.5	(165)	12.4	(110)
3	26.7	4.0	(35)	5.6	(50)	31.1	(275)	16.9	(150)
2	33.6	4.5	(40)	5.6	(50)	31.1	(275)	16.9	(150)
1	42.4	—		5.6	(50)	31.1	(275)	16.9	(150)
1/0–2/0	53.5–67.4	—		5.6	(50)	43.5	(385)	20.3	(180)

Reproduced with permission of NFPA from NFPA 70, National Electrical Code, 2023 edition. Copyright © 2022, National Fire Protection Association. For a full copy of the NFPA 70, please go to www.nfpa.org

Figure 12-26. *Informative Annex I* has torque tables for instances where manufacturer torque settings are not available.

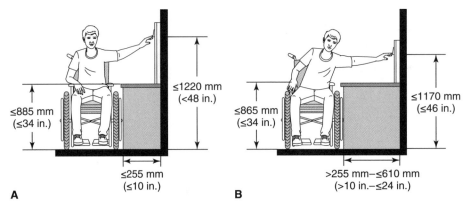

Reproduced with permission of NFPA from NFPA 70, National Electrical Code, 2023 edition. Copyright © 2022, National Fire Protection Association. For a full copy of the NFPA 70, please go to www.nfpa.org

Figure 12-27. *Informative Annex J* lists some of the applicable ADA standards that affect the electrical installation. There are several figures to clarify the standards.

Summary

- The *Chapter 9* tables are applicable as referenced by other parts of the *NEC*.
- The *Informative Annexes* are for informational purposes only and are not a part of the requirements of the *NEC*.
- The *Informative Annexes* have information to aid in understanding and application of the *National Electrical Code*.

Unit 12 Review

Name _____ Date _____ Class _____

Know and Understand

Answer the following questions based on information in this unit.

1. Chapter 9 of the *NEC* is titled _____.
   A. *General*
   B. *Tables*
   C. *Special Equipment*
   D. *Communications Systems*

2. The requirements found in *Chapter 9* _____.
   A. will supplement or modify the requirements of *Chapters 1–8*
   B. are for informational purposes only
   C. stand alone and are not subject to the requirements of *Chapters 1–7* unless specifically referenced from within *Chapter 9*.
   D. are applicable when referenced by another part of the *NEC*

3. Conductor properties are provided by _____ of *Chapter 9*.
   A. *Table 2*   C. *Table 5*
   B. *Table 4*   D. *Table 8*

4. The _____ are for informational purposes only and are not part of the requirements of the *NEC*.
   A. informative annexes   C. appendices
   B. *Chapter 9* tables   D. exceptions

5. _____ has conduit fill tables for conductors of the same size.
   A. *Informative Annex A*   C. *Informative Annex C*
   B. *Informative Annex B*   D. *Informative Annex D*

6. *Table 5A* of *Chapter 9* contains information on _____.
   A. insulated conductors
   B. insulated compact conductors
   C. bare conductors
   D. conduit fill

7. *Informative Annex A* has information on _____.
   A. accessible design
   B. administration
   C. product safety standards
   D. raceway fill

8. *Chapter 9* _____ has information on the minimum radius of conduit and tubing bends.
   A. *Table 1*   C. *Table 4*
   B. *Table 2*   D. *Table 5*

9. "Types of construction" is found in _____.
   A. *Informative Annex D*   C. *Informative Annex F*
   B. *Informative Annex E*   D. *Informative Annex G*

10. Recommended tightening torque tables are listed in _____.
    A. *Informative Annex B*   C. *Informative Annex H*
    B. *Informative Annex F*   D. *Informative Annex I*

Apply and Analyze

Answer the following questions using a copy of the National Electrical Code. *Identify the section or subsection where the answer is found.*

1. The minimum bending radius of a 3/4″ conduit using a one-shot bender is _____.
   A. 4 1/2″   C. 5 1/2″
   B. 5″   D. 6″

   NEC _____

2. A 1″ electrical metallic tubing has a nominal internal diameter of _____.
   A. 1″   C. 1.072″
   B. 1.049″   D. 1.102″

   NEC _____

3. What is the available area of a 2″ rigid metal conduit using the "Over 2 wires (40%)" column?
   A. 1.302 in^2 C. 1.342 in^2
   B. 1.307 in^2 D. 1.363 in^2

   NEC _____

4. A 14 AWG THHN conductor has an approximate area of _____.
   A. 0.0097 in^2 C. 0.0139 in^2
   B. 0.0133 in^2 D. 0.0181 in^2

   NEC _____

5. A 4/0 AWG XHHW compact conductor has an approximate area of _____.
   A. 0.2733 in^2 C. 0.3267 in^2
   B. 0.2780 in^2 D. 0.3237 in^2

   NEC _____

6. A 2 AWG Class C copper stranded conductor contains _____ strands.
   A. 7
   B. 19
   C. 37
   D. 61

   NEC _____

7. According to *Informative Annex C*, the maximum number of 12 AWG THWN conductors permitted in a 3/4 electrical metallic tubing is _____.
   A. 13
   B. 16
   C. 18
   D. 20

   NEC _____

8. _____ of the raceway articles references *Chapter 9* to determine the maximum number of conductors permitted in the raceway.
   A. *Section 3XX.22*
   B. *Section 3XX.24*
   C. *Section 3XX.26*
   D. *Section 3XX.28*

   NEC _____

9. Nipples are permitted to be filled to _____ of the total area.
   A. 31%
   B. 40%
   C. 53%
   D. 60%

   NEC _____

10. When installing one wire in a 1 1/4″ schedule 40 rigid PVC conduit, the available area is _____.
    A. 0.770 in^2
    B. 0.793 in^2
    C. 0.832 in^2
    D. 0.844 in^2

    NEC _____

Name _____ Date _____ Class _____

Critical Thinking

NEC Chapter 9 and Informative Annexes

Answer the following questions using a copy of the National Electrical Code. *Identify the section or subsection where the answer is found.*

1. Which table in *Informative Annex C* has conduit fill information for rigid metal conduit with compact conductors?

   _____

   NEC _____

2. Which product safety standard, found in *Informative Annex A*, contains listing requirements that apply to *Article 250* (Grounding and Bonding Equipment)?

   _____

   NEC _____

3. A 1″ electrical metallic tubing (EMT) has a total area of _____ in².

   _____

   NEC _____

4. The minimum bending radius for a 2″ conduit using a full shoe bender is _____″.

   _____

   NEC _____

5. Type II construction has three categories: fire-resistive, one-hour rated, and _____.

   _____

   NEC _____

6. According to ADA standards, the maximum height for a switch for a person in a wheelchair with an unobstructed side reach is _____ inches.

   _____

   NEC _____

7. A 12 AWG THHN conductor has an approximate area of _____ square inches.

   _____

   NEC _____

8. A 500 kcmil Class C copper conductor has _____ strands.

   _____

   NEC _____

9. A 24-volt dc power limited fire alarm (PLFA) source that operates at 24 volts and is inherently limited has a maximum current limitation of _____ amperes.

   _____

   NEC _____

10. When installing three conductors in a 1 1/4″ liquidtight flexible metal conduit, the available area is _____.

    _____

    NEC _____

Part 1   Layout and Structure: Learning to Use the NEC

NEC Chapter 9 and Informative Annexes

Answer the following questions using a copy of the National Electrical Code. *Identify the section or subsection where the answer is found.*

11. What size electrical metallic tubing is required for five conductors that have a combined area of 2.32 in²?

    _____

    NEC _____

12. An 8 AWG stranded conductor that is uninsulated has an overall area of _____ in².

    _____

    NEC _____

13. A 1/0 AWG aluminum compact THHW conductor has an overall area of _____ in².

    _____

    NEC _____

14. What is the area (in square inches) that nine 10 AWG THHN conductors will take up in a raceway?

    _____

    NEC _____

15. What is the approximate area (in square inches) of a 350 kcmil THHN conductor?

    _____

    NEC _____

16. What is the available space (in square inches) for 6 conductors in a 3/4″ electrical nonmetallic tubing?

    _____

    NEC _____

17. What is the direct-current resistance of a stranded uncoated 12 AWG copper conductor?

    _____

    NEC _____

18. A 1/2″ flexible metal conduit that will contain 4 conductors has a useable area of _____ in².

    _____

    NEC _____

19. According to *Informative Annex C*, how many 500 kcmil RHH conductors will fit in a 3″ intermediate metal conduit (IMC)?

    _____

    NEC _____

20. What is the maximum useable area of 20″ long 2″ schedule 40 PVC conduit that is installed between two panelboards?

    _____

    NEC _____

PART 2 | Application: Advanced NEC Topics

Sashkin/Shutterstock.com

Unit 13 Branch Circuits
Unit 14 Box Fill and Pull Boxes
Unit 15 Raceway Fill
Unit 16 Conductor Ampacity and Circuit Sizing
Unit 17 Feeders and Taps
Unit 18 Services
Unit 19 Grounding
Unit 20 Range Calculations
Unit 21 Dwelling Service Calculations

UNIT

13 Branch Circuits

Sashkin/Shutterstock.com

KEY TERMS

appliance branch circuit
arc-fault circuit interrupter (AFCI)
bathrooms
branch circuit
continuous load
grade
ground-fault circuit interrupter (GFCI)
habitable room
individual branch circuit
multiwire branch circuit

LEARNING OBJECTIVES

After completing this chapter, you will be able to:

- Define the different types of branch circuits recognized by the *NEC*.
- Identify areas where receptacles are required to have GFCI protection.
- Identify dwelling branch circuits that require AFCI protection.
- Describe the difference between continuous and noncontinuous loads.
- List the four required dwelling unit branch circuits.
- Locate dwelling unit branch circuit requirements in the *NEC*.
- Locate non-dwelling unit branch circuit requirements in the *NEC*.

Introduction

This unit covers some of the commonly used branch circuit requirements found in *Article 210*. *Article 210* of the *NEC* falls under the design classification. It is full of requirements necessary when designing the electrical system. It starts off with general branch circuit requirements that apply to all installations, and then it moves into specific requirements. Although the article is general and applies to all installations, a good portion of the article is focused on dwelling unit branch circuit, receptacle, and lighting requirements.

The intention of this unit is to get you comfortable navigating *Article 210*, introduce you to residential branch circuit requirements, and help you to practice finding requirements that relate to branch circuits.

Branch Circuits

Branch circuits are the circuit conductors between the final overcurrent device protecting the circuit and the outlet(s). The branch circuit includes the conductors from the breaker to the very last outlet in the circuit, **Figure 13-1**. The *National Electrical Code* recognizes several types of branch circuits:

- Appliance branch circuit
- General-purpose branch circuit
- Multiwire branch circuit

An *appliance branch circuit* is a branch circuit that supplies energy to one or more outlets to which appliances are to be connected and has no permanently connected luminaires that are not a part of an appliance, **Figure 13-2**.

A *general-purpose branch circuit* is a branch circuit that supplies two or more receptacles or outlets for lighting and appliances. Most of the branch circuits in dwellings are general-purpose branch circuits, **Figure 13-3**.

Goodheart-Willcox Publisher

Figure 13-2. A branch circuit that feeds equipment such as a refrigerator, dishwasher, or trash compactor is an appliance branch circuit.

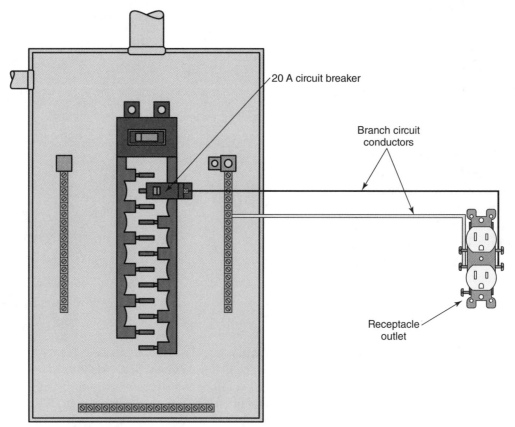

Goodheart-Willcox Publisher

Figure 13-1. The branch circuit conductors begin at the overcurrent device and end at the outlet. They can feed a single piece of equipment as is done with an individual branch circuit, or they could feed multiple outlets as is done with a general-purpose branch circuit.

Krista Abel/Shutterstock.com

Figure 13-3. Many of the branch circuits in dwellings are general-purpose branch circuits that feed two or more receptacles or lighting outlets. This bedroom has one general-purpose branch circuit that feeds the receptacles, recessed lights, paddle fan, and the patio outlets.

An *individual branch circuit* is a branch circuit that supplies only one piece of utilization equipment, **Figure 13-4**. There are many situations where we elect to install an individual branch circuit to eliminate the possibility of another load tripping the circuit, such as with refrigerators and freezers. In some cases, the *NEC* requires an individual branch circuit. *Section 422.12*, for example, requires an individual branch circuit for central heating equipment. The equipment manufacturer may also require use of an individual branch circuit.

A *multiwire branch circuit* is a branch circuit that consists of two or more ungrounded conductors that have a voltage between them, and a neutral conductor that has equal voltage between it and each ungrounded conductor of the circuit that is connected to the neutral or grounded conductor of the system, **Figure 13-5**. Multiwire branch circuits aren't as popular for general circuitry as they have been in the past. One of the main reasons for their decline in use within dwellings was the requirement for AFCI breakers. Initially, AFCI breakers would not work with multiwire branch circuits, so when AFCI protection became a requirement, multiwire branch circuits were no longer feasible. Manufacturers have since created double-pole AFCI breakers that will work with a multiwire branch circuit, but they are typically only used when a multiwire branch circuit is encountered while remodeling an existing installation.

Two types of home appliances that have multiwire branch circuits are electric ranges and dryers. They are 120/240-volt appliances that have a neutral connection. These appliances generally use the 240-volt portion of the

Goodheart-Willcox Publisher

Figure 13-4. This trash compactor is fed with an individual branch circuit. This is common practice for appliances that have larger current draws. Some manufacturers will require an individual branch circuit in the specifications or instructions.

Multiwire Branch Circuit

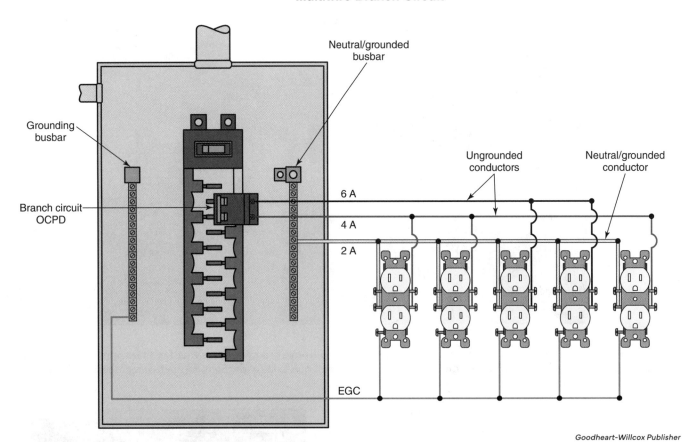

Figure 13-5. Multiwire branch circuits have two ungrounded conductors that share one grounded conductor. In a single-phase multiwire branch circuit, the grounded/neutral conductor carries the imbalance between the two ungrounded conductors. The black wire has 6 amperes, the red wire has 4 amperes, and the white wire (neutral) carries the difference of 2 amperes.

circuit for high-current items, such as heating elements, and the 120-volt portion of the circuit for lights, control circuits, and sometimes motors.

Ground-fault circuit interrupters (GFCI), **Figure 13-6**, are devices intended for the protection of personnel, that de-energize a circuit or portion thereof within an established period of time when a ground-fault current exceeds the values established for a Class A device. *Section 210.8* has general information on where GFCI protection is required. Requirements that apply to specific pieces of equipment are located within their respective areas of the *NEC*. *Article 680, Swimming Pools, Fountains, and Similar Installations* has numerous requirements for ground-fault circuit interrupters in the vicinity of items such as swimming pools and hot tubs, **Figure 13-7**.

Arc-fault circuit interrupters (AFCI), **Figure 13-8**, are devices intended to provide protection from the effects of arc faults by recognizing characteristics unique to arcing and de-energizing the circuit when an arc fault is detected. *Section 210.12* lists the areas that require AFCI protection. These requirements apply to dwellings, dorms, hotel rooms, and sleeping areas in nursing homes and limited-care facilities. If people use it for living and sleeping, it is likely to require arc-fault protection.

Ratings

The rating of a branch circuit is determined by the overcurrent device. If you have a 20-ampere branch circuit, there is a 20-ampere circuit breaker or fuse protecting the circuit from excess current, **Figure 13-9**.

In certain situations, a smaller overcurrent device than is required by the *NEC* is installed for the respective size conductors. An example is a long branch circuit. A 20-ampere branch circuit feeding receptacles would generally have 12 AWG conductors. If the receptacles are a long distance from the panelboard, it may be necessary to upsize the conductors to 10 AWG or larger, to reduce the amount of voltage drop. The 10 AWG branch circuit conductors, capable of carrying 30 amperes, would be terminated on a 20-ampere breaker, making it a 20-ampere circuit.

Another example of a situation where an undersized overcurrent device would be used is an appliance that lists a maximum overcurrent device size. This is common with central air conditioners, especially when they are upgraded. For example, a central air conditioning unit is being replaced, and the old unit was fed with 8 AWG conductors and protected by a 40-ampere breaker. The new unit

Lost_in_the_Midwest/Shutterstock.com

Figure 13-6. GFCI protection can be provided by a receptacle or a circuit breaker. A test button can be pressed to test the internal circuitry of the device. Newer GFCIs will run a self-test occasionally; if they don't pass the test they will shut off and not turn back on.

is more efficient and lists a maximum overcurrent device size of 30 amperes, **Figure 13-10**. The *NEC* permits us to use the old conductors as they are capable of safely carrying more current than the air conditioner will draw. The circuit breaker, however, must be changed from a 40-ampere unit to a 30-ampere unit to satisfy the manufacturer's maximum overcurrent device requirement.

PRO TIP — Voltage Drop

Voltage drop does not appear in the *NEC* as an enforceable requirement. It is, however, mentioned as an *Informational Note*. *Section 210.19* and *Section 215.2(A)(2)* have *Informational Notes* that suggest a maximum amount of voltage drop for the circuit/feeder to have *reasonable efficiency of operation*. It recommends a maximum voltage drop of 3% for a branch circuit and 3% for a feeder, with a combined voltage drop of 5%. Most appliances and motors will have a nominal voltage that they are designed for. If the voltage drop of the branch circuit or feeder causes the voltage at the equipment to drop below its listed voltage, it may not operate properly, efficiently, or at all. See **Figure 13-11**.

Kristina Postnikova/Shutterstock.com

Figure 13-7. There are many *Code* requirements for swimming pools and hot tubs due to electricity being in the vicinity of people who are in the water. Receptacles and sometimes lights around pools, hot tubs, etc. must be GFCI protected.

 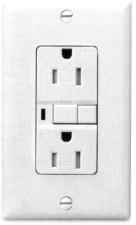

Eaton

Figure 13-8. AFCI protection is typically provided by a circuit breaker so it can protect everything downstream of the overcurrent device. AFCI receptacles are available for remodeling situations to extend conductors from an existing outlet. This is an option when it isn't feasible to install an AFCI circuit breaker in the panel, such as when a home has an old fuse panel.

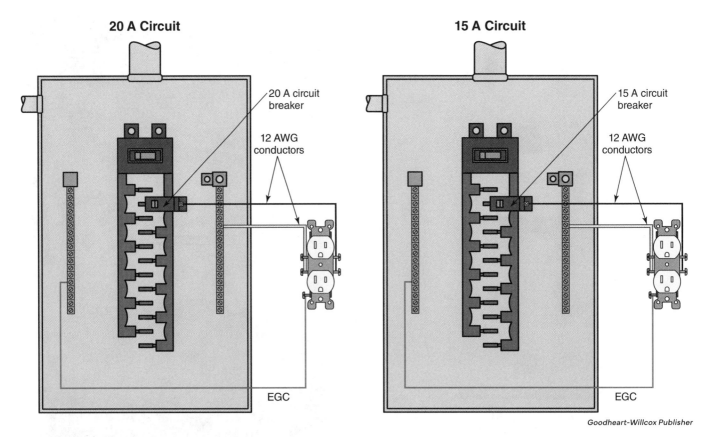

Figure 13-9. Both diagrams have 12 AWG branch circuit conductors that feed a duplex receptacle. The left diagram is a 20-ampere circuit while the right is a 15-ampere circuit. The size of the circuit is determined by the overcurrent device that is protecting the circuit.

Figure 13-10. On the nameplate of an air conditioner, the manufacturer will often list a maximum size overcurrent device that is permitted to feed the unit. The pictured nameplate lists a maximum size overcurrent device of 30 amperes.

Conductor Sizing

Section 210.19(A) has requirements for sizing the branch circuit conductors. It states that the branch circuit conductors must be large enough to carry the ampacity of the load, and if the circuit feeds a continuous load, it must be multiplied by 125%. A ***continuous load*** is a load where the maximum current is expected to continue for at least 3 hours. Lighting is an example of a continuous load: once it is turned on, it is likely to be on for three hours or more, **Figure 13-12**.

Noncontinuous loads are calculated at 100%. Many plug-in appliances, such as televisions and computers, are noncontinuous loads. Even though they are plugged in continuously and are likely to be on for three hours or more, their current draw fluctuates during use and will only draw the maximum amount of current for short periods of time.

Overcurrent Protection

Branch circuit overcurrent protection is covered in *Section 210.20*. It states that the branch circuit conductors must be protected by overcurrent devices. It has language in its subsections relating to continuous and noncontinuous loads that are similar to what is found in *Section 210.19* for conductor sizing.

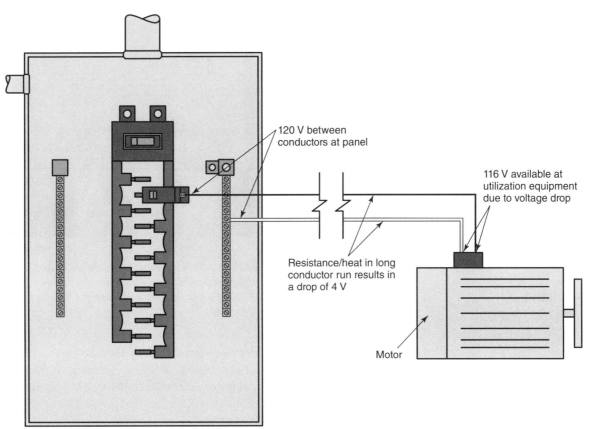

Figure 13-11. Voltage drop is due to heat losses in the conductors. The resistance of the conductor will cause it to heat up as the current is passing through. It is recommended to have a maximum of 3% voltage drop on a branch circuit. The circuit pictured has 4 volts (120 V – 116 V = 4 V) being dropped on the conductors, resulting in a 3.33% drop in voltage (4V/120V = 0.333) at the motor.

Figure 13-12. Lighting in almost any occupancy is considered a continuous load. This conference room is likely to have the lights on all day long when there is an event.

Dwelling Branch Circuits

Article 210 has many requirements that are related to dwelling units that are necessary to understand when designing the electrical system. It answers several questions. What circuits are required? Do the circuits need GFCI and/or AFCI protection? What size should the circuits be? Where are receptacles required?

Required Branch Circuits

Section 210.11, Required Branch Circuits has general information that applies to all types of occupancies but focuses on dwelling units in *210.11(C)*. There are four types of branch circuits that are identified as required in dwellings: small appliance, laundry, bathroom, and garage circuits.

Small Appliances

Section 210.11(C)(1) requires a minimum of two 20-ampere small-appliance branch circuits. The small-appliance branch circuit covers the wall and counter receptacles in the kitchen, dining room, breakfast room, pantry, and other similar rooms, **Figure 13-13**. *Section 210.52* goes on to prohibit the small-appliance branch circuits from feeding any other outlets. Kitchen lighting and fixed appliance outlets, such as the dishwasher, are included in the prohibition and cannot be on the small-appliance branch circuits. The only exception to this rule is for a kitchen clock outlet and a receptacle for a gas range.

Laundry

Section 210.11(C)(2) requires a 20-ampere laundry circuit, **Figure 13-14**. This circuit is intended for the connection of

Carla Bullock/Shutterstock.com

Figure 13-14. This laundry room will have all the receptacles, including the washing machine, on the laundry branch circuit. The laundry room lights cannot be installed on the laundry circuit.

laundry equipment, such as the washing machine and other related equipment. It is not permitted to feed the laundry room lights or any outlets outside of the laundry area.

Bathrooms

Section 210.11(C)(3) requires a 20-ampere bathroom circuit to feed the bathroom receptacles, **Figure 13-15**. It cannot feed any other outlets except for bathroom receptacles, but it is permitted to feed receptacles from multiple bathrooms. There is an exception that allows other outlets in the bathroom, such as the lights and exhaust fan, to be on the bathroom receptacle circuit provided it only feeds a single bathroom.

Garages

Section 210.11(C)(4) requires a 20-ampere branch circuit to feed the receptacles in an attached garage or a detached garage that has electric power. It is not a requirement to run power to a detached garage, so if the detached garage doesn't have a power connection, then this circuit is not required. The required garage circuit is not permitted to feed any other receptacle outlets or lights except for outdoor receptacle outlets.

Outlet Requirements and Provisions

Part III of *Article 210* is titled *Required Outlets*. It addresses specifics on where receptacles are required to be located, rooms and areas that must have lighting outlets, and any control or switching requirements.

Section 210.50(C) starts off *Part III* with a requirement that a receptacle installed for a specific appliance must be within 6′ of its intended location. The requirement is to prevent extension cords from being used to feed appliances, as

Goodheart-Willcox Publisher

Figure 13-13. Dwelling units are required to have at least two small-appliance branch circuits to feed the receptacles in the kitchen, dining, or pantry. The *NEC* provides this minimum requirement because kitchen appliances have high current draws that are used simultaneously.

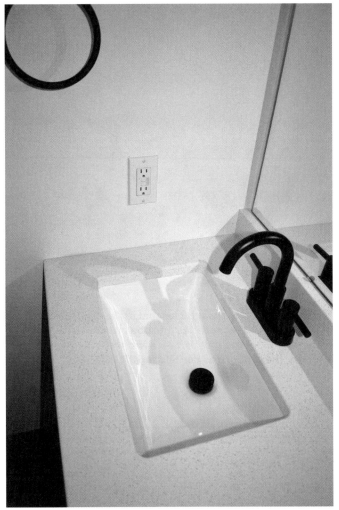

Goodheart-Willcox Publisher

Figure 13-15. The receptacles in dwelling unit bathrooms have 20-ampere branch circuits. This requirement is to prevent the circuit from tripping when high current items, such as blow dryers or curling irons, are being used.

they generally draw higher currents. Examples of appliances include a washer, dryer, range, microwave, or dishwasher.

Dwelling Receptacle Requirements

Dwelling unit receptacle outlets are generally connected to a 15- or 20-ampere branch circuit. The *NEC* requires a 20-ampere circuit for the four required dwelling branch circuits, but other than that, it is up to us to determine the layout and size of circuitry we want to install. There are advantages to both. Advantages of the 15-ampere circuit fed with 14 AWG conductors are the cable is cheaper to purchase, easier to work with/bend, and takes less space in boxes (box fill calculations). The main advantage of a 20-ampere circuit fed with 12 AWG conductors is it can carry more current than a 15-ampere circuit. If additional outlets are installed on each circuit due to the extra capacity, it could result in fewer branch circuits, which would save money.

Tamper-Resistant Receptacles

Receptacles that are below 5 1/2′ above the floor in dwelling units are required to be tamper-resistant. Tamper-resistant receptacles have a shutter behind the face of the receptacle that prevents small conductive objects, such as paperclips or keys, from being inserted into the receptacle, **Figure 13-16**. This eliminates the need to stick a plastic plug into the receptacle when small children are present. This requirement is found in *Section 406.12*, which is in the receptacle article of *Chapter 4*. Receptacles are considered equipment for general use and are installed during the trim stage of the project.

Receptacle Spacing

The general-purpose receptacles installed in most rooms in a dwelling follow the general provisions of *Section 210.52(A)*. It states that these receptacle requirements apply to the hallway, closet, living room, dining room, kitchen, bedrooms, parlor, library, sunroom, recreation room, and any similar room. Basically, this is a requirement for any habitable room, closet, and hallway. A ***habitable room*** is a room in a building for living, sleeping, eating, or cooking, but excluding bathrooms, toilet rooms, closets, hallways, storage or utility spaces, and similar areas.

Lost Mountain Studio/Shutterstock.com

Figure 13-16. Tamper-resistant receptacles have a plastic shutter that is behind the face of the device that prevents small objects from being inserted. When something is installed into the ungrounded and grounded terminal at the same time the shutter moves to the side and allows the plug to go in.

Section 210.52(A) requires any point along the wall in one of the listed areas to be within 6′ of an outlet. A lamp with a 6′ cord should be able to reach a receptacle regardless of where you decide to put it along the perimeter of the room. Within 6′ of a door or opening in the wall, there must be a receptacle, and there should be no more than 12′ between receptacles, **Figure 13-17**. With regards to narrow wall spaces, the *NEC* states that any wall space 24″ or wider requires a receptacle outlet.

Railings, the fixed portion of a patio door, and floor-to-ceiling windows in the rooms listed in *Section 210.52(A)* are considered wall space. If the railing, fixed portion of a door, or floor-to-ceiling windows are longer than 6′, the receptacle will have to be installed on the floor. For the floor receptacle to be counted as one of the required receptacle outlets, it must be within 18″ of the wall, **Figure 13-18**.

All the dwelling unit rooms that have the general receptacle spacing requirements found in *Section 210.52(A)* are required to have branch circuit AFCI protection as per *Section 210.12*. In addition to the AFCI protection, if the room is in the basement, *Section 210.8(A)(5)* requires the receptacles to have GFCI protection.

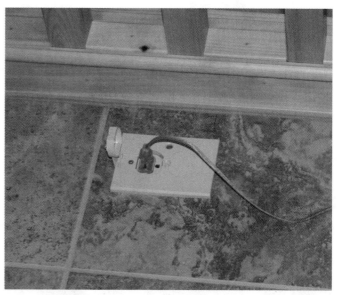

Goodheart-Willcox Publisher

Figure 13-18. Floor receptacles that are serving as one of the required wall receptacles must be installed within 18″ of the wall. Floor receptacle boxes and their covers shall be listed for the purpose as per *Section 314.27(B)*.

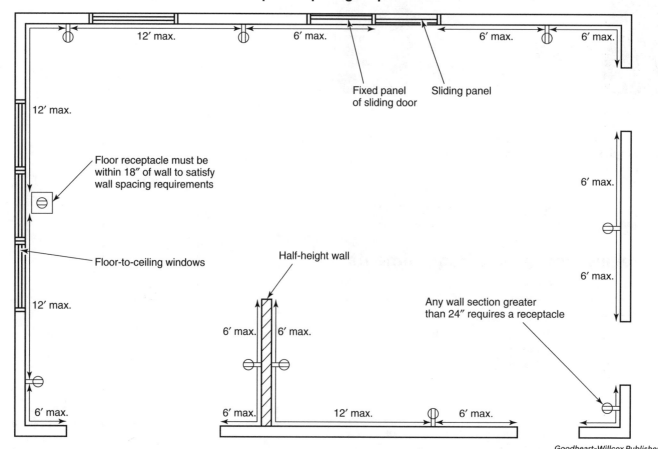

Goodheart-Willcox Publisher

Figure 13-17. The 6′ receptacle dwelling unit spacing rule requires that no point along the wall line is more than 6′ from a receptacle outlet. In addition, any wall space that is 24″ or wider is required to have an outlet.

Kitchens

The kitchen is one of the rooms fed from the small-appliance branch circuits. *Section 210.52(B)(3)* requires that the kitchen countertop receptacles must be fed by at least two different 20-ampere small-appliance branch circuits. These circuits can also feed receptacle outlets in the other rooms that are covered by the small-appliance branch circuits, but they cannot feed lighting or outlets of any type in other rooms. If a house has a second kitchen, it must have two separate small-appliance branch circuits as they are not permitted to feed more than one kitchen.

The kitchen countertop receptacles have different spacing requirements than other areas due to kitchen appliances having shorter cords. Any point along the wall shall be within 24″ of an outlet, **Figure 13-19**. That means there must be a receptacle within 24″ of the end of the countertop or any break in the counter space (sink or range), with no more than 48″ between receptacles. Receptacles are not required behind cooktops or sinks unless they are corner mounted. *NEC Figure 210.52(C)(1)* has a diagram to help determine when a receptacle is required behind a sink, **Figure 13-20**.

Requirements for islands and peninsulas have changed considerably in the last few *NEC* revisions. It is no longer a requirement to have a receptacle on an island or peninsula with a flat countertop. If the owner wants a receptacle on an island or peninsula, it will have to be mounted in or on the countertop. Because *Section 406.5(G)(1)* does not allow receptacles to be face-up on countertops, the receptacle will have to be a listed assembly that comes out of the counter or is mounted on the surface, **Figure 13-21**. Receptacles are no longer permitted on the side of the cabinet due to the chance of children pulling an appliance off the counter by the cord. If the owner decides not to have a receptacle on the island, it is still required that power is run into the cabinet and capped off in a junction box for future use.

All 120-volt kitchen circuits are required to be AFCI protected. In addition to the AFCI protection, all kitchen receptacle outlets are required to be GFCI protected.

Goodheart-Willcox Publisher

Figure 13-20. Corner sinks and ranges may need a receptacle behind them depending on the distance from the back of the sink to the corner. If it is 18″ or more, a receptacle is required.

Kitchen Counter Receptacle Spacing

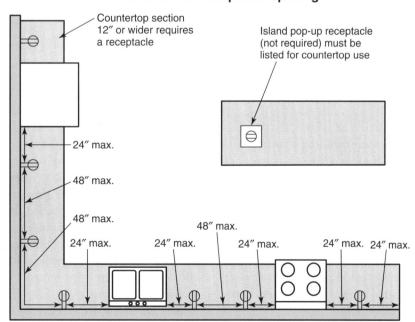

Goodheart-Willcox Publisher

Figure 13-19. The 2′ receptacle spacing requirement for kitchen countertops ensures that no point along the wall line is more than 2′ from a receptacle outlet. Small countertops that are 12″ or wider like the one next to the refrigerator are required to have a receptacle. This requirement ensures that a kitchen appliance cord will reach a receptacle regardless of where it is located on the countertop.

Goodheart-Willcox Publisher

Figure 13-21. Receptacles installed in the countertop of an island or peninsula must be a listed unit as per *Section 406.5(G)*. Consideration must be given to the location of the receptacles when the cabinets are being designed as they extend down into the cabinet below and may conflict with the top drawer.

Eaton

Figure 13-22. Dual-function breakers have become very common in dwellings. It can be difficult to determine what type of fault caused them to trip as they detect overcurrents, short circuits, ground faults, and arc faults. Many manufacturers have diagnostics built into the breaker that will indicate which type of fault caused the breaker to trip.

That includes a 120/240-volt range receptacle. A dual-function breaker provides both AFCI and GFCI protection, saving time and money when feeding circuits that require both types of protection, **Figure 13-22**.

Bathroom

A *bathroom* is defined as an area including a sink (basin) with one or more of the following: a toilet, urinal, tub, shower, bidet, or similar plumbing fixtures. The bathroom receptacle circuit is one of the 20-ampere circuits that is required in dwelling units. The bathroom receptacle circuit is permitted to feed receptacles in multiple bathrooms, but it is not permitted to feed bathroom lights, fans, or any outlet outside of the bathroom. The only exception to this rule is when a 20-ampere circuit is dedicated to one bathroom. A dedicated circuit is permitted to feed other outlets in the room, such as luminaires and bath fans.

Section 210.52(D) requires bathroom counters to have a receptacle that is adjacent to and within 36″ of the sink, **Figure 13-23**. If there are two sinks, you may need more than one receptacle serving the counter space. Having a receptacle on the sidewall of a bathroom vanity that has multiple sinks will often lead to there being more than 36″ from the receptacle to the far sink. If a receptacle can be installed in the back wall between the sinks, that will satisfy the requirement, but that may mean cutting the outlet into the mirror or backsplash. It is more common to install two receptacles for this scenario, one on each side, **Figure 13-24**. The receptacles could also be installed on the side or front of the cabinet provided they are less than twelve inches below the countertop.

Section 406.9(C) prohibits receptacles from being within 3′ of the edge of a tub or shower. This may impact where the required vanity receptacle is located. If the bathroom is very small, and it is impossible to be 3′ away from a tub or shower, the *NEC* requires the receptacle to be installed on the wall as far away from the tub/shower as possible, while still satisfying the requirements of being adjacent to and within 36″ of the sink.

Section 210.8(A)(1) requires bathroom receptacles to be GFCI protected. When an area has a tub or shower but does not meet the definition of a bathroom, any receptacle outlet within 6′ of the tub or shower must be GFCI protected. The bathroom is one of the few rooms that doesn't require AFCI protection.

Bathroom Receptacle Requirements

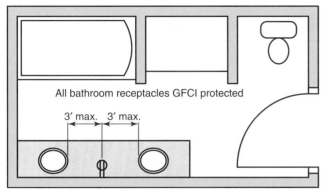

Figure 13-23. The required bathroom receptacle must be mounted on a wall adjacent to and within 36″ of the sink. Two sinks can share a single receptacle as long as it is within 36″ of each sink. However, a more common practice is to install one receptacle for each sink.

PRO TIP — Bathroom Fans

Bathroom fans are often installed to remove moisture and provide air circulation in bathrooms. It is important to read the manufacturer's instructions before determining the location of the fan. Many fans are listed to only be ceiling mounted in a horizontal position. Wall mounting a fan that is listed for ceiling mount only can put stress on the bearings that they weren't designed and listed for. Some manufacturers do have their fans listed for wall or ceiling installation, **Figure 13-25**.

If the fan is going to be mounted in the wall or ceiling space that is directly above the bathtub or shower, it will likely require GFCI protection. Nearly all the fan manufacturers require their fan to be GFCI protected when it is installed directly over a tub or shower stall.

Remember that *Section 110.3(B)* requires us to follow the manufacturer listing requirements and instructions.

Figure 13-24. Bathrooms with two sinks will often end up with two receptacles to satisfy the requirement of the receptacles being within 36″ of the sink.

Figure 13-25. Manufacturers typically require bathroom exhaust fans that are installed directly above a shower or bathtub to be GFCI protected.

Outdoor Outlets

Section 210.52(E) requires at least two exterior receptacle outlets on a dwelling unit, one in the front and one in the back. The receptacles must be mounted below 6′-6″ and be readily accessible from grade to satisfy the requirements of this section. ***Grade*** is the final elevation of the ground. When working on a project, the level of the earth around a building may not be at its final or finished grade. Ground material may be added or removed to get to the desired finished grade towards the end of the project.

If HVAC equipment is installed outside, *Section 210.63(A)* requires that a receptacle be installed within 25′ of the equipment for servicing. If one of the other required outdoor receptacles is within 25′ of the equipment, it will satisfy this requirement, **Figure 13-26**.

Exterior receptacles are not required to be AFCI protected, but they must be GFCI protected and tamper resistant. If the GFCI protection is provided by an outdoor GFCI receptacle, it must be installed in a readily accessible location.

Most outdoor receptacles are considered to be in a wet location. The exception is a receptacle installed on the underside of a soffit, and in some cases under large porch overhangs, which is considered a damp location, **Figure 13-27**. The *NEC* doesn't give any dimensional information on when an overhang changes the location from wet to damp, so this usually falls under the interpretation of the authority having jurisdiction.

Section 406.9(A) and *(B)* requires receptacles in damp and wet locations to be weather resistant. Receptacles in damp locations are permitted to have the standard weatherproof covers, while receptacles in wet locations must have a cover that ensures the receptacle is protected from the weather even when a cord is plugged in. These covers are often called "while-in-use" or simply "in-use" covers, **Figure 13-28**.

Goodheart-Willcox Publisher

Figure 13-27. Receptacles and lights on the underside of a soffit are in a damp location. The receptacle still must be weather resistant, but the old-style flip-up weatherproof covers are permitted.

Goodheart-Willcox Publisher

Figure 13-26. A disconnect with a built-in receptacle will satisfy the requirement for a disconnect within sight of the air conditioner as well as the requirement for a receptacle within 25′.

Lost_in_the_Midwest/Shutterstock.com

Figure 13-28. In-use covers maintain weatherproof integrity while the cord is plugged in.

Laundry

The laundry circuit is one of the required circuits in dwellings. It must be a 20-ampere circuit, have GFCI and AFCI protection, and is not permitted to feed anything other than laundry receptacles. The area classified as laundry can be a corner of the basement, a closet, or an entire room. The *NEC* isn't specific on what is classified as a laundry area so it is up to interpretation. The laundry circuit will include the washing machine outlet, and if it is in a room or closet, it could include the other receptacles as well.

Garages

The garage circuit is the last of the required dwelling circuits. It only applies if there is an attached garage or a detached garage with electric power. The required circuit cannot feed any other outlets except for exterior receptacle outlets. *Section 210.52(G)(1)* requires at least one receptacle to be installed in each vehicle bay, and they are not permitted to be more than 5′-6″ above the floor. A three-stall garage will be required to have at least three receptacles, one for each vehicle bay, **Figure 13-29**. Receptacles for garage door openers are permitted to be on the same circuit but are in addition to the minimum number of required wall receptacles. Garage receptacles must be protected by a readily accessible GFCI.

Hallways

Dwelling unit hallway receptacle requirements are found in *Section 210.52(H)*. If the hallway is 10′ or more in length it is required to have a receptacle. The *NEC* does not detail exactly where the receptacle is to be installed, and it does not require a second receptacle outlet for long hallways. In the case of a hallway that has a corner, an imaginary line is drawn down the center, **Figure 13-30**. If that line is 10′ or more in length, the hallway is required to have an outlet.

Foyers

Foyers, or entryways, have receptacle outlet requirements if they have an area of 60 ft^2 or more, **Figure 13-31**. If so, any wall space three feet or wider, excluding doors and windows that extend to the floor, is required to have a receptacle outlet. *Section 210.52(I)* contains these requirements.

Lighting

Dwelling unit lighting outlet requirements are found in *Section 210.70*. The *NEC* doesn't give requirements as to how many lights are required, the minimum number of lumens, etc. It simply gives minimum requirements as to what areas are required to have some form of lighting as well as its

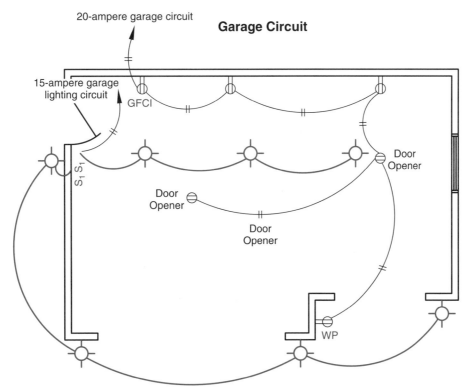

Goodheart-Willcox Publisher

Figure 13-29. The garage receptacle circuit must be dedicated to garage receptacles. The only exception is for a grade-level exterior outlet like the one between the garage doors. The GFCI can be in the electrical panel, or it can be the first receptacle in the circuit, provided it is readily accessible.

Dwelling Hallway Receptacle Requirements

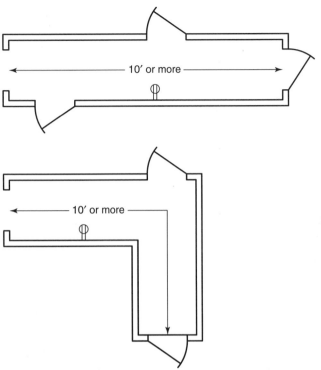

Figure 13-30. Dwelling unit hallways that are 10′ or longer are required to have a receptacle outlet. If the hallway has a turn, the centerline of the hallway is measured to determine if it is 10′ or more in length. The *NEC* doesn't say where in the hallway the receptacle shall be located, so it is up to the designer or electrician to determine the best location.

Figure 13-31. Foyers that are over 60 ft² shall have a receptacle in each wall space that is 3′ or more in width.

associated switching requirements. The rest of the lighting and switching details are left to the person designing the system. Their job is to ensure that it will meet the needs of the customer.

Habitable Rooms Kitchens and Bathrooms

Section 210.70(A)(1) requires every habitable room, kitchen, and bathroom to have a wall-controlled lighting outlet. The control device must be located on a wall near an entrance to the room. It is permitted to be controlled by a switch or automatic means, such as an occupancy sensor.

Switch-controlled receptacles can be used to control floor or table lamps, **Figure 13-32**. *Section 210.70(A)(1) Exception No. 1* allows switch-controlled receptacles to serve as the lighting outlet instead of having a ceiling or wall luminaire in habitable rooms other than kitchens and bathrooms. This is usually in the form of split-wired duplex receptacles where half of the receptacle is energized all the time, and the other half is controlled by the switch. For the receptacles to count as one of the required room receptacles in *Section 210.52(A)*, part of the receptacle must be energized all the time, regardless of the switch position. See **Figure 13-33**.

PRO TIP: Grounded Conductor in Switch Boxes

In dwelling units, a cable is commonly used as a wiring method. *Section 404.2(C)* requires that the grounded conductor be available in the switch box in habitable rooms, bathrooms, hallways, and stairways. This is to ensure that the grounded conductor is available for the future installation of occupancy sensors or timers, **Figure 13-34**. If the room is fed with three-way switching and the area being controlled is visible from the switch boxes, the grounded conductor is required in only one of the locations.

Figure 13-32. Table lamps that are plugged into split-wired receptacles allow the lamps to be turned on and off with a wall switch. Table lamps come with a six-foot cord. Using the 6′ rule when laying out receptacles will provide a place to plug the lamp in regardless of where it is placed along the wall.

Goodheart-Willcox Publisher

Figure 13-34. Timers installed in a switch box can be used to automatically turn lights on and off at predetermined times. This is common with exterior lights. Most timer switches need to have the grounded conductor in the box to power the electronics in the device.

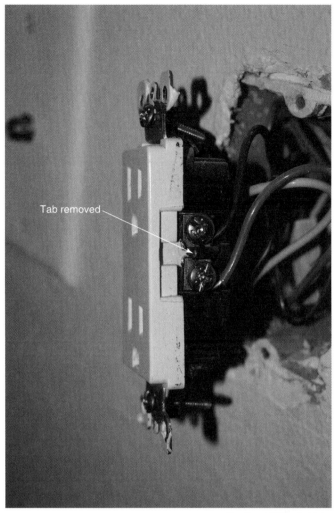

Goodheart-Willcox Publisher

Figure 13-33. When installing split-wired receptacles, the tab on the ungrounded side that connects the top and bottom terminals together is broken off so they are isolated from one another. In this picture, the black wire that connects to the top of the receptacle is energized all the time and the red wire that connects to the bottom of the receptacle is controlled by the wall switch.

David Papazian/Shutterstock.com

Figure 13-35. Walk-out doors on dwelling units are required to have an exterior light. The light can be wall mounted, recessed in the soffit, or a post light that is close.

Outside Lights

Outside lights are required at each walk-out door of a dwelling unit that has access to grade. This requirement does not apply to overhead garage doors for vehicles. *Section 210.70(A)(2)(2)* requires the required outdoor lights to have a wall-mounted switch or control device, **Figure 13-35**.

Stairway

Stairways are required to have a lighting outlet as per *Section 210.70(A)(2)(1)*. If the stairway has six risers or more, a wall switch must be installed at the top and bottom of the stair, **Figure 13-36**. Stairways that have a landing are required to have a switch on the landing if it has a door and six or more risers between levels.

Attics, Crawl Spaces, and Unfinished Basements

Lighting outlets in attics, crawl spaces, and unfinished basements are covered in *Section 210.70(C)*. It requires any of the mentioned spaces that store goods or will have equipment that requires servicing to have a switch-controlled light, **Figure 13-37**. Attics that are not meant for storage and do not have serviceable equipment are not required to have a light. This requirement applies to both dwelling and non-dwelling occupancies. Lighting outlets in crawl spaces are required to be GFCI protected by *Section 210.8(C)*.

be GFCI protected as required by *Section 210.8(E)*. This requirement applies to both dwelling and non-dwelling occupancies, **Figure 13-38**.

Non-Dwelling Branch Circuits

Non-dwelling occupancies don't have as many branch circuit requirements. The circuitry is designed around the use of the building. A manufacturing facility will have circuitry designed for the specific equipment being installed. An office building will have the branch circuits necessary to feed the cubicles and offices. While there is the flexibility to design for the use, there are a few non-dwelling areas where the *NEC* does have requirements.

Meeting Rooms

Section 210.65 addresses meeting rooms that are 1000 ft² or smaller. These rules do not apply to rooms over 1000 ft², such as auditoriums or gymnasiums, **Figure 13-39**.

The language mentions designing the electrical requirements of the space to meet the needs of the owner, but it also gives a few basic requirements. If the meeting room has dividers to make the space smaller, each space is considered separately when placing receptacles.

Fixed walls in meeting rooms shall have the minimum number of wall receptacles determined by *Section 210.52(A) (1-4)*, which is the receptacle spacing requirement used in dwellings. It allows the placement of the required number of receptacles to be determined by the installer, designer, or building owner.

Mike Higginson/Shutterstock.com

Figure 13-36. Stairways with six risers or more are required to have a switch at the top and bottom of the stair to prevent having to walk up or down the stairs in the dark to operate the switch.

Catherine.Things/Shutterstock.com

Figure 13-37. Attics with storage, or equipment that requires maintenance, will often have pull-down stairs for ease of access. The *NEC* requires a switch-controlled light in the attic.

Equipment Requiring Servicing

Section 210.63 requires a receptacle to be installed within 25′ of equipment that requires servicing. This applies to equipment inside and outside of the building. The receptacle shall

Konstantin L/Shutterstock.com

Figure 13-38. Air conditioners are an example of equipment that requires servicing and must have a receptacle within 25′. The receptacle can be in the form of an exterior receptacle, as shown between the air conditioners in the image, or it can be built into the disconnect.

Mangostar/Shutterstock.com

Figure 13-39. Meeting rooms have wall and floor receptacle requirements as per *Section 210.65*. If the room is over 215 ft², it shall have at least one receptacle on the floor that is at least 6′ from the wall and in a convenient location for plugging in items such as computers and projectors.

If the meeting room or its divided spaces are 215 ft² or more, there must be at least one floor receptacle. The floor receptacle must be at least 6′ away from the fixed walls. The idea behind this requirement is to have the floor outlet towards the middle of the room where a projector or other piece of equipment may be used.

Sign Circuit

Commercial occupancies that are accessible to pedestrians shall have an outlet for a sign at each customer entrance to the space. **Figure 13-40.** This does not apply to delivery doors or personnel doors. The required outlet must be fed with an individual 20-ampere circuit. This requirement is easy to miss as it is found in *Section 600.5*.

Goodheart-Willcox Publisher

Figure 13-40. One of the required circuits in commercial occupancies is for a business sign. The circuit is required at entrances where consumers will be entering the building or space.

Summary

- *Article 210* contains branch circuit requirements.
- The *NEC* recognizes the following types of branch circuits:
  - Appliance
  - General purpose
  - Individual
  - Multiwire
- Requirements for GFCI protection of receptacles are found in *Section 210.8*.
- Requirements for AFCI protection is found in *Section 210.12*.
- Branch circuit conductors and their overcurrent device must be rated at 125% of a continuous load.
- The following are required branch circuits in dwelling units.
  - Small appliance
  - Laundry
  - Bathroom
  - Garage
- Dwelling unit receptacle outlet requirements are found in *Section 210.52*.
- Branch circuit lighting requirements for all occupancies are found in *Section 210.70*.

Unit 13 Review

Name _____ Date _____ Class _____

Know and Understand

Answer the following questions based on information in this unit.

1. The *NEC* requires at least _____ small-appliance branch circuits.
   A. one
   B. two
   C. three
   D. four

2. A(n) _____ supplies energy to only one utilization equipment.
   A. appliance branch circuit
   B. branch circuit
   C. individual branch circuit
   D. multiwire branch circuit

3. The *NEC* recommends that the voltage drop on a circuit be limited to _____.
   A. 3%
   B. 5%
   C. 6%
   D. 8%

4. The required bathroom receptacle must be installed within _____ of the sink.
   A. 24″
   B. 36″
   C. 48″
   D. 72″

5. Foyers have receptacle requirements if its area is _____ ft² or more.
   A. 30
   B. 40
   C. 50
   D. 60

6. Which of the following is not permitted to have switched receptacles serve as the required lighting outlet?
   A. Bathroom
   B. Family room
   C. Bedroom
   D. Den

7. An outdoor air-conditioning unit is required to have a 125-volt receptacle installed within _____ for servicing.
   A. 10′
   B. 20′
   C. 25′
   D. 50′

8. Which of the following is permitted to be on the required 20-ampere garage branch circuit?
   A. Garage lights
   B. Attic light
   C. Receptacle outlet in the front soffit
   D. Grade level outdoor receptacle by the garage door

9. The maximum distance between receptacles on a kitchen countertop is _____.
   A. 12″
   B. 24″
   C. 36″
   D. 48″

10. A receptacle installed for a specific appliance must be installed within _____ of the appliance's intended location.
    A. 3′
    B. 4′
    C. 6′
    D. 8′

Apply and Analyze

Answer the following questions using a copy of the National Electrical Code. *Identify the section or subsection where the answer is found.*

1. The required receptacles installed in the vehicle bay of a garage must not be more than _____ above the floor.
   A. 48″
   B. 60″
   C. 66″
   D. 72″

   NEC _____

2. A receptacle installed to serve the kitchen countertop shall not be more than _____ above the countertop surface.
   A. 12″
   B. 16″
   C. 18″
   D. 20″

   NEC _____

3. Receptacles in which of the following locations are required to be GFCI protected?
   A. Basement bedroom
   B. Main floor living room
   C. 2nd floor hallway
   D. Dining room

   NEC _____

4. A household range that is over 8 3/4 kW shall have a minimum branch circuit rating of _____.
   A. 30 amperes
   B. 40 amperes
   C. 50 amperes
   D. 60 amperes

   NEC _____

5. Which of the following _____ is permitted to be fed with the small-appliance branch circuits?
   A. gas-fired range
   B. kitchen lighting
   C. patio receptacle
   D. living room receptacle

   NEC _____

6. Receptacles below _____ in a dwelling must be tamper-resistant.
   A. 48″
   B. 66″
   C. 72″
   D. 78″

   NEC _____

7. A 20-ampere circuit that feeds multiple receptacles shall have receptacles that are rated _____.
   A. 15-ampere
   B. 20-ampere
   C. 15-or 20-ampere
   D. 30-ampere or less

   NEC _____

8. Which of the following dwelling unit outlets is not required to have AFCI protection?
   A. Kitchen dishwasher outlet
   B. Laundry outlets
   C. Hallway lighting outlet
   D. Garage receptacle outlets

   NEC _____

9. A branch circuit feeds a piece of utilization equipment that is fastened in place as well as other lighting and receptacle outlets. The equipment that is fastened in place cannot exceed more than _____ of the branch-circuit ampere rating.
   A. 50%
   B. 75%
   C. 80%
   D. 125%

   NEC _____

10. Lighting outlets installed in a crawl space are required to be _____.
    A. 20-ampere
    B. GFCI protected
    C. AFCI protected
    D. All of the above.

    NEC _____

Name _____ Date _____ Class _____

Critical Thinking

Answer the following questions using a copy of the National Electrical Code. *Identify the section or subsection where the answer is found.*

1. The rating of a branch circuit is determined by _____.

   _____

   NEC _____

2. Guest rooms and guest suites of hotels and motels that have _____ shall have branch circuits installed to meet the rules for dwelling units.

   _____

   NEC _____

3. A dwelling unit bedroom wall space _____ or wider is required to have a receptacle.

   _____

   NEC _____

4. List five requirements for dwelling unit outdoor receptacles.

   _____

   _____

   _____

   _____

   _____

   _____

   _____

   _____

   NEC _____

5. Dwelling unit floor receptacles shall be installed within _____ of the wall to count as one of the required receptacles.

   _____

   NEC _____

6. When would you not be required to have a lighting outlet in a residential detached garage?

   _____

   NEC _____

7. Ground-fault protection of equipment shall be provided for a 277/480-volt wye system with a branch circuit disconnect rating of _____ amperes or more.

   _____

   NEC _____

8. The required receptacle outlets in show windows shall be within _____ of the top of the window. No point along the top of the window shall be more than _____ from a receptacle outlet.

   _____

   _____

   NEC _____

9. A dwelling unit has a U-shaped stairway with eight risers on each staircase and a landing in the middle. The landing contains a door. What is the minimum number of switches required to control the stair lights?

   _____

   NEC _____

10. A branch circuit overcurrent device shall not be less than _____% of a continuous load.

    _____

    NEC _____

Part 2   Application: Advanced NEC Topics

Answer the following questions using a copy of the National Electrical Code. *Identify the section or subsection where the answer is found.*

11. The _____ conductors of a multiwire branch circuit shall be grouped.

    NEC _____

12. List five kitchen appliances that are required to have GFCI protection.

    _____
    _____
    _____
    _____
    _____

    NEC _____

13. Balconies that are within _____ horizontally of a dwelling unit are required to have at least one receptacle outlet accessible from the balcony.

    NEC _____

14. Guest rooms in hotels shall have at least one lighting outlet controlled by a wall mounted control device in each _____.

    NEC _____

15. Branch circuits feeding the terminals of luminaires in dwelling units shall not exceed _____ volts.

    NEC _____

16. A dwelling unit with two kitchens requires at least _____ small appliance branch circuits.

    NEC _____

17. A corner mounted sink in a dwelling unit kitchen is required to have a receptacle behind the sink if the distance from the edge of the sink to the corner is more than _____ inches.

    NEC _____

18. The rating of any one cord and plug connected load that is plugged into a 20-ampere duplex receptacle shall not exceed _____ amperes.

    NEC _____

19. While upgrading a panelboard in a dwelling unit, the branch circuits have to be extended to reach the new panel location. When would AFCI protection be required on the extended branch circuits?

    NEC _____

20. Ground fault circuit interrupters shall be installed in a(n) _____ location.

    NEC _____

UNIT 14
Box Fill and Pull Box Calculations

Sashkin/Shutterstock.com

KEY TERMS

barrier
box volume
conduit body
device
device box
extension ring
fixture stud
hickey
junction box
lighting outlet
lighting outlet box
outlet
outlet box
plaster ring (mudring)
pull box
raised cover
short radius conduit body
tail
yoke

LEARNING OBJECTIVES

After completing this unit, you will be able to:
- Calculate the maximum number of conductors permitted in nonmetallic boxes.
- Calculate the maximum number of conductors permitted in metallic boxes.
- Perform straight pull box calculations.
- Perform angle pull box calculations.

Introduction

This unit will cover how to perform box fill and pull box calculations. There are many different sizes and types of boxes used in the electrical industry. *Article 314* has minimum standards that must be followed to ensure that a box is large enough. The minimum requirements are to ensure that conductors are not damaged during installation and that they are large enough to safely house the conductors and devices, **Figure 14-1**. This unit will present how to calculate the minimum size electrical box for various applications.

Goodheart-Willcox Publisher

Figure 14-1. Pictured is a 3-gang box at the rough-in stage of wiring a dwelling. This box will have three switches mounted in the box. It is important to keep track of the number of conductors and devices in the box to ensure it is not over maximum capacity.

Outlet Box

The *NEC* defines an **outlet** as a point in the wiring system where current is taken to supply utilization equipment. With that definition in mind, an **outlet box** is the box where power is taken to supply utilization equipment. An outlet box is a generic term that will include lighting outlet boxes as well as device boxes, **Figure 14-2**.

Lighting Outlet Box

The *NEC* defines a **lighting outlet** as an outlet intended for the direct connection of a lamp holder or luminaire. With that definition in mind, a **lighting outlet box** is a box that is intended to connect and support a luminaire, **Figure 14-3**. Lighting outlet boxes are generally round and will have 8–32 screws that connect the luminaire to the box. Boxes designed to support a luminaire should be capable of supporting at least 50 lb.

> **PRO TIP** **Finding the Correct Section**
>
> Because boxes containing conductors are found in all installations, their requirements can be found in one of the first four chapters that apply generally to all installations. More specifically, the boxes are a wiring method and are installed and sized while roughing-in. That information leads to *Chapter 3, Wiring Methods and Materials* of the *NEC*.
>
> To find the correct section using the table of contents, complete the following:
>
> 1. *Chapter 3*
> 2. *Article 314, Outlet, Device, Pull, and Junction Boxes*
> 3. *Part II, Installation*
> 4. Scan sections to find *Section 314.16, Number of Conductors in Outlet, Device, and Junction Boxes, and Conduit Bodies*
>
> To find the correct section using the index, complete the following:
>
> 1. Keyword 1—Boxes
> 2. Keyword 2—Fill calculations
> 3. Open to *Section 314.16(B)*

Electrical Boxes

Before getting started on box fill calculations, it is important to understand the terminology of the components involved. There are many different sizes and types of electrical boxes, and you must learn the differences and requirements of each.

Goodheart-Willcox Publisher

Figure 14-2. This picture of a bathroom vanity has two types of outlets. The luminaire above the mirror is an example of a lighting outlet, and the GFCI receptacle on the right side of the vanity is an example of a receptacle outlet.

Figure 14-3. Lighting outlet boxes installed in the middle of a habitable room are required to be fan rated. The *NEC* enacted this requirement due to the vast number of paddle fans installed on regular lighting boxes when consumers decided to upgrade.

Device Box

A ***device box*** is a box that will house a device such as a switch or receptacle. The *NEC* defines a ***device*** as a unit of an electrical system, other than a conductor, that carries or controls electrical energy as its principal function. Device boxes are identified by the number of gangs, or slots. A 1-gang box will support one device and a 2-gang box will support two devices, **Figure 14-4**. Device boxes are designed to use 6–32 screws to secure the device to the box.

Junction Box

A ***junction box*** is a box where conductors are spliced together, **Figure 14-5**. Sometimes outlet and device boxes will contain spliced conductors additional to the conductors necessary to feed the device or luminaire. In this situation, these boxes fit more than one box definition. For example, a box that contains spliced conductors and also houses a receptacle outlet meets the definition of an outlet box, device box, and junction box.

Pull Box

A ***pull box*** is a box that is used as a pull point in the middle of the conduit run, **Figure 14-6**. It provides a location to pull the conductors out so they can then be fed back into the conduit to complete the run. The *NEC* limits the amount of bend between pull points to four 90° bends (a total of 360°). In the case of a conduit run where there will be more than 360° of bend, it may be necessary to have a pull box in the

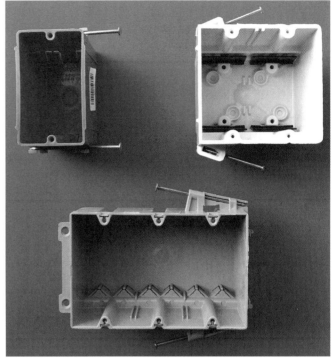

Figure 14-4. Device boxes come in many styles and are made of various types of materials. One thing they have in common is they are designed to accept the screws that come with devices such as switches and receptacles.

Figure 14-5. Junction boxes, which are used as a pull point or to splice conductors, must remain accessible in case there is a need to get back into the box to make repairs or changes to the electrical system. They must also have a cover on them to prevent accidental contact with the interior wiring and to contain any sparks or excess heat from loose connections.

middle. The purpose of limiting the amount of bend to 360° is to reduce the friction from going around too many bends, reducing the pulling forces necessary to install the conductors in the raceway. Having to use excessive force to pull the conductors into a raceway can damage the conductors.

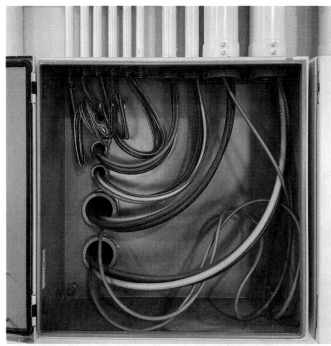

Figure 14-6. Pull boxes can provide a way of transitioning the direction of the raceway and provide a place to pull the conductors out and push them back in when there is more than 360° of conduit bend. This pull box has the raceways entering the box from the back and exiting from the top.

Box Fill Calculations— NEC Section 314.16

Box fill calculations will ensure that an electrical box has sufficient space to house the conductors, devices, and clamps that may be present in the box. A box with too many wires has conductors jammed into it, creating the potential for a short circuit, ground fault, or possible overheating of the conductor insulation. **Figure 14-7**. The box fill calculations performed by *Section 314.16* apply to boxes housing conductors 6 AWG and smaller.

Figure 14-7. Trying to contain too many conductors in a box will result in the conductors being jammed together and against the inside of the box. It is easy to damage the conductors, resulting in nicks, short circuits, and excess heat.

Box Volume Calculations

The first step in box fill calculations is to determine the box volume. The *box volume* is the amount of space in cubic inches available for conductors, devices, and clamps. For boxes under 100 in^3, the volume will either be stamped in the box, or it will be same as listed in *Table 314.16(A)*.

The materials that boxes are made from can be divided into two categories: metallic and nonmetallic, **Figure 14-8**. Although the calculations are performed in the same manner, there are slight differences that requires them to be addressed separately.

Metallic Boxes

Smaller metallic boxes are constructed in standard sizes that are listed in *Table 314.16(A)*. If a metallic box has a volume other than the standard boxes in the table, it will be stamped in the box. *Table 314.16(A)* lists each type and size metallic box as well as its volume. See **Figure 14-9**. The right side of the table lists the maximum number of conductors permitted in the box. This portion of the table assumes all the conductors are the same size and does not take into consideration any other items that may be present in the box.

Nonmetallic Boxes

Nonmetallic boxes do not have a table that lists their standard volume since they are all constructed slightly differently. They have varying wall thicknesses, are tapered differently, and may have internal clamps. All nonmetallic boxes will have their maximum volume stamped in the box, **Figure 14-10**.

Most nonmetallic boxes that are listed to be used with nonmetallic sheathed cable will have internal clamps. The only exception to that rule is nail-on 1-gang boxes that have small knockouts instead of clamps, **Figure 14-11**. Since 1-gang nonmetallic boxes do not have clamps, it is necessary to secure the cable to a framing member closer to the box

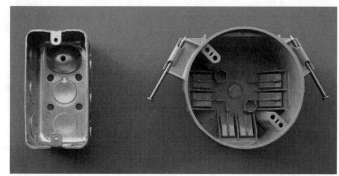

Figure 14-8. Metal boxes are generally manufactured to standard sizes corresponding to *Table 314.16(A)*. Nonmetallic boxes will have their volume stamped on the box as their internal area varies slightly by box material and manufacturer.

Table 314.16(A) Metal Boxes

Box Trade Size			Minimum Volume		Maximum Number of Conductors* (arranged by AWG size)						
mm	in.		cm³	in.³	18	16	14	12	10	8	6
100 × 32	(4 × 1¼)	round/octagonal	205	12.5	8	7	6	5	5	4	2
100 × 38	(4 × 1½)	round/octagonal	254	15.5	10	8	7	6	6	5	3
100 × 54	(4 × 2⅛)	round/octagonal	353	21.5	14	12	10	9	8	7	4
100 × 32	(4 × 1¼)	square	295	18.0	12	10	9	8	7	6	3
100 × 38	(4 × 1½)	square	344	21.0	14	12	10	9	8	7	4
100 × 54	(4 × 2⅛)	square	497	30.3	20	17	15	13	12	10	6
120 × 32	(4¹¹⁄₁₆ × 1¼)	square	418	25.5	17	14	12	11	10	8	5
120 × 38	(4¹¹⁄₁₆ × 1½)	square	484	29.5	19	16	14	13	11	9	5
120 × 54	(4¹¹⁄₁₆ × 2⅛)	square	689	42.0	28	24	21	18	16	14	8
75 × 50 × 38	(3 × 2 × 1½)	device	123	7.5	5	4	3	3	3	2	1
75 × 50 × 50	(3 × 2 × 2)	device	164	10.0	6	5	5	4	4	3	2

Reproduced with permission of NFPA from NFPA 70, National Electrical Code, 2023 edition. Copyright © 2022, National Fire Protection Association. For a full copy of the NFPA 70, please go to www.nfpa.org

Figure 14-9. *Table 314.16(A)* lists the volume of metallic boxes built to standard sizes.

Goodheart-Willcox Publisher

Figure 14-10. Nonmetallic boxes are often stamped on the inside of the box with their maximum volume. The 2-gang box pictured has an available volume of 42 in³.

than with all other boxes and enclosures. The *NEC* requires nonmetallic sheathed cable to be secured within 8″ of a 1-gang box (*Section 314.17(B)(2) Exception*). All other boxes that have a method of securing the nonmetallic sheathed cable to the box shall be secured within 12″ of the box (*Section 334.30*).

Items That Add to the Overall Volume of a Box

There are a few items that will add volume to a box, such as extension rings, plaster rings, and raised covers, **Figure 14-12**.

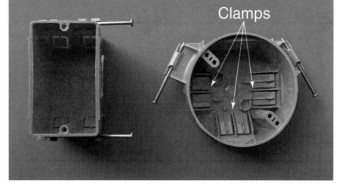

Goodheart-Willcox Publisher

Figure 14-11. The box on the left is a 1-gang device that does not have internal clamps to secure the cable to the box. The round lighting outlet box on the right has internal clamps that are molded into the box.

- *Extension rings* are added to the top of a box to increase its volume. They are the same length and width of the box but have an open back to allow access to the original box. The amount of space they add to the box will be stamped in the extension ring, or *Table 314.16* may be used for the box with similar dimensions.
- *Plaster rings*, also known as mud rings, are a component that is installed on a box (typically 4″ square) to provide an opening in the finished wall surface for the installation of a device or luminaire. Plaster rings are available to provide a 1-gang or 2-gang opening for devices and a round opening for luminaires. They come in multiple depths and may be metallic or nonmetallic. Plaster rings that are stamped with a volume will add a bit of space to the box.
- *Raised covers* provide a means of mounting devices to square metallic boxes in areas where the raceway and boxes are exposed. The amount of space they add to the box will be stamped on the cover.

Figure 14-12. Clockwise from top left: four square box, four square extension ring, 1-gang plaster ring, four square raised cover.

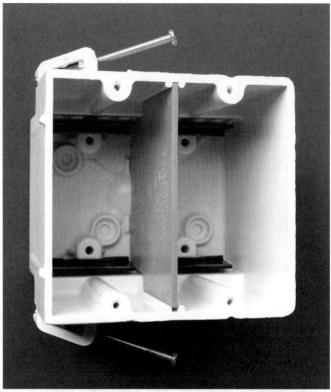

Figure 14-13. Installing a barrier into a box divides the box into individual spaces.

Items That Subtract from the Volume of a Box

Barriers will subtract from the original volume of a box. A *barrier* is a divider that is installed in a device box to separate the box into separate spaces, **Figure 14-13**. One reason for separating a box into separate spaces is for installing line voltage and data cables in the same box. A 2-gang box can have a receptacle outlet on one side and a TV jack on the other side. A barrier will separate the box into two separate spaces and keep the TV cable isolated from the power wiring.

Another reason for installing a barrier in a box is when a box contains switches with more than 300 volts potential between two devices. *NEC Section 404.8(B)* prohibits having more than 300 volts between devices in the same box without a divider. This may happen when working on a 277/480-volt system in a commercial or industrial installation.

Barriers made of nonmetallic materials are generally used with nonmetallic boxes, while metallic barriers are used with metallic boxes. Nonmetallic barriers can be used with a metallic box provided it fits properly and the box is listed to accept it. It is unusual to install a metal divider in a nonmetallic box. A metal divider is unlikely to have the correct dimensions to fit into a nonmetallic box and would have to be properly bonded to ensure it doesn't become energized. It would be cheaper and less work to use the nonmetallic divider that was designed for the box.

The barrier may be marked with the amount of volume it takes up in the box. See **Figure 14-14**. If the barrier is not

Figure 14-14. Barriers may have the amount of space they consume engraved into the unit. This barrier is listed at one cubic inch. It removes 0.5 in^3 of space from each side of the box.

marked with its volume, the *NEC* identifies a metal divider as taking up 0.5 in³ and a nonmetallic divider as taking up 1 in³. Once a divider has been installed, the box is divided into separate spaces. The overall volume of the box is reduced by the volume of the divider, and the remaining volume is divided among the spaces. For example, a 2-gang nonmetallic box has an original volume of 35 in³. A nonmetallic divider is installed in the box which reduces the overall volume of the box to 34 in³. The remaining 34 in³ is divided by 2 since the box now contains two separate spaces. Each side of the box now has a volume of 17 in³.

Box Fill Calculations

Box fill calculations begin at *Section 314.16(B)*. Box fill calculations are used to determine the maximum number of conductors that will safely fit in a box. Everything in the box is given an allowance that is equated to conductors. Nonconductor items will be given a value (allowance) that is equivalent to the amount of conductor space they will take up. For example, cable clamps take up approximately the same amount of space as one conductor, and devices take up approximately the same amount of space as two conductors.

Section 314.16(B) has five subsections that identify the items that may be in the box and describes the allowance consideration that must be given to each. The items included in box fill calculations are:

- Conductors
- Clamps
- Support fittings
- Devices
- Equipment grounding conductors

After all the items in the box have been determined, they are added together. That sum is the total number of allowances (equivalent conductors) that are in the box.

Conductors

The first item to consider is the actual conductors that enter the box. This is covered in *Section 314.16(B)(1)*. The general rule is each current-carrying conductor that originates outside the box and splices or terminates within the box counts as one conductor, **Figure 14-15**. Equipment-grounding conductors are *not* part of this calculation. They have their own calculation which will be covered later.

Each conductor that passes through the box without a termination or splice must be considered slightly differently. If a conductor passes through the box and is less than twice the minimum length, as indicated in *NEC 300.14*, it only counts as one conductor, **Figure 14-16**. Conductors will sometimes pass through a box without stopping on their way to another box in the conduit run. Passing through the box without stopping saves space in the box, the cost of a Wire-Nut®, and time spent splicing. With most boxes, the minimum conductor length in a box is 6″. The measurement

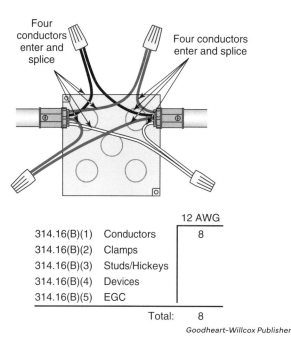

Figure 14-15. Box fill calculation for a junction box where four conductors from each raceway are spliced. The raceway is being used as the equipment grounding conductor.

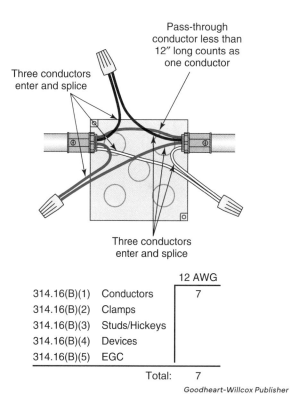

Figure 14-16. Box fill calculation for a junction box where one conductor passes through the box without stopping while the other three conductors from each raceway are spliced. The raceway is being used as the equipment grounding conductor.

is found by measuring the length of the conductor from where it leaves one conduit to where it enters the other conduit. A conductor that passes straight through a box or has a loop that is less than 12″ long counts as one conductor.

Each conductor that passes through a box without terminating or splicing and is twice the minimum length counts as two conductors, **Figure 14-17**. The reason for leaving a loop twice the minimum length is so it can be cut and spliced in with another conductor sometime in the future. If that were to happen, it would become two separate conductors that originate outside the box and splice within the box. The fact that a longer loop takes extra space and has the potential to become two conductors is why the *NEC* counts it as two conductors.

Tails, sometimes referred to as pigtails, are not counted in the box fill calculation as they do not leave the box. A *tail* is a small length of wire that extends from a wire splice to a device or luminaire. See **Figure 14-18**. The *NEC* has taken the stance that a tail does not take up enough room to be considered.

The *NEC* has an exception when it comes to fixture wires entering a box from a domed luminaire or similar canopy. If the light has a domed space, it will give more than enough room to accommodate the amount of space taken up by the fixture wires, **Figure 14-19**. This exception only applies if

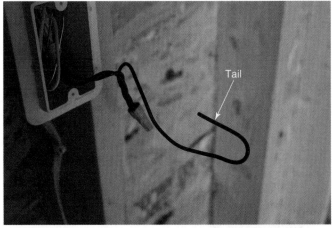

Goodheart-Willcox Publisher

Figure 14-18. A tail is a piece of wire that is spliced in with the other conductors in the box. Tails are not counted in the box fill calculation as they do not leave the box.

Goodheart-Willcox Publisher

Figure 14-19. Small fixture wires coming from a luminaire with a domed canopy are *not* counted in the box fill calculation.

there is a domed canopy, there are fewer than four fixture wires, and they are smaller than 14 gauge.

Clamps

Clamps are used to fasten cables to boxes and are covered in *Section 314.16(B)(2)*. Clamps may be on the outside of the box (external) or inside the box (internal). External clamps do not take up any space inside the box and therefore are not considered. Internal clamps count as one conductor regardless of the number of clamps present in the box. The amount of space they take up is based on the largest conductor present in the box, **Figure 14-20**.

Metallic and nonmetallic boxes designed for raceways will have knockouts that fit standard trade size raceways and will not have internal clamps. When attaching raceways to these boxes, there will be threads from the connector or

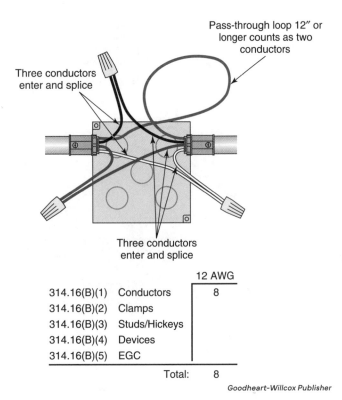

Figure 14-17. Box fill calculation for a junction box where one conductor passes through the box without stopping but is looped for future use. The other three conductors from each raceway are spliced. The raceway is being used as the equipment grounding conductor.

raceway and a locknut that are inside the box. The little bit of internal space taken by the threads and the locknut are minimal, so no allowance is necessary, **Figure 14-21**.

Support Fittings

Support fittings are installed inside a box to support a luminaire. They are not the same as a fixture strap, which mounts on the outside of the box. The two types of support fittings addressed by Section 314.16(B)(3) are fixture studs and hickeys, **Figure 14-22**. These fittings are not used very often anymore with standard luminaires, but they may still be used with exceptionally large and heavy chandeliers.

A *fixture stud* is a device that is installed inside an electrical box to support a luminaire. They are often secured directly to a framing member above the box so the weight of the fixture is supported independently of the electrical box. Fixture studs are threaded for the attachment of a hickey.

A *hickey* is a device that attaches a fixture nipple to the fixture stud. A hickey has internal threads that attach to the fixture stud on one side and the fixture nipple on the other. It has an opening that allows the wires that are traveling up the nipple to get into the box, **Figure 14-23**. With the use of a fixture stud and a hickey, there is no need for a fixture strap; they provide all the necessary support for a luminaire.

Luminaire support fittings, such as studs and hickeys, each have an allowance of one conductor. If both a fixture stud and a hickey are installed, they count as a total of two conductors. These allowances are based on the largest conductor in the box.

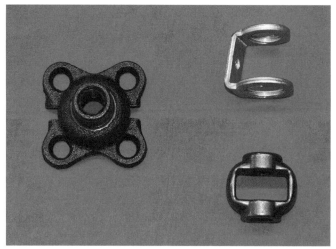

Goodheart-Willcox Publisher

Figure 14-22. A fixture stud (left) and a hickey (right).

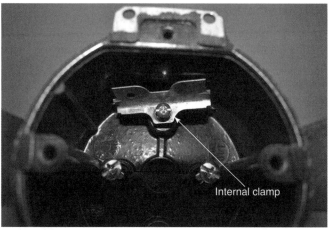

Goodheart-Willcox Publisher

Figure 14-20. This round lighting box has metallic clamps to secure nonmetallic sheathed cable to the box. They are given an allowance of one conductor, based on the largest conductor in the box.

Goodheart-Willcox Publisher

Figure 14-21. The threads and locknut that enter the box from a connector or nipple are not counted in the box fill calculation.

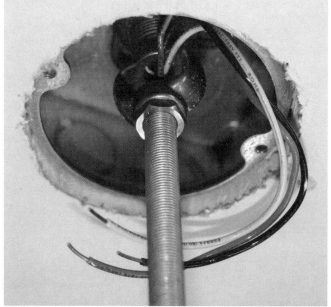

Goodheart-Willcox Publisher

Figure 14-23. This fixture stud and hickey are used to support a luminaire. The luminaire nipple threads directly into the hickey, allowing the fixture wires to enter the box.

Devices

Devices such as switches and receptacles protrude into the box and take up space, **Figure 14-24**. *Section 314.16(B)(4)* states that each yoke or strap containing a device has a volume allowance of two conductors. A *yoke* is the mounting bracket portion of a device that is used to fasten it to the box. The device allowance is based on the largest conductor attached to the device, **Figure 14-25**.

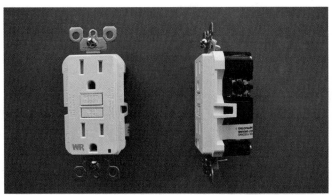

Goodheart-Willcox Publisher

Figure 14-24. The *NEC* gives devices such as GFCIs an allowance of two conductors. This is based on the largest conductor attached to the device.

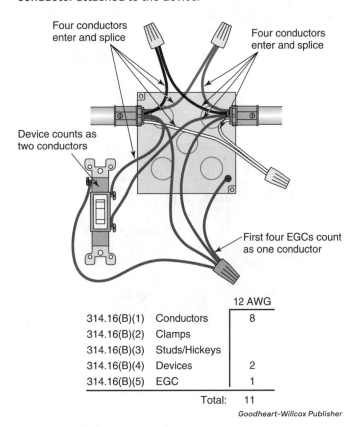

Goodheart-Willcox Publisher

Figure 14-25. Box fill calculation for a box where two conductors attach to a single pole switch and the other three conductors from each raceway are spliced. Each raceway has an equipment grounding conductor which is tailed to the box as well as the switch.

Larger devices, such as dryer and range receptacles, are typically mounted in a 2-gang box as they require more than 1-gang for mounting. The *NEC* gives larger devices a double volume allowance for each gang required for mounting. For example, a range receptacle mounted in a 2-gang box has an allowance of four conductors, **Figure 14-26**.

Equipment Grounding Conductor (EGC)

Equipment grounding conductors (EGC) are calculated separately from current carrying conductors. EGCs are covered in *Section 314.16(B)(5)*. The first four count as one conductor with each additional EGC adding 0.25 conductors. For example, two EGCs count as one conductor, four EGCs count as one conductor, and six EGCs count as 1.5 conductors. **Figure 14-27** shows a box fill calculation with an EGC. Equipment grounding conductors were counted differently prior to the 2020 edition of the *National Electrical Code*.

Allowance Table

Once the total number of conductors/allowances have been calculated, they must be equated to the amount of space they take up in the box. *Table 314.16(B)(1)* lists the equivalent amount of space that must be allotted for each conductor/allowance, **Figure 14-28**. For example, each 14 AWG conductor/allowance takes up 2 in^3, each 12 AWG conductor 2.25 in^3, and so on. It is best to add the number of conductor allowances that are in the box first, and then apply the appropriate multiplier found in *Table 314.16(B)(1)*, **Figure 14-29**.

Goodheart-Willcox Publisher

Figure 14-26. A range receptacle is an example of a device that is installed in a 2-gang box, so it has an allowance of four conductors, which is based on the largest conductor attached to the receptacle.

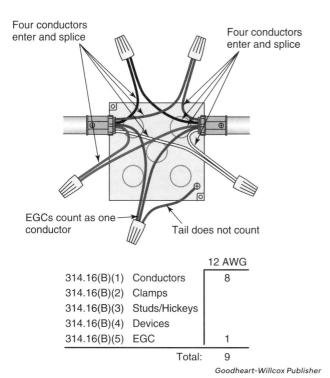

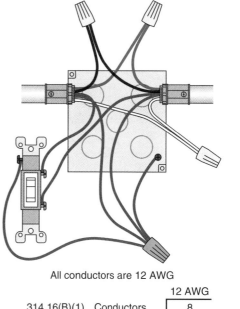

Figure 14-27. Box fill calculation for a junction box where four conductors from each raceway are spliced. Each raceway has an equipment grounding conductor that is tailed and attached to the box. The equipment grounding conductors are given an allowance of one conductor based on the largest EGC in the box.

Figure 14-29. Box fill calculation for a box where two conductors attach to a single pole switch and the other three conductors from each raceway are spliced. Each raceway has an equipment grounding conductor which is tailed to the box as well as the switch.

Table 314.16(B)(1) Volume Allowance Required per Conductor

Size of Conductor (AWG)	Free Space Within Box for Each Conductor	
	cm³	in.³
18	24.6	1.50
16	28.7	1.75
14	32.8	2.00
12	36.9	2.25
10	41.0	2.50
8	49.2	3.00
6	81.9	5.00

Reproduced with permission of NFPA from NFPA 70, National Electrical Code, 2023 edition. Copyright © 2022, National Fire Protection Association. For a full copy of the NFPA 70, please go to www.nfpa.org

Figure 14-28. Once the number of conductor allowances have been determined, *Table 314.16(B)(1)* lists the amount space that must be allotted based on the size of the conductors. The cubic inches value in the rightmost column of the table is multiplied by the number of allowances to get the total volume. If there are multiple size conductors in the box, each size conductor is calculated separately, and the results added together.

PRO TIP — Show Your Work

When calculating box fill, it is easy to make small mathematical errors if you are doing it in your head. This is especially true if you are multiplying the allowances by the *Table 314.16(B)(1)* value as you go. Showing your work can be as simple as writing on the side of a 2 × 4 stud, a piece of cardboard, or a small piece of scratch paper. Multiple conductors with different sizes can make calculations even trickier. **Figure 14-30** shows how to organize your calculations.

1. Draw a small table with a separate column for each conductor size.
2. Make rows where items present in the box are listed.
3. Tally the number of allowances in each row for all the conductor sizes present in the box.
4. Add the values for each respective conductor size together and multiply the sum by the volume found in *Table 314.16(B)(1)*.
5. The results of each column are added together to get the total.

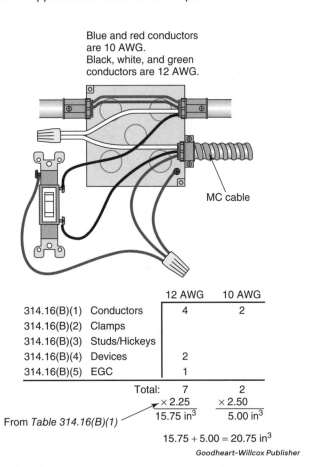

Figure 14-30. Box fill calculation with two raceways and an MC cable entering the box. The raceways contain both 12 AWG and 10 AWG conductors. The raceway is used as an equipment grounding conductor for the raceways, but the MC cable is a 12 AWG EGC that is attached to the box and tailed to the switch.

Conduit Bodies

Conduit bodies are defined as a separate portion of a conduit or tubing system that provides access through a removable cover to the interior of the system at a junction of two or more sections of the system or at a terminal point of the system. They are used to create a pull point and are used to change directions without having the sweep of a 90° bend, **Figure 14-31**. The quantity of 6 AWG and smaller conductors that are permitted to pass through a conduit body are based on the size of the raceways attached to it. This is covered in Unit 15 when performing conduit fill calculations.

Although rare, it is permitted to splice, tap, or install devices in conduit bodies if two requirements are met. First, it must be a full-size conduit body, not a short radius conduit body, **Figure 14-32**. *Short conduit bodies* are too small to house anything but the conductors passing through. Secondly, the conduit body must be stamped by the manufacturer with its available volume, which must not be exceeded, **Figure 14-33**. The calculation of the space occupied by conductors and devices within conduit bodies is calculated the same as it is for boxes.

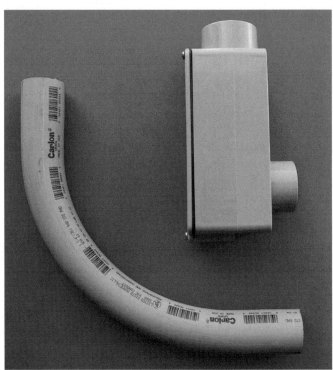

Figure 14-31. An LB type conduit body (upper right) and a 90° bend (lower left) both make a 90° change in direction. Conduit bodies have a removable cover to allow the conductors to be pulled out and fed back into the conduit in the new direction.

Figure 14-32. An SLB is a shorter conduit body that is used when there is not enough room for a standard LB. They are rarely used as they are difficult to feed conductors through.

Pull and Junction Boxes (1,000 Volts and Under)— NEC Section 314.28

Pull and junction boxes containing conductors 4 AWG and larger are calculated differently than the boxes housing smaller conductors, **Figure 14-34**. The following are applications where a large pull/junction box is used in which *Section 314.28* will apply:

- Provide a pull point in the middle of an exceptionally long conduit run.
- Provide a transition where conduits change direction.
- Provide a pull point to prevent the conduit run from having more than 360° of bend.
- Splice large conductors.
- Install power distribution blocks.

Goodheart-Willcox Publisher

Figure 14-33. Although rare, it is permissible to splice in an LB if the cubic inch capacity is not exceeded. This LB has a volume of 4.35 in³.

Goodheart-Willcox Publisher

Figure 14-34. Boxes containing conductors 4 AWG and larger are sized according to *Section 314.28*. The box pictured has six conduits with large conductors, all of which are making a straight pull.

Rather than looking at the size and number of conductors entering the box, it is sized by the raceways that enter the box. The trade size of the raceway will be used in conjunction with a multiplier to perform the calculations. There are two main categories of pull box calculations: straight pulls and angle pulls, U pulls, and splices.

Straight Pulls

With straight pulls, we are calculating the minimum distance from where the raceway enters the box to the opposite wall where it leaves, **Figure 14-35**. Multiply the trade size of the largest raceway by eight to find the minimum length (distance to the opposite wall) of the box, **Figure 14-36**. There is no minimum width other than being wide enough to install the locknut and bushing on the raceway.

Angle or U Pulls and Splices

Angle pulls, U pulls, and splices all have the same method of calculating the minimum dimensions of the box. The distance between the raceway and the opposite side of the box must be at least six times the trade size of the largest raceway. In addition to the minimum dimension requirements of the box, *Section 314.28(A)(2)* gives a minimum distance

Goodheart-Willcox Publisher

Figure 14-35. Straight pull boxes are generally used as a pull point when the conduit run will exceed 360°. The multiplier for a straight pull is eight times the trade size of the largest raceway.

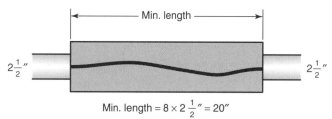

Goodheart-Willcox Publisher

Figure 14-36. A straight pull with 2 1/2″ conduit.

between raceways containing the same conductors. The raceways containing the same conductors must be separated by at least six times the largest raceway.

Angle Pull

An angle pull is where the conductors enter a box from a raceway and make a 90° turn before leaving the box, **Figure 14-37**. It is easier to tuck larger wires into a box with an angle pull than it is for a straight pull, which is why the multiplier for an angle pull is less than that of a straight pull.

Both the length and width of a box must be calculated for an angle pull, **Figure 14-38**. Start with one of the raceways and multiply its trade size times six to determine the minimum length of the box, X, **Figure 14-39**. This calculation is repeated for the other side of the box that has a raceway housing the same conductors that make the angle pull. This will determine the minimum width of the box, Y.

In addition to having a minimum length and width, it is important to remember to keep the raceways housing the same conductors at least six times the largest raceway apart, Z_s. This will determine where the raceways enter the box, and in some cases, may require a larger box than established by the length and width calculations.

The calculation for multiple raceways making angle pulls is found by taking the largest raceway in the row times six, and then adding all the other raceways in the same row, **Figure 14-40**. The resulting number is the minimum distance to the opposite side of the box. The calculation is repeated for the remaining sides of the box. All raceways

Figure 14-37. Pull boxes can provide a pull point for long conduit runs as well as change the direction of the conduit run. This box is located underground in a steam tunnel. There are several raceways that make a transition to run parallel with the tunnel.

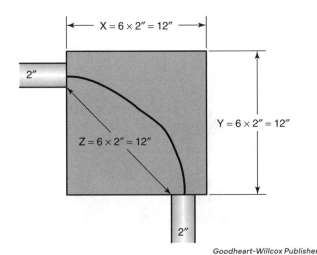

Figure 14-39. Angle pull calculation with 2″ raceways.

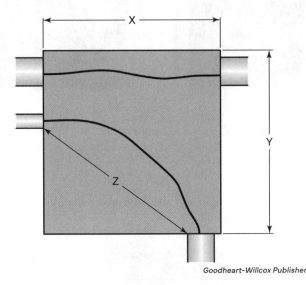

Figure 14-38. For consistency, this text will use X for the horizontal measurement, Y for the vertical measurement, and Z for the distance between raceways for all examples and activities.

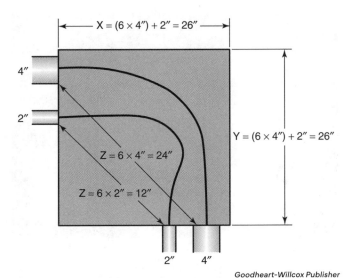

Figure 14-40. Angle pull calculation with multiple raceways on each side of the box.

containing the same conductors shall be at least six times the largest raceway apart.

If a box has multiple rows of raceways entering one side of the box, each row is calculated independently, and the largest result will determine the minimum distance to the opposite side of the box.

In the case of a box with a straight pull and an angle pull, the calculation is performed for each type of pull independently, and the largest result is used for the minimum box dimension, **Figure 14-41**.

Splice and Termination

Boxes that contain larger conductors that splice or connect to power distribution blocks shall have their minimum dimensions calculated in the same manner as angle pulls. This is covered in *Section 314.28(A)(2)*. In the case of power distribution blocks, the minimum dimensions of the box shall comply with both the manufacturer's instructions as well as the minimum wire bending space requirements found in *Section 312.6*. See **Figure 14-42**.

Angle Pull Leaving the Back of the Box

Occasionally, a raceway will enter the side of a box and exit the back of the box, **Figure 14-43**. Angle pulls with a raceway leaving the box opposite a removable cover have an additional step. The exception in *Section 314.28(A)(2)* states that the minimum distance from a removable cover to the back of the box is found using *Table 312.6(A)* for one wire per terminal based on the largest conductor in the raceway. For example, 4/0 AWG conductors that make an angle pull leaving the back of the box, opposite a removable cover, require a box that is at least 4″ deep. The minimum distance between raceways of six times the largest raceway still applies.

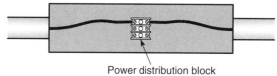

Power distribution block

Goodheart-Willcox Publisher

Figure 14-42. Power distribution blocks may be installed in a large box to splice and/or distribute the energy in multiple directions. In addition to the pull box requirements from *Section 324.28*, the minimum wire bending space from *Section 312.6* must also be followed.

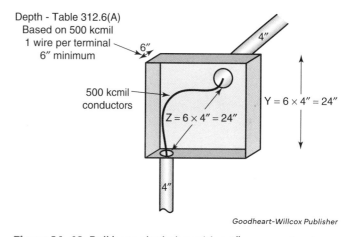

Goodheart-Willcox Publisher

Figure 14-43. Pull box calculation with a 4″ raceway leaving the back of the box opposite the cover. In this case, there is no minimum X dimension as no raceways enter the left or right side of the box.

Straight X = 8 × 3″ = 24″
Angle X = (6 × 4″) + 3″ = 27″

Goodheart-Willcox Publisher

Figure 14-41. A pull box with a 3″ straight pull and a 4″ angle pull.

Summary

- Sizing electrical boxes containing conductors 6 AWG and smaller are calculated according to *Section 314.16(A)* and *314.16(B)*.
- The volume of metallic boxes can be found in *Table 314.16(A)*.
- The volume of a nonmetallic box is marked on the box.
- When performing a box fill calculation:
  - Conductors originating outside the box and terminating or splicing in the box are given a single volume allowance.
  - Clamps are given a single volume allowance.
  - Devices are given a double volume allowance.
  - Up to four equipment grounding conductors are given a single volume allowance.

- The amount of space occupied by a conductor allowance is determined by *Table 314.16(B)(1)*.
- Sizing electrical boxes containing conductors 4 AWG and larger are calculated according to *Section 314.28*.
- When sizing straight pull boxes, eight times the largest raceway is the minimum distance to the opposite wall of the box.
- When sizing angle pull boxes, six times the largest raceway is the minimum distance to the opposite wall of the box.
- In angle pulls, raceways containing the same conductors shall be at least six times the largest raceway apart.

Unit 14 Review

Name _____ Date _____ Class _____

Know and Understand

Answer the following questions based on information in this unit.

1. A device box is manufactured to use _____ screws to attach the device to the box.
   A. 6-32   C. 10-32
   B. 8-32   D. 1/4-20

2. _____ lists the volume of standard size metallic boxes.
   A. *Table 310.16*   C. *Table 314.16(B)(1)*
   B. *Table 314.16(A)*   D. *Table 314.28*

3. When performing box fill calculations, internal clamps are allotted a _____ volume allowance based on the largest conductor in the box.
   A. single   C. triple
   B. double   D. quadruple

4. The multiplier used when calculating the size of a pull box that has a straight pull is _____ times the largest raceway.
   A. two   C. six
   B. four   D. eight

5. A non-metallic divider that is not stamped with the volume of space it takes up is allotted a value of _____.
   A. 1/2 cubic inch   C. 2 cubic inches
   B. 1 cubic inch   D. 3 cubic inches

6. The multiplier used when calculating the size of a pull box that has an angle pull is _____ times the largest raceway.
   A. two   C. six
   B. four   D. eight

7. When performing box fill calculations, a single gang device is allotted a _____ volume allowance based on the largest conductor in the box.
   A. single   C. triple
   B. double   D. quadruple

8. _____ lists the volume allowance in cubic inches for 18 AWG through 6 AWG conductors.
   A. *Table 310.16*   C. *Table 314.16(B)(1)*
   B. *Table 314.16(A)*   D. *Table 314.28*

9. The first four equipment grounding conductors are allotted a _____ volume allowance based on the largest EGC in the box.
   A. single   C. triple
   B. double   D. quadruple

10. When performing an angle pull, the minimum distance between two raceways containing the same conductor shall be separated by not be less than _____ times the largest raceway.
    A. three   C. six
    B. four   D. eight

Apply and Analyze

Answer the following questions using a copy of the National Electrical Code. *Identify the section or subsection where the answer is found.*

1. What is the volume of a 4 square box that is 1 1/2″ deep?
   A. 18 in^3 C. 29.5 in^3
   B. 21 in^3 D. 30.3 in^3

   NEC _____

2. What is the volume allowance of a 14 AWG conductor?
   A. 2 in^3 C. 2.5 in^3
   B. 2.25 in^3 D. 3 in^3

   NEC _____

3. Five 12 AWG conductors take up _____ of space in a box.
   A. 5 in^3 C. 11.25 in^3
   B. 10 in^3 D. 12.5 in^3

   NEC _____

4. Calculate the minimum volume of a 1-gang nonmetallic box that contains the following:
   - (1) 14-2 nonmetallic sheathed cable
   - (1) 14-3 nonmetallic sheathed cable
   - (1) Single-pole switch

   A. 7 in^3 C. 16 in^3
   B. 14 in^3 D. 18 in^3

   NEC _____

5. Calculate the volume of a 2-gang nonmetallic box with the following:
   - (2) 14-2 nonmetallic sheathed cables
   - (2) 14-3 nonmetallic sheathed cables
   - (2) Three-way switches

   A. 28 in³ C. 32 in³
   B. 30 in³ D. 36 in³

   NEC _____

6. Calculate the volume of a 1-gang nonmetallic box with the following:
   - (3) 12-2 non-metallic sheathed cables
   - (1) Duplex receptacle

   A. 20.25 in³ C. 24.3 in³
   B. 22 in³   D. 25 in³

   NEC _____

7. A metallic box containing a duplex receptacle has two raceways entering the box. One of the raceways has three 12 AWG conductors: black, red, and white. The other raceway has two 12 AWG conductors: red and white. All the wires are spliced or are terminated in the box. The raceway is being used as the equipment grounding conductor. What is the minimum volume box required for this situation?

   A. 13.5 in³   C. 18 in³
   B. 15.75 in³ D. 20.25 in³

   NEC _____

8. A metallic box has a 12-2 MC cable and two raceways entering the box. Each of the raceways has three 12 AWG conductors: black and white current carrying conductors and a green equipment grounding conductor. The MC cable has a black and white current carrying conductor and a green equipment grounding conductor. All the wires are spliced in the box. What is the minimum volume box required for this situation?

   A. 13.5 in³   C. 18 in³
   B. 15.75 in³ D. 20.25 in³

   NEC _____

9. A pull box contains 2 1/2" conduits that are making a straight pull. What is the minimum length of the box?

   A. 15" C. 18"
   B. 16" D. 20"

   NEC _____

10. A pull box contains two 4" conduits that are making an angle pull. What is the minimum distance to the opposite side of the box from the raceways?

    A. 24" C. 30"
    B. 28" D. 32"

    NEC _____

Name _____ Date _____ Class _____

Critical Thinking

Calculate the following box volumes using a copy of the National Electrical Code.

1. Calculate the volume of a box with the following:
   - (1) 1-gang nonmetallic box without internal clamps
   - (1) Duplex receptacle
   - (3) 14-2 with ground nonmetallic sheathed cables

   _____

2. Calculate the volume of a box with the following:
   - (1) 2-gang nonmetallic box with internal clamps
   - (2) Switches
   - (2) 14-2 with ground nonmetallic sheathed cables
   - (2) 14-3 with ground nonmetallic sheathed cables

   _____

3. Calculate the volume of a box with the following:
   - (1) 3-gang nonmetallic box with internal clamps
   - (3) Switches
   - (4) 14-2 with ground nonmetallic sheathed cables
   - (3) 14-3 with ground nonmetallic sheathed cables

   _____

4. Calculate the volume of a box with the following:
   - (1) 2-gang nonmetallic box with internal clamps
   - (1) Single-pole switch—terminated with 14 AWG
   - (1) Duplex receptacle—terminated with 12 AWG
   - (3) 14-2 NM with ground nonmetallic sheathed cables
   - (2) 12-2 NM with ground nonmetallic sheathed cables

   _____

5. Calculate the volume of a box with the following:
   - (1) Round nonmetallic ceiling box with internal clamps
   - (1) Fixture stud
   - (6) 14-2 with ground nonmetallic sheathed cables

   _____

6. Calculate the volume of a box with the following:
   - (1) Metallic box connected to electrical metallic tubing and MC cable
   - (2) 12 AWG conductors coming from one conduit—red, white
   - (3) 12 AWG conductors coming from one conduit—black, red, white
   - (3) 12-2 MC cables

   Note: All clamps are external.

   _____

7. Calculate the volume of a box with the following:
   - (1) Metallic box connected to electrical metallic tubing
   - (1) Duplex receptacle
   - (3) 12 AWG coming from one conduit
   - (5) 12 AWG coming from one conduit

   Note: Using the raceway as an equipment grounding conductor. All wires splice.

   _____

8. Calculate the volume of a box with the following:
   - (1) Round ceiling box with internal clamps
   - (1) Domed light fixture with three 18 AWG conductors
   - (2) 14-2 with ground nonmetallic sheathed cables

   _____

9. Calculate the volume of a box with the following:
   - (1) 4-gang nonmetallic box with internal clamps
   - (4) Switches
   - (5) 14-2 with ground nonmetallic sheathed cables
   - (4) 14-3 with ground nonmetallic sheathed cables

   _____

10. Calculate the volume of a box with the following:
    - (1) Metallic box connected to electrical metallic tubing
    - (2) Switches
    - (3) 12 AWG coming from one conduit
    - (5) 12 AWG coming from one conduit
    - (2) 12 AWG coming from one conduit

    Note: Using the raceway as an EGC. All wires splice.

   _____

Part 2 Application: Advanced NEC Topics

Calculate the following pull box dimensions lengths using a copy of the National Electrical Code.

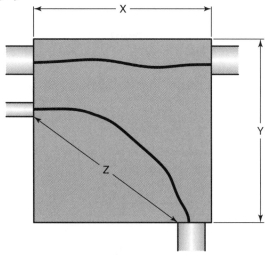

11.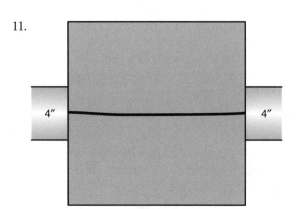

X = _____

12.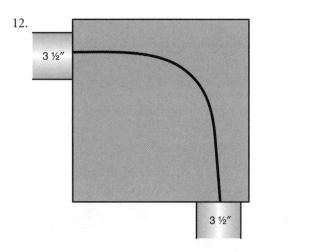

X = _____
Y = _____
Z = _____

13.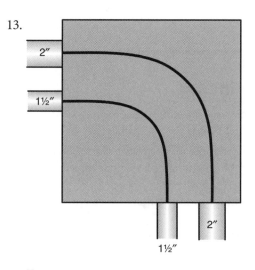

X = _____
Y = _____
Z (2") = _____
Z (1 1/2") = _____

14.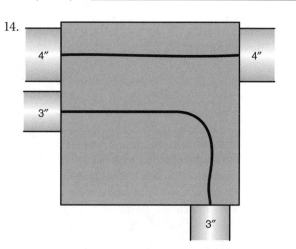

X = _____
Y = _____
Z = _____

15.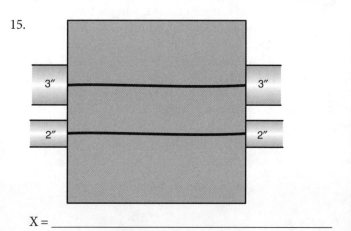

X = _____

16.

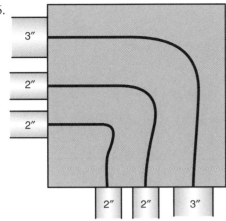

X = _____

Y = _____

Z (3″) = _____

Z (2″) = _____

17.

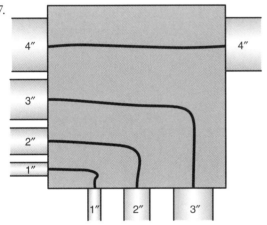

X = _____

Y = _____

Z (3″) = _____

Z (2″) = _____

Z (1″) = _____

18.

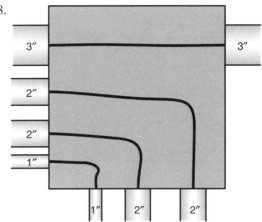

X = _____

Y = _____

Z (2″) = _____

Z (1″) = _____

19.

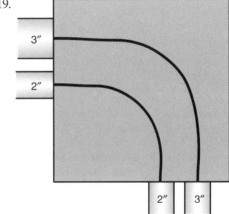

X = _____

Y = _____

Z (3″) = _____

Z (2″) = _____

20.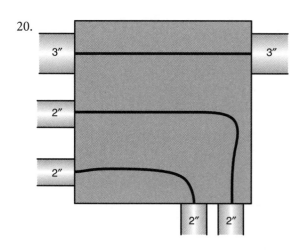

X = _____

Y = _____

Z = _____

UNIT 15
Raceway Fill Calculations

Sashkin/Shutterstock.com

KEY TERMS

auxiliary gutter
bare conductor
compact conductor
conductor
conduit
covered conductor
insulated conductor
nipple
raceway
tubing
wireway

LEARNING OBJECTIVES

After completing this unit, you will be able to:
- Identify the starting point for all raceway fill inquiries.
- Perform raceway fill calculations on circular raceways using the tables in *Chapter 9*.
- Perform raceway fill calculations on circular raceways using *Informative Annex C*.
- Calculate the maximum number of conductors permitted in noncircular raceways.
- Calculate the maximum permitted fill in noncircular raceways for splices and taps.

Introduction

The *National Electrical Code* limits the number and size of conductors that are permitted to be installed in a raceway. The main purpose of these limitations is to ensure that the conductors are not damaged during installation. Damage can occur if the conductors cross during installation (causing them to jam) or if excess pulling force is required to install them in the raceway. The problem is magnified in longer runs or when pulling through multiple bends. This unit will present how to calculate the maximum number of conductors permitted in several types of raceways.

Raceway Fill Definitions

Before beginning raceway fill calculations, it is important to understand the terminology of the components involved. The *NEC* defines a **raceway** as an enclosed channel designed expressly for holding wires, cables, or busbars, with additional functions as permitted. Typically, raceways are thought of as circular, but that is not always the case. They come in many shapes, sizes, and styles to fit the need of the application, **Figure 15-1**. As with other wiring methods, raceways must be listed and used within the limitations of the manufacturer's instructions and *NEC* requirements. Each of the raceways recognized by the *NEC* are defined within *Article 100*.

Conduit

A **conduit** is a circular raceway that houses and protects the conductors installed in an electrical system. There are many different types of conduits used in the electrical industry, such as nonmetallic, metallic, flexible, and liquidtight, **Figure 15-2**. Most conduit is capable of being bent in the field to change direction or elevation.

Goodheart-Willcox Publisher

Figure 15-1. Raceways come in many types and sizes. Raceways are chosen based on the environmental conditions, need for flexibility or protection, ease of installation, and cost.

Goodheart-Willcox Publisher

Figure 15-2. Three different types of conduits. Starting from the left: aluminum rigid metal conduit, rigid PVC conduit, and liquidtight flexible nonmetallic conduit.

Tubing

Tubing is a type of circular raceway that has a thinner wall and is therefore easier to bend and work with than some other circular raceways. There are three types of tubing recognized by the *NEC*: electrical metallic tubing (*Article 358*), electrical nonmetallic tubing (*Article 362*), and flexible metallic tubing (*Article 360*). See **Figure 15-3**.

Nipple

A **nipple** is a piece of conduit or tubing that is 24″ or less in length, **Figure 15-4**. Nipples have different fill capabilities than other raceways due to their short length.

Goodheart-Willcox Publisher

Figure 15-3. Two of the most commonly used types of tubing: electrical nonmetallic tubing (left), sometimes referred to as "Smurf tube," and electrical metallic tubing (right), the most-used raceway on commercial jobs.

Goodheart-Willcox Publisher

Figure 15-4. Any raceway that is 24″ or less in length is considered a nipple. Nipples are used to connect enclosures together.

Wireway

A *wireway* is a trough with a hinged or removable cover that houses and protects conductors and cables. It is sometimes referred to as a *gutter*. The two main types of wireways are metallic (*Article 376*) and nonmetallic (*Article 378*). They are typically installed in locations where conductors will be running back and forth between cabinets, boxes, and raceways, **Figure 15-5**.

Auxiliary Gutter

An *auxiliary gutter* is an enclosure to supplement wiring spaces at meter centers, distribution centers, switchgears, switchboards, and similar points of wiring systems. They have hinged or removable covers, and they are available in metallic and nonmetallic materials. Auxiliary gutters are very similar to wireways. In fact, the same enclosure can be classified as both. The main difference is the proximity of auxiliary gutters to the equipment it supplements and the limited length.

Conductor

A *conductor* is a wire made of copper, aluminum, or copper-clad aluminum that can carry current. It may be solid or stranded, insulated, covered, or bare. Conductors are generally installed in a raceway or are part of a cable assembly.

Bare Conductor

A *bare conductor* is defined as a conductor having no covering or electrical insulation, **Figure 15-6**. They are generally used for grounding and bonding.

Covered Conductor

A *covered conductor* is defined as a conductor encased within material of composition or thickness that is not recognized by the *Code* as electrical insulation. Covered conductors are not very common.

Insulated Conductor

An *insulated conductor* is defined as a conductor encased within material of composition and thickness that is recognized by the *Code* as electrical insulation, **Figure 15-7**. Most conductors installed in raceways are insulated. The conductor insulation is chosen based on the environment and temperature to which the conductor is exposed.

Compact Conductor

A *compact conductor* is a stranded conductor with irregularly shaped strands that have been formed together, eliminating the air spaces between the strands, **Figure 15-8**. Eliminating the space between strands allows for a conductor with a smaller overall diameter and the same current-carrying capacity.

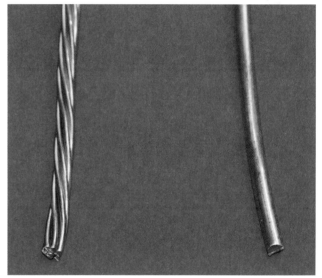

Goodheart-Willcox Publisher

Figure 15-6. Bare conductors are available in stranded and solid. The conductive material is visible.

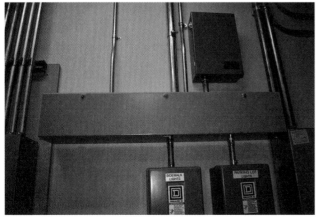

Goodheart-Willcox Publisher

Figure 15-5. This wireway contains conductors that run back and forth from between the surrounding electrical panels.

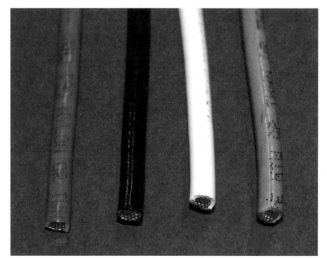

Goodheart-Willcox Publisher

Figure 15-7. Insulated conductors are available in a variety of colors and insulation types.

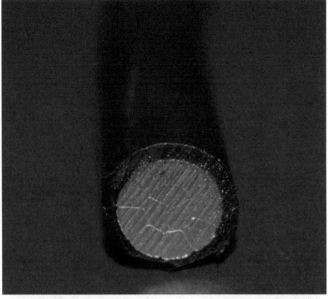

Figure 15-8. This is an aluminum compact conductor. Compact conductors have a smaller diameter than conductors with circular strands due to the air space between the strands being eliminated.

Figure 15-9. Installing more conductors than a raceway is rated for may lead to conductors jamming in corners and will require excess pulling force.

Circular Raceways

The *NEC* restricts the quantity and size of conductors that are permitted to be installed in circular raceways, **Figure 15-9**. This is to prevent damage to the conductors during installation. When being installed in a raceway, conductors tend to cross each other, which can cause them to jam if the raceway is overfilled. This is especially true when going around 90° bends. The restriction on the number of conductors also prevents excessive pulling force, which can cause damage.

Raceway Size

The *NEC* categorizes circular raceways by trade size. The inside diameter will be close to its trade size, but it is not an exact measurement. Instead, the trade size indicates the size of the knockout it is designed to enter.

Rigid metal conduit is an example of a raceway that enters directly into a knockout and is held in place with two locknuts, one inside the box and the other outside the box. Most other raceways have a connecting fitting that attaches the raceway to the box knockout, **Figure 15-10**.

Even among the same trade size, the internal diameter of raceways varies, **Figure 15-11**. For this reason, each type and size raceway have different available areas for conductors.

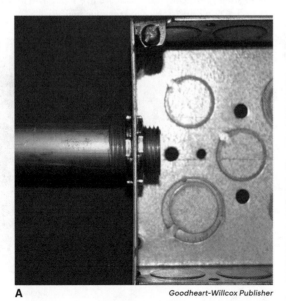

Figure 15-10. A—This rigid metal conduit is attached to the box with a locknut inside and outside of the box. B—This electrical metallic tubing attaches to the box with a set screw connector.

Goodheart-Willcox Publisher

Figure 15-11. Both raceways pictured are 2″ rigid PVC. The one on the left is Schedule 80, which has a thicker wall, and the one on the right is Schedule 40. The outside diameter of each raceway is the same because they use the same connectors and couplings. The thicker wall of Schedule 80 reduces the amount of area inside the raceway, affecting the amount of space for conductors.

Conductors

Conductors carry the current through an electrical system. They are available in many sizes, determined by the amount of current they need to carry. To keep the current flowing through the conductor and not everything else it touches, they are typically insulated. See **Figure 15-12**.

Conductor Insulation

Conductors are available with many different types of insulation. *Table 310.4(A)* lists many of the types of insulated conductors used in the electrical industry, **Figure 15-13**.

Goodheart-Willcox Publisher

Figure 15-12. Conductor insulation prevents current from leaving the conductor and traveling through other objects that it is in contact with. It is important to take care when installing the conductors to prevent damage to the insulation, which can lead to a ground fault or short circuit. The conductors in this picture were damaged when being installed in this LB which led to a ground fault to the conduit body.

Table 310.4(1) *Continued*

Trade Name	Type Letter	Maximum Operating Temperature	Application Provisions	Insulation	Thickness of Insulation			Outer Covering[1]
					AWG or kcmil	mm	mils	
Thermoset	SIS	90°C (194°F)	Switchboard and switchgear wiring only	Flame-retardant thermoset	14–10 8–2 1–4/0	0.76 1.14 1.40	30 45 55	None
Thermoplastic and fibrous outer braid	TBS	90°C (194°F)	Switchboard and switchgear wiring only	Thermoplastic	14–10 8 6–2 1–4/0	0.76 1.14 1.52 2.03	30 45 60 80	Flame-retardant, nonmetallic covering
Extended polytetra-fluoro-ethylene	TFE	250°C (482°F)	Dry locations only. Only for leads within apparatus or within raceways connected to apparatus, or as open wiring (nickel or nickel-coated copper only)	Extruded polytetra-fluoroethylene	14–10 8–2 1–4/0	0.51 0.76 1.14	20 30 45	None
Heat-resistant thermoplastic	THHN	90°C (194°F)	Dry and damp locations	Flame-retardant, heat-resistant thermoplastic	14–12 10 8–6 4–2	0.38 0.51 0.76 1.02	15 20 30 40	Nylon jacket or equivalent

Reproduced with permission of NFPA from NFPA 70, National Electrical Code, 2023 edition. Copyright © 2022, National Fire Protection Association. For a full copy of the NFPA 70, please go to www.nfpa.org

Figure 15-13. *Table 310.4(1)* has detailed information on conductor insulation and the conditions for which it is rated.

Over time, conductor insulation has advanced, with improvements in insulation quality and thickness. Although many of the types of conductors listed in *Table 310.4(A)* are not commonly used or are outdated, the *NEC* must include them all in case they are encountered.

Each type of insulation has been designed and rated for specific voltages, environmental conditions, and temperatures. Each type of insulation has a slightly different thickness, changing the overall diameter of the conductor and affecting the number of conductors that will fit in a raceway.

Circular Raceway Fill

The starting point for calculating raceway fill is the article devoted to each respective raceway. Since the raceways in *Chapter 3* follow the common numbering format, the appropriate section will be titled *3XX.22, Number of Conductors*. This is the first stop when determining the number of conductors that are permitted in a raceway. Be sure to read the Section in its entirety. If applicable, it will direct you to *Chapter 9, Table 1* where the fill calculations will begin.

Remember that *Section 90.3* states that the *Chapter 9* tables are applicable as referenced. Resist the temptation to jump right to *Chapter 9*. There may be additional information relating to raceway fill in *Section 3XX.22* of the raceway article that would be missed. Flexible metallic conduit (FMC) is an example of this. *Section 348.22* refers you to *Table 348.22* that has requirements for trade size 3/8″ FMC, **Figure 15-14**. Since 3/8″ flex is such a small raceway, having fittings that go inside the raceway instead of fittings that attach to the outside of the raceway makes a significant difference in raceway area. For all sizes other than 3/8″, the *Code* sends you back to *Chapter 9, Table 1*.

Permitted Conduit and Tubing Fill—Chapter 9, Table 1

After reading through *3XX.22*, you will be sent to *Chapter 9, Table 1*. See **Figure 15-15**. It lists the maximum permitted conductor fill as a percentage of three different conductor/cable combinations:

- **One conductor.** When installing one conductor or cable, it is permitted to be filled to 53% of the cross-sectional area. This is not very common because branch circuits will have a minimum of two conductors. An example is when running a grounding electrode conductor from the service disconnect to a grounding electrode, such as a water line.
- **Two conductors.** When installing two conductors in a raceway, it is permitted to be filled to 31% of the raceway's cross-sectional area. Two conductors have a lower permitted fill percentage due to them not being able to form into a group. They will always be side by side and the full width of two conductors. If their permitted area was too high, they would risk getting

Table 1 Percent of Cross Section of Conduit and Tubing for Conductors and Cables

Number of Conductors and/or Cables	Cross-Sectional Area (%)
1	53
2	31
Over 2	40

Reproduced with permission of NFPA from NFPA 70, National Electrical Code, 2023 edition. Copyright © 2022, National Fire Protection Association. For a full copy of the NFPA 70, please go to www.nfpa.org

Figure 15-15. *Chapter 9, Table 1* has the general fill requirements for circular raceways. The most-used row will be the one for more than two conductors.

Table 348.22 Maximum Number of Insulated Conductors in Metric Designator 12 (Trade Size 3/8) Flexible Metal Conduit (FMC)*

Size (AWG)	Types RFH-2, SF-2		Types TF, XHHW, TW		Types TFN, THHN, THWN		Types FEP, FEBP, PF, PGF	
	Fittings Inside Conduit	Fittings Outside Conduit	Fittings Inside Conduit	Fittings Outside Conduit	Fittings Inside Conduit	Fittings Outside Conduit	Fittings Inside Conduit	Fittings Outside Conduit
18	2	3	3	5	5	8	5	8
16	1	2	3	4	4	6	4	6
14	1	2	2	3	3	4	3	4
12	—	—	1	2	2	3	2	3
10	—	—	1	1	1	1	1	2

*In addition, one insulated, covered, or bare equipment grounding conductor of the same size shall be permitted.

Reproduced with permission of NFPA from NFPA 70, National Electrical Code, 2023 edition. Copyright © 2022, National Fire Protection Association. For a full copy of the NFPA 70, please go to www.nfpa.org

Figure 15-14. *Table 348.22* has additional information on 3/8″ flexible metal conduit not found in *Chapter 9*. *Table 348.22* lists the most common insulation types and the number of conductors permitted with each type of fitting.

wedged while going through bends in the conduit. More than two conductors will naturally form into a group rather than being side-by-side-by-side making them less likely to become wedged.

- **More than two conductors.** When installing more than two conductors, it is permitted to fill the raceway 40% of its cross-sectional area. This is the most common scenario.

Immediately following *Chapter 9*, *Table 1* are 10 *Notes to Tables* that provide additional information relating to raceway fill.

Note 1—Maximum Number of Conductors

The first note refers to *Informative Annex C* to find the maximum number of conductors when they are the same type and size (same cross-sectional area). This table is very useful and saves a lot of time. It has already done the math to determine the maximum number of conductors for each size raceway and type of conductor. The same information can be found by using *Table 4* and *Table 5* and performing the calculations. That method will be covered in the next section.

Informative Annex C begins with a table of contents. This table of contents is very useful as it lists the applicable table for each type of raceway and gives the page number where the table begins. It is very easy to inadvertently end up in the wrong table, so be sure to verify the correct raceway and conductor type at the top of the table, **Figure 15-16**.

The tables in *Informative Annex C* list the conductor type and size along the left side and the trade size of the raceway listed across the top, **Figure 15-17**. The maximum number of conductors will be listed at the intersection of the appropriate raceway and conductor.

Informative Annex C — Conduit, Tubing, and Cable Tray Fill Tables for Conductors and Fixture Wires of the Same Size

This informative annex is not a part of the requirements of this NFPA document but is included for informational purposes only.

Table	Page
C.1 — Electrical Metallic Tubing (EMT)	774
C.1(A)* — Electrical Metallic Tubing (EMT)	777
C.2 — Electrical Nonmetallic Tubing (ENT)	778
C.2(A)* — Electrical Nonmetallic Tubing (ENT)	781
C.3 — Flexible Metal Conduit (FMC)	782
C.3(A)* — Flexible Metal Conduit (FMC)	785
C.4 — Intermediate Metal Conduit (IMC)	786
C.4(A)* — Intermediate Metal Conduit (IMC)	789

Reproduced with permission of NFPA from NFPA 70, National Electrical Code, 2023 edition. Copyright © 2022, National Fire Protection Association. For a full copy of the NFPA 70, please go to www.nfpa.org

Figure 15-16. The table of contents is a useful tool to ensure you find the correct table.

PROCEDURE — Maximum Number of Conductors

The following is the process for determining the maximum number of 12 AWG THHN conductors that can be placed in a 3/4″ electrical metallic tubing.

1. *Section 358.22* leads us to *Chapter 9*.
2. *Note 1* leads us to *Informative Annex C* (All conductors the same size).
3. Using the table of contents for *Informative Annex C*, information for electrical metallic tubing (EMT) can be found in *Table C.1*.
4. THHN is the fourth type of conductor listed in *Table C.1*, **Figure 15-17**.
5. The 12 AWG row and 3/4 trade size column intersect at 16.

A 3/4″ electrical metallic tubing can hold sixteen 12 AWG THHN conductors.

Table C.1 *Continued*

Type	Conductor Size (AWG/kcmil)	Trade Size (Metric Designator)											
		3/8 (12)	1/2 (16)	3/4 (21)	1 (27)	1¼ (35)	1½ (41)	2 (53)	2½ (63)	3 (78)	3½ (91)	4 (103)	5 (129)	6 (155)
	1250	—	0	0	0	0	1	1	1	1	2	3	1	6
	1500	—	0	0	0	0	0	1	1	1	1	2	4	5
	1750	—	0	0	0	0	0	0	1	1	1	2	3	5
	2000	—	0	0	0	0	0	0	1	1	1	1	3	4
THHN, THWN, THWN-2	14	—	12	22	35	61	84	138	241	364	476	608	914	1290
	12	—	9	16	26	45	61	101	176	266	347	443	666	941
	10	—	5	10	16	28	38	63	111	167	219	279	420	593
	8	—	3	6	9	16	22	36	64	96	126	161	242	342
	6	—	2	4	7	12	16	26	46	69	91	116	175	247
	4	—	1	2	4	7	10	16	28	43	56	71	107	152
	3	—	1	1	3	6	8	13	24	36	47	60	91	128
	2	—	1	1	3	5	7	11	20	30	40	51	76	108
	1	—	1	1	1	4	5	8	15	22	29	37	56	80
	1/0	—	1	1	1	3	4	7	12	19	25	32	47	67
	2/0	—	0	1	1	2	3	6	10	16	20	26	40	56

Reproduced with permission of NFPA from NFPA 70, National Electrical Code, 2023 edition. Copyright © 2022, National Fire Protection Association. For a full copy of the NFPA 70, please go to www.nfpa.org

Figure 15-17. *Informative Annex C* lists the number of conductors permitted to be installed in each type and size of raceway when the conductors are all the same size.

Each of the raceways listed in *Informative Annex C* has a table for standard conductors with circular strands as well as a table for compact stranded conductors. The compact conductor table will have an *(A)* after it. For example, *Table C.1.(A)* is the table for EMT when installing compact conductors. Care must be taken to be sure you are in the correct table because compact conductors have the same types of insulation as standard conductors.

Note 6—Minimum Size Raceway

If you will have a combination of different size or type of conductors, you will have to perform a calculation to determine the minimum raceway size. *Note 6* states that *Chapter 9, Table 4* is used to determine the available space in conduit and tubing and that *Chapter 9, Table 5* is used to determine the amount of space a conductor occupies.

Table 4—Raceway Area

Note 6 states that *Table 4* is to be used to find the area of conduit or tubing, **Figure 15-18**. *Table 4* is separated into the commonly used circular raceways recognized by the *NEC*. The trade size of the raceway is listed along the left side of the table with a separate row for each size.

There are separate columns for each of the fill percentages prescribed by the *NEC*, as well as columns for raceway diameter and total area. The "Over 2 Wires 40%" is the most used column, so it is right next to the trade sizes of the raceway. The value found in that column is the amount of available space in the raceway when there are more than two wires installed. The *NEC* has already completed the math and multiplied the total area of the raceway by 40% to obtain the values.

The number at the intersection between the raceway size and the applicable column is the amount of space conductors are allowed to occupy in the raceway.

PROCEDURE — Raceway Area

The following is the process for determining the space available for conductors with a 1″ EMT with five conductors.

1. *Section 358.22 leads us to Chapter 9.*
2. *Note 6 in Chapter 9 leads us to use Table 4.*
3. The first part of *Table 4* contains requirements for electrical metallic tubing (EMT), **Figure 15-18**.
4. The 1″ trade size row and the over 2 wires column intersect at 0.346.

A 1″ EMT with five conductors has 0.346 in² of space available for conductors.

Table 5—Conductor Area

Note 6 states that *Table 5, Dimensions of Insulated Conductors and Fixture Wires* is to be used to find the area of insulated conductors, **Figure 15-19**. The left side of the table has a list of the insulated conductors that are recognized by the *NEC* for installation in raceways. Each conductor category (type of insulation) will have several rows, one for each conductor size.

The table is also divided into columns. The center two columns detail the approximate area of the conductors shown in millimeters squared as well as inches squared. This is the most commonly used column of the table.

Minimum Size Raceway

To calculate the minimum size raceway, you add up the area of the conductors, found in *Table 5*, and review *Table 4* to choose the raceway with enough space to fit the area of the conductors.

Table 4 Dimensions and Percent Area of Conduit and Tubing (Areas of Conduit or Tubing for the Combinations of Wires Permitted in Table 1, Chapter 9)

		Article 358 — Electrical Metallic Tubing (EMT)											
Metric Designator	Trade Size	Over 2 Wires 40%		60%		1 Wire 53%		2 Wires 31%		Nominal Internal Diameter		Total Area 100%	
		mm²	in.²	mm²	in.²	mm²	in.²	mm²	in.²	mm	in.	mm²	in.²
16	½	78	0.122	118	0.182	104	0.161	61	0.094	15.8	0.622	196	0.304
21	¾	137	0.213	206	0.320	182	0.283	106	0.165	20.9	0.824	343	0.533
27	1	222	0.346	333	0.519	295	0.458	172	0.268	26.6	1.049	556	0.864
35	1¼	387	0.598	581	0.897	513	0.793	300	0.464	35.1	1.380	968	1.496
41	1½	526	0.814	788	1.221	696	1.079	407	0.631	40.9	1.610	1314	2.036
53	2	866	1.342	1299	2.013	1147	1.778	671	1.040	52.5	2.067	2165	3.356
63	2½	1513	2.343	2270	3.515	2005	3.105	1173	1.816	69.4	2.731	3783	5.858
78	3	2280	3.538	3421	5.307	3022	4.688	1767	2.742	85.2	3.356	5701	8.846
91	3½	2980	4.618	4471	6.927	3949	6.119	2310	3.579	97.4	3.834	7451	11.545
103	4	3808	5.901	5712	8.852	5046	7.819	2951	4.573	110.1	4.334	9521	14.753
129	5	5220	8.085	7830	12.127	6916	10.713	4045	6.266	128.9	5.073	13050	20.212
155	6	7528	11.663	11292	17.495	9975	15.454	5834	9.039	154.8	6.093	18821	29.158

Reproduced with permission of NFPA from NFPA 70, National Electrical Code, 2023 edition. Copyright © 2022, National Fire Protection Association. For a full copy of the NFPA 70, please go to www.nfpa.org

Figure 15-18. *Chapter 9 Table 4* lists the fill requirements for each type and size of raceway.

PROCEDURE

Conductor Area

The following is the process for determining the area of a 1/0 AWG THHW conductor.

1. *Note 6* in *Chapter 9* leads us to use *Table 5*.
2. Scroll down through *Table 5* until the row containing THHW, **Figure 15-19**.
3. The 1/0 AWG THHW intersects the approximate area column at 0.2223.

A 1/0 AWG THHW conductor has an approximate area of 0.2223 in^2.

Table 5 Dimensions of Insulated Conductors and Fixture Wires

Type	Size (AWG or kcmil)	Approximate Area		Approximate Diameter	
		mm^2	in.^{2}	mm	in.
TF, TFF, XF, XFF	16	7.032	0.0109	2.997	0.118
TW, XF, XFF, THHW, THW, THW-2	14	8.968	0.0139	3.378	0.133
TW, THHW, THW, THW-2	12	11.68	0.0181	3.861	0.152
	10	15.68	0.0243	4.470	0.176
	8	28.19	0.0437	5.994	0.236
RHH*, RHW*, RHW-2*	14	13.48	0.0209	4.140	0.163
RHH*, RHW*, RHW-2*, XF, XFF	12	16.77	0.0260	4.623	0.182
Type: RHH*, RHW*, RHW-2*, THHN, THHW, THW, THW-2, TFN, TFFN, THWN, THWN-2, XF, XFF					
RHH,* RHW,* RHW-2,* XF, XFF	10	21.48	0.0333	5.232	0.206
RHH*, RHW*, RHW-2*	8	35.87	0.0556	6.756	0.266
TW, THW, THHW, THW-2, RHH*, RHW*, RHW-2*	6	46.84	0.0726	7.722	0.304
	4	62.77	0.0973	8.941	0.352
	3	73.16	0.1134	9.652	0.380
	2	86.00	0.1333	10.46	0.412
	1	122.6	0.1901	12.50	0.492
	1/0	143.4	0.2223	13.51	0.532
	2/0	169.3	0.2624	14.68	0.578
	3/0	201.1	0.3117	16.00	0.630
	4/0	239.9	0.3718	17.48	0.688

Reproduced with permission of NFPA from NFPA 70, National Electrical Code, 2023 edition. Copyright © 2022, National Fire Protection Association. For a full copy of the NFPA 70, please go to www.nfpa.org

Figure 15-19. *Chapter 9 Table 5* lists the approximate area and diameter for each size and type of conductor.

PROCEDURE

Finding the Correct Section

Because raceways are found in most installations, their requirements can be found in one of the first four chapters that apply generally to all installations. More specifically, the raceways are a wiring method and are installed and sized while roughing-in. This information leads to *Chapter 3, Wiring Methods and Materials*.

There is more than one way to find the correct section to start raceway fill calculations. Below lists the steps to find the correct sections using the table of contents and the index.

To find the correct section using the table of contents:

1. Determine where to find keywords: electrical metallic tubing, number of conductors.
2. Open table of contents to *Chapter 3*. Electrical metallic tubing is a general wiring method and installed during the rough-in of a building.
3. Locate *Article 358, Electrical Metallic Tubing (EMT)* in the table of contents.
4. Locate *Part II, Installation* in the table of contents.
5. Open *Part II* of *Article 358* and scan the sections to find *Section 358.22, Number of Conductors*.

To find the correct section using the index:

1. Determine where to find keywords: electrical metallic tubing, number of conductors.
2. Search index for electrical metallic tubing.
3. Narrow index search of electrical metallic tubing to "number of conductors."
4. Open to *Section 358.22, Number of Conductors*.

After the correct section has been found, using either the table of contents or the index, go to the tables in *Chapter 9* to complete the calculation.

- If all the same size and type conductors: *Informative Annex C*
- If multiple type or size conductors: *Table 4* to find the area of electrical metallic tubing; *Table 5* to find the area of the conductors.

EXAMPLE 15-1

Problem: The following conductors are to be installed in an electrical metallic tubing (EMT). Determine the total area of the conductors and the minimum size EMT.
- (2) 12 AWG THHN conductors
- (2) 10 AWG THHN conductors
- (3) 8 AWG THHN conductors

Table 5 Continued

Type	Size (AWG or kcmil)	Approximate Area mm²	Approximate Area in.²	Approximate Diameter mm	Approximate Diameter in.
TF, TFF, XF, XFF	16	7.032	0.0109	2.997	0.118
TW, XF, XFF, THHW, THW, THW-2	14	8.968	0.0139	3.378	0.133
TW, THHW, THW, THW-2	12	11.68	0.0181	3.861	0.152
	10	15.68	0.0243	4.470	0.176
TFN, TFFN	18	3.548	0.0055	2.134	0.084
	16	4.645	0.0072	2.438	0.096
THHN, THWN, THWN-2	14	6.258	0.0097	2.819	0.111
	12	8.581	0.0133	3.302	0.130
	10	13.61	0.0211	4.166	0.164
	8	23.61	0.0366	5.486	0.216
	6	32.71	0.0507	6.452	0.254
	4	53.16	0.0824	8.230	0.324

Reproduced with permission of NFPA from NFPA 70, National Electrical Code, 2023 edition. Copyright © 2022, National Fire Protection Association. For a full copy of the NFPA 70, please go to www.nfpa.org

Solution:
1. *Section 358.22 leads us to Chapter 9.*
2. *Note 6* leads us to *Table 5* to find the conductor areas.

$$12 \text{ AWG THHN}: 2 \times 0.0133 = 0.0266 \text{ in}^2$$
$$10 \text{ AWG THHN}: 2 \times 0.0211 = 0.0422 \text{ in}^2$$
$$8 \text{ AWG THHN}: 3 \times 0.0366 = 0.1098 \text{ in}^2$$

Adding the areas together...

$$0.0266 + 0.0422 + 0.1098 = 0.1786 \text{ in}^2$$

The total area of the conductors is 0.1786 in²

3. *Note 6* leads us to *Table 4* to determine the minimum size EMT

Table 4 Dimensions and Percent Area of Conduit and Tubing (Areas of Conduit or Tubing for the Combinations of Wires Permitted in Table 1, Chapter 9)

		Article 358 — Electrical Metallic Tubing (EMT)											
Metric Designator	Trade Size	Over 2 Wires 40% mm²	Over 2 Wires 40% in.²	60% mm²	60% in.²	1 Wire 53% mm²	1 Wire 53% in.²	2 Wires 31% mm²	2 Wires 31% in.²	Nominal Internal Diameter mm	Nominal Internal Diameter in.	Total Area 100% mm²	Total Area 100% in.²
16	½	78	0.122	118	0.182	104	0.161	61	0.094	15.8	0.622	196	0.304
21	¾	137	0.213	206	0.320	182	0.283	106	0.165	20.9	0.824	343	0.533
27	1	222	0.346	333	0.519	295	0.458	172	0.268	26.6	1.049	556	0.864
35	1¼	387	0.598	581	0.897	513	0.793	300	0.464	35.1	1.380	968	1.496

Reproduced with permission of NFPA from NFPA 70, National Electrical Code, 2023 edition. Copyright © 2022, National Fire Protection Association. For a full copy of the NFPA 70, please go to www.nfpa.org

3/4 EMT has a fill capacity of 0.213. The minimum size EMT is 3/4"

Special Conduit Fill Scenarios

The conduit fills as described in the previous section apply to complete raceway systems where the raceway leaves one box, runs a certain distance, and enters another box. That is the most common application, but there are other situations that must be addressed. The remaining *Notes to Tables* from *Chapter 9* cover some of the less-common scenarios.

Note 2—Physical Protection

Note 2 states that raceways used to protect exposed wiring from physical damage are not required to follow the fill requirements of *Table 1*. An example of this is underground service entrance cables that are being protected by Schedule 80 Rigid PVC as it emerges from underground and travels up to a meter socket, **Figure 15-20**. As long as the cable fits and can be installed without being damaged, it is permitted.

Goodheart-Willcox Publisher

Figure 15-20. Underground service entrance cable emerging from the ground is run through Schedule 80 Rigid PVC to provide physical protection for the cable.

> **PRO TIP** — **Cable Protection**
>
> Nonmetallic sheathed cable is often installed in a raceway to cover it up and offer protection. One example of this is inside kitchen cabinets. To prevent the products in the cabinet from banging into the cable, it can be installed in a raceway such as ENT, **Figure 15-21**.

Goodheart-Willcox Publisher

Figure 15-21. Electrical metallic tubing covering ENT in a cabinet will provide protection from whatever may be stored in this cabinet.

Note 4—Nipples

Note 4 contains requirements for short pieces of raceway nipples (24″ or less in length). Conduit nipples are permitted to be filled to 60%, **Figure 15-22**. Since nipples are not very long, it is easy to feed conductors into the raceway without the worry of conductors crossing or being jammed. Because of this, the *NEC* allows us to fill the conduit to a greater capacity.

Note 5—Conductors and Cables Not Listed in the NEC

Note 5 addresses conductors and cables that are not included in *Table 5*. It states that conductors and cables that are not included in *Table 5* will use their actual dimensions for calculating conduit fill. Once their dimensions have been determined, they are calculated in the same manner as all other conductors and cables.

Note 7—Conduit Fill Rounding

Note 7 details situations where it is permitted to round up to the next whole number in determining the maximum number of conductors that can be installed in a raceway. It states that when installing all the same size conductor, the next higher whole number of conductors is permitted if it results in a decimal of 0.8 or greater. This is only permitted when

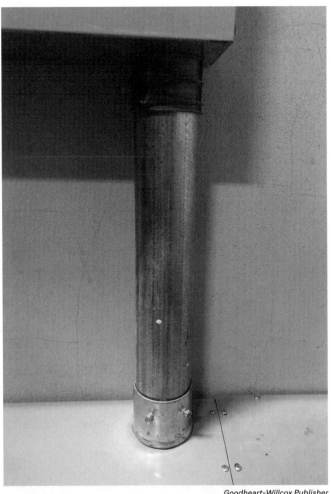

Goodheart-Willcox Publisher

Figure 15-22. An electrical metallic tubing nipple running from a metal wireway to an enclosure. Nipples are allowed to be filled to 60% of their cross-sectional area.

installing all the same conductors (overall cross-sectional area), which is also when we are permitted to use *Informative Annex C*. *Informative Annex C* has already taken this into consideration in its tables.

Note 8—Bare Conductors

Note 8 explains how to determine the cross-sectional area of bare conductors. When installing bare conductors in a raceway, *Chapter 9, Table 8* is used to determine the area of the conductors. Conductors sized 18 AWG through 8 AWG have two rows for each conductor, one for a solid conductor and the other for stranded. Under the heading of "Stranding," there is a column titled "Quantity." The quantity column represents the number of strands in the conductor. If it has 1 listed, it has only 1 strand and is considered a solid

Goodheart-Willcox Publisher

Figure 15-24. Elliptical cables, such as nonmetallic sheathed cable, are permitted to be installed in a raceway, provided the raceway is in an area that the cable is permitted. For example, the inside of a raceway installed outdoors is considered a wet location. Nonmetallic sheathed cable would not be permitted to be installed in an outdoor raceway, as it is not listed for wet locations.

Table 8 Conductor Properties

Size (AWG or kcmil)	Area		Conductors						Direct-Current Resistance at 75°C (167°F)					
			Stranding			Overall			Copper				Aluminum	
				Diameter		Diameter		Area	Uncoated		Coated			
	mm²	Circular mils	Quantity	mm	in.	mm	in.	mm²	in.²	ohm/km	ohm/kFT	ohm/km	ohm/kFT	ohm/km	ohm/kFT
18	0.823	1620	1	—	—	1.02	0.040	0.823	0.001	25.5	7.77	26.5	8.08	42.0	12.8
18	0.823	1620	7	0.39	0.015	1.16	0.046	1.06	0.002	26.1	7.95	27.7	8.45	42.8	13.1
16	1.31	2580	1	—	—	1.29	0.051	1.31	0.002	16.0	4.89	16.7	5.08	26.4	8.05
16	1.31	2580	7	0.49	0.019	1.46	0.058	1.68	0.003	16.4	4.99	17.3	5.29	26.9	8.21
14	2.08	4110	1	—	—	1.63	0.064	2.08	0.003	10.1	3.07	10.4	3.19	16.6	5.06
14	2.08	4110	7	0.62	0.024	1.85	0.073	2.68	0.004	10.3	3.14	10.7	3.26	16.9	5.17
12	3.31	6530	1	—	—	2.05	0.081	3.31	0.005	6.34	1.93	6.57	2.01	10.45	3.18
12	3.31	6530	7	0.78	0.030	2.32	0.092	4.25	0.006	6.50	1.98	6.73	2.05	10.69	3.25
10	5.261	10380	1	—	—	2.588	0.102	5.26	0.008	3.984	1.21	4.148	1.26	6.561	2.00
10	5.261	10380	7	0.98	0.038	2.95	0.116	6.76	0.011	4.070	1.24	4.226	1.29	6.679	2.04

Reproduced with permission of NFPA from NFPA 70, National Electrical Code, 2023 edition. Copyright © 2022, National Fire Protection Association. For a full copy of the NFPA 70, please go to www.nfpa.org

Figure 15-23. *Chapter 9, Table 8* has detailed information on bare conductors.

conductor. If it has 7 or more strands, it is a stranded conductor, **Figure 15-25**. Installing solid conductors in a raceway with other insulated conductors is not a common scenario.

Note 9

Note 9 addresses cables being pulled into a raceway system, **Figure 15-24**. It states that multiconductor cables are to be treated as one conductor. In the case of a multiconductor cable that is elliptical rather than circular, the area is calculated by using the longer dimension of the ellipse as the cable's diameter.

Note 10

Note 10 provides clarity on the differences between standard conductors with circular strands and compact conductors. It points out that *Table 5* is for standard (concentric-lay-stranded) conductors while *Table 5A* is for compact conductors.

| PROCEDURE | Bare Conductor Area |

The following is the process for determining the area of a 10 AWG solid bare conductor.

1. *Note 8* in *Chapter 9* leads us to use *Table 8*.
2. Scroll down through *Table 8* until the row containing 10 AWG, **Figure 15-23**.
3. Because this is a solid conductor, review the row with a quantity of 1.

A 10 AWG solid bare conductor has an area of 0.102 in^2.

Noncircular Raceways

The conductor fills for raceways that are not circular are found in *Section 3XX.22* of the applicable article just as it is for circular raceways. The raceway article will list the allowable fills as a percentage or it will identify where to go to find the information. The amount of space occupied by a conductor is found in *Chapter 9, Table 5* as it is for circular conductors.

Surface Raceways

Surface metallic and nonmetallic raceways are made in various sizes, making it difficult for the *NEC* to provide a standard way of calculating raceway fill. Instead, it defers to the manufacturer to list the maximum number of conductors permitted within their product. By going to the manufacturer's website or looking at their product literature, you can find the maximum fill requirements, **Figure 15-25**.

Wireways and Auxiliary Gutters

Metal wireways are covered in *Article 376*, and nonmetallic wireways are covered in *Article 378*, **Figure 15-26**. Auxiliary gutters are technically an enclosure rather than a raceway, but its fill requirements are the same as that of wireways. All three are permitted to be filled to 20% of their

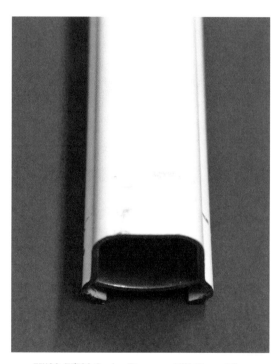

V500/V700 Series Raceway Fill Capacities

Wire Size (THHN/THWN)	OD		Number of Conductors 40% Fill	
	Inches	mm	V500 Series	V700 Series
14 AWG	0.111	2.8	7	10
12 AWG	0.130	3.3	5	7
10 AWG	0.164	4.2	3	4

Goodheart-Willcox Publisher

Figure 15-25. Wiremold® is a decorative surface metal raceway often used to add wiring in existing locations. The fill requirements will come from the manufacturer's literature.

Goodheart-Willcox Publisher

Figure 15-26. Wireways and auxiliary gutters are often the same piece of equipment. Pictured is metal wireway that is being used to move conductors back and forth between enclosures.

cross-sectional area with conductors. The area is found by multiplying the depth by the height. Once the available space has been determined, it can be divided by the area of the conductors found in *Chapter 9, Table 5* to determine the maximum number.

Unlike in circular raceways, it is permitted to splice or tap conductors in a wireway or auxiliary gutter. *Section 3XX.56* of the applicable article allows them to be filled up to 75% of their area with the combined of its conductors, taps, and splices.

EXAMPLE 15-2

Problem: How many 350 kcmil THHN conductors can be installed in a 4″ × 4″ metal wireway?

Solution: Determine the amount of area available for conductors in the metal wireway.

$$\text{Total area of the wireway} = 4'' \times 4''$$
$$= 16 \text{ in}^2$$
$$\text{Total fillable area} = 16 \text{ in}^2 \times 20\%$$
$$= 3.2 \text{ in}^2$$

The wireway has 3.2 in² of space available for conductors. Use *Table 5* to determine the approximate area of a 350 kcmil conductor.

Divide the available space by the area of the conductors.

$$3.2 \text{ in}^2 \div 0.5242 \text{ in}^2 = 6.1 \text{ in}^2$$

The 4″ × 4″ wireway can hold six 350 kcmil THHN conductors.

Table 5 *Continued*

Type	Size (AWG or kcmil)	Approximate Area mm²	Approximate Area in.²	Approximate Diameter mm	Approximate Diameter in.
Type: FEP, FEPB, PAF, PAFF, PF, PFA, PFAH, PFF, PGF, PGFF, PTF, PTFF, TFE, THHN, THWN, THWN-2, Z, ZF, ZFF, ZHF					
THHN, THWN, THWN-2	350	338.2	0.5242	20.75	0.817
	400	378.3	0.5863	21.95	0.864
	500	456.3	0.7073	24.10	0.949
	600	559.7	0.8676	26.70	1.051
	700	637.9	0.9887	28.50	1.122
	750	677.2	1.0496	29.36	1.156

Reproduced with permission of NFPA from NFPA 70, National Electrical Code, 2023 edition. Copyright © 2022, National Fire Protection Association. For a full copy of the NFPA 70, please go to www.nfpa.org

Summary

- Raceway fill calculations are covered in *Section 3XX.22* of the applicable article.
- *Informative Annex C* can be used when calculating conduit and tubing fill when the conductors are all the same size and type.
- Raceways are permitted to be filled to the following percentages:
- One conductor—53%
- Two conductors—31%
- Over two conductors—40%
- Nipples—60%
- *Chapter 9, Table 4* lists the dimensions and area of circular raceways.
- *Chapter 9, Table 5* lists the dimensions and area of conductors.
- *Chapter 9, Table 8* lists the dimensions and area of bare conductors.

Unit 15 Review

Name _____ Date _____ Class _____

Know and Understand

Answer the following questions based on information in this unit.

1. Conduit nipples are permitted to be filled to _____% of their total cross-sectional area.
   A. 31
   B. 40
   C. 53
   D. 60

2. The raceway fill requirements are found in _____ of the applicable article.
   A. *Section 3XX.2*
   B. *Section 3XX.20*
   C. *Section 3XX.22*
   D. *Section 3XX.24*

3. A(n) _____ is defined as a conductor encased within material of composition or thickness that is *not* recognized by the *Code* as electrical insulation.
   A. conductor
   B. covered conductor
   C. bare conductor
   D. insulated conductor

4. A nipple is a piece of conduit or tubing that is _____″ or less in length.
   A. 12
   B. 24
   C. 36
   D. 48

5. *Chapter 9*, _____ details the dimensions and area of conduit and tubing.
   A. *Table 1*
   B. *Table 2*
   C. *Table 4*
   D. *Table 5*

6. The number of conductors permitted in surface raceways can be found in _____.
   A. the manufacturer's literature
   B. *Chapter 9, Table 4*
   C. *Chapter 9, Table 12(A)*
   D. *Informative Annex C*

7. A conduit with five conductors is permitted to be filled to _____% of its total cross-sectional area.
   A. 31
   B. 40
   C. 53
   D. 60

8. The area of compact conductors is found in *Chapter 9*, _____.
   A. *Table 1*
   B. *Table 4*
   C. *Table 5*
   D. *Table 5A*

9. Which of the following do conduit fill requirements *not* apply to?
   A. Cables installed in short sections of a raceway for physical protection only
   B. Cables installed in a raceway
   C. A raceway with only one conductor installed
   D. Conduit nipples

10. A conduit containing two conductors is permitted to be filled to _____% of its total cross-sectional area.
    A. 31
    B. 40
    C. 53
    D. 60

Apply and Analyze

Answer the following questions using a copy of the National Electrical Code. Identify the section or subsection where the answer is found.

1. What is the total cross-sectional area of a 1″ electrical metallic tubing?
   A. 0.533 in^2
   B. 0.508 in^2
   C. 0.832 in^2
   D. 0.864 in^2

   NEC _____

2. What is the area of a 3 AWG THHW conductor?
   A. 0.0973 in^2
   B. 0.1134 in^2
   C. 0.1333 in^2
   D. 0.1521 in^2

   NEC _____

3. When installing six conductors in a 3/4″ rigid metallic conduit, how much space is available for the conductors?
   A. 0.216 in^2
   B. 0.220 in^2
   C. 0.235 in^2
   D. 0.320 in^2

   NEC _____

4. What is the area of a compact 4/0 XHHW conductor?
   A. 0.2733 in^2
   B. 0.2780 in^2
   C. 0.3237 in^2
   D. 0.3197 in^2

   NEC _____

5. How much space is available for conductors in a 1/2″ EMT nipple?
   A. 0.122 in^2
   B. 0.114 in^2
   C. 0.182 in^2
   D. 0.190 in^2

   NEC _____

6. How many 10 AWG THHN conductors are permitted in a 1″ rigid PVC conduit, Schedule 80?
   A. 12
   B. 13
   C. 15
   D. 16

   NEC _____

7. Conductors installed in a cellular metal floor raceway shall not exceed ____% of the interior cross-sectional area of the cell or header.
   A. 20
   B. 25
   C. 30
   D. 40

   NEC _____

8. A 1 5/8″ × 1 5/8″ strut-type channel raceway that is fastened together with internal joiners has _____ of space available for conductors.
   A. 0.507 in^2
   B. 0.671 in^2
   C. 0.811 in^2
   D. 2.208 in^2

   NEC _____

9. How many 6 AWG THHN compact conductors are permitted to be installed in a 1″ rigid metallic conduit?
   A. 6
   B. 7
   C. 8
   D. 9

   NEC _____

10. A Schedule 40 rigid PVC that is installed under concrete will have one 2/0 copper THWN grounding electrode conductor pulled into it. What is the minimum trade size raceway required for this installation?
    A. 1/2″
    B. 3/4″
    C. 1″
    D. 1 1/4″

    NEC _____

Name _____ Date _____ Class _____

Critical Thinking

Complete the following raceway fill calculations using a copy of the National Electrical Code. *Assume all conductors have circular strands unless otherwise noted. Identify the section or subsection used to find the answer.*

1. What is the maximum number of 6 AWG XHHW conductors that can be installed in a 1 1/4″ EMT (electrical metallic tubing)?

   _____

   NEC _____

2. What is the maximum number of 10 AWG THW conductors that can be installed in a 1″ Schedule 40 Rigid PVC raceway?

   _____

   NEC _____

3. What is the minimum size EMT (electrical metallic tubing) required for the following conductors:
   - (6) 12 AWG THHN
   - (3) 10 AWG THHN

   Total conductor area _____

   Minimum trade size raceway _____

   NEC _____

4. What is the minimum size IMC (intermediate metal conduit) required for the following conductors:
   - (4) 500 kcmil THW
   - (1) 1/0 AWG THW

   Total conductor area _____

   Minimum trade size raceway _____

   NEC _____

5. What is the minimum size FMC (flexible metal conduit) required for the following conductors:
   - (3) 4 AWG THHN
   - (1) 10 AWG THHN
   - (4) 14 AWG THHN

   Total conductor area _____

   Minimum trade size raceway _____

   NEC _____

6. What is the minimum size Schedule 40 rigid PVC required for the following conductors:
   - (2) Compact 4/0 AWG XHHW
   - (1) Compact 2/0 AWG XHHW

   Total conductor area _____

   Minimum trade size raceway _____

   NEC _____

7. What is the minimum size EMT (electrical metallic tubing) required for the following conductors:
   - (3) 12 AWG RHW (without outer cover)
   - (4) 14 AWG THW
   - (6) 14 AWG THHN

   Total conductor area _____

   Minimum trade size raceway _____

   NEC _____

8. What is the minimum size LFMC (liquidtight flexible metal conduit) required for the following conductors:
   - (3) 6 AWG THWN
   - (1) 10 AWG THWN

   Total conductor area _____

   Minimum trade size raceway _____

   NEC _____

9. How many 500 kcmil THHN conductors are permitted to be installed in an auxiliary gutter that measures 6″ × 6″?

   Available area of the auxiliary gutter _____

   Number of conductors _____

   NEC _____

10. An existing 3/4″ RMC (rigid metal conduit) has four 10 AWG TW conductors in it. How many 12 AWG THHW conductors could be added to this conduit?

    _____

    NEC _____

Complete the following raceway fill calculations using a copy of the National Electrical Code. *Assume all conductors have circular strands unless otherwise noted. Identify the section or subsection used to find the answer.*

11. What is the maximum number of 6 AWG XHHW conductors that can be installed in a 1 1/4″ schedule 80 rigid PVC?

    _____

    NEC _____

12. What is the maximum number of compact 250 kcmil THHW conductors that can be installed in a 3″ EMT (electrical metallic tubing) raceway?

    _____

    NEC _____

13. What is the minimum size ENT (electrical nonmetallic tubing) required for the following conductors:
    - (3) 8 AWG THHN
    - (1) 10 AWG THHN

    Total conductor area_____

    Minimum trade size raceway _____

    NEC _____

14. What is the minimum size RMC (rigid metal conduit) required for the following conductors:
    - (4) 4/0 AWG Compact THHW
    - (1) 2 AWG Compact THHW

    Total conductor area_____

    Minimum trade size raceway _____

    NEC _____

15. What is the minimum size LFNC-B (liquidtight flexible nonmetallic conduit) required for the following conductors:
    - (3) 6 AWG THWN
    - (1) 10 AWG THWN
    - (4) 12 AWG THWN

    Total conductor area_____

    Minimum trade size raceway _____

    NEC _____

16. What is the minimum size schedule 80 rigid PVC nipple required for the following conductors:
    - (3) 250 kcmil XHHW
    - (1) 1/0 AWG XHHW

    Total conductor area_____

    Minimum trade size raceway _____

    NEC _____

17. What is the minimum size EMT (electrical metallic tubing) required for the following conductors:
    - (6) 12 AWG RHW (with outer cover)
    - (4) 12 AWG THW
    - (8) 14 AWG THHN

    Total conductor area_____

    Minimum trade size raceway _____

    NEC _____

18. What is the minimum size FMC (flexible metal conduit) required for the following conductors:
    - (3) 2 AWG THHN
    - (2) 8 AWG Stranded and Uninsulated (Bare)

    Total conductor area_____

    Minimum trade size raceway _____

    NEC _____

19. How many 350 kcmil XHHW conductors are permitted to be installed in a metal wireway that measures 8″ × 8″?

    Available area of the wireway _____

    Total conductors _____

    NEC _____

20. What is the maximum number of 12 AWG THHN conductors permitted in a 3/8″ FMC (flexible metal conduit) that is terminated with fittings inside the conduit?

    _____

    _____

    NEC _____

UNIT 16 Conductor Ampacity

Sashkin/Shutterstock.com

KEY TERMS

adjustment factor
aluminum
ampacity
cable
copper
correction factor
insulated conductor
location, damp
location, dry
location, wet
parallel conductors

LEARNING OBJECTIVES

After completing this chapter, you will be able to:
- Utilize *NEC Table 310.4(A)* to identify conductor characteristics and ratings.
- Calculate the maximum ampacity for a given conductor based on *Table 310.16*.
- Apply conductor ampacity correction factors.
- Apply conductor adjustment factors.
- Describe the correct temperature rating for conductors based on insulation type and application.
- Calculate rooftop conductor ampacities.
- Calculate the ampacity of parallel conductors.

Introduction

Conductors have an important role in an electrical system. They carry electric current from the source of energy to the load that utilizes it. Conductors are typically insulated to contain the dangerous voltages present and prevent the conductive portion of the conductor from unintentionally contacting other objects or people.

Conductors must be sized and installed in a manner that will prevent them from exceeding their ampacity which can lead to overheating and jeopardizing the integrity of the conductor insulation. The *NEC* defines ***ampacity*** as the maximum current, in amperes, that a conductor can carry continuously under the conditions of use without exceeding its temperature rating.

This unit will describe how to calculate the amount of current a conductor can safely carry and how to determine the correct conductor size and insulation type.

General Conductor Information

Nearly all the current-carrying conductors in an electrical system will be insulated. The *NEC* defines an **insulated conductor** as a conductor that is encased within material of composition and thickness that is recognized by the *Code* as electrical insulation, **Figure 16-1**. The insulation prevents the conductive portion of the conductor from touching the raceway or box that it meets. As electricians, we need to ensure the conductor is installed in a manner that will prevent its insulation from becoming damaged. Damage can occur during installation, from excessive heat, from environmental conditions, or a number of other factors. Choosing the correct insulation and sizing the conductor appropriately will prevent the conductor from producing an electrical hazard that could lead to electric shock, arc flash, or fire.

Conductor Material

There are three types of conductors recognized by the *Code*: copper, aluminum, and copper-clad aluminum. The two most common types of conductors installed are copper and aluminum, **Figure 16-2**.

Copper

One of the advantages of using a conductor made of **copper** is that it has a lower resistance than aluminum. That means the copper conductor can carry more current than an aluminum conductor of the same size. For example, a 6 AWG THW copper conductor is rated to carry 65 amperes, while a 6 AWG THW aluminum conductor is only rated for 50 amperes. To have an aluminum conductor rated for 65 amperes, you would have to upsize to a 4 AWG.

Another advantage of copper is it has better terminating characteristics than aluminum. When exposed to temperature variations, copper expands and contracts less than aluminum. Excess expansion and contraction in terminations can lead to a loose connection.

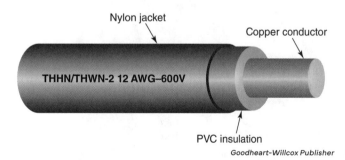

Figure 16-1. Building wire and conductors have identifying information printed on the side as required by *Section 310.8(A)*. It will list the maximum voltage, insulation type/letter designation, size, and manufacturer information.

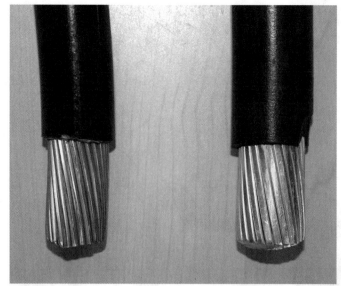

Goodheart-Willcox Publisher

Figure 16-2. The left conductor is a copper conductor with standard circular strands. The aluminum conductor on the right has compact strands.

With copper's advantages, it is the most used branch circuit conductor. The *NEC* has been written with that in mind, *Section 110.5* states that if the conductor insulation is not specified the *Code* is referring to copper conductors.

Aluminum

Aluminum's biggest advantage is its cost. It is much less expensive to purchase than copper. Even after upsizing, the price is much less than aluminum. A second advantage aluminum has over copper is its weight. An aluminum conductor weighs about half that of a copper conductor with a similar ampacity. The third advantage is it is much easier to bend and form large aluminum conductors than copper. This factors in when terminating the conductors and when pulling through pull boxes and conduit bodies. The reduction in weight and ease of bending makes the electrician's job less strenuous, leading to fewer injuries and greater productivity.

There are a few disadvantages to using aluminum conductors. The first disadvantage is having to upsize due to its higher resistance. Upsizing the conductor often means increasing the size of its raceway. A second disadvantage is that it tends to corrode or oxidize when exposed to air. This can be exacerbated when there is moisture present and large temperature fluctuations. Many parts of the country will require an antioxidizing paste to coat the aluminum as soon as it has been exposed to air. Most conductor manufactures are now using aluminum alloys that resist oxidation, but many inspectors are still requiring antioxidizing paste, **Figure 16-3**. A third disadvantage is that aluminum conductors expand and contract more than copper conductors. This becomes an issue at terminations and is magnified by large temperature variations. It is extremely important to

Figure 16-3. Antioxidizing paste is used to coat aluminum conductors after the insulation has been stripped off for termination. This will prevent oxidation when the conductor is exposed to air.

torque the terminations to the manufacturer's specifications to help prevent issues that arise from poor connections.

Copper-Clad Aluminum

A copper-clad aluminum conductor is defined as a conductor drawn from a copper-clad aluminum rod, with the copper metallurgically bonded to an aluminum core, where the copper forms a minimum of 10% of the cross-sectional area of a solid conductor or each strand of a stranded conductor.

Copper-clad aluminum conductors are treated like aluminum conductors when determining their ampacity. Their advantage is they have better terminating characteristics than an aluminum conductor and cost less than copper.

> **PRO TIP** — **Conductor vs Cable**
>
> A conductor is an individual wire. Conductors are available as bare, covered, or insulated. A *cable* is a factory assembly of conductors. Cables may have a protective sheath surrounding the conductors, but that is not always the case. For example, underground service cables and overhead service cables are conductors that are slightly twisted together to make them into a cable, **Figure 16-4**. Most cables are made of multiple conductors, however there are a few types of cables that only contain a single conductor, such as some tray cables.

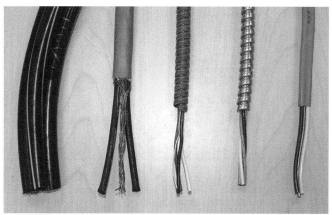

Figure 16-4. Cables are intended to be installed underground or in buildings without the need for a raceway. From left to right: underground service entrance (USE) cable, SEU service entrance type-U (SEU) cable, armored clad (AC) cable, metal-clad (MC) cable, nonmetallic (NM) sheathed cable.

Uses Permitted—Section 310.10

Conductors must have an insulation that is rated from the environmental conditions that it will be exposed to. *Section 310.10, Uses Permitted* and has requirements for various locations and conditions.

Location—Dry, Damp, Wet

Conductor locations are divided into dry, damp, and wet locations. The *NEC* defines the three locations as follows:

Location, dry. A location not normally subject to dampness or wetness. A location classified as dry may be temporarily subject to dampness or wetness, as in the case of a building under construction, **Figure 16-5A**.

Location, damp. A location protected from weather and not subject to saturation with water or other liquids but subject to moderate degrees of moisture, **Figure 16-5B**.

Location, wet. An installation underground or in concrete slabs or masonry in direct contact with the earth; in locations subject to saturation with water or other liquids, such as vehicle washing areas; and in unprotected locations exposed to weather, **Figure 16-5C**.

Conductors must be chosen that are listed for the type of location where they will be installed, even when inside a raceway. The *NEC* considers the inside of a raceway installed in a wet location or underground to be a wet location (*Section 300.5(B)* and *Section 300.9*). It is likely that water will find its way into the raceway and may even get trapped inside. For this reason, conductors installed in raceways outside or underground must be listed for wet locations.

Direct Sunlight

Sunlight will deteriorate the insulation of conductors not listed to be in direct sun. Although it is not very common to

Figure 16-5. Dry, damp, and wet locations. A—The rough-in of a commercial building. Even though the area may get wet during construction, once the building has been finished, it will no longer have moisture present, so it is considered a dry location. B—The receptacle on the underside of this soffit is considered a damp location. It will be subject to moderate degrees of moisture but will not be saturated with water. C—Equipment located outside is in a wet location. The enclosures, raceways, and fittings must be listed for use in wet locations.

find conductors in direct sunlight, there are a few applications where it is necessary. One example is overhead service conductors connecting to a service drop, **Figure 16-6**.

Direct Burial

Conductors buried directly in the ground must be rated for direct burial. Underground conductors are exposed to moisture, minerals, bugs, and sometimes rodents. Care must be taken when installing a cable directly in the ground to prevent sharp rocks and other objects from damaging the insulation. A nick in the insulation of an aluminum conductor buried in the ground will quickly deteriorate the conductor, **Figure 16-7**.

Conductor Specifications—Table 310.4(1)

Conductors installed in raceways are classified as building wire and are often referred to by their insulation type. *Table 310.4(1)* lists the different types of building wires by their trade name and letter designation, **Figure 16-8**.

Trade Name—Letter Designation

The trade name listed gives a description of the type of insulation used on the conductor. The column titled "Type Letter" is how conductors are described in the field. For example, THHN is one of the most commonly used conductors.

Figure 16-6. The service drop and overhead service conductors attached to the service drop are exposed to sunlight and must be listed accordingly.

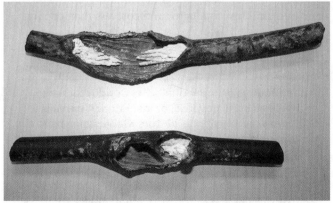

Figure 16-7. This aluminum underground service entrance (USE) cable was nicked and became deteriorated, leading to an open circuit.

Table 310.4(1) *Continued*

Trade Name	Type Letter	Maximum Operating Temperature	Application Provisions	Insulation	Thickness of Insulation			Outer Covering[1]
					AWG or kcmil	mm	mils	
Thermoset	SIS	90°C (194°F)	Switchboard and switchgear wiring only	Flame-retardant thermoset	14–10 8–2 1–4/0	0.76 1.14 1.40	30 45 55	None
Thermoplastic and fibrous outer braid	TBS	90°C (194°F)	Switchboard and switchgear wiring only	Thermoplastic	14–10 8 6–2 1–4/0	0.76 1.14 1.52 2.03	30 45 60 80	Flame-retardant, nonmetallic covering
Extended polytetra-fluoro-ethylene	TFE	250°C (482°F)	Dry locations only. Only for leads within apparatus or within raceways connected to apparatus, or as open wiring (nickel or nickel-coated copper only)	Extruded polytetra-fluoroethylene	14–10 8–2 1–4/0	0.51 0.76 1.14	20 30 45	None
Heat-resistant thermoplastic	THHN	90°C (194°F)	Dry and damp locations	Flame-retardant, heat-resistant thermoplastic	14–12 10 8–6 4–2 1–4/0 250–500 501–1000	0.38 0.51 0.76 1.02 1.27 1.52 1.78	15 20 30 40 50 60 70	Nylon jacket or equivalent
Moisture-and heat-resistant thermoplastic	THHW	75°C (167°F) 90°C (194°F)	Wet location Dry location	Flame-retardant, moisture-and heat-resistant thermoplastic	14–10 8 6–2 1–4/0 213–500 501–1000 1001–2000	0.76 1.14 1.52 2.03 2.41 2.79 3.18	30 45 60 80 95 110 125	None

Reproduced with permission of NFPA from NFPA 70, National Electrical Code, 2023 edition. Copyright © 2022, National Fire Protection Association. For a full copy of the NFPA 70, please go to www.nfpa.org

Figure 16-8. *Table 310.4(1)* has detailed information about building wire insulation. This table helps identify the temperature ratings of conductors. XHHW is rated for 90°C in dry locations but only 75°C in wet locations.

Its trade name is heat-resistant thermoplastic, but it is generally referred to as THHN.

Application Provisions

The "Application Provisions" column states if a conductor is listed for dry, damp, or wet locations. It goes on to give a short description of a few of the conductors listed for very specific applications.

Maximum Operating Temperature

The "Maximum Operating Temperature" column lists the maximum operating temperature the insulation is rated for. There are a few conductors that have more than one maximum operating temperature listed. For example, THHW lists both 75°C and 90°C, **Figure 16-8**. This is because the conductor is rated for 75°C in wet locations and 90°C in dry locations. Where the conductor will be installed will determine its maximum operating temperature.

Conductor Ampacities

Determining the minimum conductor size for a specific load is an important part of an electrician's job. Having a conductor that is too small for the load may overheat the conductor insulation and lead to a fire. Choosing a conductor that is larger than necessary is not cost effective. *Sections 310.14, 310.15,* and *310.16* detail the process of determining the maximum ampacity of conductors.

Ampacity Provisions

Section 310.14, Ampacities for Conductors Rated 0 Volts–2000 Volts starts off giving two ways to determine conductor ampacity. The first method is to go to *Section 310.15* and use the ampacity tables. The second method is a mathematical formula that can only be applied under engineering supervision. Most of the time we will use *Section 310.15* and the ampacity tables.

Ampacity Tables

Section 310.15, Ampacity Tables sends us to *Sections 310.16 through 310.21* and their corresponding tables. *Table 310.16* is the table most often used by electricians. *Section 310.17 through 310.21* are for conductors in free air and for a few high-temperature applications. *Section 310.15* also makes adjustment factors when there are more than three current-carrying conductors and correction factors when the conductor is in a high- or low-ambient temperature. Adjustment and correction factors will be discussed in greater detail later in this unit.

Section 310.16, Ampacities of Insulated Conductors in Raceways, Cables, or Earth (Directly Buried) applies to the majority of wiring done by electricians. It sends us to *Table 310.16* to find the maximum permitted current for insulated conductors, **Figure 16-9**. The values in the table are based on having not more than three current-carrying conductors in the raceway or cable. The ambient temperature that the table is based on is 30°C (86°F).

The table can be divided in half vertically, with copper conductors on the left and aluminum conductors on the right. Both halves of the table have three columns, each representing the common temperature rating of conductor insulation. The 60°C column lists the conductors with 60°C insulation, the 75°C column lists the conductor with 75°C insulation, and the 90°C column lists the conductor with 90°C insulation. There are a few conductor types that appear in more than one column. Those conductors have a dual rating where their temperature rating is based on whether it is a wet or dry location. For example, XHHW is rated for 90°C in dry and damp locations but only 75°C in wet locations. *Table 310.4(1)* can be referenced to determine the correct temperature rating.

The conductor sizes used by the electrical industry are printed along the left and right side of the table. The American Wire Gauge is used for conductors 4/0 and smaller, with kcmil being used for anything larger than 4/0. To determine the maximum ampacity of a conductor, find the intersection between the conductor size and the appropriate temperature column.

Table 310.16 shows the amount of current it will take for a conductor to reach 60°C, 75°C, or 90°C. For example, a 6 AWG copper conductor carrying 75 amperes will reach 90°C (194°F). If the current is increased to 80 amperes, the temperature of the conductor will be more than 90°C, which has the potential to deteriorate the insulation. A conductor will not reach its maximum temperature the instant it reaches its maximum current. However, if that maximum current flows for an extended time, it may reach that maximum temperature.

Table 310.16 Ampacities of Insulated Conductors with Not More Than Three Current-Carrying Conductors in Raceway, Cable, or Earth (Directly Buried)

Size AWG or kcmil	Temperature Rating of Conductor [See Table 310.4(1)]						Size AWG or kcmil
	60°C (140°F)	75°C (167°F)	90°C (194°F)	60°C (140°F)	75°C (167°F)	90°C (194°F)	
	Types TW, UF	Types RHW, THHW, THW, THWN, XHHW, XHWN, USE, ZW	Types TBS, SA, SIS, FEP, FEPB, MI, PFA, RHH, RHW-2, THHN, THHW, THW-2, THWN-2, USE-2, XHH, XHHW, XHHW-2, XHWN, XHWN-2, XHHN, Z, ZW-2	Types TW, UF	Types RHW, THHW, THW, THWN, XHHW, XHWN, USE	Types TBS, SA, SIS, THHN, THHW, THW-2, THWN-2, RHH, RHW-2, USE-2, XHH, XHHW, XHHW-2, XHWN, XHWN-2, XHHN	
	COPPER			ALUMINUM OR COPPER-CLAD ALUMINUM			
18*	—	—	14	—	—	—	—
16*	—	—	18	—	—	—	—
14*	15	20	25	—	—	—	—
12*	20	25	30	15	20	25	12*
10*	30	35	40	25	30	35	10*
8	40	50	55	35	40	45	8
6	55	65	75	40	50	55	6
4	70	85	95	55	65	75	4

Reproduced with permission of NFPA from NFPA 70, National Electrical Code, 2023 edition. Copyright © 2022, National Fire Protection Association. For a full copy of the NFPA 70, please go to www.nfpa.org

Figure 16-9. *Table 310.16* lists the allowable ampacities for all the various type and size of conductors installed in electrical installations.

EXAMPLE 16-1

Problem: Determine the maximum ampacity of a 250 kcmil TW copper conductor.

Solution:

1. The conductor is copper, so we will use the left side of Table 310.16.
2. Locate the column that includes TW insulation (60°C column).
3. Locate the 250 kcmil row and slide over to where it meets the 60°C column.

The maximum ampacity is 215 amperes.

Table 310.16 Ampacities of Insulated Conductors with Not More Than Three Current-Carrying Conductors in Raceway, Cable, or Earth (Directly Buried)

Size AWG or kcmil	Temperature Rating of Conductor [See Table 310.4(1)]						Size AWG or kcmil
	60°C (140°F)	75°C (167°F)	90°C (194°F)	60°C (140°F)	75°C (167°F)	90°C (194°F)	
	Types TW, UF	Types RHW, THHW, THW, THWN, XHHW, XHWN, USE, ZW	Types TBS, SA, SIS, FEP, FEPB, MI, PFA, RHH, RHW-2, THHN, THHW, THW-2, THWN-2, USE-2, XHH, XHHW, XHHW-2, XHWN, XHWN-2, XHHN, Z, ZW-2	Types TW, UF	Types RHW, THHW, THW, THWN, XHHW, XHWN, USE	Types TBS, SA, SIS, THHN, THHW, THW-2, THWN-2, RHH, RHW-2, USE-2, XHH, XHHW, XHHW-2, XHWN, XHWN-2, XHHN	
	COPPER			ALUMINUM OR COPPER-CLAD ALUMINUM			
18*	—	—	14	—	—	—	—
16*	—	—	18	—	—	—	—
14*	15	20	25	—	—	—	—
12*	20	25	30	15	20	25	12*
10*	30	35	40	25	30	35	10*
8	40	50	55	35	40	45	8
6	55	65	75	40	50	55	6
4	70	85	95	55	65	75	4
3	85	100	115	65	75	85	3
2	95	115	130	75	90	100	2
1	110	130	145	85	100	115	1
1/0	125	150	170	100	120	135	1/0
2/0	145	175	195	115	135	150	2/0
3/0	165	200	225	130	155	175	3/0
4/0	195	230	260	150	180	205	4/0
250	215	255	290	170	205	230	250
300	240	285	320	195	230	260	300
350	260	310	350	210	250	280	350
400	280	335	380	225	270	305	400

Reproduced with permission of NFPA from NFPA 70, National Electrical Code, 2023 edition. Copyright © 2022, National Fire Protection Association. For a full copy of the NFPA 70, please go to www.nfpa.org

Adjustment Factors— Section 310.15(C)(1)

When more than three current-carrying conductors are installed in a raceway or bundled for more than 24″, an adjustment factor must be applied. An **adjustment factor** is a multiplier applied to a conductor's maximum ampacity to compensate for the heating effect of having multiple conductors grouped together. The adjustment factor will reduce the maximum amount of current a conductor is permitted to carry. The reason for the adjustment factor is that each conductor will be giving off heat when carrying current and that heat will heat up the surrounding conductors, **Figure 16-10**. The values in *Table 310.16* have already taken into consideration the amount of heat generated and shared for three

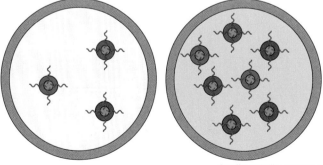

Goodheart-Willcox Publisher

Figure 16-10. A raceway with three conductors will emit some heat as current flows. The greater the number of conductors in the raceway, the greater the amount of heat being released.

conductors, but anything beyond three has the potential to exceed the conductor's temperature rating. *Section 310.15(C)* has the requirements for applying the adjustment factors and refers you to *Table 310.15(C)(1)*, **Figure 16-11**.

There are several situations that affect the application of the adjustment factors. The following are a few of the most common variables that alter the general rule of the adjustment factors:

- *Section 310.15(C)(1)(b)*: Adjustment factors do not apply to raceways less than 24″ in length (nipples). The conductors in a nipple are not near each other for a long enough distance for heat to become an issue.
- *Section 310.15(E)*: A neutral conductor in a 120/240-volt single phase system that only carries the unbalanced current of a multiwire circuit is not included as a current-carrying conductor. The neutral in a single-phase multiwire branch circuit will carry the difference between the two ungrounded (hot) conductors. If line one has 12 amperes and line two has 10 amperes, the neutral will carry two. Since it is only carrying the difference between the line conductors, the neutral current will not be contributing enough heat to impact the other conductors.
- *Section 310.15(F)*: Grounding and bonding conductors are not included as current-carrying conductors. Equipment grounding conductors should not have current in them under normal circumstances. Therefore, they do not contribute heat to the other conductors.

Table 310.15(C)(1) Adjustment Factors for More Than Three Current-Carrying Conductors

Number of Conductors*	Percent of Values in Table 310.16 Through Table 310.19 as Adjusted for Ambient Temperature if Necessary
4–6	80
7–9	70
10–20	50
21–30	45
31–40	40
41 and above	35

*Number of conductors is the total number of conductors in the raceway or cable, including spare conductors. The count shall be adjusted in accordance with 310.15(E) and (F). The count shall not include conductors that are connected to electrical components that cannot be simultaneously energized.

Reproduced with permission of NFPA from NFPA 70, National Electrical Code, 2023 edition. Copyright © 2022, National Fire Protection Association. For a full copy of the NFPA 70, please go to www.nfpa.org

Figure 16-11. *Table 310.15(C)(1)* has the adjustment factors that are applied to situations where there are more than three current-carrying conductors.

EXAMPLE 16-2

Problem: Determine the maximum ampacity of a 3/0 AWG THW aluminum conductor with five current-carrying conductors in the raceway.

Table 310.16 Ampacities of Insulated Conductors with Not More Than Three Current-Carrying Conductors in Raceway, Cable, or Earth (Directly Buried)

Size AWG or kcmil	Temperature Rating of Conductor [See Table 310.4(1)]						Size AWG or kcmil
	60°C (140°F)	75°C (167°F)	90°C (194°F)	60°C (140°F)	75°C (167°F)	90°C (194°F)	
	Types TW, UF	Types RHW, THHW, THW, THWN, XHHW, XHWN, USE, ZW	Types TBS, SA, SIS, FEP, FEPB, MI, PFA, RHH, RHW-2, THHN, THHW, THW-2, THWN-2, USE-2, XHH, XHHW, XHHW-2, XHWN, XHWN-2, XHHN, Z, ZW-2	Types TW, UF	Types RHW, THHW, THW, THWN, XHHW, XHWN, USE	Types TBS, SA, SIS, THHN, THHW, THW-2, THWN-2, RHH, RHW-2, USE-2, XHH, XHHW, XHHW-2, XHWN, XHWN-2, XHHN	
	COPPER			ALUMINUM OR COPPER-CLAD ALUMINUM			
18*	—	—	14	—	—	—	—
16*	—	—	18	—	—	—	—
14*	15	20	25	—	—	—	—
12*	20	25	30	15	20	25	12*
10*	30	35	40	25	30	35	10*
8	40	50	55	35	40	45	8
6	55	65	75	40	50	55	6
4	70	85	95	55	65	75	4
3	85	100	115	65	75	85	3
2	95	115	130	75	90	100	2
1	110	130	145	85	100	115	1
1/0	125	150	170	100	120	135	1/0
2/0	145	175	195	115	135	150	2/0
3/0	165	200	225	130	155	175	3/0
4/0	195	230	260	150	180	205	4/0

Reproduced with permission of NFPA from NFPA 70, National Electrical Code, 2023 edition. Copyright © 2022, National Fire Protection Association. For a full copy of the NFPA 70, please go to www.nfpa.org

Solution:

1. The conductor is aluminum, so we will use the right side of *Table 310.16*.
2. Locate the column that includes THW insulation (75°C column).
3. Locate the 3/0 AWG row and slide over to where it meets the 75°C column.
4. *Table 310.16* tells us that the conductor can carry 155 amperes before the adjustment factor is applied.
5. Since there are five conductors, *Table 310.15(C)(1)* instructs us to derate the maximum ampacity to 80%, **Figure 16-11**.

$$155 \text{ A} \times 0.8 = 124 \text{ A}$$

The maximum ampacity is 124 amperes.

Correction Factors

A *correction factor* is a multiplier applied to a conductor's maximum ampacity to compensate for the conductor being in an area where the ambient temperature is above or below the acceptable *NEC* value, **Figure 16-12**.

The values printed in *Table 310.16* are based on an ambient temperature of 30°C (86°F). When conductors run through an area with an ambient temperature greater than 30°C, they will not be able to dissipate heat well, which can lead to overheating the insulation. *Section 310.15(B)(1)* addresses this situation and provides correction factors in *Table 310.15(B)(1)(1)*, **Figure 16-13**. These correction factors can be applied to a conductor's maximum ampacity when being installed in an area with an ambient temperature that is above or below 30°C.

Table 310.15(B)(1)(1) has a column on the left side of the table for Celsius and a column on the right side for Fahrenheit. There are three center columns, one for each of the temperature ratings found in *Table 310.16*. The value in the appropriate column and row is multiplied by the maximum ampacity found in *Table 310.16*. This result is the maximum amount of current permitted on the conductor.

In the event of conductors installed in an area with a low ambient temperature, the correction factors in *Table 310.15(B)(1)(1)* may increase the maximum ampacity of the conductor.

EXAMPLE 16-3

Problem: Determine the maximum ampacity of a 2 AWG RWH copper conductor with three conductors in the raceway and the ambient temperature is 100°F.

Table 310.16 Ampacities of Insulated Conductors with Not More Than Three Current-Carrying Conductors in Raceway, Cable, or Earth (Directly Buried)

Size AWG or kcmil	Temperature Rating of Conductor [See Table 310.4(1)]						Size AWG or kcmil
	60°C (140°F)	75°C (167°F)	90°C (194°F)	60°C (140°F)	75°C (167°F)	90°C (194°F)	
	Types TW, UF	Types RHW, THHW, THW, THWN, XHHW, XHWN, USE, ZW	Types TBS, SA, SIS, FEP, FEPB, MI, PFA, RHH, RHW-2, THHN, THHW, THW-2, THWN-2, USE-2, XHH, XHHW, XHHW-2, XHWN, XHWN-2, XHHN, Z, ZW-2	Types TW, UF	Types RHW, THHW, THW, THWN, XHHW, XHWN, USE	Types TBS, SA, SIS, THHN, THHW, THW-2, THWN-2, RHH, RHW-2, USE-2, XHH, XHHW, XHHW-2, XHWN, XHWN-2, XHHN	
	COPPER			ALUMINUM OR COPPER-CLAD ALUMINUM			
18*	—	—	14	—	—	—	—
16*	—	—	18	—	—	—	—
14*	15	20	25	—	—	—	—
12*	20	25	30	15	20	25	12*
10*	30	35	40	25	30	35	10*
8	40	50	55	35	40	45	8
6	55	65	75	40	50	55	6
4	70	85	95	55	65	75	4
3	85	100	115	65	75	85	3
2	95	115	130	75	90	100	2
1	110	130	145	85	100	115	1
1/0	125	150	170	100	120	135	1/0
2/0	145	175	195	115	135	150	2/0

Reproduced with permission of NFPA from NFPA 70, National Electrical Code, 2023 edition. Copyright © 2022, National Fire Protection Association. For a full copy of the NFPA 70, please go to www.nfpa.org

Solution:
1. The conductor is copper, so we will use the left side of *Table 310.16*.
2. Locate the column that includes RHW insulation (75°C column).
3. Locate the 2 AWG row and slide over to where it meets the 75°C column.
4. *Table 310.16* tells us that the conductor can carry 115 amperes before any adjustment or correction factors are applied.
5. There are only three conductors in the raceway, so the adjustment factors do not apply.
6. Since the ambient temperature is 100°F, the correction factors in *Table 310.15(B)(1)(1)* apply.

Table 310.15(B)(1)(1) Ambient Temperature Correction Factors Based on 30°C (86°F)

For ambient temperatures other than 30°C (86°F), multiply the ampacities specified in the ampacity tables by the appropriate correction factor shown below.

Ambient Temperature (°C)	Temperature Rating of Conductor			Ambient Temperature (°F)
	60°C	75°C	90°C	
10 or less	1.29	1.20	1.15	50 or less
11–15	1.22	1.15	1.12	51–59
16–20	1.15	1.11	1.08	60–68
21–25	1.08	1.05	1.04	69–77
26–30	1.00	1.00	1.00	78–86
31–35	0.91	0.94	0.96	87–95
36–40	0.82	0.88	0.91	96–104
41–45	0.71	0.82	0.87	105–113
46–50	0.58	0.75	0.82	114–122

Reproduced with permission of NFPA from NFPA 70, National Electrical Code, 2023 edition. Copyright © 2022, National Fire Protection Association. For a full copy of the NFPA 70, please go to www.nfpa.org

7. The insulation is rated for 75°C, so we will use the 75°C column in *Table 310.15(B)(1)1)*. Locate the intersection between the row for 100°F and the 75°C column.
8. *Table 310.15(B)(1)(1)* instructs us to derate the maximum ampacity to 88%.

$$115 \text{ A} \times 0.88 = 101.2 \text{ A}$$

The maximum ampacity is 101.2 amperes.

Goodheart-Willcox Publisher

Figure 16-12. Boiler rooms in commercial buildings tend to be very warm, especially up near the ceiling. This is an example of where a correction factor would need to be applied to conductors run along the ceiling.

Table 310.15(B)(1)(1) Ambient Temperature Correction Factors Based on 30°C (86°F)

For ambient temperatures other than 30°C (86°F), multiply the ampacities specified in the ampacity tables by the appropriate correction factor shown below.

Ambient Temperature (°C)	Temperature Rating of Conductor			Ambient Temperature (°F)
	60°C	75°C	90°C	
10 or less	1.29	1.20	1.15	50 or less
11–15	1.22	1.15	1.12	51–59
16–20	1.15	1.11	1.08	60–68
21–25	1.08	1.05	1.04	69–77
26–30	1.00	1.00	1.00	78–86
31–35	0.91	0.94	0.96	87–95
36–40	0.82	0.88	0.91	96–104
41–45	0.71	0.82	0.87	105–113
46–50	0.58	0.75	0.82	114–122
51–55	0.41	0.67	0.76	123–131
56–60	—	0.58	0.71	132–140
61–65	—	0.47	0.65	141–149
66–70	—	0.33	0.58	150–158
71–75	—	—	0.50	159–167
76–80	—	—	0.41	168–176
81–85	—	—	0.29	177–185

Note: Table 310.15(B)(1)(1) shall be used with Table 310.16 and Table 310.17 as required.

Reproduced with permission of NFPA from NFPA 70, National Electrical Code, 2023 edition. Copyright © 2022, National Fire Protection Association. For a full copy of the NFPA 70, please go to www.nfpa.org

Figure 16-13. *Table 310.15(B)(1)(1)* lists the correction factors that are applied to conductors installed in areas with an ambient temperature that deviates from 30°C (86°F).

Rooftop Raceways and Cables

Conductors installed in the direct sun on a roof may be exposed to extreme temperatures. *Section 310.15(B)(2)* has requirements to address this scenario. If the rooftop raceway or cable is installed with 3/4″ of the roof surface, a 33°C (60°F) is added to the outdoor temperature to determine the ambient temperature, **Figure 16-14**. An *Informational Note* following *Section 310.15(B)(2)* references the ASHRAE Handbook to find the information. Once the ambient temperature has been determined, the correction factors in *Table 310.15(B)(1)(1)* are used to determine the maximum ampacity. Conductors with XHHW-2 insulation are exempt from this requirement.

High Temperature and More Than Three Conductors

In an installation with more than three current-carrying conductors and a high or low ambient temperature, both the correction factor and adjustment factor must be applied.

Goodheart-Willcox Publisher

Figure 16-14. Raceways run within 3/4″ of the roof will be dramatically warmer than the outside air temperature from the effects of the sun warming up the roof surface.

EXAMPLE 16-4

Problem: Determine the maximum ampacity of a 10 AWG THWN-2 copper conductor in a raceway that is 1/2″ above the roof surface with three conductors in the raceway. The jobsite is in a part of the country with an ambient outdoor air temperature of 102°F.

Table 310.16 Ampacities of Insulated Conductors with Not More Than Three Current-Carrying Conductors in Raceway, Cable, or Earth (Directly Buried)

	Temperature Rating of Conductor [See Table 310.4(1)]						
	60°C (140°F)	75°C (167°F)	90°C (194°F)	60°C (140°F)	75°C (167°F)	90°C (194°F)	
	Types TW, UF	Types RHW, THHW, THW, THWN, XHHW, XHWN, USE, ZW	Types TBS, SA, SIS, FEP, FEPB, MI, PFA, RHH, RHW-2, THHN, THHW, THW-2, THWN-2, USE-2, XHH, XHHW, XHHW-2, XHWN, XHWN-2, XHHN, Z, ZW-2	Types TW, UF	Types RHW, THHW, THW, THWN, XHHW, XHWN, USE	Types TBS, SA, SIS, THHN, THHW, THW-2, THWN-2, RHH, RHW-2, USE-2, XHH, XHHW, XHHW-2, XHWN, XHWN-2, XHHN	
Size AWG or kcmil	COPPER			ALUMINUM OR COPPER-CLAD ALUMINUM			Size AWG or kcmil
18*	—	—	14	—	—	—	—
16*	—	—	18	—	—	—	—
14*	15	20	25	—	—	—	—
12*	20	25	30	15	20	25	12*
10*	30	35	**40**	25	30	35	10*
8	40	50	55	35	40	45	8
6	55	65	75	40	50	55	6
4	70	85	95	55	65	75	4

Reproduced with permission of NFPA from NFPA 70, National Electrical Code, 2023 edition. Copyright © 2022, National Fire Protection Association. For a full copy of the NFPA 70, please go to www.nfpa.org

Solution:

1. The conductor is copper, so we will use the left side of *Table 310.16*.
2. The raceway is considered a wet location due to it being outside. *Table 310.4(A)* tells us that THWN-2 is listed for 90°C in dry and wet locations.
3. Locate the column that includes THWN-2 insulation (90°C column).
4. Locate the 10 AWG row and slide over to where it meets the 90°C column.
5. *Table 310.16* tells us that the conductor can carry 40 amperes before any adjustment or correction factors are applied.
6. There are only three conductors in the raceway, so the adjustment factors do not apply.
7. Since the raceway is within 3/4″ of the roof, 60°F must be added to the ambient outdoor air temperature.

$$102°F + 60°F = 162°F$$

8. The corrected ambient temperature is 162°F, so the correction factors in *Table 310.15(B)(1)(1)* apply.

Table 310.15(B)(1)(1) Ambient Temperature Correction Factors Based on 30°C (86°F)

For ambient temperatures other than 30°C (86°F), multiply the ampacities specified in the ampacity tables by the appropriate correction factor shown below.

Ambient Temperature (°C)	Temperature Rating of Conductor			Ambient Temperature (°F)
	60°C	75°C	90°C	
10 or less	1.29	1.20	1.15	50 or less
11–15	1.22	1.15	1.12	51–59
16–20	1.15	1.11	1.08	60–68
21–25	1.08	1.05	1.04	69–77
26–30	1.00	1.00	1.00	78–86
31–35	0.91	0.94	0.96	87–95
36–40	0.82	0.88	0.91	96–104
41–45	0.71	0.82	0.87	105–113
46–50	0.58	0.75	0.82	114–122
51–55	0.41	0.67	0.76	123–131
56–60	—	0.58	0.71	132–140
61–65	—	0.47	0.65	141–149
66–70	—	0.33	0.58	150–158
71–75	—	—	**0.50**	159–167
76–80	—	—	0.41	168–176

Reproduced with permission of NFPA from NFPA 70, National Electrical Code, 2023 edition. Copyright © 2022, National Fire Protection Association. For a full copy of the NFPA 70, please go to www.nfpa.org

9. The conductor insulation is rated for 90°C, so we will use the 90°C column in *Table 310.15(B)(1)(1)*.
10. Locate the intersection between the row for 162°F and the 90°C column.
11. *Table 310.15(B)(1)(1)* instructs us to derate the maximum ampacity to 50%.

$$40 \text{ A} \times 0.5 = 20 \text{ A}$$

The maximum ampacity is 20 amperes.

EXAMPLE 16-5

Problem: Determine the maximum ampacity of a 4 AWG THHW aluminum conductor with seven current-carrying conductors in the raceway and the ambient temperature is 38°C.

Table 310.16 Ampacities of Insulated Conductors with Not More Than Three Current-Carrying Conductors in Raceway, Cable, or Earth (Directly Buried)

Size AWG or kcmil	Temperature Rating of Conductor [See Table 310.4(1)]						Size AWG or kcmil
	60°C (140°F)	75°C (167°F)	90°C (194°F)	60°C (140°F)	75°C (167°F)	90°C (194°F)	
	Types TW, UF	Types RHW, THHW, THW, THWN, XHHW, XHWN, USE, ZW	Types TBS, SA, SIS, FEP, FEPB, MI, PFA, RHH, RHW-2, THHN, THHW, THW-2, THWN-2, USE-2, XHH, XHHW, XHHW-2, XHWN, XHWN-2, XHHN, Z, ZW-2	Types TW, UF	Types RHW, THHW, THW, THWN, XHHW, XHWN, USE	Types TBS, SA, SIS, THHN, THHW, THW-2, THWN-2, RHH, RHW-2, USE-2, XHH, XHHW, XHHW-2, XHWN, XHWN-2, XHHN	
	COPPER			ALUMINUM OR COPPER-CLAD ALUMINUM			
18*	—	—	14	—	—	—	—
16*	—	—	18	—	—	—	—
14*	15	20	25	—	—	—	—
12*	20	25	30	15	20	25	12*
10*	30	35	40	25	30	35	10*
8	40	50	55	35	40	45	8
6	55	65	75	40	50	55	6
4	70	85	95	55	65	75	4
3	85	100	115	65	75	85	3
2	95	115	130	75	90	100	2

Reproduced with permission of NFPA from NFPA 70, National Electrical Code, 2023 edition. Copyright © 2022, National Fire Protection Association. For a full copy of the NFPA 70, please go to www.nfpa.org

Solution:

1. The conductor is aluminum, so we will use the right side of *Table 310.16*.
2. Locate the column that includes THHW insulation (90°C column).
3. Locate the 4 AWG row and slide over to where it meets the 90°C column.
4. *Table 310.16* tells us that the conductor can carry 75 amperes before the adjustment or correction factors are applied.

Table 310.15(C)(1) Adjustment Factors for More Than Three Current-Carrying Conductors

Number of Conductors*	Percent of Values in Table 310.16 Through Table 310.19 as Adjusted for Ambient Temperature if Necessary
4–6	80
7–9	70
10–20	50
21–30	45

Reproduced with permission of NFPA from NFPA 70, National Electrical Code, 2023 edition. Copyright © 2022, National Fire Protection Association. For a full copy of the NFPA 70, please go to www.nfpa.org

5. There are seven conductors, so *Table 310.15(C)(1)* instructs us to derate the maximum ampacity to 70%.
6. The ambient temperature is 38°C, so the correction factors in *Table 310.15(B)(1)(1)* apply.

Table 310.15(B)(1)(1) Ambient Temperature Correction Factors Based on 30°C (86°F)

For ambient temperatures other than 30°C (86°F), multiply the ampacities specified in the ampacity tables by the appropriate correction factor shown below.

Ambient Temperature (°C)	Temperature Rating of Conductor			Ambient Temperature (°F)
	60°C	75°C	90°C	
10 or less	1.29	1.20	1.15	50 or less
11–15	1.22	1.15	1.12	51–59
16–20	1.15	1.11	1.08	60–68
21–25	1.08	1.05	1.04	69–77
26–30	1.00	1.00	1.00	78–86
31–35	0.91	0.94	0.96	87–95
36–40	0.82	0.88	0.91	96–104
41–45	0.71	0.82	0.87	105–113
46–50	0.58	0.75	0.82	114–122

Reproduced with permission of NFPA from NFPA 70, National Electrical Code, 2023 edition. Copyright © 2022, National Fire Protection Association. For a full copy of the NFPA 70, please go to www.nfpa.org

7. The conductor insulation is rated for 90°C, so we will use the 90°C column in *Table 310.15(B)(1)(1)*.
8. Locate the intersection between the row for 38°C and the 90°C column.
9. *Table 310.15(B)(1)(1)* instructs us to derate the maximum ampacity to 91%.

$$75 \text{ A} \times 0.7 \times 0.91 = 47.775 \text{ A}$$

The maximum ampacity is 47.78 amperes.

Weakest Link

Section 310.15(A) makes it clear that conductors must be sized in order to prevent overheating of not only the conductor but also the terminations. It sends us back to *Section 110.14(C)* to find the temperature limitations for conductor terminations. We need to keep in mind that the conductor is not the only consideration. The conductor is installed in a conduit with other conductors and terminates on both ends. If one of the terminations, the raceway, or the other conductors in the raceway have a lower maximum operating temperature, the amount of current the conductor is permitted to carry will have to be adjusted to ensure the other items are not overheated, **Figure 16-15**.

Even though a conductor may be rated for 90°C, if the termination is only rated for 60°C, then the value found in the 60°C column is the maximum amount of current a conductor is permitted to carry. The same goes for other conductors in the raceway. If a 90°C conductor is installed in a raceway with 75°C rated conductors, the final ampacity of the 90°C conductor cannot exceed the value in the 75°C column.

When applying adjustment and correction factors, the starting ampacity is based on the column appropriate for the conductor insulation. After the correction/adjustment factors have been applied, the final ampacity cannot exceed the ampacity found in the temperature column of the lowest rated device, termination, conductor, etc. that is contact with the conductor.

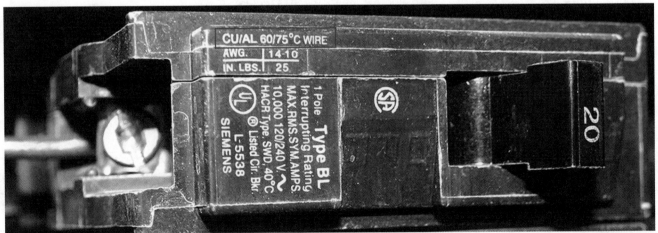

Goodheart-Willcox Publisher

Figure 16-15. Breakers and other devices will have a maximum temperature rating either printed on the equipment or found in the manufacturer literature. The terminal on this breaker is rated for up to 75°C.

EXAMPLE 16-6

Problem: Five 12 AWG THHN copper conductors are being installed in a raceway with an ambient temperature of 30°C. They connect to a breaker rated for 75°C on one end, and a switch rated for 60°C on the other.

Table 310.16 Ampacities of Insulated Conductors with Not More Than Three Current-Carrying Conductors in Raceway, Cable, or Earth (Directly Buried)

Size AWG or kcmil	Temperature Rating of Conductor [See Table 310.4(1)]						Size AWG or kcmil
	60°C (140°F)	75°C (167°F)	90°C (194°F)	60°C (140°F)	75°C (167°F)	90°C (194°F)	
	Types TW, UF	Types RHW, THHW, THW, THWN, XHHW, XHWN, USE, ZW	Types TBS, SA, SIS, FEP, FEPB, MI, PFA, RHH, RHW-2, THHN, THHW, THW-2, THWN-2, USE-2, XHH, XHHW, XHHW-2, XHWN, XHWN-2, XHHN, Z, ZW-2	Types TW, UF	Types RHW, THHW, THW, THWN, XHHW, XHWN, USE	Types TBS, SA, SIS, THHN, THHW, THW-2, THWN-2, RHH, RHW-2, USE-2, XHH, XHHW, XHHW-2, XHWN, XHWN-2, XHHN	
	COPPER			ALUMINUM OR COPPER-CLAD ALUMINUM			
18*	—	—	14	—	—	—	—
16*	—	—	18	—	—	—	—
14*	15	20	25	—	—	—	—
12*	20	25	30	15	20	25	12*
10*	30	35	40	25	30	35	10*
8	40	50	55	35	40	45	8

Reproduced with permission of NFPA from NFPA 70, National Electrical Code, 2023 edition. Copyright © 2022, National Fire Protection Association. For a full copy of the NFPA 70, please go to www.nfpa.org

Solution:
1. The conductor is copper, so we will use the left side of *Table 310.16*.
2. Locate the column that includes THHN insulation (90°C column).
3. Locate the 12 AWG row and slide over to where it meets the 90°C column.
4. *Table 310.16* tells us that the conductor can carry 30 amperes before the adjustment factor is applied.

Table 310.15(C)(1) Adjustment Factors for More Than Three Current-Carrying Conductors

Number of Conductors*	Percent of Values in Table 310.16 Through Table 310.19 as Adjusted for Ambient Temperature if Necessary
4–6	80
7–9	70
10–20	50

Reproduced with permission of NFPA from NFPA 70, National Electrical Code, 2023 edition. Copyright © 2022, National Fire Protection Association. For a full copy of the NFPA 70, please go to www.nfpa.org

5. Since there are five conductors, *Table 310.15(C)(1)* instructs us to derate the maximum ampacity to 80%.
$$30 \text{ A} \times 0.8 = 24 \text{ A}$$
6. This tells us that the maximum amount of current the conductor can carry before exceeding its 90°C rating is 24 amperes. If the conductor was permitted to carry its maximum current of 24 amperes, it will overheat the breaker, which is rated for 75°C, and the device, which is rated for 60°C.
7. Since the switch has the lowest temperature rating, which is 60°C, the maximum ampacity for the conductor cannot exceed the 60°C column. *Table 310.16* lists 20 amperes as the maximum current for a 12 AWG conductor under the 60°C column.

The maximum current is 20 amperes.

If you are not certain what the temperature ratings of the terminations are and which column to use for final ampacity, *Section 110.14(C)(1)* has a general rule to follow. It states that for circuits rated 100 amperes or less, the 60°C column is to be used, and for circuits exceeding 100 amperes, the 75°C column is to be used for maximum ampacities. An exception to the rule is if you can prove that the terminations have a higher rating. If you have a 50-ampere circuit with 90°C conductors and 75°C terminations on both sides, you can use the 75°C column rather than the 60°C column.

Parallel Conductors

Parallel conductors are two or more conductors that are electrically joined at both ends, making them equivalent to one large conductor, **Figure 16-16**. Installing conductors in parallel allows the current-carrying capacity of each conductor to be added together. Installing parallel conductors allows for larger current values that a single conductor would not be able to carry. If you attempted to use a single conductor, it would be so large that it would be very difficult, if not impossible, to work with. The *NEC* has a lot of requirements for parallel conductors to ensure they are electrically identical. *Section 310.10(G)* lists the following requirements for parallel conductors:

- 1/0 AWG or larger (general rule)
- Same length
- Same conductor material
- Same size in circular mil area
- Same insulation type
- Terminated in the same manner

Following the *NEC* requirements will result in conductors with identical electrical characteristics, causing the total current to divide evenly among the conductors. For example, two conductors installed in parallel to feed a load that draws 420 amperes will result in each conductor carrying 210 amperes.

Cable Ampacities

To determine the maximum ampacities for cables, the starting point is *Section 3XX.80* in most of the applicable cable articles. Nonmetallic sheathed (*Article 334*) cable is a good example. *Section 334.80, Ampacity* states that even though the conductors in the cable are rated for 90°C, the final ampacity shall not exceed the 60°C column in *Table 310.16*. It also addresses cables that are run together through thermal insulation. If multiple cables are run in thermal insulation without maintaining spacing, the adjustment factors in *Table 310.15(C)(1)* apply, **Figure 16-17**.

Goodheart-Willcox Publisher

Figure 16-16. Parallel conductors must be the same length, size, conductor material, etc., and they must be terminated in the same manner. Most lugs are rated for only one conductor, so lugs with multiple openings will be required for paralleled conductors.

Goodheart-Willcox Publisher

Figure 16-17. Multiple cables bundled in thermal insulation can lead to overheating. The heat generated from current flowing through the conductors is unable dissipate and the heat builds up, creating the potential to damage the conductors and could lead to a fire.

EXAMPLE 16-7

Problem: Determine the maximum ampacity of a parallel set of 300 kcmil THW copper conductors with two conductors per phase, four current-carrying conductors in each raceway, and the ambient temperature is 86°F.

Table 310.16 Ampacities of Insulated Conductors with Not More Than Three Current-Carrying Conductors in Raceway, Cable, or Earth (Directly Buried)

	Temperature Rating of Conductor [See Table 310.4(1)]						
	60°C (140°F)	75°C (167°F)	90°C (194°F)	60°C (140°F)	75°C (167°F)	90°C (194°F)	
Size AWG or kcmil	Types TW, UF	Types RHW, THHW, THW, THWN, XHHW, XHWN, USE, ZW	Types TBS, SA, SIS, FEP, FEPB, MI, PFA, RHH, RHW-2, THHN, THHW, THW-2, THWN-2, USE-2, XHH, XHHW, XHHW-2, XHWN, XHWN-2, XHHN, Z, ZW-2	Types TW, UF	Types RHW, THHW, THW, THWN, XHHW, XHWN, USE	Types TBS, SA, SIS, THHN, THHW, THW-2, THWN-2, RHH, RHW-2, USE-2, XHH, XHHW, XHHW-2, XHWN, XHWN-2, XHHN	Size AWG or kcmil
	COPPER			ALUMINUM OR COPPER-CLAD ALUMINUM			
18*	—	—	14	—	—	—	—
16*	—	—	18	—	—	—	—
14*	15	20	25	—	—	—	—
12*	20	25	30	15	20	25	12*
10*	30	35	40	25	30	35	10*
8	40	50	55	35	40	45	8
6	55	65	75	40	50	55	6
4	70	85	95	55	65	75	4
3	85	100	115	65	75	85	3
2	95	115	130	75	90	100	2
1	110	130	145	85	100	115	1
1/0	125	150	170	100	120	135	1/0
2/0	145	175	195	115	135	150	2/0
3/0	165	200	225	130	155	175	3/0
4/0	195	230	260	150	180	205	4/0
250	215	255	290	170	205	230	250
300	240	285	320	195	230	260	300
350	260	310	350	210	250	280	350
400	280	335	380	225	270	305	400

Reproduced with permission of NFPA from NFPA 70, National Electrical Code, 2023 edition. Copyright © 2022, National Fire Protection Association. For a full copy of the NFPA 70, please go to www.nfpa.org

Solution:

1. The conductor is copper, so we will use the left side of *Table 310.16*.
2. Locate the column that includes THW insulation (75°C column).
3. Locate the 300 kcmil row and slide over to where it meets the 75°C column.
4. *Table 310.16* tells us that each conductor can carry 285 amperes before the adjustment factor is applied.
5. Since there are four conductors, *Table 310.15(C)(1)* instructs us to derate the maximum ampacity to 80%.

$$285 \text{ A} \times 0.8 = 228 \text{ A}$$

Table 310.15(C)(1) Adjustment Factors for More Than Three Current-Carrying Conductors

Number of Conductors*	Percent of Values in Table 310.16 Through Table 310.19 as Adjusted for Ambient Temperature if Necessary
4–6	80
7–9	70
10–20	50

Reproduced with permission of NFPA from NFPA 70, National Electrical Code, 2023 edition. Copyright © 2022, National Fire Protection Association. For a full copy of the NFPA 70, please go to www.nfpa.org

6. Since there are two conductors per phase, each conductor can carry 228 amps.

$$228 \text{ A} \times 2 = 456 \text{ A}$$

The parallel set has a maximum ampacity of 456 amperes.

EXAMPLE 16-8

Problem: Determine the maximum ampacity of five 14-2 nonmetallic sheathed cables that are grouped together in an insulated attic.

Solution:

1. *Section 334.80* and *Section 334.112* state than nonmetallic sheathed cable is made with 90°C conductors, so the 90° column is the starting point before any adjustment or correction factors are applied. **Note:** Remember that the final ampacity cannot exceed the value in the 60°C column.

Table 310.16 Ampacities of Insulated Conductors with Not More Than Three Current-Carrying Conductors in Raceway, Cable, or Earth (Directly Buried)

Size AWG or kcmil	Temperature Rating of Conductor [See Table 310.4(1)]						Size AWG or kcmil
	60°C (140°F)	75°C (167°F)	90°C (194°F)	60°C (140°F)	75°C (167°F)	90°C (194°F)	
	Types TW, UF	Types RHW, THHW, THW, THWN, XHHW, XHWN, USE, ZW	Types TBS, SA, SIS, FEP, FEPB, MI, PFA, RHH, RHW-2, THHN, THHW, THW-2, THWN-2, USE-2, XHH, XHHW, XHHW-2, XHWN, XHWN-2, XHHN, Z, ZW-2	Types TW, UF	Types RHW, THHW, THW, THWN, XHHW, XHWN, USE	Types TBS, SA, SIS, THHN, THHW, THW-2, THWN-2, RHH, RHW-2, USE-2, XHH, XHHW, XHHW-2, XHWN, XHWN-2, XHHN	
	COPPER			ALUMINUM OR COPPER-CLAD ALUMINUM			
18*	—	—	14	—	—	—	—
16*	—	—	18	—	—	—	—
14*	15	20	**25**	—	—	—	—
12*	20	25	30	15	20	25	12*
10*	30	35	40	25	30	35	10*
8	40	50	55	35	40	45	8

Reproduced with permission of NFPA from NFPA 70, National Electrical Code, 2023 edition. Copyright © 2022, National Fire Protection Association. For a full copy of the NFPA 70, please go to www.nfpa.org

2. Nonmetallic sheathed cable has copper conductors, so we will use the left side of *Table 310.16*.
3. Locate the 14 AWG row and slide over to where it meets the 90° column.
4. *Table 310.16* tells us that the conductor can carry 25 amperes before the adjustment factor is applied.
5. Since there are ten conductors, *Table 310.15(C)(1)* instructs us to derate the maximum ampacity to 50%.

Table 310.15(C)(1) Adjustment Factors for More Than Three Current-Carrying Conductors

Number of Conductors*	Percent of Values in Table 310.16 Through Table 310.19 as Adjusted for Ambient Temperature if Necessary
4–6	80
7–9	70
10–20	50
21–30	45

Reproduced with permission of NFPA from NFPA 70, National Electrical Code, 2023 edition. Copyright © 2022, National Fire Protection Association. For a full copy of the NFPA 70, please go to www.nfpa.org

$$25 \text{ A} \times 0.5 = 12.5 \text{ A}$$

The maximum ampacity of each cable is 12.5 amperes.

| PROCEDURE | Conductor Ampacity—Finding the Correct Section |

Conductor ampacity is a general requirement. More specifically, the conductors are a wiring method and are installed and sized while roughing-in. That information leads to *Chapter 3, Wiring Methods and Materials*. There is more than one way to get into the correct section to start conductor ampacity calculations.

To find the correct section using the table of contents:

1. Determine keywords: conductors, ampacity.
2. Open the table of contents to *Chapter 3*. Conductors are a general wiring method and are installed during the rough in of a building.
3. Locate *Article 310, Conductors for General Wiring*.
4. Locate *Part III, Installation*.
5. Open to *Part III* of *Article 310* and scan the sections to find *Section 310.15, Ampacity Tables*.
6. Find the appropriate conductor ampacity in *Table 310.16* and apply the adjustment and correction factors from *Section 310.15*.

To find the correct section using the index:

1. Determine keywords: conductors, ampacity.
2. Search index for conductors.
3. Narrow index search of conductors to "ampacities."
4. Open to *Section 310.14* through *310.21* and find *Section 310.15, Ampacity Tables*.
5. Find the appropriate conductor ampacity in *Table 310.16* and apply the adjustment and correction factors from *Section 310.15*.

Summary

- *Table 310.4(A)* contains detailed information on conductor insulation.
- *Table 310.16* lists the maximum ampacities for insulated conductors.
- When four or more conductors are installed in a raceway, the maximum ampacity is reduced by the adjustment factors found in *Table 310.15(C)(1)*.
- When conductors are installed in an ambient temperature that is above or below 30°C (86°F), the correction factors found in *Table 310.15(B)(1)(1)* are multiplied by the conductor's maximum ampacity.
- When four or more conductors are installed in an area with a high ambient temperature, both the adjustment and correction factors are applied.
- Conductors installed in the sunlight within 3/4″ from the roof must be derated for the extreme temperature they will be subjected to. Type XHHW-2 conductors are exempt from this requirement.
- The starting point for determining cable ampacities is *Section 3XX.80* in most of the cable articles.

Unit 16 Review

Name _____ Date _____ Class _____

Know and Understand

Answer the following questions based on information in this unit.

1. The three conductor materials recognized by the *NEC* are copper, aluminum, and _____.
   A. steel
   B. silver
   C. gold
   D. copper-clad aluminum

2. _____ contains specific information on conductor applications and insulation.
   A. *Table 310.4(A)*
   B. *Table 310.15(B)(1)(1)*
   C. *Table 310.15(C)(1)*
   D. *Table 310.16*

3. Which conductor type is not subject to the 60°F temperature adjustment when installed in a raceway that is in direct sunlight and within 3/4″ of the roof surface?
   A. XHHW
   B. XHHW-2
   C. THHN
   D. THHN-2

4. Conductors installed in an underground raceway are considered to be in a _____ location.
   A. dry
   B. damp
   C. wet
   D. corrosive

5. Conductors _____ and smaller are sized according to the American Wire Gauge (AWG).
   A. 12
   B. 1
   C. 2/0
   D. 4/0

6. *Table 310.16* ampacities are based on not more than _____ current-carrying conductors in a raceway, cable, or buried directly in the earth.
   A. two
   B. three
   C. four
   D. seven

7. *Table 310.16* is based on an ambient temperature of _____°C.
   A. 30
   B. 60
   C. 75
   D. 90

8. Which of the following is *not* one of the parallel conductor requirements?
   A. Must be 1 AWG or larger.
   B. Must be the same length.
   C. Must be the same conductor material.
   D. Must be the same size in circular mil area.

9. _____ of the cable articles has information on ampacity.
   A. *Section 3XX.10*
   B. *Section 3XX.22*
   C. *Section 3XX.80*
   D. *Section 3XX.112*

10. _____ lists the maximum ampacities of insulated conductors.
    A. *Table 310.15(B)(1)(1)*
    B. *Table 310.16*
    C. *Table 310.15(C)(1)*
    D. *Table 310.4(A)*

Apply and Analyze

Answer the following questions using a copy of the National Electrical Code. Identify the section or subsection where the answer is found.

1. What is the maximum ampacity of a 6 AWG aluminum TW conductor? (Terminations rated for 75°C)
   A. 40 amperes
   B. 50 amperes
   C. 55 amperes
   D. 65 amperes

   NEC _____

2. What is the adjustment factor applied to a conductor's maximum ampacity when there are five current-carrying conductors in a raceway?
   A. 30%
   B. 50%
   C. 70%
   D. 80%

   NEC _____

3. What is the correction factor applied to a 75°C conductor installed in an ambient temperature of 40°C?
   A. 0.82
   B. 0.88
   C. 0.91
   D. 0.94

   NEC _____

4. What is the maximum ampacity of a 400 kcmil copper THWN conductor? (Terminations rated for 75°C)
   A. 225 amperes
   B. 270 amperes
   C. 280 amperes
   D. 335 amperes

   NEC _____

5. What is the maximum operating temperature of a TFE conductor?
   A. 60°C
   B. 75°C
   C. 90°C
   D. 250°C

   NEC _____

6. What is the maximum ampacity of a 3/0 copper XHHW conductor that is installed in a wet location? (Terminations rated for 90°C)
   A. 155 amperes
   B. 175 amperes
   C. 200 amperes
   D. 225 amperes

   NEC _____

7. Two 4/0 THW aluminum conductors are installed in parallel. What is the total ampacity of the conductors? (Terminations rated for 75°C)
   A. 180 amperes
   B. 230 amperes
   C. 360 amperes
   D. 410 amperes

   NEC _____

8. What is the maximum ampacity of a 2 AWG copper THHN conductor that is in a raceway with a total of seven current-carrying conductors? (Terminations rated for 90°C)
   A. 91 amperes
   B. 95 amperes
   C. 115 amperes
   D. 130 amperes

   NEC _____

9. What is the maximum ampacity of a 350 kcmil aluminum THWN conductor that is installed in an ambient temperature of 102°F? (Terminations rated for 90°C)
   A. 195 amperes
   B. 220 amperes
   C. 250 amperes
   D. 272.8 amperes

   NEC _____

10. What is the maximum ampacity of a 12 AWG copper THHN conductor? There are ten current-carrying conductors in the raceway and they are installed in an ambient temperature of 105° F. (Terminations rated for 90°C)
    A. 13.05 amperes
    B. 15 amperes
    C. 26.1 amperes
    D. 30 amperes

    NEC _____

Critical Thinking

Complete the following ampacity calculations using a copy of the National Electrical Code. *Identify the section or subsection used to find the answer.*

1. What is the ampacity of a 4/0 AWG THWN copper conductor? There are two conductors in the raceway, and the ambient temperature is 45°C. (Terminations rated for 75°C)

   _____

   NEC _____

2. What is the maximum ampacity of a 14 AWG THW copper conductor? There are six conductors in the raceway, and the ambient temperature is 30°C. (Terminations rated for 60°C)

   _____

   NEC _____

3. What is the maximum ampacity of a 12 AWG copper TW conductor? There are three conductors in the raceway, and the ambient temperature is 36°C. (Terminations rated for 60°C)

   _____

   NEC _____

4. What is the maximum ampacity of a 12 AWG copper THHN conductor? There are six current-carrying conductors in the raceway, and the ambient temperature is 122° F. (Terminations rated for 75°C)

   _____

   NEC _____

5. What is the maximum ampacity of a 10 AWG THW copper conductor? There are six current-carrying conductors in the raceway, and the ambient temperature is 86°F. (Terminations rated for 60°C)

   _____

   NEC _____

6. What is the maximum ampacity of a 12 AWG copper THHN conductor? There are six current-carrying conductors in the raceway, and the ambient temperature is 122°F. (Terminations rated for 90°C)

   _____

   NEC _____

7. What is the maximum ampacity of a 2/0 AWG aluminum THW conductor? There are three current-carrying conductors in the raceway, and the ambient temperature is 17°C. (Terminations rated for 90°C)

   _____

   NEC _____

8. What is the maximum ampacity of a 500 kcmil copper XHHW conductor? There are three current-carrying conductors in the raceway, the ambient temperature is 32°C, and it is a wet location. (Terminations rated for 90°C)

   _____

   NEC _____

9. What is the maximum ampacity of a 400 kcmil aluminum THWN-2 conductor? There are three current-carrying conductors in the raceway, the ambient temperature is 80°F, and it is a wet location. (Terminations rated for 90°C)

   _____

   NEC _____

10. What is the maximum ampacity of a 600 kcmil copper THHN conductor? There are four current-carrying conductors in the raceway, and the ambient temperature is 30°C. (Terminations rated for 90°C)

    _____

    NEC _____

Part 2   Application: Advanced NEC Topics

Complete the following ampacity calculations using a copy of the National Electrical Code. *Identify the section or subsection used to find the answer.*

11. What is the temperature rating of a MTW (machine tool wire) copper conductor that is installed in a wet location?

    _____

    NEC _____

12. What is the maximum ampacity of an 8 AWG RHW aluminum conductor? There are three current-carrying conductors in the raceway, and the ambient temperature is 86°F. (Terminations rated for 75°C)

    _____

    NEC _____

13. What is the maximum ampacity of a 4/0 AWG XHHW-2 aluminum conductor? There are four current-carrying conductors in the raceway, the ambient temperature is 86°F, and it is a wet location. (Terminations rated for 90°C)

    _____

    NEC _____

14. What is the maximum ampacity of an 8 AWG THHN copper conductor? There are three current-carrying conductors in the raceway, and the ambient temperature is 65°F. (Terminations rated for 90°C)

    _____

    NEC _____

15. What is the maximum ampacity of a 14 AWG TW copper conductor? There are four current-carrying conductors in the raceway, and the ambient temperature is 41°C. (Terminations rated for 60°C)

    _____

    NEC _____

16. What is the maximum ampacity of a single 10-3 AWG nonmetallic sheathed cable? There are three current-carrying conductors in the cable, and the ambient temperature is 27°C. (Terminations rated for 60°C)

    _____

    NEC _____

17. What is the maximum ampacity of a 6 AWG THWN-2 copper conductor that is in a raceway 1/2″ above a flat roof? There are three current-carrying conductors in the raceway. It is in a part of the country with an ambient outdoor air temperature of 93°F. (Terminations rated for 90°C)

    _____

    NEC _____

18. What is the maximum ampacity of a parallel set of 500 kcmil THHN copper conductors? There are two conductors per phase. There are four current-carrying conductors in each raceway, and the ambient temperature is 100°F. (Terminations rated for 90°C)

    _____

    NEC _____

19. What is the maximum ampacity of a 2 AWG XHHW aluminum conductor? There are three current-carrying conductors in the raceway, the ambient temperature is 86°F, and it is a wet location. (Terminations rated for 90°C)

    _____

    NEC _____

20. What is the maximum ampacity of a 12 AWG THHN copper conductor? There are ten current-carrying conductors in the raceway, and the ambient temperature is 86°F. (Terminations rated for 75°C)

    _____

    NEC _____

UNIT 17 Overcurrent Protection

Sashkin/Shutterstock.com

KEY TERMS
Edison-base fuse
ground fault
nonrenewable
overcurrent
overcurrent protective device (OCPD)
overload
plug fuse
short circuit
Type S fuse

LEARNING OBJECTIVES

After completing this chapter, you will be able to:
- Differentiate the three types of overcurrents: overload, short circuit, and ground fault.
- Describe plug and cartridge fuses.
- Locate fuse requirements in the *NEC*.
- Describe various types of circuit breakers.
- Locate circuit breaker requirements in the *NEC*.
- Locate the standard overcurrent device ratings in the *NEC*.
- Describe permitted locations for overcurrent protection.
- Calculate the correct overcurrent device rating for a specific load based on *NEC* requirements.

Introduction

Overcurrent protection is an important part of an electrical system, **Figure 17-1**. It provides protection against excess current that can generate heat and damage electrical components. This unit will cover some of the various types of overcurrent protective devices, identify where applicable requirements can be found in the *NEC*, and calculate the proper size overcurrent protection device. Motors, transformers, and tap conductors have additional considerations that are not covered by this unit.

Article 240, Overcurrent Protection

Article 240 is dedicated to overcurrent protection. It has requirements for the various types of overcurrent devices, restrictions on where they are located, and descriptions for how to calculate the correct size. Before going into specifics, we must first discuss the various types of overcurrents. The *NEC* defines an **overcurrent** as any current more than the rated current of equipment or the ampacity of a conductor. An overcurrent may result from an overload, short circuit, or ground fault. Current flow that is beyond what a conductor or piece of equipment is rated for may result in damage to electrical components and could lead to a fire.

Types of Overcurrents

There are three types of overcurrents identified by the *NEC*: overloads, short circuits, and ground faults.

Overload

The *NEC* defines an **overload** as operation of equipment in excess of normal, full-load rating, or of a conductor in excess of its ampacity, that would cause damage or dangerous overheating, if persisting for a sufficient length of time. The current in an overload never leaves its intended path, **Figure 17-2**.

Goodheart-Willcox Publisher

Figure 17-1. Panelboards house overcurrent devices that will shut a circuit off if it draws too much current.

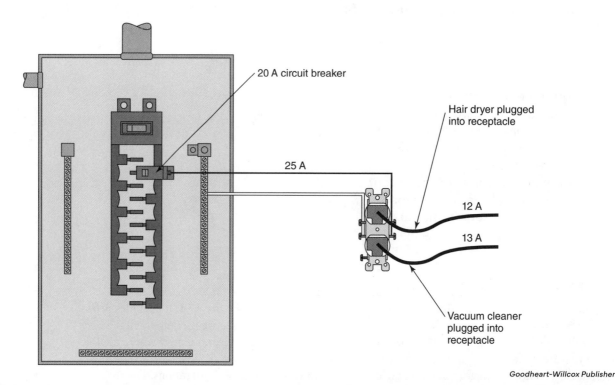

Goodheart-Willcox Publisher

Figure 17-2. Plugging too many high-current items into a circuit will cause an overload. In this picture, a vacuum and hair dryer are plugged into the same receptacle outlet. The combined current is 25 amperes, which will trip the 20-ampere breaker if it continues for too long.

When a conductor has more current flowing than it has been designed and is rated for, it can overheat the conductor insulation. Conductor insulation that has been heated beyond its maximum temperature rating may melt, become brittle, or crack. Once the insulation has been damaged, it will no longer be able to keep the conductor insulated from the other conductors, the raceway, or the enclosure that contains it. The amount of excess current and the amount of time that current is flowing will determine how much a conductor heats up. A small overcurrent for a short period of time is not likely to cause excess heating, but if a small overcurrent occurs for an extended period, it can cause damage.

PRO TIP — Motors

A motor is an example of a piece of electrical equipment that draws more current upon starting than when it is running, **Figure 17-3**. This is sometimes referred to as *inrush current*. Most motors will have an overload device, either internally or externally, that will remove the power from the motor if it draws too much current for an extended period. *Section 240.9* permits motors with this type of protection to have a larger branch circuit overcurrent protection device installed in the panelboard than would typically be required to protect the conductors. This is to prevent the breaker or fuse from tripping during startup.

Short Circuit

A *short circuit* is an excess current resulting from contact between two ungrounded conductors with a difference in potential, or an ungrounded conductor and a grounded (neutral) conductor. Conductors in this definition include insulated conductors, busbars, and conductive internal components of equipment, such as switchboards, panelboards, motor control centers, and transformers, **Figure 17-4**.

Short circuits result in a large amount of current flowing through the contacted conductive objects, often thousands of amperes. They create a parallel path around the load, leaving only the resistance of the conductor/object that has made contact to oppose the current flow. The low-resistance/-impedance path results in very high current values.

A properly sized overcurrent device functioning normally will quickly interrupt the circuit before the short circuit has the potential to cause property damage or injury. Equipment not rated for the available short circuit current or with a malfunction may result in an arc flash/blast, an extremely dangerous and life-threatening situation. Wearing the appropriate personal protective equipment is required when working around equipment that has the potential to create an arc flash/blast.

Goodheart-Willcox Publisher

Figure 17-3. Larger motors will have overload protection that will stop the motor if it starts to draw too much current.

Goodheart-Willcox Publisher

Figure 17-4. A receptacle outlet where the red ungrounded (hot) and white grounded (neutral) conductors are mistakenly connected to the same terminals on a receptacle is an example of a short circuit. This occasionally happens when people are not paying close attention to their work. When the circuit is energized, a large amount of current will travel through the circuit. This should open the breaker or fuse immediately.

PRO TIP

Branch Circuit Overcurrent Protective Devices

Branch circuit overcurrent protective devices (OCPDs) will have two current ratings, rated current and ampere-interrupting capacity (AIC), **Figure 17-5**. The rated current is the maximum amount of current the overcurrent device will allow loads from the circuit to draw. A 20-ampere breaker has been designed to safely carry up to 20 amperes. If the circuit draws over 20 amperes, the overcurrent device will open. The higher the current, the quicker the OCPD will open.

The AIC is the maximum amount of fault current the device can safely interrupt. The phrase "ampere interrupting capacity" is often shortened and simply referred to as "interrupting rating." A device with an AIC of 20,000 amperes can safely interrupt up to 20,000 amperes in the case of a ground fault or short circuit. It is important to ensure the AIC rating is not exceeded to prevent equipment failure that can lead to an explosion.

Goodheart-Willcox Publisher

Figure 17-5. Fuses and breakers have current ratings as well as ampere-interrupting capacity (AIC) ratings. This fuse has a circuit current rating of 60 amperes and an AIC rating of 200,000 amperes.

Goodheart-Willcox Publisher

Figure 17-6. A ground fault occurs when current travels on an enclosure, raceway, or grounding conductor. This figure shows where a conductor had a ground fault to the metal conduit body. The conductor insulation was damaged from pressure against the metal edge during installation. You can see where the conduit body melted a bit from the arc.

Ground Fault

The *NEC* defines a ***ground fault*** as an unintentional, electrically conducive connection between an ungrounded conductor of an electrical circuit and the normally non-current-carrying conductors, metallic enclosures, metallic raceways, metallic equipment, or earth, **Figure 17-6**.

Ground faults can result in a current flow that is small or very large, depending on where in the circuit the ground fault occurs, as well as its impedance. In the case of a ground fault with a current that is less than the value of the overcurrent device, it will not cause the breaker or fuse to open and may continue indefinitely. The danger in that situation is there will be current flowing through a grounded object not designed for current flow. Electric current traveling through metal raceways and enclosures creates a shock hazard, and if the fittings are not tight, they can arc, resulting in a fire. A ground fault with a low impedance will result in a large amount of current flowing, like a short circuit, which should quickly operate the overcurrent device.

Overcurrent Protection Devices (OCPD)

Overcurrent protection devices are designed to limit the amount of current that can flow. The *NEC* defines an ***overcurrent protection device (OCPD)*** as a device capable of providing protection for service, feeder, and branch circuits and equipment over the full range of overcurrents between its rated current and its interrupting rating. Such devices are provided with an interrupting rating appropriate to the intended use but no less than 5000 amperes.

Most overcurrent devices are installed in a panelboard or switchboard where the branch circuits and feeders receive their power. The electric service will also have overcurrent protection typically incorporated into the main service disconnect(s). The two main types of overcurrent protective devices used in the electrical industry are fuses and breakers.

Fuses

Fuses were the first type of overcurrent protective devices used to protect electrical systems, **Figure 17-7A**. Most fuses are *nonrenewable*, meaning they cannot be reset and must be replaced if they are subject to a current beyond their rating. In the case of an overcurrent, an internal link will melt, opening the circuit, **Figure 17-7B**.

Fuses have two current ratings that must be considered. The first rating, one that most people are familiar with, is the ampere rating. The ampere rating is based on the maximum amount of current expected, or the size of the circuit, such as 15 amperes, 20 amperes, or 30 amperes. Fuses also have an interrupting rating that indicates the maximum amount of current they can safely open in the case of a short circuit or ground fault.

Another consideration is voltage. All fuses will have an intended voltage they are designed for. Installing a fuse with incorrect voltage or current ratings can lead to a dangerous situation, **Figure 17-8**.

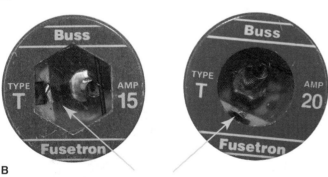

Goodheart-Willcox Publisher

Figure 17-7. **A**—Residential fuse panels contained both plug and cartridge fuses. The *NEC* does not permit plug fuses to be installed in new wiring. Some insurance companies will require a home to have a fuse panel upgraded to a panel with breakers before they will insure the home. **B**—In the middle of the looking glass on the fuse, you can see a metal link that will melt and open the circuit in the event of an overcurrent.

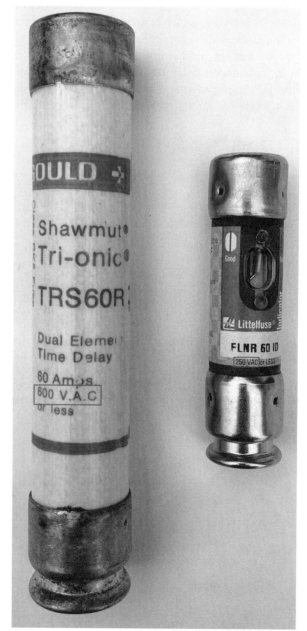

Goodheart-Willcox Publisher

Figure 17-8. In addition to verifying the correct current and interrupting ratings, the voltage of the fuses must be correct.

Plug Fuse

Plug fuses are a screw-in type fuse that are covered in *Part V* of *Article 240*. Plug fuses are no longer permitted to be installed in new construction, but they may be replaced in existing installations. There are two main types of plug fuses: Edison-base and Type S, **Figure 17-9**.

Edison-base fuses were the original style of plug fuse. Fuses with various ampacity ratings all screwed into the same screw shell. This created problems because people could install any size fuse they wanted. If a circuit had a 15-amp fuse that kept blowing due to excess current, someone could simply install a larger fuse. This practice allowed more current to flow through a conductor than it was capable of safely carrying, which would overheat the conductor insulation and could eventually lead to a fire.

Type S fuses are another type of plug fuse. *Type S fuses* screw into an adapter designed to restrict the size fuse it will accept. If a 15-amp adapter is installed, larger Type S fuses will not screw in. This prevents anyone from installing a fuse that is beyond the ampacity of the circuit. Even though Type S fuses and their adapters solve the problem of larger fuses being installed, they are only permitted for replacement in existing installations, **Figure 17-10**.

Cartridge Fuse

Cartridge fuses are covered in *Part VI* of *Article 240*. A cartridge fuse is a cylinder-shaped fuse installed in a fuse holder. Like plug fuses, they have an internal link that opens in the event of an overcurrent. Cartridge fuses are not common on residential and smaller commercial construction projects, but they are sometimes still installed in switchgear and disconnects. They may also be used for internal protection in industrial equipment and motor control centers, **Figure 17-11**.

Cartridge fuses may be nonrenewable, meaning they are thrown away if they open from an overcurrent, or renewable.

Goodheart-Willcox Publisher

Figure 17-9. The fuse on the left is a standard Edison-base plug fuse. The fuse on the right is a Type S fuse. Type S fuses have different threads for each of the ampacity ratings. This prevents a larger fuse from being installed which can lead to damage to the branch circuit conductors and may lead to a fire.

Goodheart-Willcox Publisher

Figure 17-11. Cartridge fuses come in many ampacity ratings, interrupting current ratings, sizes, and styles.

Goodheart-Willcox Publisher

Figure 17-10. A Type S fuse adapter has threads that restrict a larger fuse from being installed.

Renewable fuses can be opened so the internal link can be replaced. They are only permitted for replacement in existing installations where there is no evidence of over-fusing or tampering (*Section 240.60(D)*).

Circuit Breakers

A circuit breaker is a renewable device that opens a circuit in the event of an overload, short circuit, or ground fault beyond the current rating of the device, **Figure 17-12**. Since circuit breakers are renewable, they can be reset by turning them off and then back on again. Circuit breakers have current ratings that indicate the maximum amount of load current they are designed to carry, as well as an interrupting rating that indicates the amount of fault current they are designed to interrupt. According to *Section 240.83(C)*, a breaker must display its interrupting rating if that rating is not 5000 amperes. If the breaker does not have an interrupting rating listed, we can assume it is 5000 amperes, **Figure 17-13**.

> **PRO TIP** — **Turning On a Breaker**
>
> A breaker poses its greatest danger to electrician when it is time to turn the breaker on or reset it after it has tripped. If there is a short circuit or ground fault somewhere in the circuit when the power is turned on, thousands of amperes will flow and stress the breaker to its limit. If the breaker is sized properly and does not have a malfunction, it will just trip back off. If it is improperly sized or has a malfunction, it has the potential to explode and cause an arcing incident. To mitigate the hazard, OSHA CFR 1910.334(b)(2) and NFPA 70E 130.8(M) do not allow an overcurrent device that has opened a circuit (tripped) to be reset without first determining the cause and that it is safe to energize.
>
> To safeguard yourself when operating breakers, always wear the appropriate PPE for the hazard and stand off to the side of the panel while turning it on or off, **Figure 17-14**.

Eaton

Figure 17-12. A circuit breaker is a renewable device that opens a circuit in the case of an overload, short circuit, or ground fault that is beyond the current rating of the device.

Goodheart-Willcox Publisher

Figure 17-13. This breaker has an interrupting rating of 10,000 amperes. If a breaker has an interrupting rating that is other than 5000 amperes, it will be marked on the breaker.

Goodheart-Willcox Publisher

Figure 17-14. In addition to wearing the appropriate personal protective equipment, it is good practice to stand to the side of the enclosure when turning on breakers and disconnect switches.

Ground-Fault Circuit Interrupter (GFCI) Circuit Breaker

Ground-fault circuit interrupter (GFCI) circuit breakers provide overload and short circuit protection like a standard circuit breaker, but also provide ground-fault protection for personnel. They are designed to open the circuit in the event of a ground fault with a current of 4–6 milliamperes or more, **Figure 17-15**.

GFCI protection is required in areas with a higher risk of electric shock. This risk is higher when an energized conductor is near something that is grounded or in damp and wet locations, such as outdoors, basements, kitchens, bathrooms, and near sinks and other water sources. *Section 210.8* lists all the areas where GFCI protection for personnel is required.

Arc-Fault Circuit Interrupter (AFCI) Circuit Breaker

Arc-fault circuit interrupter (AFCI) circuit breakers provide overload and short circuit protection like a standard circuit breaker, but also monitor a circuit looking for an arcing fault that has the potential to start a fire, **Figure 17-16**. If it detects a dangerous arc, it will disconnect the circuit.

An arc fault is a fault where an arc is generating enough heat to start a fire. There are two types of arcs: series and

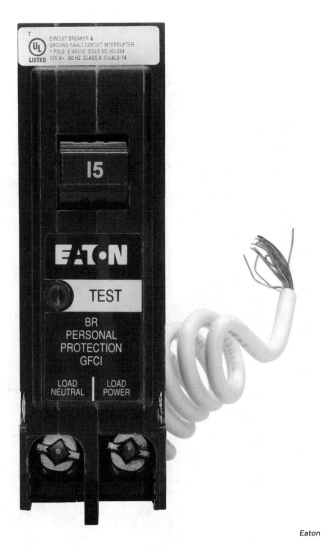

Eaton

Figure 17-15. GFCI breakers have a test button that will simulate a ground fault. If the breaker is working properly, it will trip off. This test is meant to ensure the device is working properly. In the event the test button does not trip the breaker, it should be replaced.

Eaton

Figure 17-16. Arc-fault breakers have a test button that will simulate an arcing fault. The breaker should trip off after the test button has been pushed. If it does not trip, the breaker should be replaced.

parallel. A series arc occurs when current traveling through a circuit jumps across a gap to complete a circuit. An example of a series arc is a loose connection to a device, **Figure 17-17A**. A parallel arc is when two conductors arc between one another. An example of a parallel arc is a nail or screw through a cable that presents a path for current to travel from one conductor to another, **Figure 17-17B**. AFCI breakers that monitor the circuit for both series and parallel faults are called combination arc-fault breakers. Do not confuse combination AFCIs with dual-function breakers that provide both arc-fault as well as ground-fault protection.

AFCIs are now required to be installed in most areas of dwellings. *Section 210.12* lists the requirements and locations for arc-fault circuit interrupters.

Dual-Function (AFCI and GFCI) Circuit Breaker

Dual-function circuit breakers provide both AFCI and GFCI protection. They are used in areas like kitchens, laundry rooms, and basements where the outlets are required to have both types of protection. The cost of installing a dual-function circuit breaker is cheaper than installing an AFCI breaker along with a GFCI outlet, **Figure 17-18**.

Troubleshooting what caused an arc fault or dual-function breaker to trip can be difficult. To make it a little easier, many breaker manufacturers provide built-in troubleshooting to help identify what caused the breaker to trip. Refer to the manufacturer's instructions to determine how to access the troubleshooting code. The common fault codes include overload, short circuit, ground fault, arcing fault, and self-test failure.

SWD and HID Circuit Breaker

Breakers with an SWD rating are designed and constructed to switch fluorescent lights on and off. This is sometimes the case in stores and restaurants where there are not any switches in the circuit and the circuit breakers are being used as the switch, **Figure 17-19**.

Breakers with an HID rating are designed to be used with high-intensity discharge lighting, such as metal halide,

A

B

Goodheart-Willcox Publisher

Figure 17-17. A—This receptacle has a loose connection, creating a series arc. The amount of heat generated overheated the conductor insulation, causing it to crack and melt the plastic frame of the device. **B**—*Section 300.4* of the *NEC* requires cables to be at least 1 1/4″ from the edge of the stud to protect them from screws and nails. In this picture, the exterior sheathing nails were too long and passed through the nonmetallic sheathed cable. A cable with a nail passing through it has the potential to create a parallel arc.

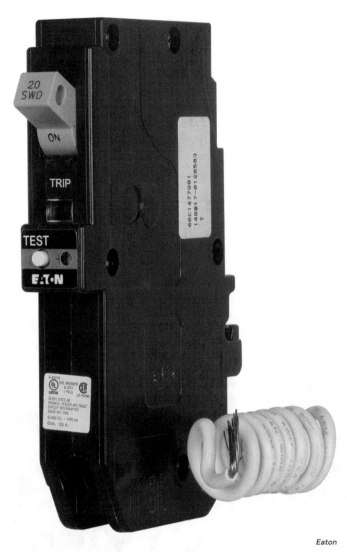

Figure 17-18. Dual-function breakers provide Class A GFCI protection as well as combination arc-fault protection.

Figure 17-19. The circuit breaker shown has a SWD and HACR designation. The HACR designation is for heating, air conditioning, and refrigeration equipment. HACR breakers are less likely to trip due to inrush current common with heating and cooling equipment. Also note that this breaker has an interrupting rating of 10,000 amperes.

high-pressure sodium, or mercury vapor lights. These breakers are designed to withstand the increased current necessary during the ignition period of HID lights. Breakers with an HID rating are also permitted to be used as a switch to turn fluorescent and HID lights on and off. *Section 240.83(D)* has the requirements for SWD and HID circuit breakers.

Standard Ampacity Ratings

Breakers and fuses installed in electrical systems come in standard sizes (ampacity ratings). *Table 240.6(A)* has a listing of the standard ampere ratings. Nonstandard size fuses and breakers are permitted by the *NEC* and are sometimes integral to a piece of equipment, but they are rarely installed in panelboards, **Figure 17-20.**

Table 240.6(A) Standard Ampere Ratings for Fuses and Inverse Time Circuit Breakers

Standard Ampere Ratings				
10	15	20	25	30
35	40	45	50	60
70	80	90	100	110
125	150	175	200	225
250	300	350	400	450
500	600	700	800	1000
1200	1600	2000	2500	3000
4000	5000	6000	—	—

Reproduced with permission of NFPA from NFPA 70, National Electrical Code, 2023 edition. Copyright © 2022, National Fire Protection Association. For a full copy of the NFPA 70, please go to www.nfpa.org

Figure 17-20. *Table 240.6(A)* lists the standard ampere ratings for fuses and breakers.

Location

There are a few ways to look at location with respect to overcurrent protection. Some things to consider are the location in the circuit, location on the wall and proximity to other items, and its location in the building. The *NEC* has several requirements for the installation of overcurrent devices and restrictions on where they are permitted to be installed. *Part II* of *Article 240* is titled *Location* and has most of these requirements.

Location in the Circuit

Section 240.21 requires that each ungrounded conductor have an overcurrent protective device. The device shall be located at the point where the conductor receives its supply, typically a panelboard, switchboard, or disconnect.

Mounting Location

Section 240.24(A) requires that overcurrent devices be mounted in a readily accessible location. They need to be easy to find and accessible so customers can get to them quickly to turn a circuit off. There is also a maximum height for overcurrent devices to ensure the majority of people will be able to reach them. The maximum height to the center of the operating handle must not exceed 6'-7" above the floor.

Section 240.33 gives an additional requirement that the enclosures containing overcurrent devices be mounted in a vertical position. With most panelboards, that means the breakers will operate horizontally (side to side). There are some smaller panelboards that will have the overcurrent devices mounted so the handle operates vertically (up and down). In this case, the breakers must be on when in the up position (*Section 240.81*).

Location in the Premises

Overcurrent protection is allowed to be installed in most areas, provided the enclosure is listed for the environment it is being installed. Although that is the general rule, there are a few areas where the *NEC* does not permit overcurrent devices:

- Areas where they are subject to physical damage (*Section 240.24(C)*)
- Vicinity of easily ignitable material, such as clothes closets (*Section 240.24(D)*)
- Dwelling unit bathrooms (*Section 240.24(E)*)
- Over steps in a stairway (*Section 240.24(F)*)

Overcurrent Protection Devices

Although overcurrent protection devices sometimes provide additional protection, such as GFCI/AFCI, their primary purpose is to protect conductors. They limit the current on a conductor to a level that is safe for the conductor and will prevent overheating. When sizing conductors, we size them to be able to carry the amount of load current. Then we size the overcurrent device to protect the conductor.

Load

Load is the equipment and appliances that consume electricity in a circuit. Examples of load include luminaires, motors, heating equipment, and appliances. There are two main types of loads recognized by the *NEC*: noncontinuous and continuous loads.

Noncontinuous Load

Noncontinuous loads are loads that will not be on for three hours or more. Most loads fall into this category. A kitchen range is an example of a noncontinuous load. Even if a person had all the burners on, as well as the oven, each heating element will cycle on and off as it reaches its set temperature. When the range initially heats up, some of the elements may be on for a bit longer, but it certainly will not be for three hours, and after reaching the desired temperature, they will cycle on and off to maintain the set temperature. Most of the items we plug in fall in the category of a noncontinuous load. Televisions, small kitchen appliances, and vacuums will be on for a short amount of time, and the amount of current they will draw will vary. Ohm's law is used to determine the maximum load, in voltamperes or watts, that is permitted on a branch circuit.

EXAMPLE 17-1

Problem: What is the maximum noncontinuous load permitted on a 120-volt, 20-ampere circuit?

Solution:

$$\text{Power (W or VA)} = \text{Voltage (V)} \times \text{Current (A)}$$
$$P = 120\text{ V} \times 20\text{ A}$$
$$= 2400\text{ W}$$

Continuous Load

A continuous load is defined as a load where the maximum current is expected to continue for three hours or more. School lighting is an example of a continuous load. The lights will be turned on in the morning and remain on the majority of the day. Most lighting is considered a continuous load as it has the potential to be on for three hours or more. *Section 210.20(A)* requires overcurrent protection devices for continuous loads to be rated at 125% of the rated load.

EXAMPLE 17-2

Problem: The lights in a classroom draw 16 amperes of current. What is the minimum size circuit required for the lights?

Solution:

$$\begin{aligned}\text{Minimum circuit size} &= \text{Load (A)} \times \text{Rating (\%)} \\ &= 16\text{ A} \times 1.25 \\ &= 20\text{ A}\end{aligned}$$

The minimum size circuit is 20 amperes.

Another way to look at it is from the perspective of the branch circuit. If a 20-ampere circuit has a continuous lighting load, what is the maximum amount of current the lights could draw? To solve that problem, divide the size of the overcurrent device by 1.25 (125%).

$$\begin{aligned}\text{Maximum load} &= \text{Overcurrent size (A)} \div \text{Rating (\%)} \\ &= 20\text{ A} \div 1.25 \\ &= 16\text{ A}\end{aligned}$$

The maximum amount of continuous load current on a 20-ampere circuit is 16 amperes.

Determining the Load

In the case of loads that are permanently connected, such as luminaires, we can determine the amount of current they will draw by looking at the nameplate and the manufacturer's literature. In situations where a branch circuit feeds receptacles, it may not be possible to determine the amount of current that will be drawn. Portable appliances, such as blenders, televisions, vacuums, and phone chargers, will be plugged into the receptacles. Each customer will have different items they want to plug in, and each item will draw a different amount of current. The *NEC* has a provision to solve this problem for nonresidential jobs. *Section 220.14(I)* allots 180 volt-amperes for each single or duplex receptacle attached to a single yoke. This gives us a number that we can use to determine the maximum number of receptacles that can be installed on a circuit. This rule does not apply to dwelling units as the receptacle outlets are considered part of the general lighting load. *Section 220.14(J)* explains which receptacle outlets in dwellings are part of the general lighting load.

EXAMPLE 17-3

Problem: What is the maximum number of receptacles that are permitted to be installed on a 20-ampere circuit in a commercial office space?

Solution: Determine the maximum circuit load.

$$\begin{aligned}\text{Power} &= \text{Voltage} \times \text{Current} \\ &= 120\text{ V} \times 20\text{ A} \\ &= 2400\text{ VA}\end{aligned}$$

Divide the maximum circuit load by 180 volt-amperes to determine how many receptacles can be placed on the circuit.

$$2400\text{ VA} \div 180\text{ VA} = 13.33\text{ receptacles}$$

The maximum number of receptacles permitted on the 20-ampere circuit is 13.

Sizing Overcurrent Devices

Sizing overcurrent devices for most equipment and appliances can be broken down into three steps:

- Step One—Determine the minimum ampacity of the circuit.
- Step Two—Determine the minimum size conductor.
- Step Three—Determine the minimum size overcurrent device.

Although these steps provide a systematic approach that applies to many applications, these steps don't have to be performed in order. For example, if the size of the OCPD has already been determined, the conductors would then be sized according to the rating of the overcurrent device. In the end, you arrive at the same result. It is also important to note that when sizing overcurrent devices for motors, transformers, and taps, there are additional requirements and considerations not covered in this unit.

Step One—Determine the Minimum Ampacity of the Circuit

The first step for sizing overcurrent devices is to determine the minimum ampacity. The method of finding this information will vary by whether you are sizing the overcurrent device for a service, feeder, or branch circuit. In the case of branch circuits, the type of circuit or piece of equipment being fed will also play a role in determining the minimum circuit ampacity.

Specific Equipment

If the circuit is feeding a specific piece of equipment, the manufacturer's specifications or the equipment nameplate will give information on the amount of energy the appliance will require. *Section 422.60* requires that a manufacturer provide a nameplate that identifies the manufacturer's name and gives the appliance's rating in volts and amperes, or volts and watts. Some appliances may also list a minimum and maximum fuse or breaker size. This is common with outdoor air conditioning units. *Section 440.22(C)* prohibits exceeding the manufacturer's maximum overcurrent device rating regardless of how large the branch circuit conductors may be, **Figure 17-21**. When sizing conductors and overcurrent devices, be sure to check the *NEC* for any specific requirements that apply to that appliance, **Figure 17-22**.

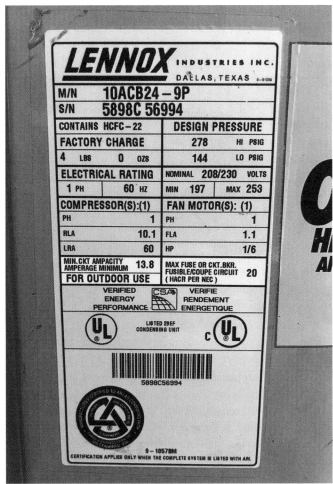

Goodheart-Willcox Publisher

Figure 17-21. The nameplate mounted on electrical equipment and appliances will identify electrical requirements. This nameplate is on an outdoor unit of an air conditioning system. In addition to the voltage requirement, it also lists the minimum circuit ampacity and maximum size overcurrent device.

General-Purpose Circuits

General-purpose circuits feed receptacle and lighting outlets that are *not* for a specific piece of equipment. For example, in a dwelling unit, we do not know what will be plugged into the bedroom receptacle outlets. Instead, we determine the minimum ampacity of the circuit we want to install. For example, you may choose to install a 15-ampere circuit in the bedroom and a 20-ampere circuit in the exercise room.

The *NEC* does have minimum requirements for a few dwelling circuits. *Section 210.11(C)* lists the required circuits in a dwelling. Included in the list are the small appliance branch circuits, laundry branch circuit(s), bathroom branch circuit(s), and the garage branch circuit(s). All of these are required to be 20-ampere circuits, most of which must be dedicated.

Service or Feeder Calculations

When determining the minimum ampacity of a service or feeder, a calculation is performed. *Article 220* describes the

Table 240.4(G) Specific Conductor Applications

Conductor	Article	Section
Air-conditioning and refrigeration equipment circuit conductors	440, Parts III, IV, VI	
Capacitor circuit conductors	460	460.8(B) and 460.25
Control and instrumentation circuit conductors (Type ITC)	335	335.9
Electric welder circuit conductors	630	630.12 and 630.32
Fire alarm system circuit conductors	760	760.43, 760.45, 760.121, and Chapter 9, Tables 12(A) and 12(B)
Motor-operated appliance circuit conductors	422, Part II	
Motor and motor-control circuit conductors	430, Parts II, III, IV, V, VI, VII	
Phase converter supply conductors	455	455.7
Remote-control, signaling, and power-limited circuit conductors	725	724.43, 724.45, 725.60, and Chapter 9, Tables 11(A) and 11(B)
Secondary tie conductors	450	450.6

Reproduced with permission of NFPA from NFPA 70, National Electrical Code, 2023 edition. Copyright © 2022, National Fire Protection Association. For a full copy of the NFPA 70, please go to www.nfpa.org

Figure 17-22. When sizing conductors and overcurrent devices, be sure to check the *NEC* to see if there are any specific requirements that apply to the appliance. *Table 240.4(G)* is a cross-reference table to overcurrent protection requirements for specific conductor applications.

process, which will be covered in Unit 21, Dwelling Service Calculations.

Step Two—Determine the Conductor Size

The second step is to determine the size of the conductors being installed and the maximum ampacity. Using the minimum circuit ampacity from step one, go to *Section 310.15* and run through the process of sizing conductors covered in Unit 16, Conductor Ampacity.

Assuming we do not have any adjustment factors or correction factors to address, we would turn to *Table 310.16*. Find the appropriate column for the type of conductor being used and slide down to the row that has an ampacity that meets or exceeds the minimum circuit ampacity from step one. *Table 310.16* will identify the minimum size conductor that can be installed as well as the maximum ampacity of the conductor.

Step Three: Determine the Correct Size Overcurrent Device

The third step is to determine the correct size overcurrent device. Using the conductor ampacity information from step two, we will turn to *Section 240.4, Protection of Conductors*. *Section 240.4* has several subsections that address the various scenarios we will encounter when sizing overcurrent devices. This section applies to all conductors except flexible cords, flexible cables, and fixture wires. The exceptions have ampacity and overcurrent requirements within their respective articles.

Section 240.4(A)

Section 240.4(A) addresses a specialized scenario where the loss of power can create a greater hazard than the danger of an overloaded circuit. One example is a fire pump. A fire pump may be the last line of defense for a building on fire. It is preferable that the fire pump draws excess current untill it destroys the motor rather than quits running while firefighters are fighting a fire and people are trying to leave a building.

Section 240.4(B)

Section 240.4(B), Overcurrent Devices Rated 800 Amperes or Less states that if an overcurrent device is rated at 800 amperes or less, the next standard size above the ampacity of the conductor is permitted to be used. We are not allowed to round up if the ampacity of the conductor corresponds directly with a standard size overcurrent device, or if it is feeding more than one receptacle for cord and plug connected portable loads. The standard ampacities of fuses and breakers can be found in *Table 240.6(A)*.

EXAMPLE 17-4

Problem: A 6 AWG copper conductor, using the 60° column, has a maximum ampacity of 55 amperes.

Solution: Since 55 amperes does not correspond to a standard size overcurrent device, the next larger size will be used which is 60 amperes.

EXAMPLE 17-5

Problem: A 2/0 THW copper conductor has a maximum ampacity of 150 amperes.

Solution: Since 150 amperes corresponds to a standard size overcurrent device, a 150-ampere breaker will be used.

Section 240.4(C)

Section 240.4(C) applies to overcurrent devices that are rated over 800 amperes. It states that if the overcurrent device is over 800 amperes, the conductor ampacity shall be equal to or greater than the rating of the overcurrent device.

EXAMPLE 17-6

Problem: Three 600 kcmil aluminum conductors installed in parallel have a combined maximum ampacity of 1020 amperes.

Solution: Since 1020 amperes does not correspond to a standard overcurrent device rating, the next lower size would be used, which is 1000 amperes.

Section 240.4(D)

Section 240.4 (D) lists the maximum size overcurrent device permitted for smaller conductors. Below is a list of the commonly used conductors this applies to.

- 14 AWG copper—15 amperes
- 12 AWG copper—20 amperes
- 10 AWG copper—30 amperes

These maximum ampacities align with the 60°C column in *Table 310.16*. Regardless of the insulation rating used, the general rule in *Section 110.14(C)(1)* states that circuits rated 100 amperes or less, which applies to conductors 14 AWG through 1 AWG, shall have their maximum ampacities based on 60°C.

PROCEDURE | Steps To Size Overcurrent Protection Devices

Step One: Determine the ampacity of the load:
- Nameplate rating(s)
- Minimum size based on *NEC* requirements
- Service or feeder calculation

Step Two: Determine the correct conductor size for the load:
- *Section 310.15* through *Table 310.16*

Step Three: Determine the correct size overcurrent protection device based on the conductor ampacity:
- *Section 240.4(B)*—Overcurrent devices 800 amperes or less
- *Section 240.4(C)*—Overcurrent devices rated over 800 amperes
- *Section 240.4(D)*—Small conductors (Up to 10 AWG copper)
- Check nameplate rating for minimum and maximum overcurrent device ratings

Summary

- There are three types of overcurrents: overloads, short circuits, and ground faults.
- *Article 240* contains requirements for overcurrent protection devices.
- *Section 240.6(A)* lists the standard size overcurrent device ratings.
- Overcurrent devices are sized to protect the circuit conductors.
- *Section 240.4* describes how to size overcurrent devices.
- The three steps to determine the minimum size overcurrent device are:
  - Determine the ampacity of the load.
  - Determine the correct conductor size for the load.
  - Determine the correct size overcurrent device based on the conductor's ampacity.

Unit 17 Review

Name _____ Date _____ Class _____

Know and Understand

Answer the following questions based on information in this unit.

1. A 20-ampere circuit with a current flow of 23 amperes is an example of a(n) _____.
   A. ground fault
   B. short circuit
   C. arcing fault
   D. overload

2. Plug fuses are permitted for _____.
   A. replacement only in existing installations
   B. residential installations only
   C. in all installations
   D. installations with a service of 200 amperes or less

3. Where a branch circuit supplies a continuous load, the rating of the overcurrent device shall not be less than _____% of the load.
   A. 80
   B. 100
   C. 125
   D. 150

4. The maximum rating of an overcurrent device for a circuit with 12 AWG copper conductors is _____ amperes.
   A. 15
   B. 20
   C. 25
   D. 30

5. An energized conductor that makes contact with the metal frame of an appliance is an example of a(n) _____.
   A. short circuit
   B. ground fault
   C. overload
   D. All of the above.

6. The *NEC* requires that an appliance nameplate list the identifying name and the appliance rating in _____.
   A. volts and amperes
   B. volts and watts
   C. volts and amperes, or volts and watts
   D. None of the above.

7. Two energized, ungrounded conductors with a difference in potential that come into contact is an example of a(n) _____.
   A. ground fault
   B. short circuit
   C. overload
   D. open circuit

8. What is the maximum size overcurrent device permitted to protect a conductor with a maximum ampacity of 53 amperes?
   A. 50 amperes
   B. 53 amperes
   C. 55 amperes
   D. 60 amperes

9. Overcurrent protection devices are permitted in all of the following locations except a _____.
   A. bedroom
   B. clothes closet
   C. hallway
   D. family room

10. The general definition for overcurrent protection states that it must be installed in a(n) _____ location.
    A. readily accessible
    B. accessible
    C. supervised
    D. dry

Apply and Analyze

Answer the following questions using a copy of the National Electrical Code. *Identify the section or subsection where the answer is found.*

1. Enclosures for overcurrent devices in damp or wet locations shall comply with _____.
   A. Section 300.14
   B. Section 110.14
   C. Section 240.32
   D. Section 312.2

   NEC _____

2. A _____-ampere plug fuse will have a hexagonal configuration in the cap.
   A. 15
   B. 20
   C. 25
   D. 30

   NEC _____

3. Which of the following is *not* a standard ampere rating of a fuse?
   A. 35 amperes
   B. 70 amperes
   C. 115 amperes
   D. 175 amperes

   NEC _____

4. Overcurrent devices shall not be installed in a location where they will be exposed to _____.
   A. physical damage
   B. low-ambient temperatures
   C. sunlight
   D. All of the above.

   NEC _____

5. When circuit breakers are installed vertically rather than horizontally, the up position _____.
   A. shall be the "off" position
   B. shall be the "on" position
   C. may be the "on" or "off" position
   D. may be the "on" or "off" position provided it is clearly marked

   NEC _____

6. Conductors protected by overcurrent devices rated over _____ amperes shall have an ampacity equal to or greater than the rating of the overcurrent device.
   A. 200   C. 600
   B. 400   D. 800

   NEC _____

7. An 18 AWG fixture wire shall be permitted to be connected to a 20-ampere branch circuit provided it is no longer than _____ feet.
   A. 10   C. 50
   B. 25   D. 100

   NEC _____

8. Overcurrent devices shall be installed in each _____ conductor at the point where it receives its supply.
   A. ungrounded   C. ungrounded and
   B. grounded   grounded
                         D. grounding

   NEC _____

9. What is the maximum size overcurrent device permitted to protect a 14 AWG copper conductor?
   A. 10 amperes   C. 20 amperes
   B. 15 amperes   D. 25 amperes

   NEC _____

10. Cartridge fuses with an interrupting rating other than _____ amperes shall be plainly marked on the fuse barrel or by a label attached to the barrel.
    A. 5000   C. 8000
    B. 6000   D. 10,000

    NEC _____

Name _____ Date _____ Class _____

Critical Thinking

Answer the following questions using a copy of the National Electrical Code. *Identify the section or subsection where the answer is found.*

1. Renewable cartridge fuses (Class H) shall be permitted to be used only for replacement in existing installations where there is _____.

   _____

   NEC _____

2. _____ circuit breakers shall not be permitted to be reconditioned.

   _____

   NEC _____

3. Cartridge fuses and fuseholders shall be classified according to _____ ranges.

   _____

   NEC _____

4. The maximum size overcurrent device for a 12 AWG aluminum conductor is _____.

   _____

   NEC _____

5. The screw shell of a plug-type fuseholder shall be connected to the _____ side of the circuit.

   _____

   NEC _____

6. What is the maximum size overcurrent device permitted to protect a 1/0 AWG copper THWN conductor?

   _____

   NEC _____

7. What is the maximum size overcurrent device permitted to protect a 6 AWG aluminum THW conductor?

   _____

   NEC _____

8. What is the maximum size overcurrent device permitted to protect a 500 kcmil copper THHN conductor?

   _____

   NEC _____

9. What is the maximum size overcurrent device permitted to protect a set of 500 kcmil copper THWN conductors with two paralleled conductors per phase?

   _____

   NEC _____

10. What is the maximum size overcurrent device permitted to protect a set of 600 kcmil aluminum XHHW conductors with three paralleled conductors per phase?

    _____

    NEC _____

Part 2   Application: Advanced NEC Topics

Answer the following questions using a copy of the National Electrical Code. *Identify the section or subsection where the answer is found.*

11. What is the maximum number of noncontinuous watts permitted on a 120-volt, 15-ampere overcurrent protective device?

    _____

    NEC _____

12. What is the maximum size overcurrent device for a 10 AWG copper THHN conductor?

    _____

    NEC _____

13. What is the maximum size overcurrent device for a 4 AWG aluminum USE cable?

    _____

    NEC _____

14. What is the maximum size overcurrent device for a 300 kcmil THHW copper conductor?

    _____

    NEC _____

15. What is the maximum overcurrent device for an 8 AWG three conductor nonmetallic sheathed cable?

    _____

    NEC _____

16. What is the maximum size overcurrent device permitted to protect a set of 350 kcmil aluminum XHHW conductors with three paralleled conductors per phase?

    _____

    NEC _____

17. What size conductors and overcurrent devices does the *NEC* require for the bathroom receptacle branch circuit in a dwelling?

    _____

    NEC _____

18. What is the minimum ampacity of a branch circuit with a continuous lighting load of 14 amperes? What is the minimum size copper THHN conductor and maximum size overcurrent device?

    _____

    _____

    _____

    NEC _____

19. An air conditioning condensing unit is being replaced. The existing conductors that fed the unit are 10 AWG and are fed with a 30-ampere breaker. The new unit has a minimum circuit ampacity of 13 amperes and lists the maximum breaker size as 25 amperes. What is the maximum size overcurrent device permitted in this installation?

    _____

    NEC _____

20. What is the minimum size copper THWN conductor and overcurrent device required for a storage type water heater in a commercial building? The nameplate rating lists it as a 20 kW, 240-volt, single-phase appliance.

    _____

    _____

    NEC _____

UNIT 18 Service

Sashkin/Shutterstock.com

KEY TERMS

overhead service conductor
service
service conductor
service drop
service-entrance conductor
service equipment
service head
service lateral
service mast
service point
surge protective device
underground service conductors

OBJECTIVES

After completing this chapter, you will be able to:

- Define the service.
- Understand the purpose of the various equipment involved in the service.
- Describe the requirement for surge-protective devices in dwellings.
- Describe the requirements for service disconnects.
- Identify the wiring methods permitted to be used as part of the service.
- Describe the two types of services.
- Locate service-related requirements in the *National Electrical Code*.

Introduction

Article 230 is dedicated to the service, **Figure 18-1**. This unit will introduce *Article 230* by discussing the various aspects and components of a service, specifically a service that feeds a building or structure, as it is the easiest to visualize. It is important to remember that electrical systems that do not have an associated building will also have a service, such as street lighting, large irrigation systems, and drilling sites.

This unit will begin with the basic service components and requirements and conclude with requirements specific to overhead and underground services. There are review questions and problems at the end of the unit that will reinforce the information and familiarize you with *Article 230* and related requirements found in other parts of the *Code*.

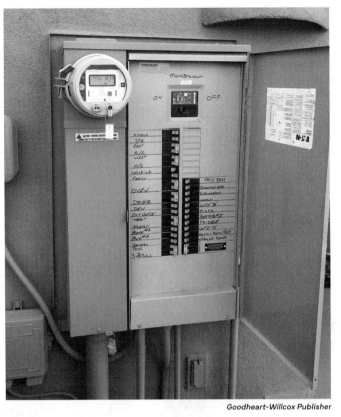

Figure 18-1. The piece of equipment in the picture contains all the service components in one unit. The left side has the utility meter socket, which is where the service point is located. The right side has the main service disconnect and the branch circuit overcurrent devices.

Service

Article 230 is an extensive article that can be difficult to navigate. The *NEC* splits it into several parts, each of which covers a different aspect of the service, **Figure 18-2**. The article starts off with a diagram that gives a graphic view of how the article is organized and acts as a cross reference to other parts of the *Code*, **Figure 18-3**.

Article 230 Services

Part I	General
Part II	Overhead Service Conductors
Part III	Underground Service Conductors
Part IV	Service Entrance Conductors
Part V	Service Equipment - General
Part VI	Service Equipment - Disconnecting Means
Part VII	Service Equipment - Overcurrent Protection

Figure 18-2. *Article 230* is divided into seven parts to cover the different aspects of the service.

Figure 18-3. *Figure 230.1* from the *National Electrical Code* has a diagram of the service with a reference to where each component is covered by the *Code*.

Understanding the terminology is crucial to interpreting the *NEC*, and this is especially true for *Article 230*. The terminology will help with understanding what is included in the service and determine where the utility responsibility ends and an electrician's begins. The *NEC* defines a ***service*** as the conductors and equipment connecting the serving utility to the wiring system of the premises served. The service starts at the service point and ends at the main service disconnect, **Figure 18-4**.

The *NEC* defines the ***service point*** as the point of connection between the facilities of the servicing utility company and the premises wiring. Everything on the utility side of the service point is the responsibility of the public utility company and is subject to the requirements of their safety standards. Everything on the customer side of the service point falls within the scope of the *National Electrical Code*.

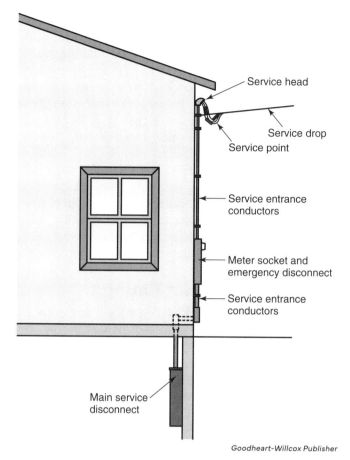

Figure 18-4. The service begins at the service point. In this diagram, the service point is where the utility service drop attaches to the service-entrance conductors. The service ends at the main service disconnect. In this diagram, the main service disconnect is located in the panelboard in the basement.

There are many factors involved in locating the service point. Some contributing factors include:
- Overhead vs underground service
- Urban or rural location
- Local codes
- Electric utility requirements

The figures in this unit identify the common service point locations. Keep in mind that service point locations will vary throughout different parts of the country and are typically determined by the local utility.

Service Equipment

The *NEC* defines ***service equipment*** as the necessary equipment, consisting of a circuit breaker(s) or switch(es) and fuse(s) and their accessories, connected to the serving utility and intended to constitute the main control and disconnect of the serving utility.

Part V of *Article 230* has general requirements for service equipment. *Section 230.66(A)* requires that all equipment used as part of the service must be marked to identify

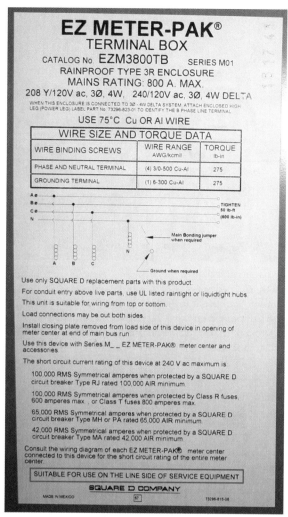

Figure 18-5. Equipment rated for use with the service is marked with a label like this one. The manufacturer lists other important information such as short circuit current ratings, torque specifications, and installation requirements.

that it is suitable for use as service equipment, **Figure 18-5**. The exception to this rule is the meter socket. Although the meter socket is not service equipment and is under the control of the electric utility, *Section 230.66(B)* still requires it to be listed and rated for the voltage and current rating of the service, **Figure 18-6**.

Surge Protective Devices (SPDs)

Section 230.67 requires a surge-protective device to be installed on dwelling unit services. The *NEC* defines a ***surge-protective device*** as a protective device for limiting transient voltage by diverting or limiting surge current; it also prevents continued flow of follow current while remaining capable of repeating these functions. In recent years, electric components such as ground-fault circuit interrupters, arc-fault circuit interrupters, LED lighting, and several types of electronic dimmers and sensors have been incorporated

Part 2   Application: Advanced NEC Topics

Goodheart-Willcox Publisher

Figure 18-6. The meter socket is not considered to be service equipment because it is under the control of the utility. Electric utilities often have very specific requirements for meter sockets that must be considered. Be sure to check with the utility for their requirements before starting the project.

Goodheart-Willcox Publisher

Figure 18-7. There are several types of surge protective devices that are available. This device snaps into the panel like a circuit breaker and has a small LED. The manufacturer recommends installing it as close to the panel feed as possible to provide maximum protection.

into residential electrical systems, **Figure 18-7.** In addition to the electronic equipment installed as part of the electrical system, homes are full of electronic appliances, such as televisions, computers, washers, and refrigerators. All these electronic components are sensitive to overvoltage.

The requirement for a surge-protective device only applies to dwellings at this point, but they can be, and often are, installed in non-dwelling installations. As sensitive electronic equipment becomes more prevalent, there will be an increased need for SPDs in all installations.

Service Disconnect

The service disconnect is covered by *Part VI* of *Article 230*. The service disconnect is a piece of service equipment that disconnects the electric source of energy from a building. The line side of the service disconnect is where the service-entrance conductors terminate, **Figure 18-8.** In most situations they are unfused, as they come directly from the electric utility. The load side of the main service disconnect is protected by the overcurrent protection that is provided as part of the disconnecting means. In many smaller installations, the service disconnect is incorporated into a panelboard, **Figure 18-9.** In the case of a panelboard, the load side of the main service disconnect has busbars where the branch circuit and feeder circuit breakers are connected.

Service Disconnect Rating

The minimum size, or rating, of the service disconnect is covered by *Section 230.79*. It states that the minimum rating of the service disconnect shall not be less than the connected load. The connected load can be found by performing a service calculation as per the requirements in *Article 220*. Residential service calculations are covered in Unit 21 of this text.

Goodheart-Willcox Publisher

Figure 18-8. On a large service, the main service disconnect may be its own section of the switchgear.

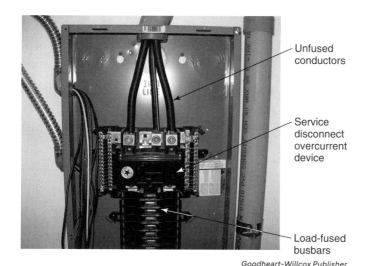

Figure 18-9. On smaller services, the main service disconnect is often incorporated into the panelboard.

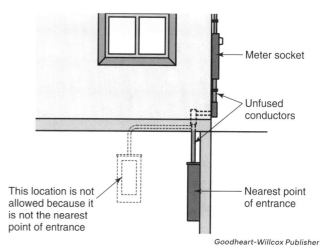

Figure 18-11. The main service disconnect must be as near as possible to where it enters the building. We want to minimize the distance the unfused conductors must travel to reach the first overcurrent device.

Although the appropriate size of the service is determined by performing a service calculation, the *NEC* gives minimum service disconnect size requirements, **Figure 18-10**.

Service Disconnect Location

Section 230.70(A)(1) requires the service disconnect to be installed in a readily accessible location at the nearest point of entrance of the service-entrance conductors, or outside of the building, **Figure 18-11**. Limiting the service disconnect to the nearest point of entrance limits the length of the service-entrance conductors that enter the building. The service-entrance conductors coming from the electric utility to the disconnect are considered unfused, so we want to minimize the distance they travel through the building. They are technically fused by the utility at the transformer but have a different purpose and a much larger fuse than the *NEC* would require for a similarly sized conductor. The utility fuse will typically allow a short circuit to burn itself open before it would clear the fault.

There are a few areas in a building where a service disconnect cannot be located. *Section 230.70(A)(2)* prohibits the installation of a service disconnect in bathrooms. The other prohibition relates to service disconnects that have overcurrent protection. *Section 240.24(D)* prohibits overcurrent protection from being installed in areas exposed to physical damage, in the vicinity of easily ignitable material, or over steps.

Service Disconnect Position

The maximum height of the service disconnect is 6'-7" to the center of the operating handle when in its highest position, **Figure 18-12**. That measurement is taken from grade level, the floor, or the platform the disconnect is mounted above. This requirement shows up in two places: *Section 240.24(A)* for overcurrent protection, and *Section 404.8(A)* for switches.

Disconnect switches and circuit breakers can be mounted in a vertical or horizontal position. If they are mounted vertically, they shall be installed so that they are energized while in the up position (*Section 404.7*). This ensures that gravity pulling down will never open the switch. Whether mounted horizontally or vertically, they must be clearly marked to indicate whether they are in the open (on) or closed (off) position (*Section 230.77*).

Number of Service Disconnects

The *NEC* has requirements that apply to service disconnects to ensure the shut off is easy to identify in the case of an emergency. Ideally there would only be one service disconnect. *Section 230.71* states that there shall only be one, but that isn't always possible. For this reason, *Section 230.71* goes on to say that additional disconnects shall be permitted provided they comply with *Section 230.71(B)*.

Up to Six Service Disconnects

Section 230.71(B) permits up to six service disconnects provided they are grouped together and are properly identified,

Minimum Rating of Service Disconnect

Section 230.79		
One-circuit installation	15 A	Section 230.79(A)
Two-circuit installation	30 A	Section 230.79(B)
One-family dwelling	100 A	Section 230.79(C)
All others	60 A	Section 230.79(D)

Figure 18-10. *Section 230.79* lists the minimum ratings of the service disconnect based on the type of application.

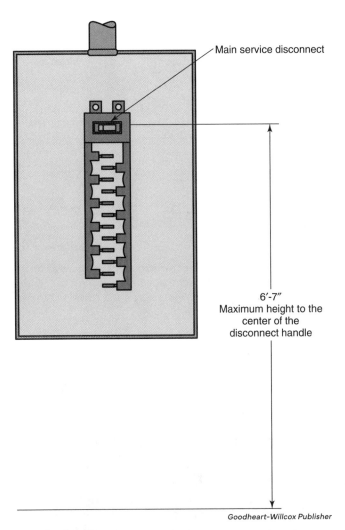

Figure 18-12. The maximum height to the center handle of the main service disconnect is 6'-7" above the floor or platform. This is to ensure that people that aren't as tall can still reach the device without needing a ladder or stool. The same goes for any other overcurrent device. If the panel is mounted so the main breaker is on the bottom, then the maximum height of one of the branch circuit overcurrent devices is 6'-7".

Figure 18-13. The grouping does not require them to be in the same enclosure, but they do have to be in close proximity to one another. Each of the disconnects must be clearly marked to indicate that it is a service disconnect and identify the load it serves.

Buildings with fire pumps, emergency systems, legally required systems, or optional standby systems are permitted to have additional disconnects beyond the six permitted by *Section 230.71*, **Figure 18-14**. They are allowed to be installed away from the rest of the service equipment to prevent inadvertent disconnection. A plaque that identifies the locations of each of the disconnects must be installed in each disconnect location so emergency personnel are able to easily locate and identify them.

Figure 18-13. This installation has three service disconnects, the main breaker in the panelboard and the two safety switches. If there is more than one main service disconnect, they must be grouped and clearly marked.

Figure 18-14. Disconnects for fire pumps are allowed to be located apart from the service equipment. Life saving devices should not inadvertently shut down in the event of a fire.

Clearances

The *NEC* requires clear working space in front of electrical equipment that will require examination, adjustment, servicing, or maintenance while energized. Disconnects, panelboards, switchboards, and motor control centers are examples of equipment that are subject to this requirement. The requirements are in place to ensure there is adequate space to be able to safely perform a task while working on or around energized equipment. The minimum spatial requirements are found in *Section 110.26*.

Depth

The minimum depth of the working space in front of a panelboard is 36″. In situations where the voltage is more than 150 volts to ground, *Table 110.26(A)(1)* lists voltages and conditions where the minimum distance shall be increased, **Figure 18-15**.

Width

The minimum width of the working space in front a panelboard is the width of the equipment or 30″, whichever is greater. The 30″ can start from the left side of the panelboard and extend to the right, it can start from the right side of the panelboard and extend to the left, or be somewhat centered, **Figure 18-16**. In the case of a panelboard that is wider than 30″, the width of the panelboard is the minimum width. When there are multiple panelboards side-by-side, the 30″ minimum clearance for each of the panelboards is permitted to overlap each other.

Height (Section 110.26(A)(3))

The minimum height in front of an electrical panel is 6′-6″ or the height of the equipment, whichever is greater. *Section 110.26(E)* describes the space above and below the panel as dedicated space. It requires that the space directly above the panel for a height of 6′ above the panel, or the ceiling height, whichever is lower, must be dedicated to electrical equipment and its related components, **Figure 18-17**. Plumbing and HVAC are not permitted to be in this space.

Illumination

Electrical equipment is required to have adequate illumination by *Section 110.26(D)*. It does not require the light to be installed within a certain distance, give a minimum level of illumination, or require the light to be dedicated to the space. It simply requires the light to be adequate, leaving the requirement up to interpretation. Since adequate has a different meaning to different people, remember that the intention is to have enough light to be able to safely work on the equipment.

Egress

For larger electrical equipment, there is an additional requirement of *egress*—a place or means of exit. In the case of an equipment malfunction that may produce an arcing incident, anyone near the equipment must be able to quickly get away. *Section 110.26(C)* details the minimum egress requirements that applies to large electrical equipment. For this section, large equipment is equipment that individually or in combination, has a rating of 1200 amperes or more,

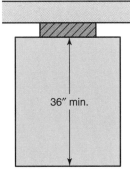

Goodheart-Willcox Publisher

Figure 18-15. There must be at least 36″ of working space in front of an electrical panel. This rule ensures there is sufficient space to safely perform the task at hand. Some electrical systems will require additional clearances. *NEC Table 110.26(A)(1)* lists the minimum requirements.

Minimum Width

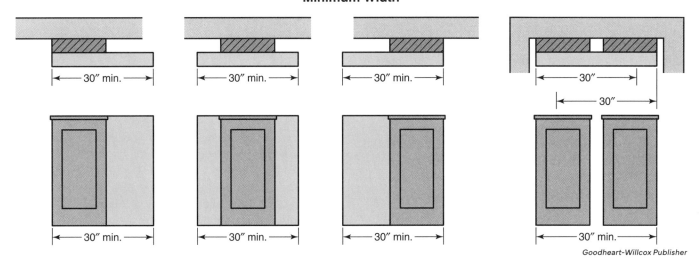

Goodheart-Willcox Publisher

Figure 18-16. The minimum width of the working space in front of a panelboard is 30 inches or the width of the equipment. If the enclosure has a door it must be able to open to a full 90-degree angle.

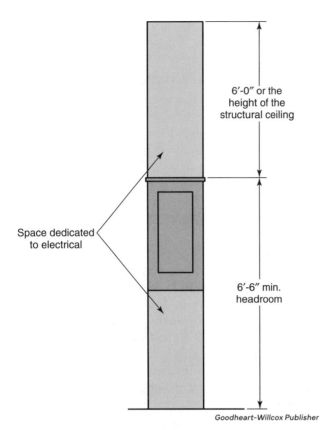

Figure 18-17. The minimum headroom in front of an electrical panel is 6'-6". The highlighted space directly above and below the panel is dedicated to electrical equipment only. Plumbing pipes, ductwork, etc. cannot be installed in that space.

and is 6' or more in width. The general rule for this scenario requires there to be two entrance/exits, one at each side, **Figure 18-18**. The exits shall be at least 24" wide and 6'-6" tall. Open equipment doors shall not impede workers' ability to enter or leave the space.

The general rule requires two entrance/exits, but there are a couple scenarios where one door will be permitted. The first is found in *Section 110.26(C)(2)(a)* and applies to a situation where there is an unobstructed egress. If there is unobstructed access to the space as in **Figure 18-19**, then only one door is required. If there is extra working space in front of the equipment, or if the working space is twice the depth that is required by *Table 110.26(A)(1)*, then only one entrance/exit is required, **Figure 18-20**.

The door(s) on a room containing electrical equipment rated 800 amperes or more must open outward and have panic hardware or listed fire exit hardware, *Section 110.26(C)(3)*. This allows a person in an emergency to be able to simply push or lean into the door hardware to open it, **Figure 18-21**.

Dwelling Emergency Disconnect

Dwelling units are required to have an emergency disconnect installed outside of the building. The purpose of this

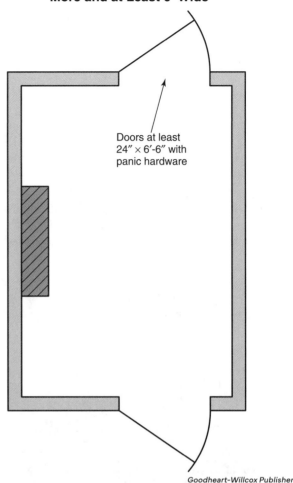

Figure 18-18. Large service equipment may require two exits so that a person can quickly and easily get out in the case of an emergency such as an arcing incident. The doors must open outward and have panic hardware so they can be opened with a push.

requirement is to provide a disconnecting means for emergency personnel, such as a firefighter needing to remove the power to a building when attempting to put out a fire. The emergency disconnect is permitted to be a nonfused disconnect that is not the service disconnect, but it must have a short circuit current rating that is greater than the available fault current. The emergency disconnect shall be appropriately marked as per *Section 230.85(E)*.

The service disconnect is permitted to satisfy the emergency disconnect requirement, provided it is mounted in a readily accessible outdoor location, **Figure 18-22**. If the service disconnect is mounted on the exterior of the building and the panelboard is mounted inside, the interior panelboard becomes a subpanel, and the grounded(neutral) conductors are separated from the equipment grounding conductors.

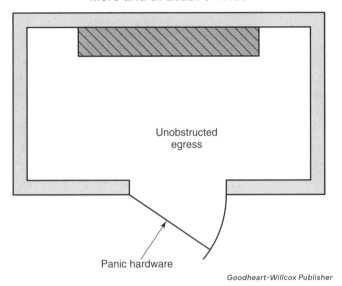

Figure 18-19. If the room has an unobstructed egress, meaning there is nothing that can block the path out, only one door would be required.

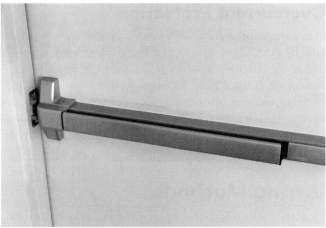

Figure 18-21. Doors with panic hardware allow for easy egress by simply pushing or leaning into the door handle.

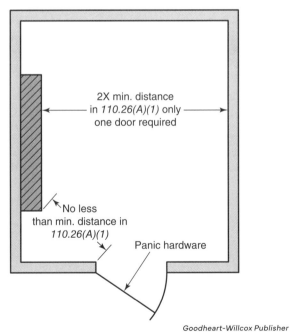

Figure 18-20. The two-door requirement can be reduced if there is twice the minimum space between the equipment and the opposite wall. If the minimum depth is 36″, there would have to be at least 72″ of working space between the equipment and the opposite wall. In addition to the extra depth, the maximum distance to the door cannot exceed 36″.

Figure 18-22. This is a meter socket main breaker assembly. The top is the meter socket portion of the assembly, and the bottom has the main breaker. This piece of equipment is the entire service as it contains the service point and the main service disconnect. Since it is outside, it also serves as the emergency disconnect.

Overcurrent Protection

Overcurrent protection associated with the service is covered by *Part VII* of *Article 230*. The overcurrent protection device must be installed in series with each ungrounded conductor. It must have an ampacity not higher than that of the service-entrance conductors. Overcurrent protection devices are not to be installed on the grounded conductor as it provides a return path to trip the ungrounded overcurrent devices in the case of a ground fault.

Wiring Methods

The wiring methods used as part of the service must be appropriately listed for the location, voltage, and ampacity, as well as being recognized as acceptable by the *NEC*. *Parts I-IV* of *Article 230* have the requirements for wiring methods associated with the service.

Service Raceways

Section 230.43 lists the raceways that are permitted to be used in the service, **Figure 18-23**. One of the most popular service raceways is rigid PVC, **Figure 18-24**. There are several reasons for its popularity: it is much less expensive than other wiring methods, easy to cut and work with, does not require bonding bushings (due to it being non-metallic), and can be used outside and underground without special fittings.

Service raceways that are installed on the outside of a building are in a wet location. *Section 230.53* requires them to be listed for use in wet locations and arranged to drain. The conductors inside the raceway must also be listed for wet locations as the interior of the raceway is also considered a wet location (*Section 300.9*).

When a raceway enters a building from outside or underground, it must be sealed to prevent the passage of moisture as per *Section 300.7(A)*. Duct seal is a common method of sealing around the conductors, **Figure 18-25**. Other sealants, such as expandable foam, are permitted to be used if they are listed for the application and will not damage the conductor insulation.

Raceways installed on the exterior of the building will be subjected to large temperature differentials, leading to expansion and contraction in the overall length of the raceway. If a raceway is unable to lengthen and shorten due to being attached to a box or conduit body on each end, it will pull out of the fittings, pull the box off the wall, or become wavy. The amount of change in temperature will vary considerably by geographic area, but all locations will have temperature variation to some degree.

Permitted Service Raceways
Section 230.43

Raceway	Stipulations	Article
Rigid polyvinyl chloride (PVC)	–	Article 352
Rigid metal conduit	–	Article 344
Intermediate metal conduit	–	Article 342
Electrical metallic tubing	–	Article 358
Electrical nonmetallic Tubing	–	Article 362
Flexible metal conduit (FMC)	Not over 6' long - with supply side bonding jumper	Article 348
Liquidtight flexible metal conduit (LFMC)	Not over 6' long - with supply side bonding jumper	Article 350
Liquidtight flexible nonmetallic conduit (LFNC)	–	Article 356
High density polyethylene conduit (HDPE)	Underground or in concrete only	Article 353
Nonmetallic Underground Conduit with Conductors (NUCC)	Underground or in concrete only	Article 354
Reinforced thermosetting resin conduit (RTRC)	–	Article 355
Metallic wireway	–	Article 376
Nonmetallic wireway	–	Article 378
Auxiliary gutters	–	Article 366
Busways	–	Article 368
Cable tray	–	Article 392

Goodheart-Willcox Publisher

Figure 18-23. Permitted service raceways as per *Section 230.43*.

Unit 18 Service

Christian Delbert/Shutterstock.com

Figure 18-24. This service has two PVC raceways extending into the ground. Each of the raceways has an expansion device to accommodate ground movement.

Goodheart-Willcox Publisher

Figure 18-25. Duct seal installed around the conductors in an LB where it transitions into a building prevents air movement that can lead to condensation.

Goodheart-Willcox Publisher

Figure 18-26. When installing an expansion fitting, the ambient temperature at the time of installation must be considered. If it is very hot, the fitting should be pushed in, as the raceway will be close to its maximum temperature/length. If it is cold, the fitting should be pulled out, as the conduit will be at its minimum length. Most expansion fittings have a mark on the inner portion that indicates the midpoint to aid in determining their proper length during installation.

If there is to be more than 1/4″ of movement, an expansion fitting will be required, **Figure 18-26**. The starting point for determining raceway expansion and contraction is in *Section 300.7(B)* and its informational notes.

Rigid PVC is the most common raceway to require expansion fittings as it expands and contracts more than metal raceways, **Figure 18-27**. *Article 352* (rigid PVC) has a table to simplify the process of determining the amount the raceway will expand/contract. The table is split into two sections with temperature change in Celsius on the left and Fahrenheit on the right. Once the amount of temperature change has been determined, the table will give the length of change for 100′ of rigid PVC. For example, 100′ of PVC installed in an area with a temperature variation of 130°F will change 5.27″ in length from its coldest temperature to its warmest temperature.

Metal raceways will also change in length, but to a lesser degree than PVC. The informational note in *Section 300.7(B)* gives a multiplier that can be used in conjunction with *Table 352.44* to determine the amount of thermal expansion for steel and aluminum conduits. It also gives the thermal

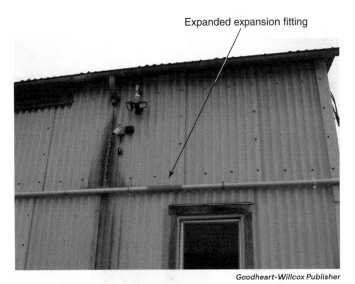

Figure 18-27. This picture was taken on a very cold day. You can see how much the expansion fitting has expanded to accommodate the shrinking of the overall length of the raceway.

Figure 18-28. Conductors like the ones leaving the service head need to be rated for wet locations and sunlight resistant. Conductor insulation that isn't sunlight resistant will deteriorate and eventually break down when exposed to the sun.

expansion coefficient if you would prefer to work through the calculation.

Service Conductors

Service conductors are the conductors from the service point to the service disconnecting means. Service conductor is an all-encompassing term that includes overhead service conductors, underground service conductors, service-entrance conductors, and service cables. It does *not* include any conductors on the utility side of the service point.

Service-entrance conductors are the service conductors between the terminals of the service equipment to the service drop, overhead service conductors, service lateral, or underground service conductors. It is possible to not have any service-entrance conductors if the service equipment is located on the outside of a building. An example is a residential service with a metersocket main disconnect assembly from a service lateral as in **Figure 18-1**.

Conductors and cables must be listed for the environment they will be subjected to. Conductors that are in a raceway underground, outside, or in any other wet location, shall be listed for wet locations (*Sections 300.5(B)* and *300.9*). Conductors that exit a service head and are in open air as they connect to service-drop or overhead service conductors, shall be listed for a wet location and be sunlight resistant (*Section 310.10*), **Figure 18-28**.

It is common to see aluminum conductors used as service conductors. There are pros and cons that go along with using aluminum conductors. Aluminum conductors are lighter, easier to bend, and cost less. A significant disadvantage of using aluminum is that it has a higher resistance and will need to be upsized as compared to a copper conductor.

Sizing

Section 230.42 lists the requirements for sizing service-entrance conductors. They must be sized to carry the maximum load to be served. The maximum load is determined by performing a service calculation as per *Article 220*. The service calculation will determine the minimum size of the main service disconnect and its associated overcurrent protection. The conductors in turn will be sized according to the overcurrent protection that is part of the main service disconnect. If there is a 200-ampere main breaker, the service conductors must be sized to be able to carry at least 200 amperes. *Table 310.16* is used to determine the number of current conductors it can safely carry. In the case of a dwelling unit, *Table 310.12(A)* is permitted to be used, **Figure 18-29**. It will allow for a smaller size conductor than is listed in *Table 310.16*. It is unlikely that a dwelling unit will draw the maximum current for an extended period, and *Table 310.12* takes that into consideration.

Types of Services

The two main types of services are overhead and underground. An overhead service has the conductors spanning from a nearby utility pole to the building. An underground service has the conductors buried underground. Overhead services were popular in the past, but most new installations will be fed from underground.

Overhead Service

An overhead service brings power to the customer overhead from a utility pole, **Figure 18-30**. The overhead cable that runs from the utility pole is called a service drop. The *NEC*

Table 310.12(A) Single-Phase Dwelling Services and Feeders

Service or Feeder Rating (Amperes)	Conductor (AWG or kcmil)	
	Copper	Aluminum or Copper-Clad Aluminum
100	4	2
110	3	1
125	2	1/0
150	1	2/0
175	1/0	3/0
200	2/0	4/0
225	3/0	250
250	4/0	300
300	250	350
350	350	500
400	400	600

Note: If no adjustment or correction factors are required, this table shall be permitted to be applied.

Reproduced with permission of NFPA from NFPA 70, National Electrical Code, 2023 edition. Copyright © 2022, National Fire Protection Association. For a full copy of the NFPA 70, please go to www.nfpa.org

Figure 18-29. *Table 310.12A* lists the minimum conductor sizes for dwelling unit services and feeders when no correction or adjustment factors are necessary, and they feed the entire load associated with the dwelling.

defines a ***service drop*** as the overhead conductors between the serving utility and the service point. The service drop does not fall under the jurisdiction of the *NEC*. It is on the utility side of the service point, so it is their responsibility and subject to their standards. Overhead distribution systems and services are still present and being maintained on older neighborhoods.

Overhead Service Conductors

Overhead service conductors are sometimes confused with the service drop. ***Overhead service conductors*** are defined as the overhead conductors between the service point and the first point of connection to the service-entrance conductors at the building or other structure. Overhead service conductors fall under the jurisdiction of the *NEC* and are installed and maintained by the electrician, but they are not very common. Old farmsteads are one place where you may run across overhead service conductors, **Figure 18-31**. The utility places a pole-mounted transformer in the yard where the service point is located. The overhead conductors running to the house and any other buildings, such as barns and workshops, are overhead service conductors. With multiple

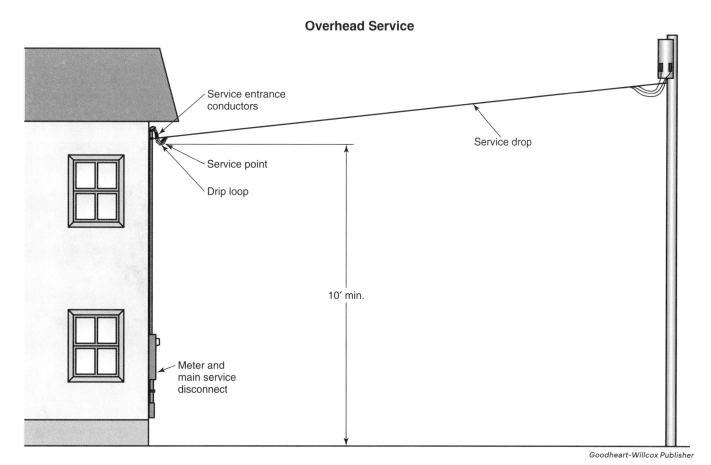

Figure 18-30. The conductors that span from the utilities pole to the building are the service-drop conductors. The service point is where their conductors connect to the service-entrance conductors. The service begins at the service point and ends at the mains service disconnect.

Overhead Service

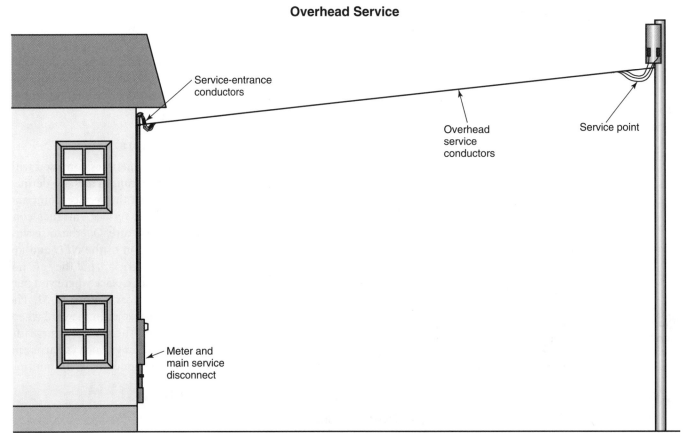

Figure 18-31. In this drawing the overhead service conductors are suspended between the pole in the yard and the house. The public utility has determined the service point is on the customer side of their pole mount transformer. Overhead service conductors are installed and maintained by the electrician. Care must be taken to ensure the minimum clearances specified in *Section 230.24(B)* are maintained.

buildings, the meter may be mounted on the pole to eliminate having an electric meter on each structure. Many farms have taken the overhead conductors down and moved them underground, eliminating the hazard of running into overhead conductors when moving large farm equipment.

Overhead Service Raceway

Raceways used with overhead services will have a service head, often referred to as a weatherhead. A *service head* is a cap that prevents snow and rain from entering the raceway, **Figure 18-32**. The attachment of the service-drop or overhead service conductors must be below the height of the service head and shall have a drip loop in the conductor. This prevents water from following the conductors back into the service head (*Section 230.54*). The bottom of the drip loop must be at least ten feet above finished grade or any platform, **Figure 18-33**.

A *service mast* is a raceway that extends through the roof for the attachment of the service-drop or overhead service conductors, **Figure 18-34**. *Section 230.28, Service Masts as Support* does not restrict the type of raceway, list a minimum size, or give a maximum length that it must extend above the

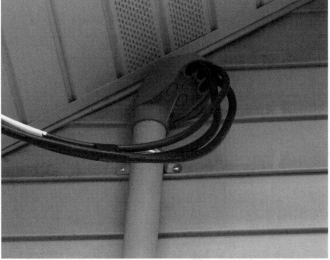

Figure 18-32. The drip loop prevents water from following the conductors into the service head. Notice how much smaller the utilities service drop is than the service-entrance conductors. It is common to see a much smaller conductor on the utility side of the service point the utility has different standards to follow.

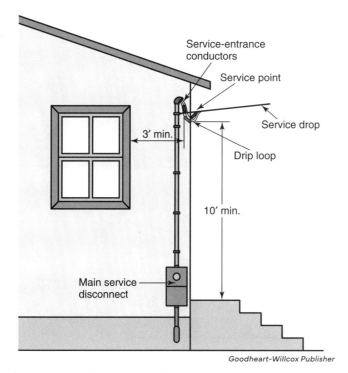

Figure 18-33. The minimum distance from the lowest point of the overhead service conductors or the drip loop, shall be at least 10′ above the ground or platforms that are below conductors.

roof. It does, however, require adequate strength or else support from guy wires or braces. This is another example of vague language that will be subject to the interpretation of the authority having jurisdiction. It is most common to see 2″ or larger rigid metallic conduit (RMC) used as a service mast. RMC is the strongest of the circular raceways. *Section 344.30(A)* gives permission for a rigid metal conduit mast to extend 36″ to the first support when going through the roof for an above-the-roof termination. If the raceway extends beyond 36″ it will require a support, such as a guy wire to ensure it has sufficient strength, **Figure 18-35**. The length of the service drop being attached to the mast and the amount the mast sticks through the roof will be used to determine if it requires additional support.

The only cables permitted to be attached to the service mast are the overhead service conductors and the utility service drop. Communication cables such as TV, telephone, and fiber optic must be attached to the building by a means other than the electric service mast as per *Sections: 800.44(C)*.

Underground Service

An underground service has a service lateral that brings the utility conductors to the service from underground, **Figure 18-36**.

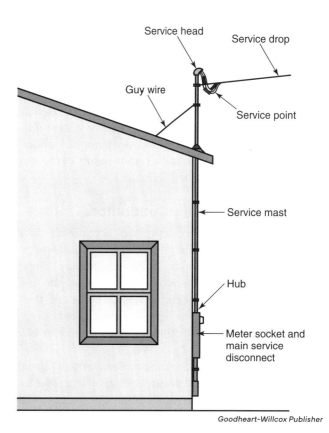

Figure 18-34. A service mast is used to support the service drop. 2″ RMC typically provides sufficient strength provided the service drop is attached within the first 24″ of where it comes through the roof.

Figure 18-35. A service mast that extends more than about 36″ above the roof will have to be guyed. The guy wire is extended opposite the service drop or will consist of multiple wires.

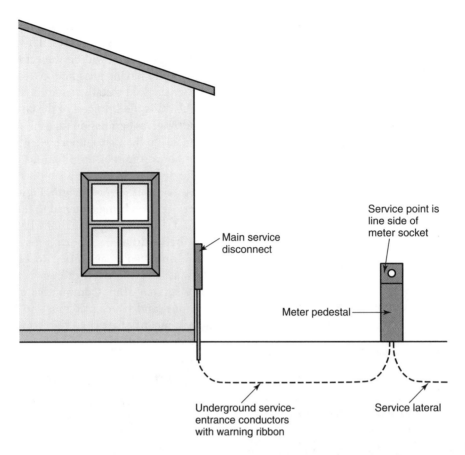

Figure 18-36. Some utilities will install a meter pedestal in the yard or will only bring power to a pole. This is more often the case in rural communities. In this case the conductors from the pedestal to the house are underground service conductors and are installed and maintained by the electrician. A warning ribbon must be installed 12″ above the conductors.

The *NEC* defines a ***service lateral*** as the underground conductors between the utility electric supply system and the service point. The service lateral does not fall under the jurisdiction of the *NEC*. It is the responsibility of the serving utility and is subject to their standards.

Underground Service Conductors

Underground service conductors will occasionally be confused with the service lateral. This is a situation where it is very important to understand the terminology. The *NEC* defines ***underground service conductors*** as the underground conductors between the service point and the first point of connection to the service-entrance conductors in a terminal box, meter, or other enclosure, inside or outside the building wall.

Rural communities where the power company installs a meter pedestal in the yard are an example of where we would install underground service conductors, **Figure 18-37**. Underground service conductors will run from the meter pedestal to the house where the main service disconnect will be located, **Figure 18-38**. The utility's responsibility ends at the pedestal, so the electrician will size and install underground service conductors.

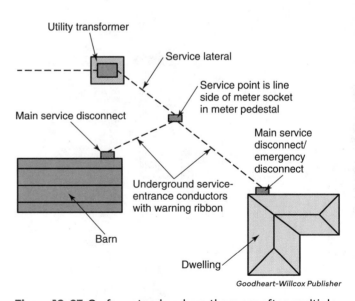

Figure 18-37. On farmsteads, where there are often multiple buildings, the utility often brings their service lateral to a pole or meter pedestal. The electrician will trench underground service conductors to the building or barn. It is the responsibility of the electrician to install and maintain underground service conductors. A warning ribbon must be installed 12″ above the conductors.

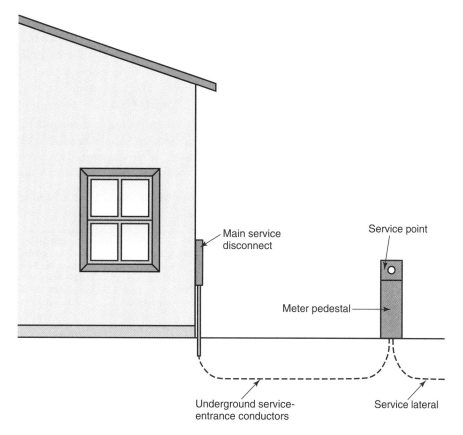

Figure 18-38. The service lateral coming from the utility falls under the utility company's jurisdiction, but the underground service-entrance conductors are the responsibility of the electrician. If there is a fault in the underground service-entrance conductors, it will be up to the electrician to locate and repair the fault.

The wiring methods that are permitted to be used as underground service conductors are listed in *Section 230.30(B)*. When trenching in underground service conductors, *Section 300.5(D)(3)* of the *NEC* requires a warning ribbon to be installed 12″ above the conductors, **Figure 18-39**.

PRO TIP: Utility Locating Services

When calling to have the underground utilities located, it is important to understand what is being located. In the United States, simply dialing 811 will connect you with the locating service in your area. They will come out and locate all the underground cables and lines that are under the control of a utility, such as electric, gas, sewer, and cable TV. They will not mark anything that is on the customer side of the service point. That means underground service conductors running from a pedestal or pole where the service point is located to a building, will not be located. It is the responsibility of the electrician to locate those cables. Other underground electrical lines that won't be marked include cables to post lights, detached garages, and sheds.

Figure 18-39. A warning ribbon must be installed 12″ above underground service-entrance conductors to alert anyone who may be digging to the conductors' presence.

Underground Service Raceway

Raceways containing service conductors that emerge from underground are covered under *Part III* of *Article 230*. The raceway may be continuous from one location to another, and contain building wire, or it may protect a directly buried cable just as it is emerging from underground.

Table 300.5(A) lists the minimum burial depths for cables and raceways, **Figure 18-40**. Depths vary by the type of raceway, location of the raceway, and type of circuit.

- An underground service cable that is trenched underground from a meter pedestal to a home must be buried at least 24″ underground. (Column 1, Row 1)
- A rigid PVC raceway containing underground service conductors that is installed in the same location must be buried at least 18″ underground. (Column 3, Row 1)

The raceways that are permitted to be used with underground service conductors are listed in *Section 230.30(B)*. The inside of an underground raceway is considered a wet

Table 300.5(A) Minimum Cover Requirements, 0 to 1000 Volts ac, 1500 Volts dc, Nominal, Burial in Millimeters (Inches)

	Type of Wiring Method or Circuit									
	Column 1 Direct Burial Cables or Conductors		Column 2 Rigid Metal Conduit or Intermediate Metal Conduit		Column 3 Electrical Metallic Tubing, Nonmetallic Raceways Listed for Direct Burial Without Concrete Encasement, or Other Approved Raceways		Column 4 Residential Branch Circuits Rated 120 Volts or Less with GFCI Protection and Maximum Overcurrent Protection of 20 Amperes		Column 5 Circuits for Control of Irrigation and Landscape Lighting Limited to Not More Than 30 Volts and Installed with Type UF or in Other Identified Cable or Raceway	
Location of Wiring Method or Circuit	mm	in.	mm	in.	mm	in.	mm	in.	mm	in.
All locations not specified below	600	24	150	6	450	18	300	12	150[1,2]	6[1,2]
In trench below 50 mm (2 in.) thick concrete or equivalent	450	18	150	6	300	12	150	6	150	6
Under a building	0 (in raceway or Type MC or Type MI cable identified for direct burial)	0	0	0	0	0	0 (in raceway or Type MC or Type MI cable identified for direct burial)	0	0 (in raceway or Type MC or Type MI cable identified for direct burial)	0
Under minimum of 102 mm (4 in.) thick concrete exterior slab with no vehicular traffic and the slab extending not less than 152 mm (6 in.) beyond the underground installation	450	18	100	4	100	4	150 (direct burial) 100 (in raceway)	6 / 4	150 (direct burial) 100 (in raceway)	6 / 4
Under streets, highways, roads, alleys, driveways, and parking lots	600	24	600	24	600	24	600	24	600	24
One- and two-family dwelling driveways and outdoor parking areas, and used only for dwelling-related purposes	450	18	450	18	450	18	300	12	450	18
In or under airport runways, including adjacent areas where trespassing is prohibited	450	18	450	18	450	18	450	18	450	18

[1]A lesser depth shall be permitted where specified in the installation instructions of a listed low-voltage lighting system.
[2]A depth of 150 mm (6 in.) shall be permitted for pool, spa, and fountain lighting, installed in a nonmetallic raceway, limited to not more than 30 volts where part of a listed low-voltage lighting system.

Notes:
1. Cover shall be defined as the shortest distance in mm (in.) measured between a point on the top surface of any direct-buried conductor, cable, conduit, or other raceway and the top surface of finished grade, concrete, or similar cover.
2. Raceways approved for burial only where concrete encased shall require a concrete envelope not less than 50 mm (2 in.) thick.
3. Lesser depths shall be permitted where cables and conductors rise for terminations or splices or where access is otherwise required.
4. Where one of the wiring method types listed in Columns 1 through 3 is used for one of the circuit types in Columns 4 and 5, the shallowest depth of burial shall be permitted.
5. Where solid rock prevents compliance with the cover depths specified in this table, the wiring shall be installed in a metal raceway, or a nonmetallic raceway permitted for direct burial. The raceways shall be covered by a minimum of 50 mm (2 in.) of concrete extending down to rock.
6. Directly buried electrical metallic tubing (EMT) shall comply with 358.10.

Reproduced with permission of NFPA from NFPA 70, National Electrical Code, 2023 edition. Copyright © 2022, National Fire Protection Association. For a full copy of the NFPA 70, please go to www.nfpa.org

Figure 18-40. Required burial depths vary by the type of raceway, location of the raceway, and type of circuit. *Table 300.5(A)* lists the minimum requirements.

location, so any conductors inside the raceway must be listed accordingly.

Raceways containing service conductors that emerge from underground are addressed in *Section 230.32*, which then sends us to *Section 300.5* for the requirements. The *NEC* sends us to *Article 300* because this requirement applies to all raceways emerging from the ground, not just service raceways. Rather than printing the information multiple times, they put it in the general requirements portion of *Chapter 3* so it applies generally to all installations.

Exterior raceways that emerge from underground are considered subject to physical damage, such as from a lawn mower, rake, yard trimmer, or baseball. *Section 300.5(D)(4)* requires that raceways exposed to physical damage shall be EMT (not permitted underground in some parts of the country), RMC, IMC, RTRC-XW, or Schedule 80 PVC. Schedule 40 PVC has a thinner wall, so it does not have the strength to withstand physical damage. Schedule 80 on the other hand has a thicker wall and is recognized by the *NEC* as being strong enough to protect conductors that emerge from grade.

When a trench, hole, or the area around buildings is backfilled, the ground will settle and compact over time. This can cause problems for raceways and cables that emerge from underground. (*Section 300.5(J)*). It is a good practice to leave extra cable in meter sockets, cabinets, or junction boxes to allow the conductors to pull downward as they settle. For a directly buried cable, putting an S loop in the cable before going up the raceway can offer relief to earth movement, **Figure 18-41**.

A raceway that emerges from underground will also require attention to prevent it from pulling out of the fittings or pulling boxes and meter sockets off the wall. A slip meter riser or an expansion coupling can be used to accommodate the raceway being pulled downward, **Figure 18-42**. The slip meter riser attaches to the meter socket with another raceway being inserted into the bottom. Enough of the internal raceway must be inserted so that it is not pulled out. As the ground settles, it pulls the interior raceway down, leaving the riser attached to the meter socket.

Goodheart-Willcox Publisher

Figure 18-42. A slip meter riser allows the internal conduit that extends to the ground to slip down as it settles. The conduit can also slide up if the ground heaves due to frost.

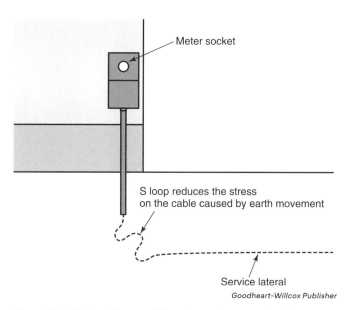

Goodheart-Willcox Publisher

Figure 18-41. Installing an S loop in underground cables before they emerge from the ground can help with the cable being pulled down due to ground settling. As the ground settles the S gets straightened out rather than pulling down.

Summary

- *Article 230* has requirements for services.
- Service equipment must be identified as suitable for use as service equipment.
- Dwellings are required to have whole-house surge-protective devices.
- The rating of the service disconnect shall be not less than the rated load.
- Service disconnects shall be readily accessible.
- The service may consist of up to six service disconnects, provided they are grouped and clearly marked.
- The clearance requirements of *Section 110.26* apply to service equipment.
- Dwellings are required to have an emergency disconnect that removes power to the building in the case of an emergency.
- The service drop is the overhead cable that is installed and maintained by the utility, while the overhead service conductors are the overhead cable installed and maintained by the electrician.
- The service lateral is the underground cable that is installed and maintained by the utility, while the underground service conductors are the underground cable installed and maintained by the electrician.

Unit 18 Review

Name _____ Date _____ Class _____

Know and Understand

Answer the following questions based on information in this unit.

1. _____ is dedicated to the electric service.
   A. *Article 220*
   B. *Article 230*
   C. *Article 240*
   D. *Article 310*
2. The service disconnect must be in a _____ location.
   A. dry
   B. exposed
   C. accessible
   D. readily accessible
3. With an overhead service, the minimum height from grade to the bottom of the drip loop is _____.
   A. 6′-7′
   B. 8′
   C. 10′
   D. 12′
4. The maximum height of the service disconnect to the center of the operating handle is _____.
   A. 6′-0″
   B. 6′-6″
   C. 6′-7″
   D. 6′-8″
5. The underground cable that the utility runs from their transformer to the service point is the _____.
   A. service lateral
   B. service drop
   C. underground service conductors
   D. underground feeder cable
6. Underground service conductors trenched across a yard are required to have a warning ribbon installed _____ above the conductors.
   A. 3″
   B. 6″
   C. 8″
   D. 12″
7. The service conductors installed in a raceway on the exterior of a building shall be rated for _____ locations.
   A. dry
   B. damp
   C. wet
   D. hazardous
8. The *NEC* permits _____ service disconnect(s).
   A. only one
   B. up to two
   C. up to three
   D. up to six
9. The service disconnect with a 200-ampere main breaker is permitted in which of the following locations?
   A. Bedroom
   B. Stairway
   C. Clothes closet
   D. Bathroom
10. The minimum depth of the working space in front of a 120/240-volt, single-phase panel is _____.
    A. 24″
    B. 30″
    C. 36″
    D. 48″

Apply and Analyze

Answer the following questions using a copy of the National Electrical Code. *Identify the section or subsection where the answer is found.*

1. Overhead service conductors shall not be smaller than _____ AWG copper.
   A. 8
   B. 6
   C. 4
   D. 3

   NEC _____

2. Which of the following is not a permitted wiring method for underground service conductors?
   A. RMC conduit
   B. PVC conduit
   C. USE cable
   D. UF cable

   NEC _____

3. A meter socket has a 2″ RMC mast that leaves the top of the meter assembly and goes through the roof for the attachment of the service drop. The _____ that connects the mast to the meter socket must be identified for use with service-entrance equipment.
   A. connector
   B. hub
   C. terminal adapter
   D. female adapter

   NEC _____

4. The minimum burial depth of underground service-entrance conductors installed in rigid PVC conduit that runs underneath an alley is _____.
   A. 4″
   B. 12″
   C. 18″
   D. 24″

   NEC _____

5. The minimum headroom in front of the service equipment shall be _____ or the height of equipment, whichever is greater.
   A. 6′-0″
   B. 6′-6″
   C. 6′-7″
   D. 8′-0″

   NEC _____

6. Ground-fault protection for equipment shall be installed on wye service with 277 volts to ground if the disconnect rating is _____ or more.
   A. 400 amperes
   B. 600 amperes
   C. 800 amperes
   D. 1000 amperes

   NEC _____

7. The service conductors leaving a service head to attach to the service drop, must be at least _____ away from the sides and bottom of windows that open.
   A. 24″
   B. 36″
   C. 48″
   D. 60″

   NEC _____

8. Service conductors are considered outside of a building when installed underneath a concrete slab that is at least _____ thick.
   A. 2″
   B. 3″
   C. 4″
   D. 6″

   NEC _____

9. Service conductors with a higher voltage to ground, such as a 4-wire delta, shall have the phase with the higher voltage to ground durably and permanently marked with an outer finish that is _____ in color.
   A. pink
   B. red
   C. orange
   D. yellow

   NEC _____

10. Overhead service conductors with a voltage of 120 volts to ground shall have a clearance of at least _____ above residential driveways.
    A. 10′
    B. 12′
    C. 15′
    D. 18′

    NEC _____

Name _____ Date _____ Class _____

Critical Thinking

Answer the following questions using a copy of the National Electrical Code. *Identify the section or subsection where the answer is found.*

1. Service equipment shall be enclosed so that _____ are not exposed to accidental contact.

   _____

   NEC _____

2. Schedule _____ PVC conduit is permitted to protect service-entrance cables that may be subject to physical damage.

   _____

   NEC _____

3. Meter sockets are not considered service equipment but shall be _____ for the voltage and current rating of the service.

   _____

   NEC _____

4. Vegetation such as trees shall not be used for the support of _____ or service equipment.

   _____

   NEC _____

5. Services supplying dwelling units shall be provided with a _____ or _____ surge protective device.

   _____

   NEC _____

6. Services requiring ground-fault protection shall be performance tested at the time of installation by _____.

   _____

   NEC _____

7. The service disconnecting means shall clearly indicate whether it is in the _____ or _____ position.

   _____

   NEC _____

8. The grounded service-entrance conductor shall not be smaller than the minimum size required by Section _____.

   _____

   NEC _____

9. What is the minimum height of an overhead service-entrance conductor over public street?

   _____

   NEC _____

10. Each service disconnect shall disconnect all _____ conductors from the premises wiring.

    _____

    NEC _____

Part 2   Application: Advanced NEC Topics

Answer the following questions using a copy of the National Electrical Code. *Identify the section or subsection where the answer is found.*

11. What is the minimum vertical clearance of an overhead service-entrance conductor over water or a swimming pool?

    _____

    NEC _____

12. What is the minimum size copper service-entrance conductor for a 100-ampere dwelling service?

    _____

    NEC _____

13. All service equipment rated at 1000 volts or less shall be marked to identify it as being _____.

    _____

    NEC _____

14. Underground service-entrance conductors shall not be smaller than _____ aluminum.

    _____

    NEC _____

15. What is the minimum size aluminum service-entrance conductors for a 200-ampere service for a commercial building? (Terminals rated for 75°C)

    _____

    NEC _____

16. A non-dwelling service with equipment rated at 1000 amperes or more will require an arc flash hazard warning label that is in accordance with applicable industry practice and shall include _____.

    _____

    _____

    _____

    NEC _____

17. A service mast supporting a service drop shall not have a _____ between the service head and the last point of securement to the building.

    _____

    NEC _____

18. Service-entrance cables shall be supported within _____ of the service head and meter socket, and at intervals not exceeding _____.

    _____

    _____

    NEC _____

19. Dwelling units are required to have a(n) _____ in a readily accessible outdoor location to disconnect the power from the building in case of an emergency.

    _____

    NEC _____

20. The service disconnect shall have a rating not less than _____.

    _____

    NEC _____

UNIT 19 Grounding and Bonding

Sashkin/Shutterstock.com

OBJECTIVES

After completing this chapter, you will be able to:
- Describe the purpose of grounding.
- Identify the grounding electrodes recognized by the *NEC*.
- Determine the size of the grounding electrode conductor.
- Identify equipment required to have a connection to an equipment grounding conductor.
- List the recognized types of equipment grounding conductors.
- Size the equipment grounding conductor.
- Describe the purpose of the main bonding jumper.
- Determine the size of a wire-type main bonding jumper.

KEY TERMS

bonding
effective ground-fault current path
equipment grounding conductor (EGC)
ground
grounded
grounded conductor
ground fault
grounding
grounding electrode
grounding electrode conductor (GEC)
grounding electrode system
intersystem bonding termination
main bonding jumper (MBJ)
separately derived system
solidly grounded
system bonding jumper (SBJ)

Introduction

Article 250 of the *National Electrical Code* is dedicated to grounding, **Figure 19-1**. There are entire books devoted to grounding and bonding alone. This unit is intended only to be an introduction to grounding and bonding principles and the most commonly used applications, as well as the related requirements in *Article 250*.

This unit splits *Article 250* into three grounding and bonding topics: grounding electrode system, equipment grounding, and bonding. *Article 250* has a lot of terminology that can create

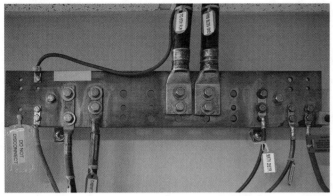

Figure 19-1. A telecommunications grounding bar.

confusion, so it is essential to reference the definitions and pay close attention to the terms. There are review questions and problems at the end of the unit to reinforce the topics covered and help you to become familiar with *Article 250*.

Grounding Fundamentals

Article 250 is an extensive article that can be difficult to navigate. The *NEC* splits it into several parts, each covering a different aspect of grounding. Refer back to **Figure 5-28** to review the parts and sections contained in *Article 250*. To help provide some clarification, the *NEC* provides a diagram, *Figure 250.1*, that identifies how the parts of *Article 250* relate to one another, **Figure 19-2**.

There many terms relating to grounding that are similar but have different meanings. Some of the *NEC* definitions have changed in recent editions, creating confusion when some electricians, who have been using the older terminology for years, use terms interchangeably. We will start with fundamental grounding definitions.

Ground is defined as the earth. A connection to the ground is made at the main service disconnect, bringing the earth to the same potential as the grounded components in the electrical system.

Solidly grounded is the state of a system being connected to ground without inserting any resistor or impedance device.

A *ground fault* is an unintentional, electrically conductive connection between an ungrounded conductor of an electrical circuit and the normally non-current-carrying conductors, metallic enclosures, metallic raceways, metallic equipment, or earth. If a conductor shorts out against a metal enclosure, creating a ground fault, we want to ensure a low impedance path back to the grounded conductor to facilitate the operation of the overcurrent device, **Figure 19-3**.

Grounding is the act of connecting a system to the ground or to a conductive body that extends the ground connection. When something is referred to as being *grounded*, it means that it is electrically connected to the earth.

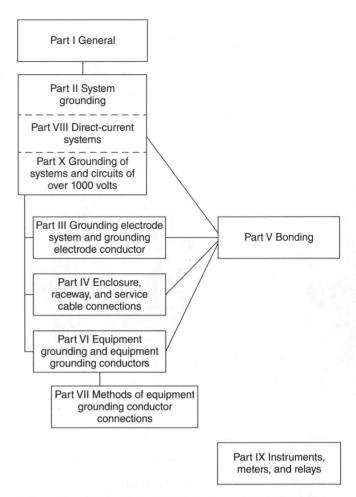

Informational Note Figure 250.1   Grounding and Bonding.

Reproduced with permission of NFPA from NFPA 70, National Electrical Code, 2023 edition. Copyright © 2022, National Fire Protection Association. For a full copy of the NFPA 70, please go to www.nfpa.org

Figure 19-2. *NEC Figure 250.1* is a diagram of *Article 250*'s parts and how they are connected.

A *grounded conductor* is a system or circuit conductor that is intentionally grounded. The grounded conductor is a conductor that will have current flow under most circumstances. In 120/240-volt, single-phase systems and 120/208- or 277/480-volt systems, the grounded conductor and the neutral refer to the same conductor, **Figure 19-4**.

> **PRO TIP** — **Grounded Conductor vs Grounding Conductor**
>
> Care must be taken to distinguish between a grounded conductor and a grounding conductor (equipment grounding conductor). The grounded conductor is a current-carrying conductor that will have current flowing during normal operation. The equipment grounding conductor provides a low impedance current path for a ground-fault current, so it quickly operates the overcurrent device. It should not have a current flow during normal operation.

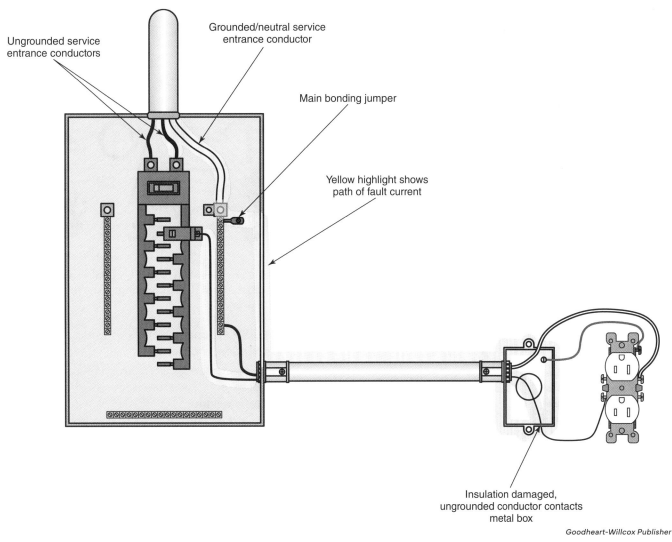

Figure 19-3. A ground fault occurs when an ungrounded conductor shorts to a grounded metal object. In this illustration, a branch circuit shorts out to a metal device box. The fault current travels through the box, conduit, panel enclosure, main bonding jumper, and then to the grounded service-entrance conductor.

Most electrical systems are grounded, but not all. Some industrial facilities have ungrounded systems so they can continue to operate in the event of a single fault to ground. The first fault to ground will not shut down the system instruments that monitor the system, but will notify qualified personnel of the fault. This allows enough time for an orderly shutdown that will cause the least amount of interruption and danger, rather than shutting down instantly when a fault occurs. A second fault can result in catastrophic failure, equipment damage, and significant downtime. For that reason, if a ground fault occurs, the people responsible for the system must find and correct it in a timely manner. This unit will not discuss ungrounded systems any further and will focus on grounded systems as they are far more common.

Purpose of Grounding

There are several different aspects to grounding and reasons that we ground. The purpose of grounding and bonding is spelled out in *Section 250.4(A)*.

- Limit the voltage to ground from lighting, line surges, or unintentional contact with higher voltage lines.
- Stabilize the voltage to earth during normal operation.
- Limit the voltage to ground on non-current-carrying metal structural components and equipment.
- Create a low-impedance ground-fault current path to quickly trip the overcurrent device in the case of a ground fault on a metallic piece of equipment.

We ground electrical systems to dissipate unintentional voltage spikes. We bond and ground metallic components to

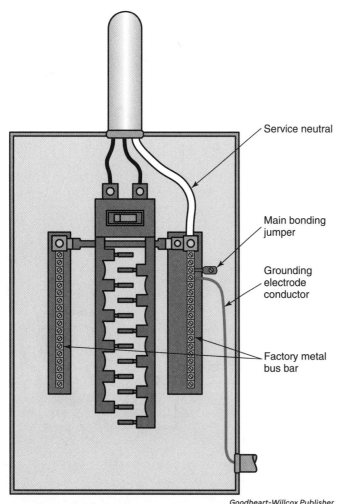

Figure 19-4. The grounded conductor is the conductor that is intentionally grounded. The main bonding jumper connects the grounded conductor to the metal enclosure and provides a path for ground-fault currents to return to the grounded service-entrance conductor.

prevent arcing and electric shock. We ensure there is a low-impedance path from metal objects to ground to open the overcurrent device quickly. All these measures ensure the metallic components of the building and electrical system are free from fire and electric shock hazards.

Grounding and Bonding Connections

All connections to grounding equipment shall be made in accordance with the requirements of *Section 250.8*, **Figure 19-5**. If the connections are exposed to physical damage, they shall be protected as per *Section 250.10*. According to *Section 250.12*, when attaching to a surface that has a nonconductive coating, such as paint, the coating shall be removed unless the fitting ensures adequate electrical continuity.

Figure 19-5. Components used for grounding and bonding must be listed and installed according to the manufacturer's recommendations.

Section 250.8(A) permits the following methods of establishing a grounding connection:

- Pressure connectors listed for grounding and bonding
- Terminal bars
- Exothermic welding
- Machine screw-type fastener that engages no less than two threads or is secured with a nut
- Thread forming machine screws that engage no less than two threads
- Connections that are part of a listed assembly
- Other listed means

Grounding Electrode System

A grounding electrode system connects the grounded portion of the electrical system to the ground (the earth). *Part III* of *Article 250* is dedicated to the grounding electrode system. It consists of a grounding electrode, grounding electrode conductor, and any protection that may be necessary to protect the grounding electrode conductor from damage, **Figure 19-6**. The grounding electrode conductor connects the grounding electrode(s) to the system grounded conductor.

Grounding Electrode

The *NEC* defines a **grounding electrode** as a conducting object through which a direct connection to earth is established. Several types of grounding electrodes are used to connect electrical systems to the earth. Some of the grounding electrodes recognized by the *NEC* are already present on the jobsite due to other building systems and construction methods, and others are installed by the electrician. *Section 250.50* states that all grounding electrodes present shall be connected to form the **grounding electrode system**.

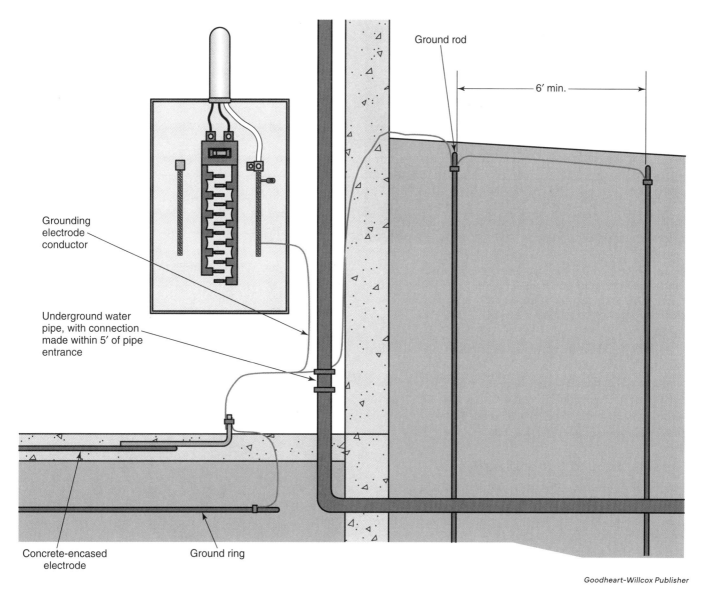

Figure 19-6. The *NEC* requires that every grounding electrode present on a site shall be used. This diagram has two ground rods, a concrete-encased electrode, a metal underground water pipe, and a ground ring.

The *NEC* has two sections covering grounding electrodes: *Section 250.52* and *Section 250.53*. *Section 250.52* covers the various types of grounding electrodes, while *Section 250.53* has installation requirements for each of the respective grounding electrodes. The following are the grounding electrodes recognized by the *NEC*.

Metal Underground Water Pipe

Metal underground water pipes are covered in *Section 250.52(A)(1)* and *Section 250.53(D)*. It is common to have a metal underground water pipe (typically copper) bringing the water from the city's water supply to the building. Provided the pipe is in contact with the earth for at least 10′, it shall be used as a grounding electrode. The water meter itself cannot be relied upon to be a part of the grounding path as it has the potential to have nonconductive fittings or be altered someday, which would open the connection to the electrode. The grounding electrode conductor is connected across the water meter to ensure continuity, **Figure 19-7**. Having the jumper across the water meter will also satisfy the requirement of bonding the interior metal water piping as per *Section 250.104(A)*, provided the building has metal water piping. It is becoming more common to see plastic water lines run throughout buildings. If there is a manifold to split the water line, it will be required to be bonded.

The metal underground water pipe must be supplemented with at least one other grounding electrode. This is because an underground break in the pipe could be repaired with a plastic piece of pipe that is nonconductive. Depending on where the repair or replacement happens, there may no longer be enough contact with the earth to have a sufficient connection to ground.

Figure 19-7. The grounding electrode conductor must be bonded to both sides of the water meter to ensure the path is maintained if the water meter was removed or a new one with nonmetallic fittings was installed.

Buildings fed from rural water supplies or individual wells often have a plastic water line. Because they are nonconductive, plastic water lines are not a grounding electrode. If the plastic is switched over to metal in the interior of the building, it will need to be bonded as per *Section 250.104(A)*. Bonding is covered later in this unit.

Metal In-Ground Support Structure

The metal in-ground support structure, which is the metal structural steel of a building, is a grounding electrode if it is in contact with the earth for ten feet or more, with or without concrete encasement, **Figure 19-8**. These structures are covered in *Section 250.52(A)(2)*. Large steel structures will often have metal pilings driven into the ground or underground steel columns encased in concrete. Both situations will qualify the steel structures to be used as grounding electrodes, provided they are in contact with the earth for 10′ or more.

In the case where the structural steel is not in contact with the earth for 10′ or more, it must be bonded to the electrical system as per *Section 250.104(C)*. Bonding is covered in greater detail later in this unit.

Concrete Encased Electrode

The reinforcing bars (rebar) present in a concrete footing, slab, or foundation wall are considered a grounding electrode, **Figure 19-9**. These are covered in *Section 250.52(A)(3)*. Concrete encased electrodes are sometimes referred to as

Figure 19-9. Reinforcing bars (rebar) in concrete foundations that have direct contact with the soil are considered a grounding electrode. The total length of the rebar must be at least 20′ long, including pieces that are tied together using tie wires or some other effective means. Most footings and walls will have hundreds of feet of rebar tied together. This picture shows a foundation wall that is formed up and ready for concrete to be poured in. After the concrete has been poured and set up, the forms will be removed.

Figure 19-8. Structural steel buildings that have at least 10′ of steel in contact with the earth are grounding electrodes. Since the structure is bolted together throughout, the entire metal portion of the structure is effectively grounded.

Ufer grounds. The term "Ufer" comes from Herbert Ufer, who did the early research into using concrete-encased electrodes as a connection to the earth. The connection of the grounding electrode conductor to the rebar is permitted to be in the concrete, provided the fitting is listed accordingly. The concrete crew will usually stub the rebar up, so it sticks out through the floor, **Figure 19-10**.

Reinforcing bars in concrete are considered a concrete-encased electrode provided the following requirements are met:

- The concrete must be in direct contact with the earth. If the concrete is wrapped in plastic or foam, it is no longer a grounding electrode as it does not directly contact the ground.
- The rebar must be bare or electrically coated.
- The rebar must be at least 1/2″ in diameter.
- The rebar must be at least 20′ or more in length. Shorter pieces are permitted, provided they are connected using tie wires or some other effective means and have a combined total length of 20′.

If there is no rebar in the concrete, or the rebar does not qualify as a grounding electrode, it is still possible to have a concrete-encased electrode. Installing a bare copper conductor directly into the slab will also serve as a concrete-encased electrode. The conductor shall be at least 4 AWG copper and must be at least 20′ long. All the same rules for the concrete being in direct contact with the earth apply.

Ground Ring

A ground ring is a bare copper conductor installed directly in the earth that encircles the building or structure, **Figure 19-11**.

Ground rings are covered in *Section 250.52(A)(4)* and *Section 250.53 (F)* and must meet the following requirements:

- Bare copper conductor 2 AWG or larger.
- A minimum length of 20′.
- Installed at least 30″ below the earth's surface.

Ground rings are common around telecommunication sites and transformer pads.

Rod and Pipe Electrodes

Ground rods and pipes are driven into the ground to serve as grounding electrodes. They are covered in *Section 250.52(A)(5)* and *Section 250.53(A)*. Ground rods are more commonly used than pipes, **Figure 19-12**. Ground rods are available in different materials, but copper is the most common. The other types of ground rods available are

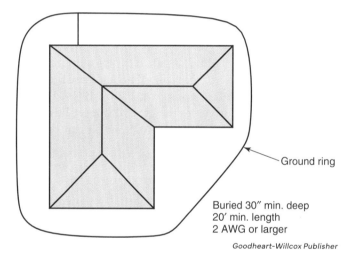

Goodheart-Willcox Publisher

Figure 19-11. A ground ring is a bare copper conductor that encircles a building.

Goodheart-Willcox Publisher

Figure 19-10. The concrete crew will typically stub a piece of rebar up from the footings so the rebar is accessible for the attachment of the grounding electrode conductor.

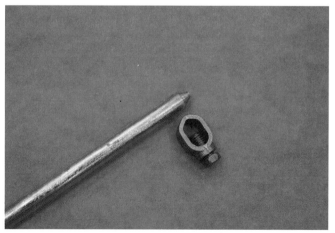

Goodheart-Willcox Publisher

Figure 19-12. Ground rods have one pointed end to make them easier to drive into the ground. The clamp used to attach the grounding electrode conductor to a ground rod must be listed for direct burial because the termination will be underground.

galvanized and stainless steel. A copper ground rod is not solid copper; it is a steel rod with a copper coating.

Ground rods must meet the following requirements:
- Shall be at least 5/8″ in diameter, unless listed. (1/2″ listed ground rods are very common)
- Shall not be less than 8′ in length.
- Shall be in contact with the earth for at least 8′.

Ground pipe type electrodes must meet the following requirements:
- Shall be at least 3/4″ (trade size) diameter.
- Shall be made of a corrosion-resistant conductive material or have a coating that protects against corrosion.
- Shall be in contact with the earth for at least 8′.

If solid rock is encountered when driving a ground rod or pipe, it is permitted to be installed at an angle up to 45°. If rock is in the way of installing it at a 45° angle, it is permissible to bury it horizontally at least 30″ below ground level, **Figure 19-13**.

Ground rods and pipes must be augmented with an electrode (rod pipe or plate) unless it can be proven that the resistance to ground is 25 ohms or less. It is typically cheaper to drive an additional rod/pipe rather than purchase the meter necessary to take the measurement, **Figure 19-14**. When a second rod/pipe is driven in the ground it must be at least 6′ away from the other electrode to ensure it can properly disperse current into the earth.

Plate Electrodes

Plate electrodes consist of a metal plate that is buried directly in the ground, **Figure 19-15**. Plate electrodes must meet the following requirements:
- Shall have at least 2 ft² of surface area exposed to the soil.
- Shall be at least 1/4″ thick if made of steel, or at least 0.06″ thick if made of a nonferrous material such as copper.
- Shall be buried at least 30″ deep.

Goodheart-Willcox Publisher

Figure 19-14. A tester can be used to measure the resistance of the ground rod to earth. Pictured is a clamp-on tester that induces a current into the conductor in order to measure its resistance.

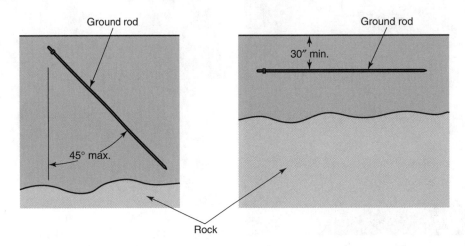

Goodheart-Willcox Publisher

Figure 19-13. In areas of the country where solid rock prevents vertical burial of a ground rod, the *NEC* allows the ground rod to be driven at a 45° angle. If the ground rod is driven at a 45° angle and it still runs into solid rock, then the ground rod is permitted to be buried horizontally, but it must be buried at least 30″ below the surface.

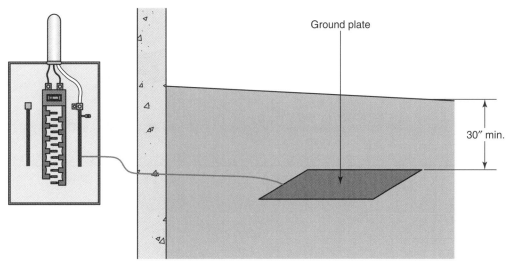

Figure 19-15. Ground plates must be buried at least 30″ below the earth's surface.

Plate electrodes must be augmented by an additional electrode (rod, pipe, or plate) unless it can be proven that the resistance to ground is 25 ohms or less. As with ground rods, the second electrode must be at least 6′ away from the first one.

Other Local Metal Underground Systems or Structures

If there are other metal underground piping systems, such as underground tanks or well casings, that aren't already bonded to an existing water piping system, they shall also be used as a grounding electrode.

Not Permitted as Grounding Electrodes

There are a few systems that would perform well as a grounding electrode that we cannot use for safety reasons. The first example is a metal underground gas pipe. We do not want to use a gas pipe as an electrode due to the hazard of a ground current on a pipe with gas in it. The other example is the rebar and metal structures around pools and hot tubs. Having a ground current on the metal components around a pool could create a shock hazard for people in the area.

Grounding Electrode Conductor

The *NEC* defines a ***grounding electrode conductor*** as a conductor used to connect the system grounded conductor or the equipment to a grounding electrode or to a point on the grounding electrode system. The general rule in *Section 250.64(C)* states that the grounding electrode conductor must be continuous without any splices or connections, however, splicing the grounding electrode conductor is permitted by irreversible compression connectors or exothermic welding, **Figure 19-16**.

Figure 19-16. These two stranded-copper conductors are connected by exothermic welding. The conductors are melted together, making the connection irreversible. Dies are available to connect conductors in various configurations.

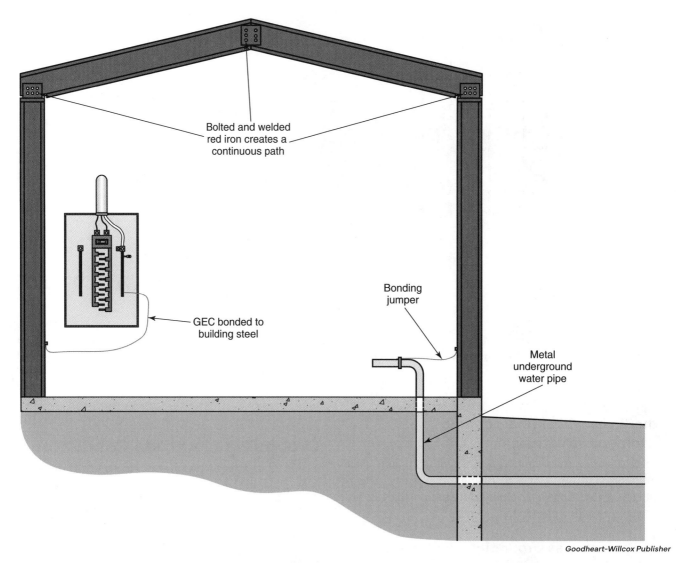

Figure 19-17. Grounding electrodes are permitted to extend the grounding electrode conductor. In this figure, the structural steel is connected to the service by the grounding electrode conductor. Rather than running a grounding electrode conductor from the service to the water pipe, a bonding jumper is used to attach the metal underground water pipe to the structural steel on the other side of the structure.

A grounding electrode is permitted to extend the grounding electrode to an additional electrode. For example, structural steel can be used to extend the grounding electrode conductor to an additional electrode provided it has bolted, riveted, or welded connections that are continuous between the connection points, **Figure 19-17**.

Grounding Electrode Conductor Sizing

The grounding electrode conductor is sized to carry the current necessary to quickly interrupt the overcurrent protection devices of the service or separately derived system. The conductor size is based on the size of the ungrounded conductors of the service or separately derived system.

Section 250.66 sends us to *Table 250.66* to determine the minimum size, **Figure 19-18**. The left half of the table has rows that correspond with the size of the ungrounded conductors and the right half of the table lists the minimum grounding electrode conductor size. The ungrounded conductor side of the table has two columns, one for copper and one for aluminum. The grounding electrode conductor side of the table has information for both copper and aluminum conductors.

Section 250.66(A), (B), and *(C)* have additional information on sizing the grounding electrode conductor when it connects to a ground rod, pipe, or plate, a concrete-encased electrode, or a ground ring.

Table 250.66 Grounding Electrode Conductor for Alternating-Current Systems

Size of Largest Ungrounded Conductor or Equivalent Area for Parallel Conductors (AWG/kcmil)		Size of Grounding Electrode Conductor (AWG/kcmil)	
Copper	Aluminum or Copper-Clad Aluminum	Copper	Aluminum or Copper-Clad Aluminum
2 or smaller	1/0 or smaller	8	6
1 or 1/0	2/0 or 3/0	6	4
2/0 or 3/0	4/0 or 250	4	2
Over 3/0 through 350	Over 250 through 500	2	1/0
Over 350 through 600	Over 500 through 900	1/0	3/0
Over 600 through 1100	Over 900 through 1750	2/0	4/0
Over 1100	Over 1750	3/0	250

Notes:
1. If multiple sets of service-entrance conductors connect directly to a service drop, set of overhead service conductors, set of underground service conductors, or service lateral, the equivalent size of the largest service-entrance conductor shall be determined by the largest sum of the areas of the corresponding conductors of each set.
2. If there are no service-entrance conductors, the grounding electrode conductor size shall be determined by the equivalent size of the largest service-entrance conductor required for the load to be served.
3. See installation restrictions in 250.64.

Reproduced with permission of NFPA from NFPA 70, National Electrical Code, 2023 edition. Copyright © 2022, National Fire Protection Association. For a full copy of the NFPA 70, please go to www.nfpa.org

Figure 19-18. *Table 250.66* is used to size the grounding electrode conductor.

Section 250.66(A) applies to the grounding electrode conductor connecting to a ground rod. It states that the grounding electrode conductor that is run to a ground rod is not required to be any larger than a 6 AWG copper. It doesn't matter if *Table 250.66* calls for a larger conductor, it never has to be larger than a 6 AWG copper. The reason for a limit to the size of the GEC is that a ground rod is limited in the amount of current it can carry to the earth, and a 6 AWG copper is sufficient to carry that amount of current.

EXAMPLE 19-1

Problem: Determine the minimum size grounding electrode conductor for a service that has 250 kcmil copper ungrounded service-entrance conductors.

Solution:

Table 250.66 Grounding Electrode Conductor for Alternating-Current Systems

Size of Largest Ungrounded Conductor or Equivalent Area for Parallel Conductors (AWG/kcmil)		Size of Grounding Electrode Conductor (AWG/kcmil)	
Copper	Aluminum or Copper-Clad Aluminum	Copper	Aluminum or Copper-Clad Aluminum
2 or smaller	1/0 or smaller	8	6
1 or 1/0	2/0 or 3/0	6	4
2/0 or 3/0	4/0 or 250	4	2
Over 3/0 through 350	Over 250 through 500	**2**	**1/0**
Over 350 through 600	Over 500 through 900	1/0	3/0

Reproduced with permission of NFPA from NFPA 70, National Electrical Code, 2023 edition. Copyright © 2022, National Fire Protection Association. For a full copy of the NFPA 70, please go to www.nfpa.org

The minimum size grounding electrode conductor is 2 AWG copper, or 1/0 AWG aluminum.

EXAMPLE 19-2

Problem: Determine the minimum size grounding electrode conductor for a service that has two paralleled 350 kcmil aluminum ungrounded service-entrance conductors per phase. The paralleled conductors have a combined circular mil area of 700 kcmil per phase.

Solution:

Table 250.66 Grounding Electrode Conductor for Alternating-Current Systems

Size of Largest Ungrounded Conductor or Equivalent Area for Parallel Conductors (AWG/kcmil)		Size of Grounding Electrode Conductor (AWG/kcmil)	
Copper	Aluminum or Copper-Clad Aluminum	Copper	Aluminum or Copper-Clad Aluminum
2 or smaller	1/0 or smaller	8	6
1 or 1/0	2/0 or 3/0	6	4
2/0 or 3/0	4/0 or 250	4	2
Over 3/0 through 350	Over 250 through 500	2	1/0
Over 350 through 600	Over 500 through 900	**1/0**	**3/0**
Over 600 through 1100	Over 900 through 1750	2/0	4/0

Reproduced with permission of NFPA from NFPA 70, National Electrical Code, 2023 edition. Copyright © 2022, National Fire Protection Association. For a full copy of the NFPA 70, please go to www.nfpa.org

The minimum size grounding electrode conductor is a 1/0 AWG copper or 3/0 AWG aluminum.

Section 250.66(B) applies to the grounding electrode conductor that connects to a concrete-encased electrode. It states that the grounding electrode conductor that is run to a concrete-encased electrode is not required to be any larger than a 4 AWG copper. It doesn't matter if *Table 250.66* calls for a larger conductor, it never has to be larger than a 4 AWG copper. The reason for a limit to the size of the GEC is that a concrete-encased electrode is limited in the amount of current it can pass to the earth, and a 4 AWG copper is sufficient to carry that amount of current.

Section 250.66(C) applies to the grounding electrode conductor that connects to a ground ring. It states that the grounding electrode conductor run to a ground ring is not required to be any larger than the conductor size of the ground ring.

Grounding Electrode Conductor Protection

Grounding electrode conductors are permitted to be run in raceways, through framing members, or exposed on the surface, **Figure 19-19**. If they are exposed, they may require some form of protection, depending on the size of the grounding electrode conductor, and whether it will be subjected to physical damage. *Section 250.64(B)* describes when physical protection is required.

- If the grounding electrode conductor is not subjected to physical damage and is 6 AWG or larger, it can be run on the surface of the structure without protection.
- If the grounding electrode conductor is 6 AWG or larger and subjected to physical damage, it must be protected with one of the methods listed in *Section 250.64(B)(2)*.
- A grounding electrode conductor that is 8 AWG or smaller must be protected by one of the methods listed in *Section 250.64(B)(3)*.

Smaller services may only require an 8 AWG grounding electrode conductor, but it is common to upsize to a 6 AWG to avoid having to install it in a raceway or protect it by some other means. The extra cost of installing a larger conductor is often less than the cost of material and labor to provide protection.

EXAMPLE 19-3

Problem: Determine the minimum size grounding electrode conductor for a service that has 4/0 aluminum service-entrance conductors run to a metal underground water pipe.

Solution:

Table 250.66 Grounding Electrode Conductor for Alternating-Current Systems

Size of Largest Ungrounded Conductor or Equivalent Area for Parallel Conductors (AWG/kcmil)		Size of Grounding Electrode Conductor (AWG/kcmil)	
Copper	Aluminum or Copper-Clad Aluminum	Copper	Aluminum or Copper-Clad Aluminum
2 or smaller	1/0 or smaller	8	6
1 or 1/0	2/0 or 3/0	6	4
2/0 or 3/0	4/0 or 250	4	2
Over 3/0 through 350	Over 250 through 500	2	1/0

Reproduced with permission of NFPA from NFPA 70, National Electrical Code, 2023 edition. Copyright © 2022, National Fire Protection Association. For a full copy of the NFPA 70, please go to www.nfpa.org

The minimum size grounding electrode conductor connecting run to a metal underground water pipe is 4 AWG copper or 2 AWG aluminum.

If ground rods are used to supplement the underground water pipe, the minimum size grounding electrode conductor that would be required for the ground rods is a 6 AWG copper. (Since the supplemental grounding electrode conductor is connecting to ground rods, an aluminum conductor would not be permitted as it cannot be installed within 18″ of the earth.)

EXAMPLE 19-4

Problem: Determine the minimum size grounding electrode conductor for a service that has two parallel sets of 500 kcmil copper conductors.

Solution:

Table 250.66 Grounding Electrode Conductor for Alternating-Current Systems

Size of Largest Ungrounded Conductor or Equivalent Area for Parallel Conductors (AWG/kcmil)		Size of Grounding Electrode Conductor (AWG/kcmil)	
Copper	Aluminum or Copper-Clad Aluminum	Copper	Aluminum or Copper-Clad Aluminum
2 or smaller	1/0 or smaller	8	6
1 or 1/0	2/0 or 3/0	6	4
2/0 or 3/0	4/0 or 250	4	2
Over 3/0 through 350	Over 250 through 500	2	1/0
Over 350 through 600	Over 500 through 900	1/0	3/0
Over 600 through 1100	Over 900 through 1750	2/0	4/0
Over 1100	Over 1750	3/0	250

Reproduced with permission of NFPA from NFPA 70, National Electrical Code, 2023 edition. Copyright © 2022, National Fire Protection Association. For a full copy of the NFPA 70, please go to www.nfpa.org

The paralleled conductors have a combined circular mil area of 1000 kcmil per phase. The minimum size grounding electrode conductor connecting to a metal in-ground support structure that qualifies as a grounding electrode is 2/0 AWG copper or 4/0 AWG aluminum. If the building also has a concrete-encased electrode, the maximum size grounding electrode conductor required to connect to the rebar is 4 AWG.

Section 250.64(E) has requirements that apply when installing the grounding electrode conductor in a raceway. If nonmetallic or nonferrous metal raceways are used, there are no special bonding requirements. If a ferrous raceway is used, it will have to be electrically continuous by bonding to the raceway at both ends, **Figure 19-20**. This creates a parallel path so the fault current can travel through the conductor as well as the raceway. If a ferrous raceway is not bonded to the grounding electrode conductor that passes through it, the large current involved with a fault or lightning strike will cause the raceway to act as a choke that will limit the current. The raceway may be bonded to the grounding electrode conductor itself, the enclosure it is attached to, or the grounding electrode.

Equipment Grounding and Equipment Grounding Conductors

Nearly all metal raceways, equipment, appliances, and fixtures are required to be grounded. The primary reason is to prevent their metallic components from becoming energized due to a ground fault that would create a shock hazard. The goal of equipment grounding is to create an effective ground-fault current path that will ensure the overcurrent device quickly opens in the event of a ground fault. The *NEC* defines the *effective ground-fault current path* as an

Goodheart-Willcox Publisher

Figure 19-19. A 6 AWG or larger grounding electrode conductor is permitted to be run through the floor joists without any protection, provided it is not in an area where it will be subjected to physical damage.

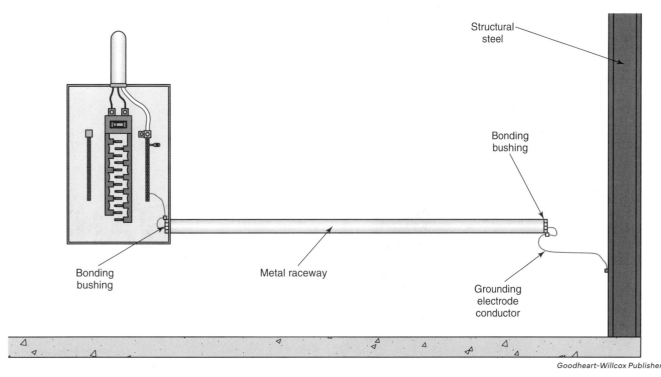

Goodheart-Willcox Publisher

Figure 19-20. A grounding electrode conductor that is installed in a metallic raceway is required to be bonded to both sides of the raceway.

intentionally constructed, low-impedance electrically conductive path designed and intended to carry current under ground-fault conditions from the point of a ground fault on a wiring system to the electrical supply source and that facilitates the operation of the overcurrent protective device or ground-fault detectors. *Part VI* of *Article 250* covers equipment grounding and equipment grounding conductors.

PRO TIP: Ground Fault on Ungrounded Branch Circuit

The equipment grounding conductor is intended to prevent conductive objects and appliances from becoming energized, **Figure 19-21**. The dishwasher pictured is in an older house and was connected to an existing ungrounded circuit. The metal frame of the dishwasher became energized when an exposed conductive portion of the ungrounded conductor made contact with the frame. This caused the entire shell of the appliance to become energized as shown by the noncontact voltage detector that is lit up. The owner received an electric shock when touching the dishwasher and the kitchen faucet, so they called an electrician. If there had been an equipment grounding conductor connected to the appliance, it would have tripped the overcurrent device when the ungrounded conductor contacted the frame of the dishwasher.

Goodheart-Willcox Publisher

Figure 19-21. This noncontact voltage detector is showing that the frame and front panel of this dishwasher have become energized.

Equipment Grounding

Equipment is grounded by connecting it to an equipment grounding conductor that is ultimately connected to the grounded point of the service. Most pieces of equipment and appliances that have exposed noncurrent carrying metal components are required to be connected to an equipment grounding conductor.

The first four sections in *Part VI* of the *Article 250* address equipment grounding. *Sections 250.109* and *250.110* have the general requirements for grounding electrical equipment and components. *Section 250.112* lists grounding requirements for specific pieces of electrical equipment that are fixed in place.

Section 250.114 lists grounding requirements for cord and plug equipment. For example, refrigerators, freezers, washers, dryers, electric ranges, aquarium equipment, electric lawnmowers, and snowblowers, **Figure 19-22**. This requirement isn't likely to cause any issues with new installations as they will have an equipment grounding conductor run with all the branch circuits that will be present in each outlet box. Older installations, on the other hand, may not always have an equipment grounding conductor in each outlet box. Many homes built prior to 1960 had knob and tube wiring, or ungrounded nonmetallic cables, so there was not an equipment grounding conductor present in the boxes. Consideration must be given as to where the appliances found in *Section 250.114* will be located to ensure there is an equipment grounding conductor present in the outlet box.

Section 250.116 has requirements for grounding nonelectrical equipment. Many times, the nonelectrical equipment being addressed is part of an assembly where a portion of it is electrically driven. For example, the metal structure associated with a crane or hoist or nonelectric elevator cars, **Figure 19-23**.

Vlad Kochelaevskiy/Shutterstock.com

Figure 19-22. Most of the major appliances installed in dwellings are required to have an equipment grounding conductor.

Figure 19-23. The metal structure of a crane is required to be grounded.

Equipment Grounding Conductor (EGC)

The first thing that generally comes to mind when thinking about an equipment grounding conductor is a wire-type conductor, but that isn't the only equipment grounding conductor recognized by the *NEC*. The *NEC* definition of an *equipment grounding conductor (EGC)* is a conductive path that is part of an effective ground-fault current path and connects normally non-current-carrying metal parts of equipment together and to the system grounded conductor or to the grounding electrode conductor, or both. *Section 250.118* lists the types of equipment grounding conductors recognized by the *Code*. They can be broken down into two categories: metallic raceways/boxes and wire-type conductors.

Raceway System

Metal raceways that are not flexible, such as RMC, IMC, and EMT, may be used as the equipment grounding conductor, **Figure 19-24**. Flexible raceways may be permitted to serve as the equipment grounding conductor, but there are restrictions in *Section 250.118* that will apply. The most common limitation to using a flexible metal raceway is a maximum length, which is 6′. There are also restrictions on the size of the overcurrent device ahead of it. If the flexible raceway is installed to allow for flexibility due to vibration or movement, such as a motor, then the raceway itself is not permitted to serve as the EGC and a separate wire-type EGC is required.

When using the raceway as the equipment grounding conductor, the fittings must be listed for use with that raceway and shall be made tight by the appropriate tools. When using a raceway as an equipment grounding conductor, the integrity of the ground-fault current path relies on the proper installation of the raceway and its fittings.

The metal jacket of raceways, such as AC cable and MI cable, may be used as the equipment grounding conductor. In order for the jacket to qualify as an EGC, the fittings have to be rated appropriately and properly installed.

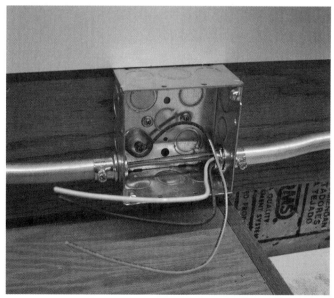

Figure 19-24. Raceways that are installed with listed fittings that are properly secured can be used as the equipment grounding conductor. The receptacle pictured is grounded through a ground tail secured to the box. The ground screw attaching the tail to the box is a listed grounding screw and has two threads in contact with the box.

> **PRO TIP**
>
> ### Insulated Equipment Grounding Conductor
>
> It is becoming more common to encounter project specifications that require an insulated wire-type equipment grounding conductor to be installed in a raceway, along with the branch circuit conductors, regardless of the type of raceway that is installed, **Figure 19-25**. In this situation, insulated wire-type equipment grounding conductors will be installed from the panelboard to the outlet(s). When the equipment grounding conductor is spliced within a box, it is required to be bonded to the box. This bonds the wire-type equipment grounding conductor to the box and metal raceway, creating multiple paths for a ground-fault current to get back to the grounded point of the service. Having multiple paths provides extra assurance there is a low-impedance path for ground-fault current. Be sure to check the specifications for the job, as although it may be permitted by the *NEC* to use the raceway as the EGC, the specifications may require a wire-type EGC to be installed in the raceway.

Wire Type

Wire-type equipment grounding conductors are covered in *Section 250.119*. They are permitted to be bare, covered, or insulated. Insulated or covered equipment grounding conductors that are 6 AWG and smaller shall have an outer finish that is green or green with yellow stripes. Since larger conductors may not be readily available in green, the *NEC* allows us to reidentify insulated conductors that are 4 AWG and larger to indicate they are being used as equipment grounding conductors, **Figure 19-26**. The reidentification may be accomplished by stripping the insulation completely off, or by some type of covering such as green tape. When reidentifying the EGC by stripping the insulation off, it must be completely removed where it is exposed in boxes or cabinets. When using a covering such as tape or paint, it must be green, encircle the conductor, and it must be identified at each box or cabinet.

Goodheart-Willcox Publisher

Figure 19-25. Equipment grounding conductors that are installed in a raceway are required to be bonded to the box when the conductors splice or terminate. The green equipment grounding conductors will be attached to the box at the threaded hole in the dimpled portion on the back of the box. The ground screw hole is dimpled so that the screw won't run into a solid wall when it is surface mounted.

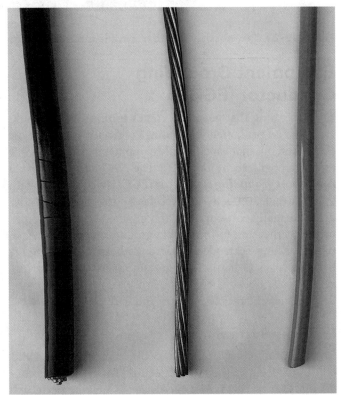

Goodheart-Willcox Publisher

Figure 19-26. Equipment grounding conductors may be bare or insulated. Bare equipment grounding conductors are often found in cables such as nonmetallic-sheathed cable. Smaller equipment grounding conductors must have an outer finish that is green or green with a yellow stripe. Larger equipment grounding conductors, 4 AWG and larger, are permitted to be reidentified with green tape provided it goes all the way around the conductor.

Table 250.122 Minimum Size Equipment Grounding Conductors for Grounding Raceway and Equipment

Rating or Setting of Automatic Overcurrent Device in Circuit Ahead of Equipment, Conduit, etc., Not Exceeding (Amperes)	Size (AWG or kcmil)	
	Copper	Aluminum or Copper-Clad Aluminum*
15	14	12
20	12	10
60	10	8
100	8	6
200	6	4
300	4	2
400	3	1
500	2	1/0
600	1	2/0
800	1/0	3/0
1000	2/0	4/0
1200	3/0	250
1600	4/0	350
2000	250	400
2500	350	600
3000	400	600
4000	500	750
5000	700	1250
6000	800	1250

Note: Where necessary to comply with 250.4(A)(5) or (B)(4), the equipment grounding conductor shall be sized larger than given in this table.
*See installation restrictions in 250.120.

Reproduced with permission of NFPA from NFPA 70, National Electrical Code, 2023 edition. Copyright © 2022, National Fire Protection Association. For a full copy of the NFPA 70, please go to www.nfpa.org

Figure 19-27. *Table 250.122* is used to size the equipment grounding conductors.

Equipment Grounding Conductor Sizing

Section 250.122 has general information about sizing the equipment grounding conductor. The equipment grounding conductor is sized according to *Table 250.122*, **Figure 19-27**. It has the rating of the overcurrent device on the left side of the table, and the minimum size EGC in both copper and aluminum on the right side of the table. The minimum size of the EGC is based on the size of the overcurrent device that protects the circuit.

EXAMPLE 19-5

Problem: Determine the minimum size equipment grounding conductor for a branch circuit protected by a 20-ampere overcurrent protective device.

Solution:

Table 250.122 Minimum Size Equipment Grounding Conductors for Grounding Raceway and Equipment

Rating or Setting of Automatic Overcurrent Device in Circuit Ahead of Equipment, Conduit, etc., Not Exceeding (Amperes)	Size (AWG or kcmil)	
	Copper	Aluminum or Copper-Clad Aluminum*
15	14	12
20	12	10
60	10	8
100	8	6

Reproduced with permission of NFPA from NFPA 70, National Electrical Code, 2023 edition. Copyright © 2022, National Fire Protection Association. For a full copy of the NFPA 70, please go to www.nfpa.org

The minimum size equipment grounding conductor is a 12 AWG copper or 10 AWG aluminum.

The table doesn't list every size of overcurrent device. If your OCPD isn't listed, you will pick the row with the next larger size OCPD to determine the minimum size equipment grounding conductor.

EXAMPLE 19-6

Problem: Determine the minimum size equipment grounding conductor for a branch circuit protected by a 50-ampere overcurrent protective device.

Solution:

Table 250.122 Minimum Size Equipment Grounding Conductors for Grounding Raceway and Equipment

Rating or Setting of Automatic Overcurrent Device in Circuit Ahead of Equipment, Conduit, etc., Not Exceeding (Amperes)	Size (AWG or kcmil)	
	Copper	Aluminum or Copper-Clad Aluminum*
15	14	12
20	12	10
60	10	8
100	8	6

Reproduced with permission of NFPA from NFPA 70, National Electrical Code, 2023 edition. Copyright © 2022, National Fire Protection Association. For a full copy of the NFPA 70, please go to www.nfpa.org

Because a 50-ampere device is not listed, jump up to the next-largest size of 60 amperes. The minimum size equipment grounding conductor is a 10 AWG copper or 8 AWG aluminum.

Raceways will often have multiple circuits in them. *Section 250.122(B)* allows one equipment grounding conductor to be installed for all the contained circuits. If there are multiple sizes of overcurrent devices protecting the conductors, the size of the equipment grounding conductor is based on the largest overcurrent protective device.

Bonding

Bonding is defined as the act of connecting metal components to establish electrical continuity and conductivity. Bonding metal components together brings them to the same potential so that touching two separate metal portions of the system will not result in an electric shock. The term bonding (bonded) is often interchanged with grounding (grounded) in the field. Although they are different, in most installations they work together to ensure the same result, which is to have metal components electrically connected to prevent a difference in potential. With a grounded system, in addition to components being connected, they are also connected to the grounded point in the electrical system. This provides a low-impedance path to facilitate the operation of the overcurrent device of the circuit that has a fault to ground. This minimizes the length of time that the metal components are energized. As indicated by *NEC Figure 250.1*, bonding is interconnected with several of the parts in *Article 250*, **Figure 19-28**.

Bonding Enclosures and Equipment

Metal enclosures, equipment, raceways, and fittings are all required to be bonded together as per *Section 250.96*. For example, use of the correct fittings for the type of raceway, tightening locknuts, and setscrews is required. Similar language shows up in *Section 300.10*, requiring electrical continuity of metal raceways and enclosures. This ensures that all components have the continuity to safely pass any fault current imposed on them by a ground fault to quickly open the overcurrent device. If there is paint or some other nonconductive coating, it must be removed to ensure good contact. Having equipment grounding conductors that are bonded to the boxes and enclosers does not relax this requirement.

Bonding Service Equipment

The service has some additional bonding requirements since it has unfused conductors before the first overcurrent device. These requirements are found in *Section 250.92*. Like the rest of the electrical system, all the metallic service components, such as raceways, enclosures, and framework, are required to be bonded together and connected to the grounded conductor. The additional requirements are found on the supply side of the service disconnect. Concentric and eccentric knockouts, reducing washers, and standard locknuts are not sufficient to bond a service raceway to an enclosure. A bonding locknut or bushing with a correctly sized bonding conductor shall be used to bond raceways on the supply side of the main service disconnect to the enclosures, **Figure 19-29**. Raceways installed wrench tight into hubs do not require any additional bonding.

These bonding requirements are one of the main reasons that electricians choose to use nonmetallic raceways for service conductors. Since nonmetallic raceways are not conductive, the need to bond the raceway is eliminated, saving time and money.

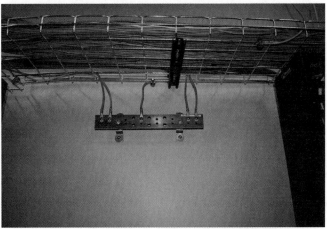

Goodheart-Willcox Publisher

Figure 19-28. The wire mesh cable tray pictured has the bonding conductors connecting the various sections of cable tray back to a bonding/grounding bar.

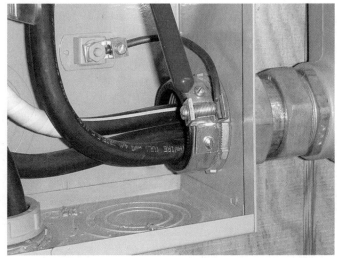

Goodheart-Willcox Publisher

Figure 19-29. A bonding bushing has a lug for the connection of a bonding conductor. It has set screws that dig into the threads of the raceway or connector to ensure a good connection.

Main Bonding Jumper

The component that connects the metal enclosure of the service disconnect and its associated metal components to the grounded conductor is the main bonding jumper, **Figure 19-30**. Main bonding jumpers are discussed in *Section 250.24*. The *NEC* defines the ***main bonding jumper (MBJ)*** as the connection between the grounded circuit conductor and the equipment grounding conductor, or the supply-side bonding jumper, or both, at the service. *Section 250.24(C)* tells us that the main bonding jumper is used to connect the equipment grounding conductors and the frame of the enclosure to the grounded conductor. It then sends us to *Section 250.28* to describe how that is to be accomplished.

The main bonding jumper is permitted to be a wire, bus, screw, or some other suitable conductor. It must be large enough to carry the available current and provide a low-impedance path to open the overcurrent device in the event of a ground fault. Panelboard manufacturers often supply a green screw or a molded conductor that connects the grounded busbar to the enclosure. Being provided by the manufacturer for that particular panelboard, it has been designed and tested to assure it is capable of carrying the fault current necessary to open the overcurrent device.

Sizing the Main Bonding Jumper

If a main bonding jumper is not supplied by the service equipment manufacturer, or it is lost, a wire can be used. *Section 250.28(D)(1)* sends us to *Table 250.102(C)(1)* to determine its minimum size, **Figure 19-31**. *Table 250.102(C)(1)* is set up like *Table 250.66* for sizing the grounding electrode conductor. The size of the service-entrance conductors is on the left side of the table and the minimum size of the main bonding jumper is on the right side of the table. The minimum size of the MBJ is based on the size of the service-entrance conductors.

Goodheart-Willcox Publisher

Figure 19-30. This service panel has a screw-type main bonding jumper that connects the grounded/neutral conductor to the metal enclosure.

Table 250.102(C)(1) Grounded Conductor, Main Bonding Jumper, System Bonding Jumper, and Supply-Side Bonding Jumper for Alternating-Current Systems

Size of Largest Ungrounded Conductor or Equivalent Area for Parallel Conductors (AWG/kcmil)		Size of Grounded Conductor or Bonding Jumper (AWG/kcmil)	
Copper	Aluminum or Copper-Clad Aluminum	Copper	Aluminum or Copper-Clad Aluminum
2 or smaller	1/0 or smaller	8	6
1 or 1/0	2/0 or 3/0	6	4
2/0 or 3/0	4/0 or 250	4	2
Over 3/0 through 350	Over 250 through 500	2	1/0
Over 350 through 600	Over 500 through 900	1/0	3/0
Over 600 through 1100	Over 900 through 1750	2/0	4/0
Over 1100	Over 1750	See Notes 1 and 2.	

Reproduced with permission of NFPA from NFPA 70, National Electrical Code, 2023 edition. Copyright © 2022, National Fire Protection Association. For a full copy of the NFPA 70, please go to www.nfpa.org

Figure 19-31. *NEC Table 102(C)(1)* is used to size main bonding jumpers and system bonding jumpers. It also is used to determine the minimum size grounded conductor.

EXAMPLE 19-7

Problem: Determine the minimum size main bonding jumper for a service that has 4/0 AWG aluminum ungrounded service-entrance conductors.

Solution:

Table 250.102(C)(1) Grounded Conductor, Main Bonding Jumper, System Bonding Jumper, and Supply-Side Bonding Jumper for Alternating-Current Systems

Size of Largest Ungrounded Conductor or Equivalent Area for Parallel Conductors (AWG/kcmil)		Size of Grounded Conductor or Bonding Jumper (AWG/kcmil)	
Copper	Aluminum or Copper-Clad Aluminum	Copper	Aluminum or Copper-Clad Aluminum
2 or smaller	1/0 or smaller	8	6
1 or 1/0	2/0 or 3/0	6	4
2/0 or 3/0	4/0 or 250	4	2
Over 3/0 through 350	Over 250 through 500	2	1/0
Over 600 through 1100	Over 900 through 1750	2/0	4/0
Over 1100	Over 1750	See Notes 1 and 2.	

Notes:
1. If the circular mil area of ungrounded supply conductors that are connected in parallel is larger than 1100 kcmil copper or 1750 kcmil aluminum, the grounded conductor or bonding jumper shall have an area not less than 12½ percent of the area of the largest ungrounded supply conductor or equivalent area for parallel supply conductors. The grounded conductor or bonding jumper shall not be required to be larger than the largest ungrounded conductor or set of ungrounded conductors.
2. If the circular mil area of ungrounded supply conductors that are connected in parallel is larger than 1100 kcmil copper or 1750 kcmil aluminum and if the ungrounded supply conductors and the bonding

Reproduced with permission of NFPA from NFPA 70, National Electrical Code, 2023 edition. Copyright © 2022, National Fire Protection Association. For a full copy of the NFPA 70, please go to www.nfpa.org

The minimum size main bonding jumper is a 4 AWG copper, or 2 AWG aluminum.

EXAMPLE 19-8

Problem: Determine the minimum size main bonding jumper for a service that has four parallel sets of 500 kcmil copper conductors. The paralleled conductors have a combined circular mil area of 2000 kcmil per phase.

Solution:

Note 1 from *Table 250.102(C)(1)* directs us to multiply the combined area by 12.5%.

$$2000 \text{ kcmil} \times 0.125 = 250 \text{ kcmil}$$

The minimum size main bonding jumper is 250 kcmil copper.

PRO TIP — Swimming Pools

Swimming pools create a unique hazard because pool users' bare skin will contact water, concrete, soil, and all of the pool components, such as handrails, slides, and diving boards, **Figure 19-32**. Keeping in mind that there are often electrically operated pool components, such as pumps, lights, and motors for pool covers, that are in the vicinity of the pool, special consideration must be taken to ensure there is no voltage potential between any of the objects.

Since a swimming pool and its associated components are considered special equipment, the bonding requirements are found in *Article 680*. *Section 680.26, Equipotential Bonding* requires all metal components in and around a swimming pool and the pool water, to be bonded together. This will reduce the possibility of a voltage gradient that could create a lethal electric shock.

Goodheart-Willcox Publisher

Figure 19-32. In order to avoid voltage gradients, all metal objects in and around the pool are bonded together.

Bonding Communication Systems

The *NEC* requires an intersystem bonding termination device to be installed for bonding communication systems, **Figure 19-33**. The *NEC* definition of an **intersystem bonding termination** is a device that provides a means for connecting intersystem bonding conductors for communications systems to the grounding electrode system. Communication systems, such as a satellite dish, antenna, telephone landline, or cable TV, all require a connection to ground. The intersystem bonding termination device provides a place for them to connect to the grounding electrode system.

Section 250.94 lists the requirements that apply to the intersystem bonding termination device. It is permitted to be located outside of the building or inside near the service equipment. It must meet the following requirements:

- Remain accessible for connection and inspection.
- Have at least three spaces for communication connections.
- Not interfere with the opening of equipment.
- If not attached directly to equipment, it must be connected to the grounding electrode system with a 6 AWG or larger copper conductor.

Bonding Piping Systems and Structural Metal

If a metal piping system or building steel comes into contact with an energized conductor, it could energize the entire piping system or the metal structure of the building, creating a very dangerous situation. For this reason, piping systems and structural metal must be bonded together and connected to ground. If the water line and structural steel are considered grounding electrodes, they are part of the grounding electrode system and will already have been connected to the grounded point of the electrical system through the grounding electrode conductor. In the case that the water line or building steel does not qualify as a grounding electrode, the interior metal water piping and building steel will have to be bonded to the grounded point of the electrical system. This applies to other piping systems such as gas piping as well.

Interior Water Piping

Buildings that have metal interior water piping but are fed from the city or rural area by a plastic water line must be bonded. *Section 250.104(A)* requires the interior metal water piping to be bonded to the service equipment enclosure, the grounded conductor at the service, the grounding electrode conductor (if large enough), or a grounding electrode (if its grounding electrode conductor is large enough). Grounding electrodes such as ground rods and concrete-encased electrodes are examples where the grounding electrode conductor may not be large enough to serve as the bonding jumper. The size of the bonding conductor required to bond the water line is found in *Table 250.102(C)(1)*. It is based on the size of the ungrounded service-entrance conductors.

Goodheart-Willcox Publisher

Figure 19-33. Intersystem bonding termination devices provide a place for communication systems to bond to our grounding electrode system.

EXAMPLE 19-9

Problem: Determine the minimum size bonding conductor for a building has 4/0 AWG copper ungrounded service-entrance conductors.

Solution:

Table 250.102(C)(1) Grounded Conductor, Main Bonding Jumper, System Bonding Jumper, and Supply-Side Bonding Jumper for Alternating-Current Systems

Size of Largest Ungrounded Conductor or Equivalent Area for Parallel Conductors (AWG/kcmil)		Size of Grounded Conductor or Bonding Jumper (AWG/kcmil)	
Copper	Aluminum or Copper-Clad Aluminum	Copper	Aluminum or Copper-Clad Aluminum
2 or smaller	1/0 or smaller	8	6
1 or 1/0	2/0 or 3/0	6	4
2/0 or 3/0	4/0 or 250	4	2
Over 3/0 through 350	Over 250 through 500	2	1/0
Over 350 through 600	Over 500 through 900	1/0	3/0
Over 600 through 1100	Over 900 through 1750	2/0	4/0
Over 1100	Over 1750	See Notes 1 and 2.	

Reproduced with permission of NFPA from NFPA 70, National Electrical Code, 2023 edition. Copyright © 2022, National Fire Protection Association. For a full copy of the NFPA 70, please go to www.nfpa.org

The minimum size bonding conductor is a 2 AWG copper or 1/0 AWG aluminum.

Building Steel

Buildings with structural steel that doesn't go into the earth, and is therefore not considered a grounding electrode, will need to be bonded. *Section 250.104(C)* requires the structural building steel to be connected to the service equipment enclosure, the grounded conductor at the service, the grounding electrode conductor (if large enough), or a grounding electrode (if its grounding electrode conductor is large enough). Grounding electrodes such as ground rods and a concrete-encased electrode are examples where the grounding electrode conductor may not be large enough to serve as the bonding jumper. The size of the bonding conductor is found in *Table 250.102(C)(1)*. It is based on the size of the ungrounded service-entrance conductors.

EXAMPLE 19-10

Problem: Determine the minimum size bonding conductor for a service has four parallel sets of 400 kcmil copper conductors. The paralleled conductors have a combined circular mil area of 1600 kcmil per phase.

Solution: *Note 1* from *Table 250.102(C)(1)* directs us to multiply the combined area by 12.5%.

$$1600 \text{ kcmil} \times 0.125 = 200 \text{ kcmil}$$

Since 200 kcmil doesn't correspond to a standard-sized conductor in "kcmil", *Chapter 9 Table 8* can be used to determine the minimum conductor size in AWG. A 4/0 AWG conductor has a circular mil area of 211,600 circular mils which is larger than the minimum 200 kcmil calculated by *Note 1* of *Table 250.102(C)(1)*.

The minimum size bonding jumper is 4/0 AWG copper.

Metal Gas Piping Systems

Gas piping and other metal piping systems are not permitted to be used as a grounding electrode, but they must be bonded to the grounded point of the service to ensure they cannot become energized. *Section 250.104(B)* states that the conductor that bonds the piping system shall be sized by *Table 250.122* based on the circuit that is likely to energize it. In the case of a gas line, it is connected to an appliance, such as a furnace, that is fed with an electric branch circuit containing an equipment grounding conductor, **Figure 19-34**. The frame of the furnace is grounded through the equipment grounding conductor, and the gas line is connected to the frame of the appliance, satisfying the *NEC* requirement.

It is common to have a manifold where several gas lines split off to feed each of the gas appliances. The manifold will be connected to the equipment grounding conductor through the gas line feeding the appliances. Be sure to read the manufacturer's instructions for the manifold as some manufacturers will require a minimum size conductor to bond the manifold.

Goodheart-Willcox Publisher

Figure 19-34. Appliances such as gas water heaters and furnaces have a mechanical connection to the gas line. The gas line is generally bonded through the appliance's equipment grounding conductor, because that is the branch circuit that is likely to energize it if there is a fault.

System Bonding Jumper

A system bonding jumper is used to bond the equipment grounding conductors and the metal enclosure to the grounded conductor in a separately derived system. The *NEC* defines the ***system bonding jumper (SBJ)*** as the connection between the grounded circuit conductor and the supply-side bonding jumper, or the equipment grounding conductor, or both, at a separately derived system. The system bonding jumper has the same function in a separately derived system that the main bonding jumper has with the service. *Section 250.30* has requirements that apply to the supply bonding jumper.

A ***separately derived system*** is an electrical source, other than a service, having no direct connection(s) to circuit conductors of any other electrical source other than those established by grounding and bonding connections. A common use for a separately derived system for commercial and industrial buildings is on the secondary side of transformers, **Figure 19-35**. A transformer will change the supply voltage, either up or down, to accommodate the voltage for specific loads. Buildings with a 277/480-volt wye service will often have transformers that step the voltage down to 120/208 wye for receptacles and other loads that require the lower voltage.

Goodheart-Willcox Publisher

Figure 19-35. A transformer is an example of a separately derived system. A transformer has a system bonding jumper that provides a low impedance path back to the transformer so the overcurrent device will open in the event of a ground fault.

Summary

- *Article 250* of the *National Electrical Code* has requirements for grounding.
- Fittings used for grounding and bonding must be listed for the use and installed according to the manufacturer's instructions.
- The grounding electrodes recognized by the *NEC* are:
  - Metal underground water pipe
  - Metal in-ground support structure
  - Concrete-encased electrode
  - Ground ring
  - Rod and pipe electrodes
  - Plate electrodes
- Grounding electrode conductors are sized according to *Table 250.66* based on the size of the ungrounded service-entrance conductors.
- Equipment used to connect the grounding electrode conductor to the grounding electrode must be listed for the purpose.
- Metal raceways are permitted to be used as an equipment grounding conductor provided that they are installed according to *Section 250.118*.
- Equipment grounding conductors may be bare, green, or green with a yellow stripe.
- Equipment grounding conductors are sized according to *Table 250.122* based on the overcurrent protection device of the circuit conductors.
- The main bonding jumper provides a low-impedance path for ground-fault current to return to the grounded conductor and open the overcurrent protection device.
- The main bonding jumper is sized by *Table 250.102(C)(1)* according to the size of the ungrounded service-entrance conductors.

Unit 19 Review

Name _____ Date _____ Class _____

Know and Understand

Answer the following questions based on information in this unit.

1. _____ is dedicated to grounding and bonding.
   A. *Article 220*
   B. *Article 230*
   C. *Article 240*
   D. *Article 250*

2. _____ is to connect together to establish electrical continuity and conductivity.
   A. Grounding
   B. Bonding
   C. Earthing
   D. Splicing

3. A ground ring must be installed at least _____ below the surface of the earth.
   A. 30″
   B. 48″
   C. 60″
   D. 72″

4. Machine screw fasteners used for grounding and bonding must engage at least _____ threads.
   A. 2
   B. 3
   C. 4
   D. 5

5. Which of the following is *not* one of the purposes of grounding?
   A. Limit the voltage to ground from lighting and line surges.
   B. Stabilize the voltage to earth under normal operation.
   C. Limit the voltage to ground on non-current carrying metal components.
   D. Provide a high-impedance fault current path.

6. Equipment grounding conductors that are _____ AWG or smaller are required to have an outer finish that is green.
   A. 8
   B. 6
   C. 4
   D. 2

7. The _____ is the system or circuit conductor that is intentionally grounded.
   A. equipment grounding conductor
   B. grounding electrode conductor
   C. ungrounded conductor
   D. grounded conductor

8. The maximum size equipment grounding conductor required for a ground rod is _____ AWG.
   A. 2
   B. 4
   C. 6
   D. 8

9. Flexible metal conduit is *not* permitted to be used as an equipment grounding path if it is _____.
   A. installed to allow for flexibility
   B. longer than four feet long
   C. made of aluminum
   D. All of the above.

10. _____ is used to size the main bonding jumper.
    A. *Table 250.3*
    B. *Table 250.66*
    C. *Table 250.102(C)(1)*
    D. *Table 250.122*

Apply and Analyze

Answer the following questions using a copy of the National Electrical Code. *Identify the section or subsection where the answer is found.*

1. Grounded systems shall have a(n) _____ main bonding jumper that connected the equipment grounding conductors and the service disconnect enclosure to the grounded conductor.
   A. copper
   B. unspliced
   C. screw-type
   D. solid

   NEC _____

2. A ground ring shall encircle the building and be at least _____ long.
   A. 20′
   B. 23′
   C. 30′
   D. 100′

   NEC _____

3. The minimum size equipment grounding conductor required for a branch circuit protected by a 20-ampere breaker is _____.
   A. 16 AWG
   B. 14 AWG
   C. 12 AWG
   D. 10 AWG

   NEC _____

4. The insulated equipment grounding conductor installed to reduce electromagnetic interference to a(n) _____ is not required to be connected to the box and is permitted to run directly to the appropriate panelboard.
   A. isolated ground receptacle
   B. grounded receptacle
   C. grounded switch
   D. grounded luminaire

   NEC _____

5. Aluminum grounding electrode conductors on the exterior of buildings shall not be terminated within _____ of the earth.
   A. 6″
   B. 12″
   C. 18″
   D. 24″

   NEC _____

6. A service with 2/0 copper ungrounded conductors shall have a _____ copper grounding electrode conductor that is connected to a water pipe grounding electrode.
   A. 10 AWG
   B. 8 AWG
   C. 6 AWG
   D. 4 AWG

   NEC _____

7. A ground rod with a resistance of _____ or less to earth will not require a supplemental electrode.
   A. 10 ohms
   B. 25 ohms
   C. 50 ohms
   D. 100 ohms

   NEC _____

8. Which conductor is to be grounded in a single-phase three-wire system?
   A. Neutral conductor
   B. Phase conductor
   C. Any conductor as long as it is identified
   D. None of the above.

   NEC _____

9. When multiple ground rods are driven in the ground, they must be at least _____ apart.
   A. 4′
   B. 5′
   C. 6′
   D. 8′

   NEC _____

10. A _____ or larger grounding electrode conductor that is not exposed to physical damage is permitted to be run along the surface of the building without metal covering or protection.
    A. 8 AWG
    B. 6 AWG
    C. 4 AWG
    D. 2 AWG

Name _____ Date _____ Class _____

Critical Thinking

Complete the following ampacity calculations using a copy of the National Electrical Code. *Identify the section or subsection used to find the answer.*

1. Rod, pipe, and plate electrodes shall be installed below _____.

   _____

   NEC _____

2. Connections to grounding and bonding equipment that rely on _____ shall not be used.

   _____

   NEC _____

3. A single equipment grounding conductor is permitted to be installed for multiple circuits that are installed in the same _____.

   _____

   NEC _____

4. A concrete-encased electrode shall be permitted to be multiple pieces of rebar connected together with wire ties provided the overall length is at least _____ long.

   _____

   NEC _____

5. Expansion fittings shall be made _____ by equipment bonding jumpers or other means.

   _____

   NEC _____

6. The continuity of the metal water pipe grounding electrode shall not rely on water meters, _____, and similar equipment.

   _____

   NEC _____

7. A grounding electrode is not required at a separate building or structure if it is only fed with a(n) _____ that contains an equipment grounding conductor.

   _____

   NEC _____

8. A _____ used as a grounding electrode, shall be supplemented with an additional electrode regardless of its resistance to ground.

   _____

   NEC _____

9. List four residential cord- and plug-connected pieces of equipment that are required to have an equipment grounding conductor that connects all of its non-current carrying metal components to ground.

   _____

   NEC _____

10. Equipment grounding conductor continuity shall not be interrupted by the removal of a _____.

    _____

    NEC _____

Complete the following ampacity calculations using a copy of the National Electrical Code. *Identify the section or subsection used to find the answer.*

11. Determine the minimum size copper grounding electrode conductor required for a concrete-encased electrode. The ungrounded service-entrance conductors are 500 kcmil copper.

    _____

    NEC _____

12. Determine the minimum size copper grounding electrode conductor required for a metal underground water pipe. The ungrounded service-entrance conductors are 4/0 AWG aluminum.

    _____

    NEC _____

13. Determine the minimum size copper grounding electrode conductor required for two ground rods that are installed six feet apart. The ungrounded service-entrance conductors are three paralleled 500 kcmil copper conductors per phase.

    _____

    NEC _____

14. Determine the minimum size aluminum grounding electrode conductor required for building steel that meets the requirements to be a metal in-ground support structure grounding electrode. The ungrounded service-entrance conductors are two paralleled 500 kcmil aluminum conductors per phase.

    _____

    NEC _____

15. Determine the minimum size copper grounding electrode conductor required for a concrete-encased electrode. The ungrounded service-entrance conductors are 2 AWG aluminum.

    _____

    NEC _____

16. Determine the minimum size copper grounding electrode conductor required for a 2 AWG ground ring. The ungrounded service-entrance conductors are 750 kcmil aluminum conductors.

    _____

    NEC _____

17. Determine the minimum size copper grounding electrode conductor required for a metal underground water pipe. The ungrounded service-entrance conductors are 250 kcmil copper conductors.

    _____

    NEC _____

18. Determine the minimum size copper grounding electrode conductor required for a metal underground water pipe. The ungrounded service-entrance conductors are 2 AWG copper conductors.

    _____

    NEC _____

19. Determine the minimum size copper grounding electrode conductor required for building steel that meets the requirements to be a metal in-ground support structure grounding electrode. The ungrounded service-entrance conductors are two paralleled 300 kcmil copper conductors per phase.

    _____

    NEC _____

20. Determine the minimum size copper grounding electrode conductor required for a concrete-encased electrode. The service disconnect is rated at 200 amperes and it is on a dwelling.

    _____

    NEC _____

Complete the following ampacity calculations using a copy of the National Electrical Code. *Identify the section or subsection used to find the answer.*

21. Determine the minimum size copper main bonding jumper for a service with 4/0 aluminum ungrounded service-entrance conductors.

    _____

    NEC _____

22. Determine the minimum size copper main bonding jumper for a service with four paralleled 500 kcmil copper ungrounded service-entrance conductors per phase.

    _____

    NEC _____

23. A building has copper interior water piping that is connected to a plastic water line that feeds the building from the rural water supplier. The ungrounded service-entrance conductors are 500 kcmil copper. Determine the minimum size copper bonding conductor required to bond the metal water piping to the service equipment.

    _____

    NEC _____

24. A building has structural steel that is not in contact with the earth. The ungrounded service-entrance conductors are four paralleled 500 kcmil copper ungrounded service-entrance conductors per phase. Determine the minimum size copper bonding conductor required to bond the metal structural steel.

    _____

    NEC _____

25. Determine the minimum size copper equipment grounding conductor for a branch circuit protected by a 15-ampere breaker.

    _____

    NEC _____

26. Determine the minimum size copper equipment grounding conductor for a branch circuit protected by a 100-ampere fuse.

    _____

    NEC _____

27. Determine the minimum size aluminum equipment grounding conductor for a branch circuit protected by a 400-ampere fuse.

    _____

    NEC _____

28. Determine the minimum size copper equipment grounding conductor for a branch circuit protected by a 50-ampere breaker.

    _____

    NEC _____

29. A nonmetallic raceway has two 20-amp branch circuits and one 30-amp branch circuits in it. What is the minimum number and size of equipment grounding conductors required in this installation?

    _____

    _____

    NEC _____

30. A feeder has two paralleled 500 kcmil aluminum conductors per phase. The conductors are split up into two nonmetallic raceways. The paralleled conductors are protected by a 600-ampere overcurrent protective device. What is the minimum size equipment grounding conductor that is required to be installed in each of the raceways?

    _____

    NEC _____

UNIT 20
Dwelling Unit Cooking Appliance Calculations

Sashkin/Shutterstock.com

KEY TERMS

counter-mounted cooking unit (cooktop)
demand factor
load
oven
range

LEARNING OBJECTIVES

- Describe the difference between a range, cooktop, and oven.
- Calculate the service demand for the cooking appliances in dwellings.
- Calculate the branch circuit demand for a dwelling unit cooking appliance.
- Calculate the minimum size branch circuit conductors required for a cooking appliance.
- Calculate the required branch circuit overcurrent protective device for a given cooking appliance.

Introduction

The focus of this chapter is residential cooking appliance calculations, often referred to as range calculations. Range calculations are one of the steps involved when performing a service or feeder calculation. Although range calculations are a part of calculating the size of an electrical service, which is covered in Unit 21, there are many scenarios, and a unit dedicated to calculating the cooking appliance demand factors is necessary. This unit will also cover demand factors that may be applied to range branch circuits.

Most single-family range calculations are simple, with only one cooking appliance. When performing service or feeder calculations for multifamily dwellings, there are many cooking appliances, making the calculations a bit more complex. This unit will discuss the straightforward single-family calculations, as well as the more complex multifamily calculations.

Definitions

Before performing cooking appliance calculations, you must understand applicable definitions. Without understanding the proper terminology, it is impossible to properly apply the *Code*.

- *Demand factor* is the ratio of the maximum demand of a system, or part of a system, to the total connected load of a system or the part of the system under consideration.
- A *counter-mounted cooking unit (cooktop)* is a cooking appliance designed for mounting in or on a counter and consisting of one or more heating elements, internal wiring, and built-in or mountable controls, **Figure 20-1**.
- An *oven* is a permanently installed compartment with a door designed to cook food. It will often have a heating element on the top and bottom of the heating compartment, **Figure 20-2**.
- A *range* is a cooktop and an oven built in one unit, **Figure 20-3**.
- A *load* is the portion of an electrical system that converts electrical energy into a useful form of energy such as light, heat, and rotation.

Household Cooking Equipment Demand Factors

When performing a service or feeder calculation, the *NEC* allows a demand factor to be applied to household cooking appliances, such as ranges, ovens, and cooktops. The reason for the demand factor is that the heating elements in these appliances frequently cycle on and off, so they are rarely all on at the same time. Even if all four burners are on and the oven is being used, each element will be independently

Goodheart-Willcox Publisher

Figure 20-2. Pictured is a double oven installed in a cabinet with a warming drawer mounted in the space below the oven. Even though it looks like the warming drawer is part of the oven, it is a separate appliance with its own branch circuit. A double oven has two oven compartments, each of which has heating elements.

NavinTar/Shutterstock.com

Figure 20-1. This cooktop is mounted on the counter.

ApoGapo/Shutterstock.com

Figure 20-3. A range has both the cooktop and the oven built into one appliance for convenience and saving space.

turned on as it cools and off as it reaches its desired temperature. Additionally, when looking at multifamily dwellings, it is unlikely that every unit will be using its cooking appliance at full capacity at the same time. Therefore, the greater the number of appliances, the greater the demand factor applied.

Throughout this unit, the examples and problems will be based on performing a service calculation so it will be referred to as "service demand" for consistency. The calculation is performed the same way for feeders.

> **PRO TIP** — **Review NEC Structure**
>
> Calculations are performed when designing an electrical installation, which is addressed in *Chapter 2* of the *NEC*, **Figure 20-4**. More specifically, load calculations are addressed in *Article 220, Branch Circuit, Feeder, and Service Calculations*. *Section 220.55* applies to household cooking appliances.

Article 220.55 states that when calculating the load for household cooking appliances over 1 3/4 kW, *Table 220.55* will be used. Appliances under 1 3/4 kW, such as most countertop appliances, are considered small appliances and are not calculated using *Table 220.55*, **Figure 20-5**. It is also worth noting that the *NEC* states that kW is considered equivalent to kVA for these calculations.

The key to understanding *Table 220.55* is reading the table and the notes carefully. This starts with the heading on the table, **Figure 20-6**. It states that the table's demand factors apply to household cooking appliances rated over 1 3/4 kW. It goes on to say that Column C may be used in all cases except as otherwise permitted in *Note 3*. Column C will provide a possible answer in almost all scenarios.

Goodheart-Willcox Publisher

Figure 20-5. Small household cooking appliances are not included in *Table 220.55*.

CODE ARRANGEMENT

- Chapter 1: General
- Chapter 2: Wiring and Protection
- Chapter 3: Wiring Methods and Materials
- Chapter 4: Equipment for General Use

} Applies generally to all electrical installations

- Chapter 5: Special Occupancies
- Chapter 6: Special Equipment
- Chapter 7: Special Conditions

} Supplements or modifies chapters 1 through 7

- Chapter 8: Communications Systems

} Chapter 8 is not subject to the requirements of chapters 1 through 7 except where the requirements are specifically referenced in chapter 8

- Chapter 9: Tables

} Applicable as referenced

- Informative Annex A through Informative Annex J

} Informational only; not mandatory

Goodheart-Willcox Publisher

Figure 20-4. Having a thorough understanding of the layout of the *National Electrical Code* will make finding information much easier.

Table 220.55 Demand Factors and Loads for Household Electric Ranges, Wall-Mounted Ovens, Counter-Mounted Cooking Units, and Other Household Cooking Appliances over 1¾ kW Rating (Column C to be used in all cases except as otherwise permitted in Note 3.)

Number of Appliances	Demand Factor (%) (See Notes)		Column C Maximum Demand (kW) (See Notes) (Not over 12 kW Rating)
	Column A (Less than 3½ kW Rating)	Column B (3½ kW through 8¾ kW Rating)	
1	80	80	8
2	75	65	11
3	70	55	14
4	66	50	17
5	62	45	20
6	59	43	21
7	56	40	22
8	53	36	23
9	51	35	24
10	49	34	25
11	47	32	26
12	45	32	27
13	43	32	28
14	41	32	29
15	40	32	30
16	39	28	31
17	38	28	32
18	37	28	33
19	36	28	34
20	35	28	35
21	34	26	36
22	33	26	37
23	32	26	38
24	31	26	39
25	30	26	40
26–30	30	24	15 kW + 1 kW for each range
31–40	30	22	
41–50	30	20	25 kW + ¾ kW for each range
51–60	30	18	
61 and over	30	16	

Notes:
1. *Over 12 kW through 27 kW ranges all of same rating.* For ranges individually rated more than 12 kW but not more than 27 kW, the maximum demand in Column C shall be increased 5 percent for each additional kilowatt of rating or major fraction thereof by which the rating of individual ranges exceeds 12 kW.
2. *Over 8¾ kW through 27 kW ranges of unequal ratings.* For ranges individually rated more than 8¾ kW and of different ratings, but none exceeding 27 kW, an average value of rating shall be calculated by adding together the ratings of all ranges to obtain the total connected load (using 12 kW for any range rated less than 12 kW) and dividing by the total number of ranges. Then the maximum demand in Column C shall be increased 5 percent for each kilowatt or major fraction thereof by which this average value exceeds 12 kW.
3. *Over 1¾ kW through 8¾ kW.* In lieu of the method provided in Column C, adding the nameplate ratings of all household cooking appliances rated more than 1¾ kW but not more than 8¾ kW and multiplying the sum by the demand factors specified in Column A or Column B for the given number of appliances shall be permitted. Where the rating of cooking appliances falls under both Column A and Column B, the demand factors for each column shall be applied to the appliances for that column, and the results added together.
4. Calculating the branch-circuit load for one range in accordance with Table 220.55 shall be permitted.
5. The branch-circuit load for one wall-mounted oven or one counter-mounted cooking unit shall be the nameplate rating of the appliance.
6. The branch-circuit load for a counter-mounted cooking unit and not more than two wall-mounted ovens, all supplied from a single branch circuit and located in the same room, shall be calculated by adding the nameplate rating of the individual appliances and treating this total as equivalent to one range.
7. This table shall also apply to household cooking appliances rated over 1¾ kW and used in instructional programs.

Reproduced with permission of NFPA from NFPA 70, National Electrical Code, 2023 edition. Copyright © 2022, National Fire Protection Association. For a full copy of the NFPA 70, please go to www.nfpa.org

Figure 20-6. *Table 220.55* of the *NEC* is used when calculating household cooking equipment demand factors.

Table 220.55 is divided into four columns:
- The far-left column indicates the number of appliances.
- Column A gives a demand factor percentage for appliances less than 3 1/2 kW.
- Column B gives a demand factor percentage for appliances between 3 1/2 kW and 8 3/4 kW.
- Column C gives a demand quantity in kilowatts (kW). Note that columns A and B give a demand percentage, while column C gives a value in kW.

> **PRO TIP**
>
> ## Instructional Programs
>
> Table 220.55 of the NEC also applies to instructional programs, such as those found in high schools, **Figure 20-7**. Household electric ranges are used to provide a more realistic experience for students. Since the elements cycle to maintain the temperature, they only draw their maximum rated current for a short period of time. Applying a demand factor provides a more realistic load calculation.

B Brown/Shutterstock.com

Figure 20-7. Family and consumer science classrooms often contain the same appliances found in a standard home, such as cooktops, ovens, and washing machines.

Column C may be applied to all scenarios with the help of *Notes 1* and *2*, **Figure 20-6**. The top of Column C states that it applies to appliances with a maximum rating of 12 kW. If an appliance has a rating larger than 12 kW, then either *Note 1* or *2* will apply. This is discussed later in this unit. Assuming our appliance is rated 12 kW or less, we go to the appropriate row for the number of appliances, slide over to Column C, and that will give us the demand in kW. Column C is the most-used portion of *Table 220.55*; most single-family dwellings have a range that is rated at 12 kW or less.

EXAMPLE 20-1

The following problems will all use Column C to calculate the service demand. Find the given number of appliances on the left side of the table and slide over to Column C to find the demand.

Problem: What is the service demand for one 10.5 kW range?

Solution: For one appliance, Column C reads 8 kW. The service demand is 8 kW.

Problem: What is the service demand for four 11 kW ranges?

Solution: For four appliances, Column C reads 17 kW. The service demand is 17 kW.

Problem: What is the service demand for forty 12 kW ranges?

Solution: For 40 appliances, Column C reads 15 kW + 1 kW for each range.

$$15 \text{ kW} + 40 \text{ kW} = 55 \text{ kW}$$

The service demand is 55 kW.

In the situation where the cooking appliances are rated over 12 kW, either *Note 1* or *2* will apply. *Note 3* applies to cooking appliances with ratings between 1 3/4 kW and 8 3/4 kW. *Note 1* applies to a single range, or multiple ranges that have the same rating. For each kW the individual rating of the appliance exceeds 12, the number in Column C is increased 5%. If the rating of a range results in a major fraction, it is rounded up. If not, it is rounded down. A 13.6 kW range (round up to 14 kW) is two more than 12, so the number in Column C will be increased by 10%.

$$8 \text{ kW} \times 110\% = 8.8 \text{ kW}$$

EXAMPLE 20-2

The following problems will all use Column C to calculate the service demand. Since the value of the ranges exceeds 12 kW, *Note 1* must be applied.

Problem: What is the service demand for one 15 kW range?

Solution: For one appliance, Column C reads 8 kW.

$$15 \text{ kW} - 12 \text{ kW} = 3$$
$$3 \times 5\% = 15\%$$
$$8 \text{ kW} \times 115\% = 9.2 \text{ kW}$$

The service demand is 9.2 kW.

Problem: What is the service demand for three 13 kW ranges?

Solution: For three appliances, Column C reads 14 kW.

$$13 \text{ kW} - 12 \text{ kW} = 1$$
$$1 \times 5\% = 5\%$$
$$14 \text{ kW} \times 105\% = 14.7 \text{ kW}$$

The service demand is 14.7 kW.

Problem: What is the service demand for twenty 13.6 kW ranges?

Solution: For twenty appliances, Column C reads 35 kW. According to *Note 1*, 13.6 is a major fraction, so the average size is considered 14 kW.

$$14 \text{ kW} - 12 \text{ kW} = 2$$
$$2 \times 5\% = 10\%$$
$$35 \text{ kW} \times 110\% = 38.5 \text{ kW}$$

The service demand is 38.5 kW.

Note 2 is used when there are multiple appliances with different ratings. Start by finding the average size of the appliances. When finding the average value, 12 kW is used for any of the appliances that have a value that is less than 12 kW.

Even if a range has a nameplate rating of 10 kW, 12 kW is used when finding the average value. *Note 2* states that if the average value of the ranges ends in a decimal, a major fraction will round up. An average range value of 13.5 kW would round up to an average value of 14 kW, while an average range value of 13.4 would round down to 13 kW.

EXAMPLE 20-3

The following problem will use Column C to calculate the service demand. Since the value of the ranges exceed 12 kW, but all have different values, *Note 2* must be applied.

Problem: What is the service demand for three ranges: 10 kW, 12 kW, and 14 kW?

Solution: Column C for 3 appliances reads 14 kW.

$$12 \text{ kW} + 12 \text{ kW} + 14 \text{ kW} = 38 \text{ kW}$$
$$38 \text{ kW} \div 3 = 12.67 \text{ or } 13 \text{ kW (average value)}$$
$$13 \text{ kW} - 12 \text{ kW} = 1$$
$$1 \times 5\% = 5\%$$
$$14 \text{ kW} \times 105\% = 14.7 \text{ kW}$$

The service demand is 14.7 kW

Note 3 is typically used with appliances such as counter-mounted cooking units and wall-mounted ovens. It applies to appliances rated 1 3/4 kW through 8 3/4 kW. Instead of using Column C, it allows the use of Column A or Column B, whichever fits the nameplate rating of the appliance. Column C can still be used but may not give the lowest final number. Whichever calculation gives you the lowest answer should be used.

Column A is for appliances that have a rating less than 3 1/2 kW, while Column B is for appliances rated 3 1/2 kW through 8 3/4 kW, **Figure 20-6**. One of the things about *Table 220.55* that tends to cause confusion is that Column A and Column B are set up differently than Column C. Column C gives a demand number in kW, while Columns A and B give a demand factor percentage that must be multiplied by the nameplate rating of the appliance. For example, to find the demand factor for two ovens, each rated at 7 kW, Column B gives a demand factor percentage of 65% for two appliances.

$$2 \text{ appliances} \times 7 \text{ kW} \times 65\% = 9.1 \text{ kW}$$

Column C reads 11 kW for two appliances. Using the smaller of the two, 9.1 kW would be the final demand used in the calculation.

EXAMPLE 20-4

The following problems are permitted to use either Columns A and B or Column C. Both calculations should be performed and the lowest value in kW should be used. Note: Columns A and B give a multiplier to be applied to the value of the cooking appliances while Column C gives a value in kW.

Problem: What is the service demand for one 3-kW cooktop?

Solution: Column A for 1 appliance gives an 80% multiplier.

$$3 \text{ kW} \times 80\% = 2.4 \text{ kW}$$

Column C gives a value of 8 kW. Using the lower of the two values, the service demand is 2.4 kW.

Problem: What is the service demand for one 6 kW cooktop and two 7.5 kW ovens?

Solution: Find the total value of appliances by adding their ratings.

$$6 \text{ kW} + 7.5 \text{ kW} + 7.5 \text{ kW} = 21 \text{ kW}$$

For three appliances, Column B gives a 55% multiplier.

$$21 \text{ kW} \times 55\% = 11.55 \text{ kW}$$

Column C gives a value of 14 kW. Using the lower of the two values, the service demand is 11.55 kW.

Problem: What is the service demand for twenty 8-kW ovens?

Solution: Find the total value of appliances by adding their ratings.

$$20 \times 8 \text{ kW} = 160 \text{ kW}$$

For twenty appliances, Column B gives a 28% multiplier.

$$160 \text{ kW} \times 28\% = 44.8 \text{ kW}$$

Column C gives a value of 35 kW. Using the lower of the two values, the service demand is 35 kW.

When the appliance values fall under both Column A and B, the answer is found using the appropriate column for the appliance, but the row is based on the total number of appliances. So, if there are six total appliances: three 3 kW and three 6 kW, the appropriate column would be used for the respective appliance, but the row is based on six appliances, not three. Then the results are added together.

EXAMPLE 20-5

The following problem has appliances that fall under both Column A and B. Calculations are done for the appliances within the respective column and the results added together. Column C may also be used.

Problem: What is the service demand for three 3 kW cooktops and three 6 kW ovens?

Solution: Calculate the service demands for each type of appliance.

$$\text{Total of Column A appliances} = 3 \times 3 \text{ kW}$$
$$= 9 \text{ kW}$$

For six appliances, Column A gives a 59% multiplier.

$$9 \text{ kW} \times 59\% = 5.31 \text{ kW}$$
$$\text{Total of Column B appliances} = 3 \times 6 \text{ kW}$$
$$= 18 \text{ kW}$$

For six appliances, Column B gives a 43% multiplier.

$$18 \text{ kW} \times 43\% = 7.74 \text{ kW}$$

Add totals together to find the total service demand.

$$5.31 \text{ kW} + 7.74 \text{ kW} = 13.05 \text{ kW}$$

Column C gives a value of 21 kW. Using the lower of the two values, the service demand is 13.05 kW.

> **PRO TIP** — **Commercial Cooking Equipment Calculations**
>
> This unit is dedicated to dwelling unit calculations. Commercial cooking equipment calculations for non-dwelling units, such as restaurants, are calculated using *NEC Table 220.56*.

Note 4 takes a different direction than the rest of *Table 220.55*, **Figure 20-6**. While the majority of the table is used for service and feeder calculations, *Note 4* is used when sizing a branch circuit. It states that the branch circuit for one range is calculated according to the table, meaning a demand factor may be applied. This situation often results in using Column C for one appliance. We can apply a demand factor to a range because it is unlikely that the appliance will draw its full nameplate rating for more than a short period of time. This allows us to size the conductors more appropriately for the actual load and save on wire cost. For example, a 12 kW range will require a branch circuit that is rated for 8 kW.

The rating reduction allowed for ranges is not allowed for ovens and cooktops. *Note 5* states that the branch circuit load for ovens and cooktops shall be the nameplate rating. With a range, there are at least four burners and two heating elements all cycling on and off at different intervals. With a stand-alone oven or cooktop, there is a greater chance it will draw the full nameplate rating for an extended period of time. For this reason, we need to ensure the branch circuit is sized to handle the full nameplate rating of the appliance. For example, a 5 kW cooktop will require a branch circuit rated for at least 5 kW.

Note 6 applies to a cooktop and up to two ovens, all fed from the same circuit. In this situation, the nameplate ratings of all the appliances are added together and treated as one range. Column C can then be used to size the branch circuit. Before planning to combine a cooktop and oven onto the same branch circuit, be sure to check the appliance manufacture requirements/instructions. Many cooking appliance manufactures require a dedicated (individual) branch circuit for their appliances.

> **EXAMPLE 20-6**
>
> When a single branch circuit feeds multiple cooking appliances, *Note 4* allows us to add the values together and treat them as a single appliance. Column C for one appliance will be used to find the branch circuit demand.
>
> **Problem:** What is the branch circuit demand for one 5 kW cooktop and one 7 kW oven?
>
> **Solution:** Add the appliance values together and treat as one single appliance.
>
> $$5 \text{ kW} + 7 \text{ kW} = 12 \text{ kW}$$
>
> For one appliance, Column C gives a value of 8 kW. Using the lower of the two values, the branch circuit demand is 8 kW.

Household Cooking Appliance Neutral Calculation

Most substantial cooking appliances, such as cooktops, ovens, and ranges, are 120/240 loads. That means they use both 120-volt and 240-volt power. Typically, the heating portions of the appliance operate at 240 V, while the controls and light operate on 120 V. Since the high-current portions of the appliances operate at 240 volts, the neutral will carry far less current than the ungrounded/hot conductors. For this reason, *Section 220.61(B)* allows an additional demand factor of 70% to the calculation of the neutral for cooktops, ovens, and ranges when performing service or feeder calculations.

> **EXAMPLE 20-7**
>
> *NEC 220.61(B)(1)* allows a residential cooking appliance neutral to receive an additional 70% demand factor. This is in addition to any demand factors applied by *NEC Table 220.55*.
>
> **Problem:** What is the service neutral demand for one 12 kW range?
>
> **Solution:** For one appliance, Column C gives a value of 8 kW. *Section 220.61(B)* allows an additional demand factor of 70%.
>
> $$8 \text{ kW} \times 70\% = 5.6 \text{ kW}$$
>
> The service neutral demand for the range is 5.6 kW.

Sizing Branch Circuit Conductors

Once the branch circuit demand has been determined, we are ready to size the conductors. There are a few requirements that must be considered. The first is found in *Section 210.19(C)*, which provides branch circuit information for ranges, **Figure 20-8**. It states that the branch circuit conductors must not have an ampacity less than the rating of the branch circuit, and not less than the maximum load to be served. It goes on to state there must be a minimum circuit size of 40 amps for a range that is 8 3/4 KW or greater. *Exception No. 2* goes on to say that if the branch circuit was calculated according to Column C, it is permissible to derate the neutral to 70% of the ungrounded branch circuit conductors, but it shall not be smaller than a #10 copper. Although

(C) Household Ranges and Cooking Appliances. Branch-circuit conductors supplying household ranges, wall-mounted ovens, counter-mounted cooking units, and other household cooking appliances shall have an ampacity not less than the rating of the branch circuit and not less than the maximum load to be served. For ranges of 8¾ kW or more rating, the minimum branch-circuit rating shall be 40 amperes.

Exception No. 1: Conductors tapped from a branch circuit not exceeding 50 amperes supplying electric ranges, wall-mounted electric ovens, and counter-mounted electric cooking units shall have an ampacity of not less than 20 amperes and shall be sufficient for the load to be served. These tap conductors include any conductors that are a part of the leads supplied with the appliance that are smaller than the branch-circuit conductors. The taps shall not be longer than necessary for servicing the appliance.

Exception No. 2: The neutral conductor of a 3-wire branch circuit supplying a household electric range, a wall-mounted oven, or a counter-mounted cooking unit shall be permitted to be smaller than the ungrounded conductors where the maximum demand of a range of 8¾ kW or more rating has been calculated according to Column C of Table 220.55, but such conductor shall have an ampacity of not less than 70 percent of the branch-circuit rating and shall not be smaller than 10 AWG.

Reproduced with permission of NFPA from NFPA 70, National Electrical Code, 2023 edition. Copyright © 2022, National Fire Protection Association. For a full copy of the NFPA 70, please go to www.nfpa.org

Figure 20-8. The first requirement to be considered when sizing branch circuit conductors is in *Section 210.19(C)*. This section provides branch circuit information for ranges along with some exceptions.

derating the neutral is possible, it would be unusual to have a neutral smaller than the ungrounded conductors. Most dwellings are wired with cables, which are made with all three current-carrying conductors of the same size.

The second requirement to be considered when sizing conductors is found in *Section 110.14C, Temperature Limitations*, **Figure 20-9**. It states that unless we can prove that the terminations are listed for a higher temperature, we must use the lowest rating of any component in the circuit (weakest link). The *NEC*'s general rule states that for circuits 100 A or less, the 60°C column of *Table 310.16* must be used for the final conductor ampacity. If you can prove that the terminations as well as the conductors have a temp rating higher than 60°C, that temperature may be used to find the conductor ampacity.

To size the conductors, *Table 310.16* is used, **Figure 20-10**. The left half of the table is based on copper conductors,

(C) Temperature Limitations. The temperature rating associated with the ampacity of a conductor shall be selected and coordinated so as not to exceed the lowest temperature rating of any connected termination, conductor, or device. Conductors with temperature ratings higher than specified for terminations shall be permitted to be used for ampacity adjustment, correction, or both.

Reproduced with permission of NFPA from NFPA 70, National Electrical Code, 2023 edition. Copyright © 2022, National Fire Protection Association. For a full copy of the NFPA 70, please go to www.nfpa.org

Figure 20-9. The second requirement to be considered when sizing branch circuit conductors is in *Section 110.14(C)*, which provides information on temperature limitations.

Table 310.16 Ampacities of Insulated Conductors with Not More Than Three Current-Carrying Conductors in Raceway, Cable, or Earth (Directly Buried)

	Temperature Rating of Conductor [See Table 310.4(1)]						
	60°C (140°F)	75°C (167°F)	90°C (194°F)	60°C (140°F)	75°C (167°F)	90°C (194°F)	
	Types TW, UF	Types RHW, THHW, THW, THWN, XHHW, XHWN, USE, ZW	Types TBS, SA, SIS, FEP, FEPB, MI, PFA, RHH, RHW-2, THHN, THHW, THW-2, THWN-2, USE-2, XHH, XHHW, XHHW-2, XHWN, XHWN-2, XHHN, Z, ZW-2	Types TW, UF	Types RHW, THHW, THW, THWN, XHHW, XHWN, USE	Types TBS, SA, SIS, THHN, THHW, THW-2, THWN-2, RHH, RHW-2, USE-2, XHH, XHHW, XHHW-2, XHWN, XHWN-2, XHHN	
Size AWG or kcmil	COPPER			ALUMINUM OR COPPER-CLAD ALUMINUM			Size AWG or kcmil
18*	—	—	14	—	—	—	—
16*	—	—	18	—	—	—	—
14*	15	20	25	—	—	—	—
12*	20	25	30	15	20	25	12*
10*	30	35	40	25	30	35	10*
8	40	50	55	35	40	45	8
6	55	65	75	40	50	55	6
4	70	85	95	55	65	75	4
3	85	100	115	65	75	85	3
2	95	115	130	75	90	100	2
1	110	130	145	85	100	115	1
1/0	125	150	170	100	120	135	1/0
2/0	145	175	195	115	135	150	2/0

Reproduced with permission of NFPA from NFPA 70, National Electrical Code, 2023 edition. Copyright © 2022, National Fire Protection Association. For a full copy of the NFPA 70, please go to www.nfpa.org

Figure 20-10. *Table 310.16* provides information required to size conductors.

and the right half of the table is for aluminum conductors. While copper nonmetallic sheathed cable is often used for branch circuit wiring in homes, it would not be unusual to save money by installing an aluminum SE (service entrance) cable to feed the range.

When sizing conductors, the first step is to determine the ampacity. To determine the ampacity, divide the branch circuit demand by the voltage. For example, 8000 W divided by 240 V is 33.34 A. The next step is to use *Table 310.16* to determine the minimum size conductor that can safely carry the expected current. First, we will determine the minimum size nonmetallic sheathed cable (copper) using 33.34 A as the minimum ampacity. The 60°C copper column states that an 8 AWG copper cable is capable of carrying 40 A. This can safely handle the 33.34 A and meets the minimum circuit size specified in *Section 210.19(C)*. An 8 AWG nonmetallic sheathed cable is permitted to feed the range.

Performing the same calculation for an aluminum SE cable is a bit more complicated. Use the 60°C column on the right side of the table to find the conductor with an ampacity of at least 33.34 A. The table states that an 8 AWG aluminum can carry 35 A. This is where it gets a little tricky. Remember that *Section 210.19(C)* gave a minimum circuit ampacity of 40 A; we must use a 6 AWG aluminum.

Branch Circuit Overcurrent Protective Device (OCPD)

Overcurrent protective devices for ranges are sized according to *Section 240.4(D)*. It states that unless the range ampacity matches up with a standard size overprotective device, the next standard size above the ampacity can be used. If the ampacity of the conductor is 40 A as in the previous example, 40 A is a standard size so the OCPD will be a 40 A. Had the ampacity been 46 A, the next standard size would have been a 50-amp OCPD.

EXAMPLE 20-8

Problem: Calculate the minimum size overcurrent protective device required for a 12 kW range fed with NM cable.

Solution:

1. Calculate the branch circuit demand according to *Table 220.55*. For one appliance, Column C gives a value of 8 kW.
2. Determine the minimum circuit ampacity by dividing the branch circuit demand by the voltage.
    $$8000 \text{ W} \div 240 \text{ V} = 33.33 \text{ A}$$
3. Determine the conductor type, size, and maximum ampacity using the 60°C column of *Table 310.16*.
    $$40 \text{ A} = 8 \text{ AWG Copper}$$
4. Using *Section 240.4(B)*, we determine the overcurrent protective device rating is 40A. The minimum size overcurrent protective device required is 40 A.

Summary

- A range is an appliance that contains both an oven and a cooktop in one unit.
- A demand factor takes into consideration the fact that appliances such as ranges will not draw the full nameplate current for an extended period due to heating elements cycling on and off.
- *NEC Table 220.55* applies demand factors to household cooking appliances over 1 3/4 kW.
- Column A of *Table 220.55* applies to cooking equipment less than 3 1/2 kW and gives a demand factor percentage to be multiplied by the appliance nameplate rating.
- Column B of *Table 220.55* applies to cooking equipment 3 1/2 kW through 8 3/4 kW and gives a demand factor percentage to be multiplied by the appliance nameplate rating.
- Column C of *Table 220.55* applies to cooking equipment over 8 kW and gives a demand factor in kW.
- *Note 4* of *Table 220.55* applies to branch circuit sizing.
- A demand factor may be applied when sizing the branch circuit conductors for a range.
- When sizing the branch circuit conductors for an oven or cooktop, the nameplate rating must be used.

Unit 20 Review

Name _____ Date _____ Class _____

Know and Understand

Answer the following questions based on information in this unit.

1. A _____ is a counter-mounted cooking unit and oven built in one unit.
   A. cooktop
   B. convection oven
   C. microwave oven
   D. range

2. Kilowatts and _____ are considered equivalent when performing range calculations.
   A. amperes
   B. volts
   C. kilovolt-amperes
   D. ohms

3. *NEC Table 220.55* applies to household kitchen equipment _____ and greater.
   A. 1 3/4 kW
   B. 3 1/2 kW
   C. 8 3/4 kW
   D. 12 kW

4. _____ of *NEC Table 220.55* gives the maximum demand in kW.
   A. Column A
   B. Column B
   C. Column C
   D. Column D

5. _____ of *Table 220.55* is used to determine cooking appliance branch circuit load.
   A. *Note 1*
   B. *Note 2*
   C. *Note 3*
   D. *Note 4*

6. The values in Column C of *Table 220.55* are for kitchen appliances up to _____.
   A. 1 3/4 kW
   B. 3 1/2 kW
   C. 8 3/4 kW
   D. 12 kW

7. When calculating the demand for a 14 kW range, the value found in Column C of *Table 220.55* shall be increased by _____.
   A. 5%
   B. 10%
   C. 15%
   D. 20%

8. The branch circuit load for a 5 kW cooktop is _____.
   A. 4 kW
   B. 5 kW
   C. 6.25 kW
   D. 8 kW

9. It is permitted to apply a demand factor of _____ to the neutral conductor of a range.
   A. 70%
   B. 80%
   C. 100%
   D. 125%

10. _____ shall be permitted to be used when calculating the demand factor for non-dwelling kitchen equipment.
    A. *Table 220.54*
    B. *Table 220.55*
    C. *Table 220.56*
    D. *Table 220.84*

Apply and Analyze

Answer the following questions using a copy of the National Electrical Code. Identify the section or subsection where the answer is found.

1. What is the service demand for a 15 kW range?
   A. 8 kW
   B. 8.4 kW
   C. 8.8 kW
   D. 9.2 kW

2. What is the service demand for a 9.4 kW range?
   A. 8 kW
   B. 8.4 kW
   C. 8.8 kW
   D. 9.2 kW

3. What is the service demand for fifteen 13 kW ranges?
   A. 30 kW
   B. 31.5 kW
   C. 36.5 kW
   D. 40 kW

4. What is the service demand for a 7 kW oven?
   A. 5 kW
   B. 5.6 kW
   C. 6.2 kW
   D. 7 kW

5. What is the service demand for six 3 kW cooktops?
   A. 8 kW
   B. 9.2 kW
   C. 10.62 kW
   D. 12 kW

6. What is the service demand for two 14 kW ranges and three 15 kW ranges?
   A. 20 kW
   B. 22 kW
   C. 23 kW
   D. 25 kW

7. What is the service neutral demand for a 12 kW range?
   A. 5 kW
   B. 5.6 kW
   C. 6.2 kW
   D. 8 kW

8. What is the branch circuit load for a 6 kW cooktop?
   A. 4.8 kW
   B. 5 kW
   C. 5.4 kW
   D. 6 kW

9. What is the minimum size branch circuit conductor for a 11 kW range? (Copper nonmetallic sheathed cable)
   A. 6 AWG
   B. 8 AWG
   C. 10 AWG
   D. 12 AWG

10. What is the minimum size OCPD (overcurrent protective device) for a 12 kW range?
    A. 40 amperes
    B. 45 amperes
    C. 50 amperes
    D. 60 amperes

Name _____ Date _____ Class _____

Critical Thinking

Complete the following service demand calculations using a copy of the National Electrical Code. *Identify the section or subsection used to find the answer.*

1. What is the service demand for a 13.2 kW range?

   _____

   NEC _____

2. What is the service demand for a 9 kW range?

   _____

   NEC _____

3. What is the service demand for a 15.8 kW range?

   _____

   NEC _____

4. What is the service demand for twelve 13.7 kW ranges?

   _____

   NEC _____

5. What is the service demand for thirty 14 kW ranges?

   _____

   NEC _____

6. What is the service demand for sixty 14 kW ranges?

   _____

   NEC _____

7. What is the service demand for a 4 kW cooktop?

   _____

   NEC _____

8. What is the service demand for two 5 kW cooktops?

   _____

   NEC _____

9. What is the service demand for twenty-five 3 kW cooktops and twenty-five 6 kW ovens?

   _____

   NEC _____

10. What is the service demand for fifteen 3.5 kW ranges?

    _____

    NEC _____

11. What is the service demand for the following?
    - 5 – 9 kW ranges
    - 4 – 12 kW ranges
    - 7 – 15.6 kW ranges

    _____

    NEC _____

12. What is the service neutral demand for six 12 kW ranges?

    _____

    NEC _____

13. What is the service neutral demand for a 3 kW cooktop?

    _____

    NEC _____

14. What is the service neutral demand for the following?
    - (12) 11 kW ranges
    - (12) 12 kW ranges

    _____

    NEC _____

Part 2   Application: Advanced NEC Topics

15. What is the branch circuit load for a 5 kW cooktop and a 6 kW oven that are in the same room and fed from the same circuit?

    _____

    NEC _____

16. What is the branch circuit load for a 10 kW range?

    _____

    NEC _____

17. What is the branch circuit load for a 5 kW oven?

    _____

    NEC _____

18. What is the minimum size branch circuit conductor for a 14 kW range? (Aluminum SER cable)

    _____

    NEC _____

19. What is the minimum size branch circuit conductor for a 7.5 kW double oven? (Copper nonmetallic sheathed cable)

    _____

    NEC _____

20. What is the minimum size OCPD for a 3.5 kW cooktop?

    _____

    NEC _____

Complete the following service demand calculations using a copy of the National Electrical Code. *Identify the section or subsection used to find the answer.*

21. What is the service demand for an 11 kW range?

    _____

    NEC _____

22. What is the service demand for three 12 kW ranges?

    _____

    NEC _____

23. What is the service demand for eighteen 8 kW ranges?

    _____

    NEC _____

24. What is the service demand for fifty 12 kW ranges?

    _____

    NEC _____

25. What is the service demand for one hundred 13 kW ranges?

    _____

    NEC _____

26. What is the service demand for a 2 kW cooktop?

    _____

    NEC _____

27. What is the service demand for five 3.5 kW cooktops?

    _____

    NEC _____

28. What is the service demand for twenty 7 kW cooktops?

    _____

    NEC _____

29. What is the service demand for the following?
    - 11 kW range
    - 13 kW range
    - 16 kW range

    _____

    NEC _____

30. What is the service demand for the following?
    - 12 - 7 kW ovens
    - 12 - 6 kW cooktops

    _____

    NEC _____

Part 2   Application: Advanced NEC Topics

Complete the following service demand calculations using a copy of the National Electrical Code. *Identify the section or subsection used to find the answer.*

31. What is the service demand for the following?
    - 6 - 10 kW ranges
    - 8 - 11 kW ranges
    - 12 - 14 kW ranges

    _____

    NEC _____

32. What is the service neutral demand for twenty-four 12.6 kW ranges?

    _____

    NEC _____

33. What is the service neutral demand for the following?
    - (20) 9 kW ranges
    - (15) 13 kW ranges
    - (15) 14 kW ranges

    _____

    NEC _____

34. What is the branch circuit load for a 7 kW oven?

    _____

    NEC _____

35. What is the branch circuit load for a 13 kW range?

    _____

    NEC _____

36. What is the branch circuit load for a 3.5 kW cooktop and a 7 kW double oven that are in the same room and fed from the same circuit?

    _____

    NEC _____

37. What is the minimum size branch circuit neutral conductor for 12 kW range? (Copper THHN)

    _____

    NEC _____

38. What is the minimum size branch circuit conductor for a 5 kW cooktop? (Copper nonmetallic-sheathed cable)

    _____

    NEC _____

39. What is the minimum size OCPD for a 14.8 kW range?

    _____

    NEC _____

40. What is the minimum size OCPD for a 5.5 kW double oven?

    _____

    NEC _____

UNIT 21 Dwelling Service Calculations

Sashkin/Shutterstock.com

KEY TERMS

demand factor
general unit load
inductive load
load
noncoincident load
optional method (service calculation)
resistive load
standard method (service calculation)

OBJECTIVES

After completing this chapter, you will be able to:
- Determine the general unit load for a dwelling unit.
- Determine which dwelling unit loads are included in a service calculation.
- Apply the demand factors found in *Article 220* to a service calculation.
- Determine the minimum size service by using the standard service calculation.
- Determine the minimum size service by using the optional service calculation.
- Determine the minimum size service disconnect.
- Determine the minimum size service entrance conductors.

Introduction

The purpose of a service calculation is to determine the minimum size of electric service required for an installation. Although this unit will focus on and refer to service calculations, feeder calculations are performed in the same manner.

Article 220 of the *National Electrical Code* contains the requirements for these calculations. Although it is considered good practice to ensure there is extra capacity for future expansion, the *NEC* only includes what is present at the time of installation when performing the calculation, **Figure 21-1**.

Goodheart-Willcox Publisher

Figure 21-1. This meter assembly has the utility meter on the top of the enclosure and the main service disconnect on the bottom. The disconnect means has overcurrent protection rated at 200 amperes, which means this home has a 200-ampere service.

Figure 21-2. The parts of *Article 220*.

The *NEC* provides two methods for performing a service calculation on a dwelling, the standard method and the optional method. This unit will introduce a systematic approach to performing both calculation methods.

Article 220

Article 220, Branch-Circuit, Feeder, and Service Load Calculations contains the majority of the service calculation information. *Article 220* is broken into five parts, **Figure 21-2**.

- *Part I, General*, is general and gives general information that applies to calculations.
- *Part II, Branch-Circuit Load Calculations* provides values for branch circuit calculations. These values will also be used by *Part III* when performing service or feeder calculations.
- *Part III, Feeder and Service Load Calculations* can be used to calculate the service for all occupancies except for farms. It sends us back to *Part II* to acquire some of the values necessary to complete the calculation.
- *Part IV, Optional Feeder and Service Load Calculations* provides an optional method of performing a service calculation for dwelling units, schools, and restaurants.
- *Part V, Farm Load Calculations* uses values and calculations from *Parts II* and *III* but provides additional demand factors specific to farms.
- *Part V, Health Care Facilities* uses values and calculations from *Part II* and *III* but provides additional demand factors specific to health care facilities.

General Requirements

Before beginning the actual calculations, we must understand the general requirements that will be applied to service calculations. This is covered in *Part I* of *Article 220*.

Values

To have accurate and consistent calculations, *Section 220.5* gives us guidance on voltages, rounding, and power values.

Voltages

It is common to hear people refer to a 120-volt source as 110 volts or 115 volts, and a 240-volt source as 220 volts or 230 volts. They could be referring to nominal voltage, but most of the time, they are simply repeating voltage values/terminology they have heard others use. Using a voltage that differs from the actual voltage in a calculation will result in current and power values that are slightly off.

To ensure uniform interpretation and accurate calculations, *Section 220.5(A)* lists the voltages to use when applying *Article 220*. There are several voltages listed that correspond with the common nominal voltage systems that are provided by an electric utility. Since this unit is focused on dwelling units, the nominal voltage system that will be used throughout is the 120/240-volt single-phase system.

When performing a calculation for large multifamily dwelling units, such as apartments, you would typically encounter a 120/208-volt wye-connected, 3-phase service that feeds the building.

Rounding

Section 220.5(B) gives a rule for dealing with a fraction of an ampere. It states that a value of 0.5 or larger is rounded up to the nearest whole number, while decimals that are less than 0.5 are rounded down to the nearest whole number. With respect to service calculations, this rounding is not implemented until the very end of the calculation when the minimum total ampacity of the service has been determined.

Watts, Volt-Amperes, and Horsepower

Electrical loads may be rated in watts or volt-amperes. A load is a component in an electrical system that uses electricity to perform a function. The *NEC* seems to interchange the terms watts and volt-amperes, which can cause confusion. Understanding the difference between true power (watts) and reactive power (volt-amperes) begins with differentiating between resistive and reactive loads.

A resistive load has a current flow in phase with the applied voltage and is rated in watts. Resistive loads include heating elements and incandescent lighting. A reactive load has a current flow that is either leading or lagging the applied voltage and is rated in volt-amperes. Reactive loads may be inductive, such as a transformer or motor, or capacitive, which are typically found only in a capacitor.

Appliances that have a combination of resistive and reactive loads are usually rated in volt-amperes due to the current being out of phase with the applied voltage. Examples include dishwashers, washing machines, and dryers, as they have resistive heating loads along with inductive motor loads.

The *NEC* uses both watts and volt-amperes when performing calculations. For the purpose of calculating the minimum size service or feeder, watts and volt-amperes can be interchanged without impacting calculation accuracy.

PRO TIP — Motor Values

Motors are often rated in horsepower. Although a value of 746 watts is generally thought of as being equivalent to one horsepower, that is not how the *NEC* calculates current values associated with motors. 746 watts per horsepower does not take into consideration electrical and mechanical losses that are associated with electric motors, such as heat, friction, and power factor. *Articles 430* (for motors) and *440* (for air-conditioning units) describe how to determine the minimum ampacity associated with these loads.

Sample Problems in the Annex

The *NEC* provides a few sample calculations in *Annex D*. The examples can be helpful to give a framework for setting up a calculation, but they aren't very thorough, **Figure 21-3**. Many potential questions are left unanswered.

Setting Up a Service Calculation

It is best to use a systematic approach to service calculations. Compartmentalizing each aspect of the calculation will reduce the chance of error by keeping the information well organized. It is extremely important to show your work. There is a practice sheet in the appendix of this text that can be used as a guide when performing these calculations.

Example D1 (a) One-Family Dwelling

The dwelling has a floor area of 1500 ft^2, exclusive of an unfinished cellar not adaptable for future use, unfinished attic, and open porches. Appliances are a 12-kW range and a 5.5-kW, 240-V dryer. Assume range and dryer kW ratings equivalent to kVA ratings in accordance with 220.54 and 220.55.

Calculated Load (see 220.40)

General Lighting Load 1500 ft^2 at 3 VA/ft^2 = 4500 VA

Minimum Number of Branch Circuits Required [see 210.11(A)]

General Lighting Load: 4500 VA ÷ 120 V = 38 A

    This requires three 15-A, 2-wire or two 20-A, 2-wire circuits.

    Small-Appliance Load: Two 2-wire, 20-A circuits [see 210.11(C)(1)]

    Laundry Load: One 2-wire, 20-A circuit [see 210.11(C)(2)]

    Bathroom Branch Circuit: One 2-wire, 20-A circuit (no additional load calculation is required for this circuit) [see 210.11(C)(3)]

Reproduced with permission of NFPA from NFPA 70, National Electrical Code, 2023 edition. Copyright © 2022, National Fire Protection Association. For a full copy of the NFPA 70, please go to www.nfpa.org

Figure 21-3. *Informative Annex D* in the back of the *NEC* is titled *Examples*. It has several service calculation examples as well as many other types of sample calculations.

Although it is a good idea to complete service calculations without the use of a guide sheet, using one will help you build an organized routine that will eventually lead to less of a need for the worksheet.

Demand Factors

Demand factors are used in various parts of a service calculation. The *NEC* defines a ***demand factor*** as the ratio of the maximum demand of a system, or part of a system, to the total connected load of a system or the part of the system under consideration. Demand factors take into consideration the fact that not everything will be consuming energy at the same time. Although a dwelling may have twenty lights, three TVs, and two garage door openers, they will not all be used at the same time. Demand factors give a more realistic picture as to the amount of energy that will be used.

Dwelling Scenario

The following dwelling unit scenario will be used for the sample calculations within this unit. It includes many of the common items found in homes to give a realistic example of a service calculation. It will be solved using both the standard method as well as the optional method to highlight the differences in the two calculations.

A two-story house with a basement and an attached garage. House area is 1440 square feet per floor. The garage area is 1080 square feet. The dwelling contains the following circuits and appliances:

- (3) small-appliance branch circuits
- (1) laundry branch circuit
- 1200-volt-ampere dishwasher (120 V)
- 1/2 hp garbage disposer (120 V)
- 1500-volt-ampere microwave above the range (120 V)
- (2) 1/2 hp garage door openers (120 V)
- 1/3 hp sump pump (120 V)
- 1000-volt-ampere gas furnace (120 V)
- 4500-watt electric water heater (240 V)
- 4500-volt-ampere dryer (120/240 V)
- 11.6-kilowatt range (120/240 V)
- 7632-volt-ampere (31.8 amperes) air conditioner (240 V)
- 9-kilowatt car charger (120/240 V)

Standard Method

The standard method of performing a service calculation can be used in all applications. It looks at the loads from the worst-case scenario to ensure that the service size will be sufficient. In the case of residential calculations, the standard method usually results in a larger number than the optional method.

The standard method begins in *Section 220.40*, where it states that the service shall not be less than the sum of the branch circuit loads found in *Part II* after applying the demand factors of *Part III*. To simplify the calculation, this textbook breaks the standard method of performing a service calculation into eight steps. Each step of the calculation is completed independently with the value found in each step being added together to determine the minimum calculated load, **Figure 21-4**.

Step 1–General Lighting Load

The first step of the service calculation is referred to as the general lighting portion of the calculation. It takes the general unit, small-appliance branch circuit, and laundry branch circuit loads and applies a demand factor to all three.

General Unit

When calculating the general unit load for dwellings, *Section 220.41* states that it includes all general illuminations as well as general-purpose receptacles. This includes all the general-use receptacles in habitable rooms, hallways, bathrooms, the garage, and outdoors. The small-appliance and laundry branch circuits, as well as built-in appliances such as dishwashers and garage door openers, are not included in the general unit load.

To find the general unit load, multiply the useable area of the dwelling unit by 3 volt-amperes per square foot. This gives a value that takes into consideration an average use of lighting and portable plug-in appliances in a typical home.

When finding the useable area of the dwelling unit, outside dimensions are used, *Section 220.5(C)*. Starting with the 2023 edition of the *NEC*, the garage is included in this calculation. Unfinished spaces that are adaptable for future use as a habitable room are also included in the useable square footage. Open porches, decks, crawl spaces, and areas not adaptable for future use are omitted from this calculation.

A two-story home with a basement has three levels that need to be included. If each level is 1000 square feet, the total useable area of the house is 3000 square feet. To find the general lighting load, multiply the total area of 3000 square feet by 3 volt-amperes. The resulting 9000 volt-amperes is the general unit load.

Small-Appliance Branch Circuits

Section 220.52(A) covers small-appliance branch circuit calculations. Each small-appliance branch circuit is calculated at 1500 volt-amperes. *Section 210.11(C)(1)* requires a minimum of two small-appliance branch circuits, but many homes will have more than two. In the case of a dwelling with three small-appliance branch circuits, multiply the three small-appliance circuits by 1500 volt-amperes to get the total small-appliance branch circuit load of 4500 volt-amperes.

Service Calculation "Standard Method" Dwelling Unit (Single Family)			Ungrounded		Neutral
Step 1	**General Lighting, Small Appliance, Laundry**				
Dwelling area "sq ft"	_____ × 3 VA =	_____			
Small appliance	_____ × 1500 VA =	_____			
Laundry	_____ × 1500 VA =	_____			
	Total	_____			
First 3000 at 100%		_____			
3001 to 120,000 × 35%	_____ × 0.35	_____			
Remainder over 120,000 × 25%	_____ × 0.25				
Step 2	**Fixed Appliances**				
Appliance	Ungrounded	Neutral			
	_____	_____			
	_____	_____			
	_____	_____			
	_____	_____			
	_____	_____			
	_____	_____			
	_____	_____			
Total	_____	_____			
Four or more × 0.75	_____	_____			
Step 3	**Dryer**				
5000 VA or the nameplate rating - neutral 70% of ungrounded		_____			
Step 4	**Range**				
Range calculation - neutral 70% of ungrounded		_____			
Step 5	**Heat/Air Conditioning**				
Include the larger of the two/noncoincidence load					
	Ungrounded	Neutral			
	_____	_____			
Step 6	**Electric Vehicle Supply Equipment**				
7200 VA or the nameplate rating					
Step 7	**Other Loads**				
	Ungrounded	Neutral			
	_____	_____			
	_____	_____			
Total	_____				
Step 8	**Largest Motor**				
	Ungrounded	Neutral			
Largest motor × 0.25	_____	_____			
Total in Volt-Amperes		Total	_____		_____
Total Amperes = Total VA/240 V		Total	_____		_____

Goodheart-Willcox Publisher

Figure 21-4. Template for performing a dwelling unit service calculation using the standard method.

Laundry Branch Circuit

The laundry branch circuit is calculated in the same manner as the small appliance branch circuits. *Section 220.52(B)* has us multiply the 120-volt laundry circuit(s) that are present by 1500 VA per circuit. Most residential installations will only have one laundry circuit, but some larger homes and multifamily dwellings will have multiple laundry rooms. In this case, each laundry circuit present is calculated at 1500 volt-amperes.

General Lighting Demand Factor

The general unit, small-appliance branch circuit and the laundry branch circuit loads are added together to get the total general lighting load. Permission to combine the small appliance and laundry loads with the general lighting load is found in *Section 220.52(A)* and *(B)*.

Table 220.45 provides the demand factor that is applied to the general lighting load, **Figure 21-5**. The first 3000 volt-amperes are taken at 100% with the remaining VA up to 120,000 being taken at 35%. It is rare for a single-family dwelling unit to have a general lighting value over 120,000 volt-amperes, but it is possible for some large multifamily dwelling units. In the case of a multifamily dwelling where the combined general lighting load is more than 120,000 volt-amperes, the amount over 120,000 is taken at 25%.

Since the general unit load, small-appliance, and laundry branch circuits are all 120-volt circuits with a neutral conductor, the value that is found in this calculation applies to both the ungrounded as well as the grounded conductors, **Figure 21-6**.

PRO TIP — Appliance Voltage Ratings

It can be hard to know the type of connection an appliance will have if only the power value is given. For that reason, in this unit the voltages of appliances are listed after their power ratings.

If the appliance rating is followed by (120 V), it is a 120-volt appliance with a ungrounded and neutral connection. That means there will be current flow on the ungrounded conductor as well as the neutral conductor. The rating of the appliance in watts or volt-amperes is included in the ungrounded column as well as the neutral column of the standard calculation template.

An appliance rating that is followed by (240 V) is a 240-volt appliance with a connection to two ungrounded conductors. This appliance does not have a neutral connection, so the appliance value in watts or volt-amperes is only included in the ungrounded column of the standard calculation template. The neutral space can be left blank or a zero entered for this appliance.

An appliance rating that is followed by (120/240 V) has a connection to both ungrounded conductors as well as the neutral conductor. The appliance's rated value will be included in both the ungrounded as well as the neutral column of the standard calculation template. Some appliances, such as dryers and ranges, will have the neutral value multiplied by 70% as permitted by *Section 220.61(B)*.

Table 220.45 Lighting Load Demand Factors

Type of Occupancy	Portion of Lighting Load to Which Demand Factor Applies (Volt-Amperes)	Demand Factor (%)
Dwelling units	First 3000 at	100
	From 3001 to 120,000 at	35
	Remainder over 120,000 at	25
Hotels and motels, including apartment houses without provision for cooking by tenants*	First 20,000 or less at	60
	From 20,001 to 100,000 at	50
	Remainder over 100,000 at	35
Warehouses (storage)	First 12,500 or less at	100
	Remainder over 12,500 at	50
All others	Total volt-amperes	100

*The demand factors of this table shall not apply to the calculated load of feeders or services supplying areas in hotels and motels where the entire lighting is likely to be used at one time, as in ballrooms or dining rooms.

Reproduced with permission of NFPA from NFPA 70, National Electrical Code, 2023 edition. Copyright © 2022, National Fire Protection Association. For a full copy of the NFPA 70, please go to www.nfpa.org

Figure 21-5. *Table 220.45* provides lighting load demand factors that can be applied to service and feeder calculations.

Step 2–Fastened in Place Appliances

Section 220.53 applies to fastened-in-place appliances that are rated at 500 watts and greater, and motors rated at 1/4 horsepower and greater. For the appliances to qualify for this section, they must be fastened in place by some means. It may be a connection to a plumbing pipe, as in the case with a dishwasher or water heater, fastened to a cabinet like a trash compactor or an above-the-range microwave; or it

Step 1	General Lighting, Small Appliance, Laundry		Ungrounded	Neutral
Dwelling area "sq ft"	5400 × 3 VA =	16200		
Small appliance	3 × 1500 VA =	4500		
Laundry	1 × 1500 VA =	1500		
	Total	22,200		
First 3000 at 100%		3000		
3001 to 120,000 × 35%	19200 × 0.35	6720		
Remainder over 120,000 × 25%	× 0.25		9720 VA	9720 VA

Goodheart-Willcox Publisher

Figure 21-6. Step 1 of performing a service calculation using the standard method provides the load for general lighting, small appliances, and laundry.

may be fastened to the building, like a garage door opener, **Figure 21-7**. This is one of the areas of a service calculation where interpretation will affect which items are considered "fastened in place appliances." It is good practice to discuss this with your authority having jurisdiction to ensure you are calculating in the same manner.

The following are not considered fixed appliances by the *NEC*: household electric cooking equipment that is fastened in place, clothes dryers, space-heating equipment, air-conditioning equipment, or electric vehicle supply equipment.

Horsepower Ratings

The electric motors found in homes are often an integral component of an appliance. Appliances such as garbage disposers and sump pumps, which are simply an electric motor mounted within a housing, are typically rated in horsepower. *Section 220.14(C)* sends us back to *Article 430* to determine the full-load current of the motor. There is some jumping around within *Article 430*, but we will ultimately end up at *Table 430.248* (single-phase motors) to find the current values [*220.14(C)* →*430.22* →*430.6(A)(1)* → *Table 430.248*]. See **Figure 21-8**.

When using *Table 430.248*, choose the column that has a nominal voltage matching the applied voltage. The current indicated by the table will then be multiplied by the actual voltage, which will be 120 or 240 as indicated by *Section 220.5(A)*, to determine the volt-amperes of the motor.

Example: 1/3 hp sump (120 V)

$$120 \text{ V} \times 7.2 \text{ A} = 864 \text{ VA}$$

If there are four or more appliances fastened in place, a demand factor of 75% is permitted to be applied. The ungrounded conductors are counted separately from the grounded/neutral for this part of the calculation because some appliances are rated at 240 volts and will not have a neutral. Water heaters are an example of 240-volt appliances

Lost_in_the_Midwest/Shutterstock.com

Figure 21-7. Dishwashers are an example of a fastened in place appliance as they are connected to plumbing lines as well as being fastened to the countertop.

Table 430.248 Full-Load Currents in Amperes, Single-Phase Alternating-Current Motors
The following values of full-load currents are for motors running at usual speeds and motors with normal torque characteristics. The voltages listed are rated motor voltages. The currents listed shall be permitted for system voltage ranges of 110 to 120 and 220 to 240 volts.

Horsepower	115 Volts	200 Volts	208 Volts	230 Volts
1/6	4.4	2.5	2.4	2.2
1/4	5.8	3.3	3.2	2.9
1/3	7.2	4.1	4.0	3.6
1/2	9.8	5.6	5.4	4.9
3/4	13.8	7.9	7.6	6.9
1	16	9.2	8.8	8.0
1 1/2	20	11.5	11.0	10
2	24	13.8	13.2	12
3	34	19.6	18.7	17
5	56	32.2	30.8	28
7 1/2	80	46.0	44.0	40
10	100	57.5	55.0	50

Reproduced with permission of NFPA from NFPA 70, National Electrical Code, 2023 edition. Copyright © 2022, National Fire Protection Association. For a full copy of the NFPA 70, please go to www.nfpa.org

Figure 21-8. *Table 430.248* provides the full-load current values for single-phase alternating-current motors. The column with the nominal voltage closest to the actual voltage is used. The ampere value given by the column is multiplied by the actual voltage to determine the volt-amperes that are entered into the service calculation.

that do not have a neutral connection. Each column should be counted individually. If there are four appliances in that column, the 75% demand factor may be applied. See **Figure 21-9**.

Step 3–Dryer

The amount included in a service calculation for an electric dryer is covered by *Section 220.54*. It states that we shall include 5000 volt-amperes or the nameplate rating, whichever is greater.

In the case of a home or multifamily dwelling with multiple dryers, *Table 220.54* has a demand factor that can be applied, **Figure 21-10**.

Since an electric dryer is a 120/240 V load that has two ungrounded conductors as well as a neutral conductor, we also must include a value in the neutral portion of the calculation. *Section 220.61(B)* permits the neutral to have a demand factor of 70% of the ungrounded conductors. A dryer with 5000 VA in the ungrounded portion of the calculation will have 3500 VA for the neutral (5000 × 0.7 = 3500). See **Figure 21-11**.

Step 4–Household Cooking Appliances

The amount included in a service calculation for household cooking appliances is determined by *Section 220.55*. Although this can be complicated for very large homes and

Step 2	Fixed Appliances		Ungrounded	Neutral
Appliance	Ungrounded	Neutral		
Dishwasher	1200	1200		
Garbage Disposer	1176	1176		
Microwave	1500	1500		
Garage Door 1	1176	1176		
Garage Door 2	1176	1176		
Sump Pump	864	864		
Water Heater	4500	0		
Gas Furnace Blower	1000	1000		
Total	12,592	8092		
Four or more × 0.75	9444	6069	9444 VA	6069 VA

Goodheart-Willcox Publisher

Figure 21-9. Step 2 of performing a service calculation using the standard method provides the load for fixed appliances.

Table 220.54 Demand Factors for Household Electric Clothes Dryers

Number of Dryers	Demand Factor (%)
1–4	100
5	85
6	75
7	65
8	60
9	55
10	50
11	47
12–23	47% minus 1% for each dryer exceeding 11
24–42	35% minus 0.5% for each dryer exceeding 23
43 and over	25%

Reproduced with permission of NFPA from NFPA 70, National Electrical Code, 2023 edition. Copyright © 2022, National Fire Protection Association. For a full copy of the NFPA 70, please go to www.nfpa.org

Figure 21-10. *Table 220.54* has demand factors that apply to household electric ranges. This table is most often used with multifamily dwelling units as they have multiple laundry rooms.

multifamily dwellings, most single-family dwellings are straightforward. Unit 20 is dedicated to household cooking appliances and goes into greater depth.

Table 220.55 applies to cooking appliances that are rated at 1750 watts or more. Column C is applicable in all cases, although it may not provide the smallest value for smaller cooking appliances. When it comes to household ranges, Column C is used almost exclusively. Provided the range is 12,000 watts or less, Column C has the demand factor already applied and gives a final value that can be entered into the service calculation. Find the number of cooking appliances along the left column and slide over to the value in Column C, **Figure 21-12**.

Like the electric dryer, most ranges, ovens, and cooktops are 120/240 V appliances that have two ungrounded conductors as well as a neutral conductor. *Section 220.61(B)* permits the neutral to have a demand factor of 70% of the ungrounded conductors. A range with 8000 VA in the ungrounded portion of the calculation will have 5600 VA for the neutral (8000 × 0.7 = 5,600). See **Figure 21-13**.

For more in-depth range calculations, refer back to Unit 20 for further explanation.

Step 5–Heat/AC

Step 5 of a dwelling unit service calculation is used to calculate electric space-heating and air-conditioning loads. These two items will be considered independently.

Section 220.51 states that the electric space-heating loads are to be considered at 100%. Simply add all the electric heating loads at their nameplate value to determine the heating load. *Section 220.50(B)* sends us to *Part IV* of *Article 440* to determine the air-conditioning load.

Noncoincidence loads are loads that are unlikely to be used simultaneously. Heating and air-conditioning units that are installed in dwellings are considered noncoincidence loads. *Section 220.60* permits the smaller of the two to be omitted from the service calculation since they are unlikely to be on at the same time.

Since it is common for central heating equipment and air conditioners to run on 240-volt power, there may not be any value included for the neutral in this portion of

Step 3	Dryer	Ungrounded	Neutral
	5000 VA or the nameplate rating - neutral 70% of ungrounded	5000 VA	3500 VA

Goodheart-Willcox Publisher

Figure 21-11. Step 3 of performing a service calculation using the standard method provides the load for dryers.

Table 220.55 Demand Factors and Loads for Household Electric Ranges, Wall-Mounted Ovens, Counter-Mounted Cooking Units, and Other Household Cooking Appliances over 1¾ kW Rating (Column C to be used in all cases except as otherwise permitted in Note 3.)

Number of Appliances	Demand Factor (%) (See Notes)		Column C Maximum Demand (kW) (See Notes) (Not over 12 kW Rating)
	Column A (Less than 3½ kW Rating)	Column B (3½ kW through 8¾ kW Rating)	
1	80	80	8
2	75	65	11
3	70	55	14
4	66	50	17
5	62	45	20
6	59	43	21
7	56	40	22
8	53	36	23
9	51	35	24
10	49	34	25

Reproduced with permission of NFPA from NFPA 70, National Electrical Code, 2023 edition. Copyright © 2022, National Fire Protection Association. For a full copy of the NFPA 70, please go to www.nfpa.org

Figure 21-12. *NEC Table 220.55* Column C applies in all cases.

Step 4	Range	Ungrounded	Neutral
Range calculation - neutral 70% of ungrounded		8000 VA	5600 VA

Goodheart-Willcox Publisher

Figure 21-13. Step 4 of performing a service calculation using the standard method provides the load for ranges.

Step 5	Heat/Air Conditioning		Ungrounded	Neutral
Include the larger of the two/noncoincidence load				
	Ungrounded	Neutral		
Electric Heat	0	0		
Air Conditioning	7632	0	7632 VA	0 VA

Goodheart-Willcox Publisher

Figure 21-14. Step 5 of performing a service calculation using the standard method provides the load for heating and air conditioning.

the calculation. The example lists the air conditioner as a 240-volt load, signifying that it does not have a neutral connection. See **Figure 21-14**.

PRO TIP

Temperate and Overcurrent Devices

In some of the northern states, extreme cold temperatures will result in the heating system running for extended periods of time. This can result in excessive heating of the overcurrent devices, specifically the overcurrent device that is integral to the main service disconnect. To help prevent overheating and ensure the service size is sufficient, some areas will implement requirements in addition to the *National Electrical Code*. North Dakota, for example, has a state standard that requires heating loads to be increased by 25% when performing a service calculation due to sustained cold winter temperatures in the state. It is important to be familiar with the local and state requirements in your area.

Step 6 – Electric Vehicle Supply Equipment

The load for electric vehicle supply equipment (EVSE), or car chargers, is covered by *Section 220.57*, **Figure 21-15**. It states

Herr Loeffler/Shutterstock.com

Figure 21-15. In a service calculation for a dwelling, the load rating for electric vehicle supply equipment is 7200 volt-amperes or the nameplate rating, whichever is greater.

that the value inserted for the EVSE shall be the nameplate rating or 7200 volt-amperes, whichever is greater. Since the example has an electric vehicle charger that is rated at 9000 volt-amperes, that is the value entered into the calculation. The *NEC* does not provide a demand factor that can be applied to EVSE, so it is entered at its full value. Many chargers have a neutral connection. The example lists the car charger as a 120/240-volt load, indicating it has a neutral connection. Since *Section 220.61* does not include electric vehicle supply equipment in the neutral loads that are permitted to be derated, it must be included at its nameplate value. See **Figure 21-16**.

Step 7–Other Loads

Any loads that do not fit into one of the previous categories are considered other loads. This is one of the areas where there may be differences in interpretation as to if these loads are other, part of the fixed appliances, or part of the general lighting load. There is no demand factor applied to loads in the section.

A common interpretation is that if it appears in *NEC Chapters 5* or *6*, it belongs in the other load category. Examples include power connection to an RV, swimming pool equipment, hot tubs, elevators, dumbwaiters, stair chair lifts, and electric welders. Since the interpretation of the *NEC* plays a large role in this section, is important to check with your authority having jurisdiction. They may permit some of this equipment to be included with the fastened in place appliances and the demand factor applied. See **Figure 21-17**.

Step 8–Largest Motor

Dwelling unit calculations will normally have the motors included as a part of an appliance or another piece of equipment so it will likely already have been counted by another part of the service calculation. *Section 220.50* requires the largest motor to be included in the calculation at 125%. Since we have already included the motors in another part of the calculation, we just need to add the extra 25%. If the largest motor is the air conditioner, which was omitted due to the heating load being larger, *Section 220.60* requires us to keep it in consideration while determining the largest motor. Look through the calculation at the ungrounded and neutral conductors independently. In Step 8, 25% of the largest motor load from each column is entered. The largest ungrounded motor load was the air conditioner, while the largest motor load on the neutral was one of the 1/2 hp fastened in place appliances. See **Figure 21-18**.

Putting it Together

To finish the service calculation, add up the totals from each of the sections. This will provide the minimum number of volt-amperes required for the service. The next step is to divide the total volt-amperes by 240 V to determine the minimum ampacity. This will be done for both the ungrounded as well as the grounded values. See **Figure 21-19**.

Step 6	Electric Vehicle Supply Equipment	Ungrounded	Neutral
7200 VA or the nameplate rating		9000 VA	9000 VA

Goodheart-Willcox Publisher

Figure 21-16. Step 6 of performing a service calculation using the standard method provides the load for electric vehicle supply equipment.

Step 7	Other Loads		Ungrounded	Neutral
	Ungrounded	Neutral		
Hot Tub	10800	10800		
Total	10800	10800	10800 VA	10800 VA

Goodheart-Willcox Publisher

Figure 21-17. Step 7 of performing a service calculation using the standard method provides the load for any loads not included in the previous categories.

Step 8	Largest Motor		Ungrounded	Neutral
	Ungrounded	Neutral		
Largest motor × 0.25	7632 × 0.25	1176 × 0.25	1908 VA	294 VA

Goodheart-Willcox Publisher

Figure 21-18. Step 8 of performing a service calculation using the standard method provides the load for motors.

Total in Volt-Amperes	Total	61504 VA	44983 VA
Total Amperes = Total VA/240 V	Total	256 A	187 A

Goodheart-Willcox Publisher

Figure 21-19. The final step of performing a service calculation using the standard method is to add all the loads and divide by 240 V.

Minimum Size

Once the minimum service size in amperes has been calculated, the minimum size disconnect and service-entrance conductors can be determined. The service disconnect shall have a rating not less than the minimum number of total amperes as determined by the calculation (*Section 230.79*). The standard calculation had a total ampere rating of 256 amperes. Overcurrent devices, disconnects, and panelboards are only available in standard sizes. Since 256 amperes is not a standard size, the next larger will be required, which is rated at 300 amperes. (Note that this is not a common size for panels used in dwellings, but for the sake of determining the conductor size for this scenario, we will continue with the calculation.) The more likely solution is multiple panelboards.

> **PRO TIP** **Panelboards**
>
> Panelboards are available in many sizes but are only readily available and cost-effective in certain standard sizes. The most common panelboard size used in dwellings is 200 amperes. With so many 200-ampere panelboards manufactured and sold, economies of scale make them the most cost-effective choice. The standard service calculation example in this unit requires a minimum service size of 256 amperes. It is likely more cost-effective to install two 200-ampere panels, making the service rated at 400 amperes, than it would be to order a 300-ampere panelboard. This also provides extra capacity for future expansion.
>
> In this situation, each 200-ampere panelboard would be fed with 4/0 aluminum or 3/0 copper service entrance conductors as per *Table 310.12*.

Service-entrance conductors that feed the entire load of a dwelling unit may be sized according to *Section 310.12*. *Section 310.12(A)* states that if no adjustment or correction factors need to be applied, *Table 310.12* can be used to determine the minimum size of service entrance conductors. According to the table, the 300-ampere service, as determined by the standard method, will have 250 kcmil copper or 350 kcmil aluminum ungrounded service entrance conductors.

Table 310.16 can also be used to find the minimum size conductors, but it will designate a larger conductor. *Section 310.12* allows for smaller conductors due to it being unlikely that all the dwelling unit loads will be drawing energy at the same time.

Grounded Service Entrance Conductor (Neutral)

It is permitted to have a grounded (neutral) service entrance conductor that is smaller than the ungrounded conductors, provided it is large enough to carry the calculated load. According to the sample calculation, the neutral shall have a minimum ampacity of 187 amperes. The 75°C column of *Table 310.16* is used to determine the minimum neutral conductor size. It gives a minimum value of 3/0 copper or 250 kcmil aluminum. Even though it is permitted to reduce the size of the neutral service entrance conductor, many electricians will choose to install one that is the same size as the ungrounded conductors.

In addition to being able to carry the minimum calculated load, *Section 250.24(C)* requires the neutral service entrance conductor to be large enough to safely conduct any current imposed on it from a short circuit or ground fault. This ensures that fault current will be large enough to quickly open the main service overcurrent device.

Section 250.24(C) sends us to *Table 250.102(C)(1)* to determine this minimum size. The table uses the size of the ungrounded service entrance conductors to determine the minimum size grounded/neutral conductor. With 250 kcmil copper ungrounded conductors, *Table 250.102(C)(1)* gives us a minimum grounded/neutral conductor of 2 copper or 1/0 aluminum.

The service calculation will require a larger grounded/neutral conductor in almost all situations, but it is important to keep *Section 250.24(C)* in mind if the neutral current is small or has been heavily derated. The larger of the two values will be the minimum ampacity of the neutral service entrance conductor.

Optional Method

Part IV of *Article 220* provides an alternate method of performing a dwelling service calculation that is referred to as the optional method, **Figure 21-20**. The optional method of calculating a service and feeder is permitted to be used for dwelling units, schools, restaurants, and existing buildings that have past load data available.

The optional method of calculating a single-family dwelling service is covered by *Section 220.82*. The minimum size service this can be applied to is 100 amperes. The optional method is less complicated than the standard method and will typically result in a smaller value than the standard method.

The optional method is only available for ungrounded conductors. If you would like to downsize the neutral conductor, the standard method must be used to determine its minimum size. Installing a neutral conductor that is the same size as the ungrounded conductors makes the neutral calculation unnecessary.

There are only two steps to calculating a dwelling service using the optional method. The first step involves adding the general loads and applying a demand factor. The second step involves calculating the heating and air-conditioning load.

Service Calculation "Optional Method" Dwelling Unit (Single Family)		
Step 1	**General Lighting, Small Appliance, Laundry**	**Ungrounded**
Dwelling Area "sq ft" _____ × 3 VA =		
Small Appliance _____ × 1500 VA =		
Laundry _____ × 1500 VA =		
	Total	
Appliances (Nameplate Rating)		
Appliance		
_____	_____	
_____	_____	
_____	_____	
_____	_____	
_____	_____	
_____	_____	
_____	_____	
_____	_____	
_____	_____	
_____	_____	
_____	_____	
_____	_____	
_____	_____	
_____	_____	
	Total	
General Load Demand Factor		
Sum of General lighting and appliances:		
First 10,000 at 100%	10,000	
Remainder at 40% 0.4 × _____		
Total		
Step 2	**Heat/Air Conditioning**	
Include the largest of the following:		
100% of Air Conditioning		
100% of Heat Pump (Without Supplemental Electric)		
100% of Heat Pump and 65% of Supplemental Electric)		
65% of Electric Space Heating (Less Than Four Controllers)		
40% of Electric Space Heating (More Than Four Controllers)		
100% of Electric Thermal Storage System		
Total in Volt-Amperes	Total	
Total Amperes = Total VA/240V	Total	

Figure 21-20. Template for performing a dwelling unit service calculation using the optional method.

Step 1–General Loads

Section 220.82(A) takes all the general loads, except for heating and cooling, and adds them together. If the service was also calculated using the standard method, all these values will be readily available and can be transferred quickly into the optional method. Note: The appliances are included at their nameplate rating when performing the optional calculation.

The following are included as general loads:
- General lighting–3 VA per square foot
- Small-appliance branch circuits–1500 VA each
- Laundry circuits–1500 VA each
- Fixed appliances–nameplate rating
- Dryer–nameplate rating
- Range–nameplate rating
- Car charger and other loads–nameplate rating

All of these values are added together. The first 10,000-volt-amperes are taken at 100% with the remaining taken at 40%. See **Figure 21-21**.

Step 2–Heating and Air-Conditioning Load

Section 220.82(C) is used to calculate the heating and air-conditioning load. There are six scenarios covered. Be sure to read the *NEC* for each of the scenarios to understand any special considerations given within the language of the subsections.

- Air-conditioning loads at 100%
- Heat pumps without supplemental electric heat at 100%
- Heat pumps with supplemental electric heat
  - Heat pump at 100%
  - Supplemental heat at 65%
- Fixed electric space heating with less than four separately controlled units at 65%
- Fixed electric space heating with four or more separately controlled units at 40%
- Electrical thermal storage heat at 100%

Review the totals of the heating loads with the appropriate demand factor. The largest of the Heat/AC loads is inserted into the second part of the optional service calculation. See **Figure 21-22**.

Putting it Together

To finish the optional service calculation, you take the general load value from Step 1 and add it to the heating/ac load from Step 2 to determine the minimum size of the service in volt-amperes. The next step is to divide the total volt-amperes by 240 V to determine the minimum ampacity. See **Figure 21-23**.

Step 1	General Lighting, Small Appliance, Laundry		Ungrounded
Dwelling Area "sq ft" 5400 × 3 VA =		16,200	
Small Appliance 3 × 1500 VA =		4500	
Laundry 1 × 1500 VA =		1500	
	Total	22,200	
Appliances (Nameplate Rating)			
Appliance			
Dishwasher		1200	
Garbage Disposer		1176	
Microwave		1500	
Garage Door 1		1176	
Garage Door 2		1176	
Sump Pump		864	
Water Heater		4500	
Gas Furnace		1000	
Dryer		4500	
Range		11,600	
Elct Vehicle Supply Equip		9000	
Hot Tub		10,800	
	Total	47,124	
General Load Demand Factor			
Sum of General lighting and appliances:		70,692	
First 10,000 at 100%		10,000	
Remainder at 40%	0.4 × 60,692	24,276.8	
Total		34,277	34,277 VA

Goodheart-Willcox Publisher

Figure 21-21. Step 1 of performing a service calculation using the optional method provides the load for general lighting, small appliances, and laundry.

Step 2	Heat/Air Conditioning	
Include the largest of the following:		
100% of Air Conditioning	7632	
100% of Heat Pump (Without Supplemental Electric)		
100% of Heat Pump and 65% of Supplemental Electric)		
65% of Electric Space Heating (Less Than Four Controllers)		
40% of Electric Space Heating (More Than Four Controllers)		
100% of Electric Thermal Storage System		7632 VA

Goodheart-Willcox Publisher

Figure 21-22. Step 2 of performing a service calculation using the optional method provides the load for heating and air conditioning.

Total in Volt-Amperes	Total	41909 VA
Total Amperes = Total VA/240V	Total	175 A

Goodheart-Willcox Publisher

Figure 21-23. The final step of performing a service calculation using the optional method is to add all the loads and divide by 240 V.

Minimum Size

Once the minimum service size in amperes has been determined, we can then size the disconnect and service entrance conductors. The service disconnect shall have a rating not less than the minimum number of total amperes as determined by the calculation (*Section 230.79*). The optional calculation has a total rating of 175 amperes, so a 200-ampere disconnect is the minimum size.

Service entrance conductors that feed the entire load of a dwelling unit may be sized according to *Section 310.12*.

According to *Table 310.12(A)*, the 200-ampere service, as determined by the optional method, will have 2/0 copper or 4/0 aluminum service entrance conductors.

The grounded/neutral would be the same size as the ungrounded conductors for this calculation. If a person was interested in trying to downsize the neutral of a service calculation, the standard method would have to be used to determine the minimum size.

Summary

- *Article 220* of the *National Electrical Code* is the starting point for service calculations.
- The general unit load of a dwelling includes lighting and general-use receptacle outlets.
- The standard method of performing a service calculation can be used in all applications.
  - Small-appliance and laundry branch circuits are included at 1500 VA each and are added to the general unit load to make up the general lighting load.
  - The first 10,000 VA of the general lighting load is taken at 100% with the remaining taken at 35%.
  - The demand factor that may be applied to four or more fastened in place appliances is 75%.
  - Electric clothes dryers are included at 5000 VA or the nameplate rating, whichever is greater.
  - Electric range demand factors are found in *Table 220.55*.
  - The neutral for a dryer or electric range can be reduced to 70% of the ungrounded conductors.
  - Electric vehicle supply equipment is included at 7200 volt-amperes or the nameplate rating, whichever is greater.
  - Electric heat and air conditioning are noncoincident loads and so the larger of the two is included in the calculation at 100% of its nameplate value.
- The optional method is only available for ungrounded conductors.
  - The general lighting load and the nameplate rating of all the appliances except for heating and air-conditioning equipment are added together. The first 10,000 VA are taken at 100% with a demand factor of 40% being applied to the remaining balance.
  - The heating load is determined by *Section 220.82(C)*
- The total calculated load determined by the service calculation of either calculation method is divided by 240 V to determine the minimum ampacity of the service.

Unit 21 Review

Name _____ Date _____ Class _____

Know and Understand

Answer the following questions based on information in this unit.

1. When performing a service calculation, each small-appliance branch circuit is counted as _____.
   A. zero, it has already been included with the general lighting load
   B. 1200 VA
   C. 1500 VA
   D. 5000 VA

2. When calculating the minimum unit load for a dwelling, the applicable square footage is multiplied by _____.
   A. 3 volt-amperes per square foot
   B. 3.5 volt-amperes per square foot
   C. 3.5 volt-amperes per square meter
   D. 5 volt-amperes per square meter

3. When performing a service calculation, each laundry branch circuit is counted as _____.
   A. zero, it has already been included with the general lighting load
   B. 1200 VA
   C. 1500 VA
   D. 5000 VA

4. When performing a service calculation, the dryer circuit is counted as _____ or the nameplate rating, whichever is greater.
   A. 1200 VA
   B. 1500 VA
   C. 4500 VA
   D. 5000 VA

5. When performing a service calculation, a demand factor is applied to the fixed appliances if there are _____ or more appliances.
   A. 3
   B. 4
   C. 5
   D. 6

6. A demand factor of _____ may be applied to the neutral load of a dryer when performing a service calculation.
   A. 50%
   B. 60%
   C. 65%
   D. 70%

7. When using the optional method to perform a service calculation, the first 10,000 VA of the general loads are figured at 100% with a demand factor of _____ being applied to the rest.
   A. 40%
   B. 50%
   C. 60%
   D. 70%

8. With the standard method of performing a service calculation, the heating load is calculated at _____.
   A. 70%
   B. 80%
   C. 100%
   D. 125%

9. Loads that are unlikely to be used simultaneously are considered _____.
   A. non-simultaneous
   B. noncoincident loads
   C. nonconcurrent
   D. separate

10. Which of the following is *not* counted when determining the square footage of a dwelling unit to calculate the general lighting load of a service calculation?
    A. Open porch
    B. Finished basement
    C. Bathroom
    D. Bonus room

Apply and Analyze

Answer the following questions using a copy of the National Electrical Code. *Identify the section or subsection where the answer is found.*

1. What is the minimum unit load of a home that has a usable area of 2200 square feet?
   A. 1500 VA C. 4400 VA
   B. 2200 VA D. 6600 VA

   NEC _____

2. What is entered in a service calculation for a 4800-watt dryer when using the standard method?
   A. 4500 watts C. 5000 watts
   B. 4800 watts D. 5250 watts

   NEC _____

3. What is entered in a service calculation for four small-appliance branch circuits?
   A. 1500 VA C. 4500 VA
   B. 3000 VA D. 6000 VA

   NEC _____

4. What is entered in a service calculation for a 12,000-watt range when using the optional method?
   A. 8000 watts C. 12,000 watts
   B. 9600 watts D. 15,000 watts

   NEC _____

5. What value is entered into a service calculation for a 1/3 hp garbage disposer?
   A. 864 volt-amperes C. 696 volt-amperes
   B. 828 volt-amperes D. 249 volt-amperes

   NEC _____

6. What value is entered in the optional service calculation for ten 1500-watt baseboard heaters, each of which is controlled by its own thermostat?
   A. 6000 watts C. 15000 watts
   B. 9750 watts D. 18750 watts

   NEC _____

7. What value is entered into a service calculation for the following fastened in place appliances when using the standard method?
   - 1500-watt microwave mounted to the cabinets
   - 1200-volt-ampere dishwasher
   - 1/2 hp garbage disposer
   - 1350-volt-ampere trash compactor

   A. 3000 volt-amperes C. 5226 volt-amperes
   B. 3919 volt-amperes D. 6533 volt-amperes

   NEC _____

8. What value is entered in the service calculation for an EV charger with a nameplate rating of 5000 volt-amperes when using the standard method?
   A. 5000 volt-amperes C. 7200 volt-amperes
   B. 6250 volt-amperes D. 8000 volt-amperes

   NEC _____

9. What value is entered in an optional service calculation for a 14,000-watt range?
   A. 8000 watts C. 11,200 watts
   B. 9800 watts D. 14,000 watts

   NEC _____

10. What neutral value is entered in the service calculation for a 5500-volt-ampere dryer when using the standard method?
    A. 3500 volt-amperes C. 5000 volt-amperes
    B. 3850 volt-amperes D. 5500 volt-amperes

    NEC _____

Critical Thinking

Answer the following questions using a copy of the National Electrical Code.

1. Using the standard calculation, determine the general lighting load (Step 1) after applying any relevant demand factors. The dwelling contains the following:
   - House–2200 square feet of useable space
   - Garage–800 square feet
   - (2) small-appliance branch circuits
   - (1) laundry branch circuit

   Ungrounded: _____

   Neutral: _____

2. Using the standard calculation, determine the general lighting load (Step 1) after applying any relevant demand factors. The dwelling contains the following:
   - House–2500 square feet of useable space
   - Garage–750 square feet
   - (3) small-appliance branch circuits
   - (1) laundry branch circuit

   Ungrounded: _____

   Neutral: _____

3. Using the standard calculation, determine the general lighting load (Step 1) after applying any relevant demand factors. The dwelling contains the following:
   - House–7200 square feet of useable space
   - Garage–1200 square feet
   - (4) small-appliance branch circuits
   - (2) laundry branch circuit

   Ungrounded: _____

   Neutral: _____

4. Using the standard calculation, determine the fastened-in-place appliance load (Step 2) after applying any relevant demand factors. The dwelling contains the following:
   - 1500 VA fastened microwave (120V)
   - 1/2 hp garbage disposer (120V)
   - 1200 VA dishwasher (120V)

   Ungrounded: _____

   Neutral: _____

5. Using the standard calculation, determine the fastened in place appliance load (Step 2) after applying any relevant demand factors. The dwelling contains the following:
   - 1350 VA dishwasher (120V)
   - 1/3 hp garbage disposer (120V)
   - 1/2 hp garage door opener (120V)
   - 4500 W water heater (240 V)

   Ungrounded: _____

   Neutral: _____

6. Using the standard calculation, determine the fastened-in-place appliance load (Step 2) after applying any relevant demand factors. The dwelling contains the following:
   - 1400VA fastened microwave (120V)
   - 1/2 hp garbage disposer (120V)
   - 3/4 hp sump pump (120V)
   - 4500 W water heater (240 V)
   - 1200 VA dishwasher
   - 1/4 hp trash compactor
   - 1/2 hp garage door opener
   - 1100 VA gas furnace

   Ungrounded: _____

   Neutral: _____

7. Using the standard calculation, determine the dryer load (Step 3) after applying any relevant demand factors. The dryer is rated at 5500 VA (120/240V).

   Ungrounded: _____

   Neutral: _____

8. Using the standard calculation, determine the range load (Step 4) after applying any relevant demand factors. The range is rated at 11,000 W (120/240).

   Ungrounded: _____

   Neutral: _____

9. Determine the calculated load for Step 1 of the optional service calculation after applying any relevant demand factors. The dwelling contains the following:
   - House–2800 square feet of useable space
   - Garage–800 square feet
   - (2) small-appliance branch circuits
   - (1) laundry branch circuit
   - 1350 VA dishwasher (120V)
   - 1/3 hp garbage disposer (120V)
   - 1/2 hp garage door opener (120V)
   - 4500 W water heater (240 V)
   - 4500 VA dryer
   - 12000 VA range
   - 8200 VA car charger

   Ungrounded: _____

10. Determine the value for each of the following heating/air-conditioning loads after any demand factors from the optional service calculation have been applied.
    A. 6484 VA air conditioner
    B. 7680 VA heat pump without supplemental heating
    C. 5600 VA heat pump with 10 KW of supplemental electric heat. The heat pump and space heating may run simultaneously
    D. 25 kW electric furnace controlled by one thermostat
    E. (6) 1500 W baseboard heaters and (4) 1000 W baseboard heaters all of which are controlled by individual thermostats
    F. 10 kW electric thermal storage heat in a concrete floor

    Ungrounded: _____

    Ungrounded: _____

    Ungrounded: _____

    Ungrounded: _____

    Ungrounded: _____

    Ungrounded: _____

Answer the following questions using a copy of the National Electrical Code.

11. Using the standard method, complete a service calculation for a dwelling containing the following:
    - House–1200 square feet of useable space
    - Garage–500 square feet
    - (2) small-appliance branch circuits
    - (1) laundry branch circuit
    - 1500 VA dishwasher (120 V)
    - 1/3 hp garage door opener (120 V)
    - 1000 VA gas furnace (120 V)
    - 4500 VA dryer (120/240 V)
    - 11.5 kW range (120/240 V)
    - 2600 VA air-conditioning compressor (240 V)

Step 1	Ungrounded _____	Neutral _____
Step 2	Ungrounded _____	Neutral _____
Step 3	Ungrounded _____	Neutral _____
Step 4	Ungrounded _____	Neutral _____
Step 5	Ungrounded _____	Neutral _____
Step 6	Ungrounded _____	Neutral _____
Step 7	Ungrounded _____	Neutral _____
Step 8	Ungrounded _____	Neutral _____
Total VA	Ungrounded _____	Neutral _____
Total Amperes	Ungrounded _____	Neutral _____

12. Using the standard method, complete a service calculation for a dwelling containing the following:
    - House–1500 square feet of useable space
    - Garage–600 square feet
    - (3) small-appliance branch circuits
    - (1) laundry branch circuit
    - 1300 VA dishwasher (120 V)
    - 1/3 hp garbage disposer (120 V)
    - 1/2 hp trash compactor (120 V)
    - 1200 VA microwave (above range) (120 V)
    - 4000 W electric water heater (240 V)
    - 5000 VA dryer (120/240 V)
    - 12 kW range (120/240 V)
    - 10 KW electric space heating with one thermostat (240 V)
    - 6720 VA air-conditioning compressor (240 V)

Step 1	Ungrounded _____	Neutral _____
Step 2	Ungrounded _____	Neutral _____
Step 3	Ungrounded _____	Neutral _____
Step 4	Ungrounded _____	Neutral _____
Step 5	Ungrounded _____	Neutral _____
Step 6	Ungrounded _____	Neutral _____
Step 7	Ungrounded _____	Neutral _____
Step 8	Ungrounded _____	Neutral _____
Total VA	Ungrounded _____	Neutral _____
Total Amperes	Ungrounded _____	Neutral _____

13. Using the standard method, complete a service calculation for a dwelling containing the following:
    - House–2500 square feet of useable space
    - Garage–1000 square feet
    - (3) small-appliance branch circuits
    - (1) laundry branch circuit
    - 1200 VA dishwasher (120 V)
    - 1/2 hp garbage disposer (120 V)
    - 1/2 hp sump pump (120 V)
    - (2) 1/3 hp garage door openers (120 V)
    - 1300 VA microwave (above range) (120 V)
    - 4500 W electric water heater (240 V)
    - 5500 VA dryer (120/240 V)
    - 10 kW range (120/240 V)
    - 6400 VA heat-pump compressor with 15000 KW supplemental electric space heating (will run simultaneously) (240 V)
    - 6000 VA car charger

Step 1	Ungrounded _____	Neutral _____
Step 2	Ungrounded _____	Neutral _____
Step 3	Ungrounded _____	Neutral _____
Step 4	Ungrounded _____	Neutral _____
Step 5	Ungrounded _____	Neutral _____
Step 6	Ungrounded _____	Neutral _____
Step 7	Ungrounded _____	Neutral _____
Step 8	Ungrounded _____	Neutral _____
Total VA	Ungrounded _____	Neutral _____
Total Amperes	Ungrounded _____	Neutral _____

14. Using the standard method, complete a service calculation for a dwelling containing the following:
    - House–4500 square feet of useable space
    - Garage–1200 square feet
    - (4) small-appliance branch circuits
    - (1) laundry branch circuit
    - 1350 VA dishwasher (120 V)
    - 1/3 hp garbage disposer (120 V)
    - Three 1/2 hp garage door opener (120 V)
    - 1200 VA gas furnace (120 V)
    - 5000 VA dryer (120/240 V)
    - 13 kW range (120/240 V)
    - 6720 VA air-conditioning compressor (240 V)
    - 7200 VA car charger

Step 1	Ungrounded _____	Neutral _____
Step 2	Ungrounded _____	Neutral _____
Step 3	Ungrounded _____	Neutral _____
Step 4	Ungrounded _____	Neutral _____
Step 5	Ungrounded _____	Neutral _____
Step 6	Ungrounded _____	Neutral _____
Step 7	Ungrounded _____	Neutral _____
Step 8	Ungrounded _____	Neutral _____
Total VA	Ungrounded _____	Neutral _____
Total Amperes	Ungrounded _____	Neutral _____

15. Using the standard method, complete a service calculation for a dwelling containing the following:
    - House–7500 square feet of useable space
    - Garage–2000 square feet
    - (6) small-appliance branch circuits
    - (2) laundry branch circuit
    - (2) 1500 VA dishwashers (120 V)
    - (2) 1/2 hp garbage disposer
    - (2) 1200 microwaves (above range) (120 V)
    - (4) 1/2 hp garage door opener (120 V)
    - (2) 1000 VA gas furnaces (120 V)
    - (2) 5000 VA dryers (120/240 V)
    - (2) 12 kW ranges (120/240 V)
    - (2) 6720 VA air-conditioning compressors (240 V)
    - 10,800 VA car charger (120/240 V)
    - 11,280 VA hot tub (120/240 V)

    Step 1   Ungrounded _____ Neutral _____
    Step 2   Ungrounded _____ Neutral _____
    Step 3   Ungrounded _____ Neutral _____
    Step 4   Ungrounded _____ Neutral _____
    Step 5   Ungrounded _____ Neutral _____
    Step 6   Ungrounded _____ Neutral _____
    Step 7   Ungrounded _____ Neutral _____
    Step 8   Ungrounded _____ Neutral _____
    Total VA   Ungrounded _____ Neutral _____
    Total Amperes Ungrounded _____ Neutral _____

16. Using the optional method, complete a service calculation for a dwelling containing the following:
    - House–1200 square feet of useable space
    - Garage–500 square feet
    - (2) small-appliance branch circuits
    - (1) laundry branch circuit
    - 1500 VA dishwasher (120 V)
    - 1/3 hp garage door opener (120 V)
    - 1000 VA gas furnace (120 V)
    - 4500 VA dryer (120/240 V)
    - 11.5 kW range (120/240 V)
    - 2600 VA air-conditioning compressor (240 V)

    Step 1   Ungrounded _____
    Step 2   Ungrounded _____
    Total VA   Ungrounded _____
    Total Amperes Ungrounded _____

17. Using the optional method, complete a service calculation for a dwelling containing the following:
    - House–1500 square feet of useable space
    - Garage–600 square feet
    - (3) small-appliance branch circuits
    - (1) laundry branch circuit
    - 1300 VA dishwasher (120 V)
    - 1/3 hp garbage disposer (120 V)
    - 1/2 hp trash compactor (120 V)
    - 1200 VA microwave (above range) (120 V)
    - 4000 W electric water heater (240 V)
    - 5000 VA dryer (120/240 V)
    - 12 kW range (120/240 V)
    - 10 KW electric space heating (one thermostat) (240 V)
    - 6720 VA air-conditioning compressor (240 V)

    Step 1   Ungrounded _____
    Step 2   Ungrounded _____
    Total VA   Ungrounded _____
    Total Amperes Ungrounded _____

18. Using the Optional method, complete a service calculation for a dwelling containing the following:
    - House–2500 square feet of useable space
    - Garage–1000 square feet
    - (3) small-appliance branch circuits
    - (1) laundry branch circuit
    - 1200 VA dishwasher (120 V)
    - 1/2 hp garbage disposer (120 V)
    - 1/2 hp sump pump (120 V)
    - (2) 1/3 hp garage door openers (120 V)
    - 1300 VA microwave (above range) (120 V)
    - 4500 W electric water heater (240 V)
    - 5500 VA dryer (120/240 V)
    - 10 kW range (120/240 V)
    - 6400 VA heat-pump compressor with 15000 KW supplemental electric space heating (will run simultaneously) (240 V)
    - 6000 VA car charger

    Answers:
    Step 1   Ungrounded _____
    Step 2   Ungrounded _____
    Total VA   Ungrounded _____
    Total Amperes Ungrounded _____

19. Using the optional method, complete a service calculation for a dwelling containing the following:
    - House–4500 square feet of useable space
    - Garage–1200 square feet
    - (4) small-appliance branch circuits
    - (1) laundry branch circuit
    - 1350 VA dishwasher (120 V)
    - 1/3 hp garbage disposer (120 V)
    - (3) 1/2 hp garage door opener (120 V)
    - 1200 VA gas furnace (120 V)
    - 5000 VA dryer (120/240 V)
    - 13 kW range (120/240 V)
    - 6720 VA air-conditioning compressor (240 V)
    - 7200 VA car charger

    Step 1   Ungrounded _____
    Step 2   Ungrounded _____
    Total VA   Ungrounded _____
    Total Amperes   Ungrounded _____

20. Using the optional method, complete a service calculation for a dwelling containing the following:
    - House–7500 square feet of useable space
    - Garage–2000 square feet
    - (6) small-appliance branch circuits
    - (2) laundry branch circuits
    - (2) 1500 VA dishwashers (120 V)
    - (2) 1/2 hp garbage disposers
    - (2) 1200 microwaves (above range) (120 V)
    - (4) 1/2 hp garage door opener (120 V)
    - (2) 1000 VA gas furnaces (120 V)
    - (2) 5000 VA dryers (120/240 V)
    - (2) 12 kW Ranges (120/240 V)
    - (2) 6720 VA air-conditioning compressors (240 V)
    - 10,800 VA car charger (120/240 V)
    - 11,280 VA hot tub (120/240 V)

    Step 1   Ungrounded _____
    Step 2   Ungrounded _____
    Total VA   Ungrounded _____
    Total Amperes   Ungrounded _____

Appendix

Math Review **418**
    Why Math? 418
    Calculators 418
    Whole Numbers 419
    Fractions 420
    Reading a Ruler 421
    Decimal Fractions 422
    Converting Fractions to Decimals 423
    Converting Decimals to Fractions 424
    Equations 424
    Area Measure 424
    Volume Measure 426
    Exponents and Roots 428
    Working with Right Angles 428
    Computing Averages 429
    Percent and Percentage 429
    Ohm's Law Calculations 429
Service Calculation–Standard Method Template **431**
Service Calculation–Optional Method Template **432**

Math Review

Why Math?

All technical trades involve using math for numerous tasks. Industrial maintenance technicians use math to make precise measurements and convert units. They must be able to read and interpret prints. Perhaps most importantly, technicians must be able to determine if equipment is maintained and operated within specific tolerances. Some maintenance tasks require other specialized or more advanced math, but all technicians require an understanding of what is presented here.

Calculators

Calculators can be a great time-saver, but you should not rely on a calculator to replace your knowledge of basic math. Knowing some basic math and using common-sense observation can help prevent big errors when using a calculator.

For example, the formula for the area of a circle requires multiplying π (pronounced "pi" and roughly equal to 3.1416) by the radius of the circle squared (multiplied by itself). Say the radius is 5″. The area is found by multiplying 5″ × 5″ (5″ squared), then multiplying the result by 3.1416. If you know 5 × 5 (written 5^2) is 25, and 3 × 25 is 75, then you know the answer is slightly more than 75. Now you can check to see if the answer you find on your calculator is close. The correct answer is 78.54 in^2, as expected. It is important to do operations in the correct order when using a calculator. Using this example, if you multiply 3.1416 × 5 and then multiply that by itself, you get 246.7413—not even close! Knowing the correct order of operations on paper will help you enter numbers in the correct order on your calculator.

> **TECH TIP — Check Your Calculation**
>
> Always compare your answer with what might be a reasonable answer. This is the most important tip for calculator use. If the answer does not look reasonable, it is probably not correct.

There are several types of calculators commonly used in technical trades. A *general calculator* allows the user to perform basic math functions. *Basic calculators* have a memory function that allows the user to store the result of a calculation so it can be recalled and used in a subsequent calculation. Most also have a % key. When a number is shown in the display, pressing the % key allows the user to then press a number key to calculate that percent of the displayed number. The √ key is for finding the square root of a number. The ± key changes the value of a displayed number from positive to negative or negative to positive.

Scientific calculators are more advanced, with the ability to perform many trigonometric functions and other advanced operations, **Figure A-1**. Scientific calculators can be useful when working on electrical systems or other

Goodheart-Willcox Publisher

Figure A-1.

systems that include angles in calculations, such as load or welding angles.

Another type of calculator that can be useful for trades workers is the *construction calculator*. Most construction calculators allow the user to convert between metric and US Customary measurements. Construction calculators vary by manufacturer and model. All calculators come with instructions that explain their functions.

Whole Numbers

Whole numbers are simply numbers without fractions or decimal points, such as 1, 2, 3, 4, etc. Adding, subtracting, multiplying, and dividing whole numbers primarily requires memorizing a few math facts.

Adding and Subtracting Whole Numbers

For example, adding this column of whole numbers requires memorizing the *sum* of 3 + 5 and the sum of 8 + 2.

$$\begin{array}{r} 3 \\ 5 \\ + 2 \\ \hline 10 \end{array}$$

The same type of memorization of math facts is required to subtract whole numbers. We know the result of subtracting 12 from 37 is 25, because we know 2 from 7 is 5 and 1 from 3 is 2.

$$\begin{array}{r} 37 \\ - 12 \\ \hline 25 \end{array}$$

The key to both addition and subtraction is to line up the columns of digits correctly. Whole numbers should be aligned on the right.

In subtraction, if the number being subtracted (the number on the bottom) is larger than the number it is being subtracted from (the number on the top), borrow 10 from the next digit to the left and add it to the one on the right. Write small numerals above the column to help you keep track. Consider this example:

$$\begin{array}{r} {}^{2}\!3\!{}^{16}\!\!\!\!\!/ \\ - 19 \\ \hline 17 \end{array}$$

Multiplying Whole Numbers

Multiplication of whole numbers requires memorization of a multiplication table. The only way to get 6 × 5 = 30 is to know that multiplication fact or to add 6 + 6 + 6 + 6 + 6. Longhand addition quickly becomes tedious for bigger multiplication problems. To multiply numbers whose values are 10 or more, align the digits representing 0 through 9 (the 1s digit) in the right-hand column. Then multiply the top row by the 1s digit in the second row:

$$\begin{array}{r} {}^{10s}\searrow\;\swarrow{}^{1s} \\ 31 \\ \times 15 \\ \hline 155 \end{array}$$

Next, multiply the top row by the 10s digit in the second row. Because you multiplied by the 10s digit, the *product* (the result of multiplication) is written with its right-most digit in the 10s column:

$$\begin{array}{r} 31 \\ \times 15 \\ \hline 155 \\ 31 \end{array}$$

If the problem has more digits in the second row, repeat the above steps for each digit and write the products in rows beneath one another. Be sure to record the right-most digit in each row in the column for the place it represents: 100s, 1000s, etc.

When all multiplication is complete, add the products just as you would for a simple addition problem. The result is the product (answer) of the multiplication problem.

$$\begin{array}{r} 31 \\ \times 15 \\ \hline 155 \\ 31 \\ \hline 465 \end{array}$$

Dividing Whole Numbers

Division of whole numbers is simply the reverse of multiplication, but the problem must be set up differently. The *dividend* (the number being divided) is written inside the division symbol. The *divisor* (the number the dividend will be divided by) is written to the left of the symbol:

divisor ⟶ 7)28 ⟵ dividend

From the multiplication table, we know 7 × 4 = 28. So, if 28 is divided into 7 parts, each part will have 4, or 28 ÷ 7 = 4. The answer to a division problem, in this case 4, is the *quotient*. It is written above the division symbol and above the 1s place of the 28:

4 ⟵ quotient
7)28

When the dividend is large compared to the divisor, the process is divided into steps, as follows:

4)320

The 4 goes into 32 eight times. Write the 8 above the 2 (the right column of the 32). Now, multiply 4 × 8, which is 32. Write the 32 beneath the 32 in the division symbol.

$$\begin{array}{r} 8 \\ 4{\overline{\smash{\big)}\,320}} \\ 32 \end{array}$$

Subtract the product of your multiplication (32) from the number above it (32). Because 4 goes into 32 exactly 8 times, the numbers are the same, so the result of your subtraction is 0.

$$\begin{array}{r} 8 \\ 4\overline{)320} \\ \underline{32} \\ 0 \end{array}$$

Drop the next digit to the right in the dividend (0 in this case) down beside the result of your subtraction. That makes the number at the bottom 00. The 4 will not go into 0 (or 00), so the quotient of that step is 0. The quotient (answer) of the division problem is 80, meaning 320 can be divided by 4 80 times.

$$\begin{array}{r} 80 \\ 4\overline{)320} \\ \underline{32} \\ 00 \end{array}$$

If there are more places under the division symbol, keep doing the same division, multiplication, subtraction, and drop-down for each digit, moving to the right. Consider the following example:

$$\begin{array}{r} 102 \\ 6\overline{)616} \\ \underline{6} \\ 01 \\ \underline{0} \\ 016 \\ \underline{12} \\ 4 \leftarrow \text{remainder} \end{array}$$

If the last number produced by the drop-down cannot be divided evenly by the divisor, that number is called the *remainder*. In the example above, the quotient is 102 with a remainder of 4.

Practice A-1

Test your skills with the following problems. Do *not* use a calculator.

A. $\begin{array}{r} 342 \\ +\ 16 \end{array}$

B. $\begin{array}{r} 79 \\ +29 \end{array}$

C. $\begin{array}{r} 68 \\ -\ 13 \end{array}$

D. $\begin{array}{r} 124 \\ -\ 35 \end{array}$

E. $\begin{array}{r} 18 \\ \times\ 4 \end{array}$

F. $\begin{array}{r} 213 \\ \times\ 24 \end{array}$

G. $3\overline{)36}$

H. $7\overline{)214}$

Fractions

A fraction is a part of something larger. If there are three machine bolts in a pound, each machine bolt weighs one-third of one pound. One-third can be written as follows:

$\dfrac{1}{3}$ ← numerator
← denominator

The number above the fraction bar is called the *numerator*. It indicates how many parts are in the fraction, in this case 1 machine bolt. The number below the fraction bar is the *denominator*. The denominator indicates how many parts are in the whole, in this case 3 machine bolts in the whole pound. If there are 50 machine bolts in a carton and you take 7 out, you have taken

$$\frac{7}{50}$$

of the machine bolts.

Equivalent Fractions

If two fractions represent the same value, they are said to be *equivalent fractions*. For example, 1/3 and 2/6 are equivalent fractions because they both represent one-third of the whole. If both the numerator and denominator of a fraction are multiplied by the same amount, the result is an equivalent fraction.

$$\frac{1 \times 2}{3 \times 2} = \frac{2}{6}$$

Adding Fractions

Fractions must have a *common denominator* in order to be added. If the denominator of one of the fractions is 8, the other fraction must be written as an equivalent fraction with a denominator of 8. For example, to add 3/4 and 1/8, write 3/4 as an equivalent fraction with a denominator of 8.

$$\frac{3}{4} \times \frac{2}{2} = \frac{6}{8}$$

When both fractions have the same denominator, add the numerators.

$$\frac{6}{8} + \frac{1}{8} = \frac{7}{8}$$

Adding 6 eighths plus 1 eighth gives a total of 7 eighths.

A common denominator can be found by multiplying all the denominators in a problem. Consider the following example:

$$\frac{1}{3} + \frac{2}{5} + \frac{9}{14}$$
$$3 \times 5 \times 14 = 210$$
$$\frac{70}{210} + \frac{84}{210} + \frac{135}{210} = \frac{289}{210}$$

In this example, the numerator is larger than the denominator. This is because the value of the fraction is greater than 1. The numerator of 289 is 79 parts larger than the whole of 210 parts. The same number could be written as follows:

$$1\frac{79}{210}$$

This is called a *mixed number* because it is made up of a whole number plus a fraction.

Subtracting Fractions

Subtracting fractions is similar to adding them. Both fractions must have a common denominator; then the numerators are subtracted just like whole numbers.

$$\frac{2}{3} - \frac{1}{4} = ?$$

Find common denominators and subtract the numerators.

$$\frac{8}{12} - \frac{3}{12} = \frac{5}{12}$$

Multiplying Fractions

To multiply fractions, multiply the numerators; then multiply the denominators.

$$\frac{3}{4} \times \frac{2}{5} = \frac{3 \times 2}{4 \times 5} = \frac{6}{20}$$

To make the result easier to work with, always reduce it to its lowest terms. That is, write it as an equivalent fraction with the lowest possible denominator. For example, both 6 and 20 can be divided by 2 to make an equivalent fraction with a smaller denominator.

$$\frac{6 \div 2}{20 \div 2} = \frac{3}{10}$$

To multiply a fraction by a mixed number, first change the mixed number to a fraction. Then multiply as common fractions. Reduce the product to its lowest terms.

$$\frac{2}{3} \times 4\frac{1}{2} = \frac{2}{3} \times \frac{9}{2}$$
$$\frac{2}{3} \times \frac{9}{2} = \frac{18}{6}$$
$$\frac{18}{6} = \frac{3}{1} = 3$$

Dividing Fractions

To divide fractions, invert the divisor (swap the numerator and denominator); then multiply as common fractions. It may help you to remember "keep it, change it, flip it"—keep the first fraction as it is, change the division to multiplication, and flip the second fraction.

$$\frac{3}{4} \div \frac{2}{3} = \frac{3}{4} \times \frac{3}{2}$$
$$\frac{3 \times 3}{4 \times 2} = \frac{9}{8}$$

In this example, the quotient is a fraction with a larger numerator than denominator. This indicates the value is greater than 1. To express this in its simplest form, convert it to a mixed number.

$$\frac{9}{8} = \frac{8+1}{8} = 1\frac{1}{8}$$

Sometimes it is necessary to combine operations in a single problem. Some problems also involve more than one type of unit, such as inches and feet. The first step in solving more complex problems is to decide which operation should be done first. It often helps to write the problem in such a way that it states the order of operations. Next, convert everything to the same unit(s). Where operations include adding or subtracting fractions, convert them to their least common denominators. Now solve the problem, doing all operations in the planned order. Finally, convert the units to whatever makes sense for the problem. For example, you would not write fine measurements in square feet.

Work through the following example:

How long is the shaded space in this drawing?

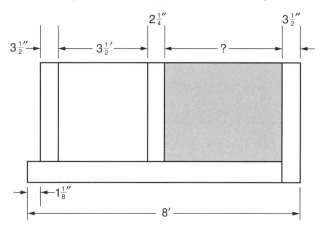

1. Convert the overall length to inches.   96″
2. Subtract $1\frac{1}{8}″$. $95\frac{7}{8}″$
3. Convert $3\frac{1}{2}′$ to inches.   42″
4. Add dimensions at top. $3\frac{1}{2}″ + 42″ + 2\frac{1}{4}″ + 3\frac{1}{2}″ = 51\frac{1}{4}″$
5. Subtract $51\frac{1}{4}″$ from $95\frac{7}{8}″$. $95\frac{7}{8}″ - 51\frac{2}{8}″ = 44\frac{5}{8}″$

Practice A-2

A. $\frac{1}{4} + \frac{5}{8}$
B. $\frac{3}{4} - \frac{1}{3}$
C. $\frac{1}{2} \times \frac{3}{4}$
D. $\frac{1}{3} \div \frac{1}{2}$
E. $\frac{2}{3} + \frac{7}{12}$
F. $\frac{13}{16} - \frac{2}{5}$
G. $\frac{2}{3} \times 2\frac{1}{2}$
H. $1\frac{1}{4} \div \frac{2}{5}$

Reading a Ruler

Measuring devices, such as rulers and tape measures, may be marked for measuring inches and fractions of an inch; meters, centimeters, and millimeters; feet, inches and tenths of an inch; or any other system. The measuring system used to divide the spaces on a measuring device is called the *scale*. The most common linear (in a line) scale in construction uses yards, feet, inches, and fractions of an inch. There are 3 feet in 1 yard and 12 inches in 1 foot. Inches are most often divided into halves, fourths, eighths, and sixteenths. See **Figure A-2**.

On finer measuring devices, the longest marks on the scale may indicate inches, **Figure A-3**. The inches on a measuring device may be divided into eighths, sixteenths, or even thirty-seconds. The second-longest marks on the scale represent halves, the next-longest represent fourths, and so on. The first step in reading the scale is determining what the smallest marks on the scale represent. Count down from the whole inch to the halves, then the quarters, eighths, sixteenths, and thirty-seconds, if they are used. Then count the number of marks from the last inch mark to the mark you are reading.

Appendix Math Review

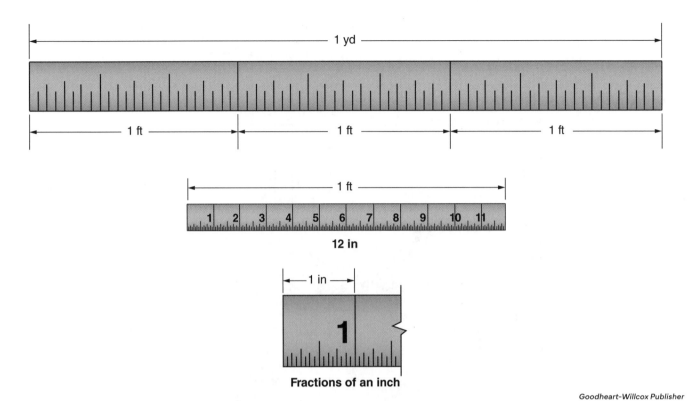

Figure A-2.

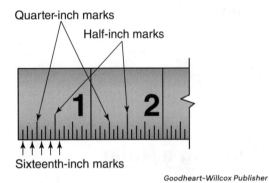

Figure A-3.

Practice A-3

What measurements are represented by the letters in **Figure A-4**?

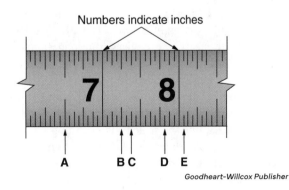

Figure A-4.

Decimal Fractions

Decimal fractions are commonly called *decimals*. Decimal fractions are fractions whose denominators are multiples of 10. If the denominator is 10, the fraction is tenths. If the denominator is 100, the fraction is hundredths.

Decimal fractions are often written on a single line with a dot separating digits. The dot between the whole number and the decimal fraction is the *decimal point*. Every place to the left of the decimal point increases the value of the digit in that place tenfold. That is why the second place to the left of the decimal point is called the tens place, the third place to the left is the hundreds place, etc. Moving to the right of the decimal point, the place values decrease tenfold. A decimal fraction of $\frac{5}{10}$ can be written as 0.5. A decimal fraction of $\frac{12}{1000}$ can be written as 0.012.

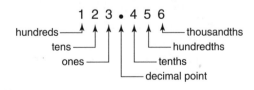

The value of the number in the example above is one hundred twenty-three and four hundred fifty-six thousandths.

Adding and Subtracting Decimals

To add decimals, line up the decimal points in a column, add the numbers, and put the decimal point in the result in the decimal point column.

$$\begin{array}{r} 1.4 \\ 19.2 \\ + \ 31.7 \\ \hline 52.3 \end{array}$$

Subtracting decimals is similar. Line up the decimal points in the problem and the answer, and subtract as usual.

$$\begin{array}{r} 27.74 \\ - \ 2.23 \\ \hline 25.51 \end{array}$$

If there are more decimal places in the number being subtracted than in the number it is being subtracted from, zeros can be added to the right without affecting the value of the number.

$$\begin{array}{r} 5.70 \quad \leftarrow \text{added zero} \\ - \ 2.02 \\ \hline 3.68 \end{array}$$

Multiplying Decimals

Decimals are multiplied the same as whole numbers, except for the placement of the decimal point in the product (answer). Add the number of decimal places to the right of the decimal point in both the number being multiplied and the number it is being multiplied by. The decimal point should be placed that many places to the left in the product.

$$\begin{array}{r} 12.25 \ \leftarrow \text{two decimal places to the right} \\ \times \ 3.75 \ \leftarrow \text{two decimal places to the right} \\ \hline 6125 \quad \text{(total of four decimal places)} \\ 8575 \\ 3675 \\ \hline 45.9375 \ \leftarrow \text{decimal point is four places to the left} \end{array}$$

Dividing Decimals

Dividing decimals is like dividing whole numbers, except for keeping track of the placement of the decimal point. As a reminder, the number being divided is the dividend, the number it is divided by is the divisor, and the answer is the quotient. To start the division problem, move the decimal point in the divisor all the way to the right. Move the decimal point in the dividend the same number of places to the right. Add zeros to the right of the dividend, if necessary. Divide as you would for whole numbers.

divisor ⟶ .4⟌20 ⟵ dividend
4.⟌200.   move decimal points

$$\begin{array}{r} 50 \\ 4\overline{)200} \\ \underline{20} \\ 00 \end{array}$$

Practice A-4

Test your skills with the following problems.

A. $\begin{array}{r} 2.12 \\ 17.01 \\ + \ 9.05 \end{array}$

B. $\begin{array}{r} 34.09 \\ 12.125 \\ + \ 2.899 \end{array}$

C. $\begin{array}{r} 18.48 \\ - \ 12.25 \end{array}$

D. $\begin{array}{r} 134.02 \\ - \ \ \ 8.14 \end{array}$

E. $\begin{array}{r} 5.25 \\ \times \ \ \ 5 \end{array}$

F. $\begin{array}{r} 15.34 \\ \times \ 6.25 \end{array}$

G. $35 \div .07$

H. $2.25 \div .25$

Converting Fractions to Decimals

To change a common fraction to a decimal fraction, divide the numerator by the denominator. For example, change 1/4 to a decimal fraction.

$$\begin{array}{r} .25 \\ 4\overline{)1.00} \\ \underline{8} \\ 20 \end{array}$$

$\frac{1}{4} = .25$

Sometimes the division yields a number with a repeating decimal, as in converting 2/3 to a decimal.

$$\begin{array}{r} .666 \\ 3\overline{)2.000} \\ \underline{1\ 8} \\ 20 \\ \underline{18} \\ 20 \end{array}$$

These numbers should be rounded off to the desired number of places. When rounding off, the last digit should be increased by 1 if the next digit is 5 or more. If the next digit is less than 5, the last digit used stays the same. In the above example, round the answer to two places. The second digit is rounded up to 7 because .666 is closer to .67 than it is to .66.

To convert a mixed number as a decimal, keep the whole number as is and convert the fractional part as above. Express $12\frac{3}{4}$ as a decimal.

$$12 + 4\overline{)3.00}^{\ .75} = 12.75$$
$$\underline{2\ 8}$$
$$20$$

Practice A-5

Convert these fractions to decimals and round the answers to three places.

A. $\frac{1}{3}$ B. $\frac{22}{7}$ C. $\frac{10}{15}$

Converting Decimals to Fractions

To convert a decimal to a fraction, drop the decimal point and write the given number as the numerator. The denominator will be 10, 100, 1000, or 1 with as many 0s as there were places in the decimal number.

$$.42 = \frac{42}{100} \text{ or } .125 = \frac{125}{1000}$$

Equations

An *equation* is a mathematical statement that two things have the same or equal value. An equation can be thought of as a mathematical sentence. The words of the sentence are mathematical values called *terms*. An equation is always written with an equal sign (=). For example, 3 + 4 = 7. In that statement, 3, 4, and 7 are terms. The statement says 3 plus 4 has the same value as 7. Many useful formulas are stated as equations. Equations can be used to find the value of one unknown term when the other values in the equation are known.

For example, if you know a truck is loaded with 10 bundles of shingles weighing 80 lb each and an unknown weight of sheet metal, and the total load is 1000 lb, you can find the weight of the metal with the following equation:

$$(80 \text{ lb} \times 10) + \text{weight of metal} = 1000 \text{ lb}$$

The first term, (80 lb × 10), represents the total weight of the shingles. It is enclosed in parentheses to indicate it is a single term that should be computed before the rest of the equation. Whenever a mathematical term is enclosed in parentheses, that computation should be done first. Now, write the equation with the shingle weight computed:

$$800 \text{ lb} + \text{weight of metal} = 1000 \text{ lb}$$

When a mathematical operation is done on one side of an equation, the equation remains a true statement if the same thing is done on the other side of the equal sign. If we subtract 800 lb from both sides of our equation, it is still a true equation.

$$800 \text{ lb} + \text{weight of metal} - 800 \text{ lb} = 1000 \text{ lb} - 800 \text{ lb}$$
$$\text{weight of metal} = 200 \text{ lb}$$

Practice A-6

Find the unknown value in each equation.

A. cost = ($.60 − $.04) × 5

B. $X = \frac{3}{4} + 20$

C. 240 lb = 2 × weight of crate

D. $\frac{1}{4} \div \frac{2}{3} = Y$

Area Measure

The area of a surface is always measured in units of square inches, square meters, square feet, etc. When a number is squared, it is multiplied by itself. For example, 3 squared is 9. Square units are written with a superscript 2, indicating that it is units × units.

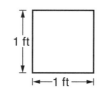

This square is 1′ × 1′ or 1 square foot.
1 ft²

This rectangle is made up of 2 squares that are 1 square foot each. It is 1′ × 2′ or 2 square feet.
2 ft²

Squares and Rectangles

The area of a square or rectangle is the number of units it is wide multiplied by the number of units it is long. The width and length must be expressed in the same units. For example, to find the area of this rectangle, convert all feet to inches, then multiply.

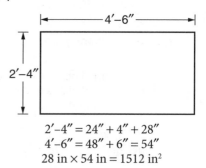

$$2'\text{-}4'' = 24'' + 4'' = 28''$$
$$4'\text{-}6'' = 48'' + 6'' = 54''$$
$$28 \text{ in} \times 54 \text{ in} = 1512 \text{ in}^2$$

A square foot is 12″ × 12″, or 144 in². If an area is given as a large number of square inches (in²), it can be converted to square feet (ft²) by dividing by 144. Consider the example above where the area of the rectangle is 1512 in². Divide by 144 to find the area in ft².

$$\frac{1512 \text{ in}^2}{144} = 10.5 \text{ ft}^2$$

Triangles

To find the area of any triangle, multiply the height times 1/2 the base.

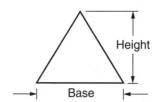 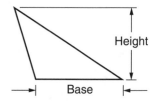

Find the area of this triangle.

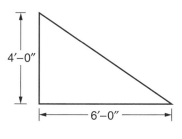

$$4 \text{ ft} \times \frac{6 \text{ft}}{2} = 4 \text{ ft} \times 3 \text{ ft} = 12 \text{ ft}^2$$

Another way to achieve the same result is to multiply the base times the height, then divide by 2.

$$4 \text{ ft} \times 6 \text{ ft} = 24 \text{ ft}^2$$

$$\frac{24 \text{ ft}^2}{2} = 12 \text{ ft}^2$$

Some figures may be made up of squares, rectangles, and triangles of varying sizes. To find the area of such a figure, break it into its various parts and find the area of each part, then add those areas.

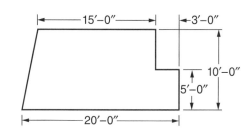

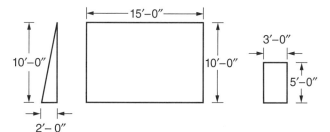

Triangle
$10 \times 2 = 20$
$20 \div 2 = 10 \text{ ft}^2$

Rectangle
$10 \times 15 = 150 \text{ ft}^2$

Rectangle
$3 \times 5 = 15 \text{ ft}^2$

$$10 \text{ ft}^2 + 150 \text{ ft}^2 + 15 \text{ ft}^2 = 175 \text{ ft}^2$$

Practice A-7

Find the areas of the figures that follow.

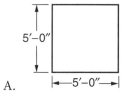

A.

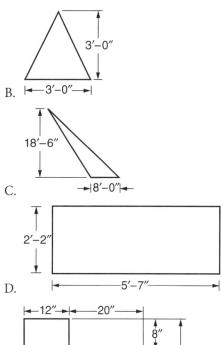

B.

C.

D.

E.

Circumference and Area of a Circle

The distance from a circle's center point to its outer edge is its *radius*. The total distance across a circle through its center point is its *diameter*.

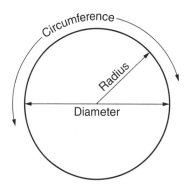

Many calculations involving circles or parts of circles use a constant of approximately 22/7, or 3.1416. The Greek letter π (pi) is used to represent this constant. It is a constant because it never changes, regardless of the dimensions of the circle.

The *circumference* of a circle is its perimeter. To find the circumference of a circle, multiply the diameter by π. This is the same as multiplying the radius by 2 and multiplying that product by π. Practice by finding the circumference of a circle with a diameter of 8′.

Appendix Math Review

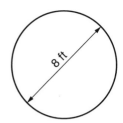

circumference = π × diameter
circumference = 3.1416 × 8′
circumference = 25.1328′

The area of a circle is found by multiplying π by the radius squared (the radius times the radius). Find the area of a circle with a radius of 3″.

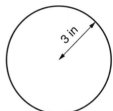

area = π × radius^2
area = 3.1416 × (3″)^2
area = 28.2744 in^2

Notice that in both examples using π, the answer is rounded to four decimal places. That is because π was rounded to four places. The answer cannot be accurate to more decimal places than the values used to calculate it.

Sometimes shapes encountered in the workplace are semicircles or quarter-circles. The areas and perimeters of these shapes can be found using the formulas for circles and dividing the result by 2 for a semicircle or 4 for a quarter-circle. For example, find the perimeter of the semicircular shape that follows.

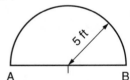

diameter = 2 × 5 ft = 10 ft
circumference = π × 10 ft = 31.416 ft
circumference of semicircular portion = $\frac{\text{circumference}}{2}$ = 15.708 ft
line AB = 10 ft

perimeter of semicircle = 10 ft + 15.708 ft = 25.708 ft

Practice A-8

Find the perimeter and area of each of these figures.

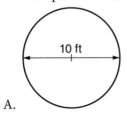

A.

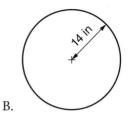

B.

C. Quarter-circle

Volume Measure

The volume of a solid is always measured in units of cubic inches, cubic meters, cubic feet, etc. When a number is cubed, that means it is multiplied by itself, then by itself again. For example, 3 cubed is 27 (3 × 3 × 3). Cubic units are written with a superscript 3, indicating units × units × units.

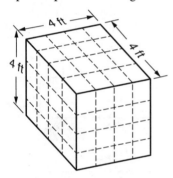

4 ft × 4 ft × 4 ft = 64 cubic feet, or 64 ft^3

This cube is made up of 64 individual cubes, each measuring 1 foot by 1 foot by 1 foot.

As long as a solid (a three-dimensional shape) has the same size and cross-sectional shape throughout its depth, its volume can be found by multiplying the area of one surface by the depth from that surface. To find the volume of a *cube* (all edges are the same size) or a rectangular solid (a rectangle with a third dimension), multiply the length, width, and height. For a 3-inch cube,

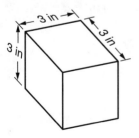

volume = 3 in × 3 in × 3 in = 27 in^3

For a rectangular solid,

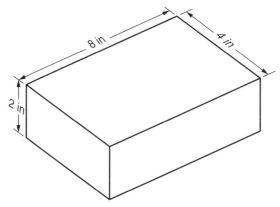

volume = 4 in × 8 in × 2 in = 64 in³

A solid with two opposite triangular faces is called a *triangular prism*. A solid with two circular faces is a *cylinder*. The volume of a triangular prism or a cylinder is found by multiplying the area of its face by its height.

Consider a triangular prism:

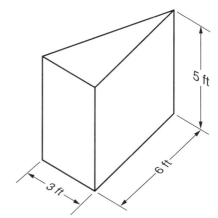

area of face = 1/2 × 3 ft × 6 ft = 9 ft²
volume = 9 ft² × 5 ft = 45 ft²

Consider a cylinder:

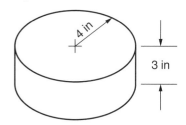

area of face = π × 4 in × 4 in = 50.27 in²
volume = 50.27 in² × 3 in = 150.81 ft³

Practice A-9

Find the volume of each solid.

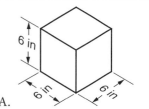

A.

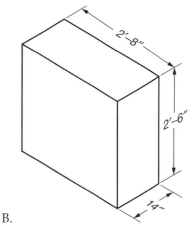

B.

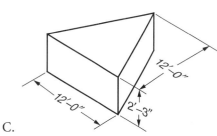

C.

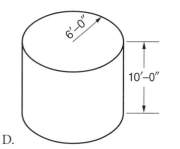

D.

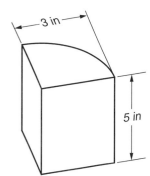

E. Quarter-cylinder

Exponents and Roots

When a number is squared or cubed, the little superscript number written to the right is called an *exponent*. For example, in the number 10^2, the exponent is 2, indicating the number is 10 multiplied by itself. Another way of saying this is "10 to the second power." If the number were to be $10 \times 10 \times 10$, it could be written as 10^3 and it could be called "10 cubed" or "10 to the third power." Exponents of 2 and 3 have names—squared and cubed, respectively—because they are the exponents used with area and volume measure. Higher exponents are only referred to as powers. For example, 10^5 is read as "10 to the 5th power." It is easier than saying "$10 \times 10 \times 10 \times 10 \times 10$." Both forms of that number equal 100,000.

Roots are the opposite of exponents. When you take the square root of a number, you are asking, "What number multiplied by itself is equal to this number?" For example, 4^2 equals 16. That means the square root of 16 is 4. Most square roots are not whole numbers. To find a square root, use the $\sqrt{}$ button on your calculator.

Practice A-10

A. What is 12 squared?
B. What is 8.5 cubed?
C. Calculate 4 to the 6th power.
D. What are two other ways to write $3 \times 3 \times 3 \times 3$?
E. What is the square root of 25?
F. Calculate $\sqrt{215}$.

Working with Right Angles

A *right angle* measures 90°. A triangle having a right angle is called a *right triangle*. It will be helpful to know a few terms associated with right triangles, **Figure A-11**.

A triangle can only have one 90° corner. The sum total of all three angles in a triangle is always 180°, so if one angle is 90°, the other two must add up to 90° together. *Hypotenuse* is a special term for the longest side of a triangle. The hypotenuse will always be across from the largest angle in a triangle.

The *Pythagorean theorem* is a principle that makes right triangles convenient to work with. Named after the Greek mathematician Pythagoras, the Pythagorean theorem states that the sum of the squares of the sides of a right triangle is equal to the square of the hypotenuse. To help keep track of the terms used in the Pythagorean theorem, it is common to label the two sides a and b and the hypotenuse c. Then the theorem can be stated as an equation:

$$a^2 + b^2 = c^2$$

If the lengths of the two sides of a right triangle are known, the Pythagorean theorem can be used to find the length of the hypotenuse. Consider a right triangle with sides equal to 3 ft and 6 ft:

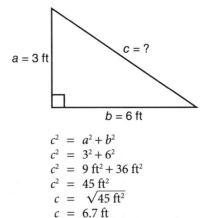

$$\begin{aligned} c^2 &= a^2 + b^2 \\ c^2 &= 3^2 + 6^2 \\ c^2 &= 9 \text{ ft}^2 + 36 \text{ ft}^2 \\ c^2 &= 45 \text{ ft}^2 \\ c &= \sqrt{45 \text{ ft}^2} \\ c &= 6.7 \text{ ft} \end{aligned}$$

The Pythagorean theorem can be used to find the length of any side of a right triangle if the other two are known. For example, if side b and the hypotenuse, c, are known, $a^2 + b^2 = c^2$ can be rearranged to $a^2 = c^2 - b^2$.

- $a^2 + b^2 = c^2$
- The equation stays in balance if you do the same thing on both sides of the equal sign.
- Subtract b^2 from both sides of the equation.
- $a^2 + b^2 - b^2 = c^2 - b^2$
- $a^2 = c^2 - b^2$

Practice with the following example:

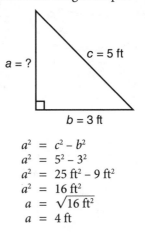

$$\begin{aligned} a^2 &= c^2 - b^2 \\ a^2 &= 5^2 - 3^2 \\ a^2 &= 25 \text{ ft}^2 - 9 \text{ ft}^2 \\ a^2 &= 16 \text{ ft}^2 \\ a &= \sqrt{16 \text{ ft}^2} \\ a &= 4 \text{ ft} \end{aligned}$$

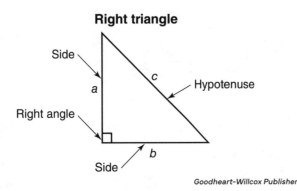

Figure A-11.

The same can be done to find side *b* when side *a* and the hypotenuse are known. Other formulas can be derived from the Pythagorean theorem, such as:

$$c = \sqrt{(a^2 + b^2)}$$
$$a = \sqrt{(c^2 - b^2)}$$
$$b = \sqrt{(c^2 - a^2)}$$

Computing Averages

An average is a typical value of one unit in a group of units. For example, if four windows have areas of 12.0 ft², 11.2 ft², 11.2 ft², and 14.5 ft², the average size of one of those windows is 12.2 ft². The average is computed by adding all of the units and dividing that sum by the number of units in the group.

```
12.0 ft²
11.2 ft²
11.2 ft²
14.5 ft²
48.9 ft²
```

```
     12.22
4)48.90
     4
     08
      8
      09
       8
       10
```

The quotient should be rounded off to the same number of decimal places as is used in the problem.

Average window size is 12.2 ft²

Practice A-11

Compute the averages of these groups.
A. 14, 14.4, 14.5, 15
B. 80 lb, 83 lb, 88 lb, 79.5 lb, 81.6 lb, 84 lb
C. 11 cubic yards, 13 cubic yards, 11.5 cubic yards, 12 cubic yards, 12.8 cubic yards

Percent and Percentage

A *percent* is one part in a hundred. One penny is 1% of a dollar. Twenty-five cents is 25% of a dollar. On the other hand, 25 cents is 50% of a half-dollar. If the half-dollar were divided into 100 parts of 1/2 cent each and the quarter were also divided into 1/2-cent increments, the quarter would equal 50 of those 1/2-cent increments.

Think of percent as hundredths. To find a given percentage of an amount, multiply the amount times the desired number of hundredths. For example, say you want to find 12% of $4.40 (12% is 0.12 times the whole):

$4.40 × 0.12 = $0.528, or 53 cents

Percent is sometimes interchanged with *percentage* in common usage, but there is a slight difference. *Percent* should be used when a specific number is used, such as 15% of the labor force. *Percentage* should be used when no specific number is used with the term, such as a large percentage of our homes are green.

Using your knowledge of equations and solving for an unknown, you can calculate the whole if you know the percentage. For instance, what was the total spent on tools if $22.00 were spent on sales tax and the tax rate is 8%?

1. Write an equation with the facts you know.

$22.00 = 8% × total

2. Write percent as hundredths.

$22.00 = 0.08 × total

3. Divide each side of the equation by 0.08.

Total = $275

Practice A-12

A. What is 10% of 225?
B. What is 65% of $350.50?
C. If merchandise cost $22.25 and the bill comes to $23.14, what percentage was added for sales tax?
D. If a 500-gallon tank is 60% full of water, how much water does it contain?

Ohm's Law Calculations

When performing Ohm's law calculations, all values must be converted into a base unit before the calculation can be performed. Some important prefixes you may use include the following:

- Milli: one thousandth, or 1/1000, or 0.001, or $1/(10^3)$ (also 10^{-3})
- Base unit: ones unit, or 10^0
- Kilo: one thousand, or 1000, or 10^3
- Mega: one million, or 1,000,000, or 10^6

When converting from one prefix to another, first examine the placement value (the exponent of the 10), then examine the final placement value. Take the difference between these two values and move the decimal point either left or right, depending on the movement. For example:

750 ohms = x kilo-ohms (often shortened to kilohms)

The base unit is 750 (10^0), while kilohms is in 10^3, a difference of 3 (3 − 0 = 3). The decimal point is moved three points to the left, thus 750 becomes 0.750 kilohms. Because we moved down the list of prefixes (from ones to kilo), the decimal was moved to the left.

Let's try another example:

1.2 mega-ohms (often shortened to megohms) = x ohms

When converting 1.2 megohms (10^6) to the base unit (10^0), there is a difference of six. Because we moved up the list of prefixes (from a larger unit to a smaller unit), the decimal point must move six places to the right. As a result, 1.2 megohms equals 1,200,000 ohms.

This same principle applies across all conversions: resistance, current, and voltage. There are other prefixes, but they are not normally used in electrical calculations. For instance, you would not say 1.25 centi-ohms; you would simply say 12.5 ohms.

Appendix   Math Review

Service Calculation "Standard Method" Dwelling Unit (Single Family)			
Step 1 **General Lighting, Small Appliance, Laundry**		**Ungrounded**	**Neutral**
Dwelling area "sq ft" _____ × 3 VA = _____			
Small appliance _____ × 1500 VA = _____			
Laundry _____ × 1500 VA = _____			
Total _____			
First 3,000 at 100% _____ 3,001 to 120,000 × 35% _____ × 0.35 _____ Remainder over 120,000 × 25% _____ × 0.25		_____	
Step 2 **Fixed Appliances**			
Appliance	Ungrounded Neutral		
_____	_____ _____		
_____	_____ _____		
_____	_____ _____		
_____	_____ _____		
_____	_____ _____		
_____	_____ _____		
_____	_____ _____		
Total Four or more × 0.75	_____ _____		
Step 3 **Dryer**			
5000 VA or the nameplate rating - neutral 70% of ungrounded		_____	_____
Step 4 **Range**			
Range calculation - neutral 70% of ungrounded		_____	_____
Step 5 **Heat/Air Conditioning**			
Include the larger of the two/nonconcidence load			
	Ungrounded Neutral		
_____	_____ _____		
_____	_____ _____		
Step 6 **Electric Vehicle Supply Equipment**			
7200 VA or the nameplate rating		_____	
Step 7 **Other Loads**			
	Ungrounded Neutral		
_____	_____ _____		
_____	_____ _____		
Total			
Step 8 **Largest Motor**			
	Ungrounded Neutral		
Largest motor × 0.25	_____ _____		
Total in Volt-Amperes Total		_____	_____
Total Amperes = Total VA/240 V Total		_____	_____

Goodheart-Willcox Publisher

Service Calculation "Optional Method" Dwelling Unit (Single Family)

		Ungrounded
Step 1 — General Lighting, Small Appliance, Laundry		

Dwelling Area "sq ft" _____ × 3 VA = _____

Small Appliance _____ × 1500 VA = _____

Laundry _____ × 1500 VA = _____

Total _____

Appliances (Nameplate Rating)

Appliance

_____   _____
_____   _____
_____   _____
_____   _____
_____   _____
_____   _____
_____   _____
_____   _____
_____   _____
_____   _____
_____   _____
_____   _____
_____   _____
_____   _____
_____   _____

Total _____

General Load Demand Factor

Sum of General lighting and appliances: _____

First 10,000 at 100%		10,000
Remainder at 40%	0.4 × _____	_____
Total		_____

Step 2 — Heat/Air Conditioning

Include the largest of the following:

100% of Air Conditioning _____
100% of Heat Pump (Without Supplemental Electric) _____
100% of Heat Pump and 65% of Supplemental Electric) _____
65% of Electric Space Heating (Less Than Four Controllers) _____
40% of Electric Space Heating (More Than Four Controllers) _____
100% of Electric Thermal Storage System _____

| **Total in Volt-Amperes** | **Total** _____ | _____ |
| **Total Amperes = Total VA/240V** | **Total** _____ | _____ |

Goodheart-Willcox Publisher

Glossary

A

adjustable speed drive. A piece of power conversion equipment that provides a means of adjusting the speed of an electric motor. (7)

adjustment factor. A multiplier applied to a conductor's maximum ampacity to compensate for the heating effect of having multiple conductors grouped together. (16)

aluminum. A conductive material whose benefits are that it is low cost, lightweight, and easy to bend. However, this material has a higher resistance than some materials and tends to corrode or oxidize when exposed to air. (16)

ampacity. The maximum current, in amperes, that a conductor can carry continuously under the conditions of use without exceeding its temperature rating. (16)

appliance. A piece of utilization equipment, generally other than industrial, that is normally built in standardized sizes or types and is installed or connected as a unit to perform one or more functions. (7)

appliance branch circuit. A branch circuit that supplies energy to one or more outlets to which appliances are to be connected and that has no permanently connected luminaires that are not a part of an appliance. (13)

arc blast. An explosion associated with an arcing incident with extremely high temperatures, blinding light, loud noise, toxic gases, and shrapnel. (4)

arc-fault circuit interrupter (AFCI). A device intended to provide protection from the effects of arc faults by recognizing characteristics unique to arcing and de-energizing the circuit when an arc fault is detected. (13)

arc flash. A rapid release of energy that occurs when two metal components with different electrical potentials come into contact. (4)

article. A chapter subdivision that covers a specific topic of the *NEC*. (2)

artificially made body of water. A body of water that has been constructed or modified to fit some decorative or commercial purpose. (9)

assembly occupancy. An area that is designed or intended for 100 or more persons to gather. (8)

audio signal processing equipment. A piece of electrically operated equipment that produces and/or processes electronic signals that, when appropriately amplified and reproduced by a loudspeaker, produce an acoustic signal within the range of normal human hearing. (9)

auxiliary gutter. An enclosure to supplement wiring spaces at meter centers, distribution centers, switchgear, switchboards, and similar points of wiring systems. (15)

B

bare conductor. A conductor having no covering or electrical insulation. (15)

barrier. A divider that is installed in a device box to separate the box into separate spaces. (14)

bathroom. An area including a sink (basin) with one or more of the following: a toilet, a urinal, a tub, a shower, a bidet, or similar plumbing fixtures. (13)

bonding. The act of connecting metal components to establish electrical continuity and conductivity. (5, 19)

box volume. The amount of space in cubic inches available for conductors, devices, and clamps. (14)

branch circuit. The circuit conductors between the final overcurrent device protecting the circuit and the outlet(s). (5, 13)

C

cabinet. An enclosure that is designed for either surface mounting or flush mounting and is provided with a frame, mat, or trim in which a swinging door or doors are or can be hung. (6)

cable. A factory assembly of more than one conductor. Permitted to be installed in their listed environment without the need to be contained in a raceway. (6, 16)

chapter. The major subdivisions of the *NEC*. (2)

circular mil. The unit of area for a circle with a diameter of one mil. (12)

Class 1 circuit. The portion of the wiring system between the load side of the overcurrent device or power-limited supply and the connected equipment. (10)

Note: The number in parentheses following each definition indicates the unit in which the term can be found.

Class 2 circuit. The portion of the wiring system between the load side of a Class 2 power source and the connected equipment. Due to its power limitations, a Class 2 circuit considers safety from a fire initiation standpoint and provides acceptable protection from electric shock. (10)

Class 3 circuit. The portion of the wiring system between the load side of a Class 3 power source and the connected equipment. Due to its power limitations, a Class 3 circuit considers safety from a fire initiation standpoint. Since a Class 3 circuit permits higher levels of voltage than a Class 2 circuit, additional safeguards are specified to provide protection from an electric shock hazard that could be encountered. (10)

Class I, Division 1 location. A class/division classification describing an area where flammable gases or vapors may exist under normal operating conditions. (8)

Class I, Division 2 location. A class/division classification describing an area where flammable gases or vapors are stored or handled but are confined within containers and therefore not present under normal conditions. (8)

Class II, Division 1 location. A class/division classification describing an area where combustible dusts may exist in quantities sufficient to produce ignition under normal operating conditions. (8)

Class II, Division 2 location. A class/division classification describing an area where combustible dusts may be present but not in quantities sufficient to cause ignition under normal conditions. (8)

Class III, Division 1 location. A class/division classification describing an area where ignitable fibers or flyings are handled or manufactured and may exist under normal operating conditions. (8)

Class III, Division 2 location. A class/division classification describing an area where ignitable fibers are stored or handled and should not be present under normal conditions. (8)

code-making panel (CMP). A group of individuals from varied backgrounds that are responsible for considering and voting upon public input and public comments that relate to a specific portion of the *NEC*. (1)

combustible dust. Dust particles that are 500 microns or smaller and present a fire or explosion hazard when dispersed and ignited in the air. (8)

communication circuit. A metallic, fiber, or wireless circuit that provides voice/data (and associated power) for communications-related services between communication equipment. (11)

compact conductor. A stranded conductor with irregularly shaped strands that have been formed together, eliminating the air spaces between the strands. (12, 15)

conductor. A wire made of copper, aluminum, or copper-clad aluminum that can carry current. (15)

conduit. A circular raceway that houses and protects the conductors installed in an electrical system. (15)

conduit body. A separate portion of a conduit or tubing system that provides access through a removable cover(s) to the interior of the system at a junction of two or more sections of the system or at a terminal point of the system. (6, 14)

confined space. An area that has limited means of entrance and egress. (4)

consensus standard. A standard that takes the viewpoints of a group of individuals from varied backgrounds into consideration during the standards development process. (1)

continuous load. A load where the maximum current is expected to continue for 3 hours or more. (13)

controller. A device or group of devices that serves to govern in some predetermined manner the electric power delivered to the apparatus to which it is connected. (7)

copper. A conductive material whose benefits include that it has a lower resistance than some materials, can carry more current, and operates better under temperature variations. However, this material is expensive and heavier than other materials. (16)

correction factor. A multiplier applied to a conductor's maximum ampacity to compensate for the conductor being in an area where the ambient temperature is above or below the acceptable *NEC* value. (16)

counter-mounted cooking unit (cooktop). A cooking appliance designed for mounting in or on a counter and consisting of one or more heating elements, internal wiring, and built-in or mountable controls. (20)

covered conductor. A conductor encased within material of composition or thickness that is not recognized by the *Code* as electrical insulation. (15)

critical operations power system (COPS). A power system for facilities or parts of facilities that require continuous operation for the reasons of public safety, emergency management, national security, or business continuity. (10)

cutout box. An enclosure designed for surface mounting that has swinging doors or covers secured directly to and telescoping with the walls of the enclosure. (6)

D

demand factor. The ratio of the maximum demand of a system, or part of a system, to the total connected load of a system or the part of the system under consideration. (20, 21)

device. A unit of an electrical system, other than a conductor, that carries or controls electrical energy as its principal function. (14)

device box. A box that will house a device such as a switch or receptacle. (14)

E

Edison-base fuse. A screw-in type of fuse that is no longer permitted to be installed in new installations. Its screw shell does not restrict the size of fuse installed. (17)

effective ground-fault current path. An intentionally constructed, low-impedance electrically conductive path designed and intended to carry current under ground-fault conditions from the point of a ground fault on a wiring system to the electrical supply source and that facilitates the operation of the overcurrent protective device or ground-fault detectors. (19)

egress. The act of leaving a space. (4)

electric sign. A fixed, stationary, or portable self-contained electrically operated and/or electrically illuminated utilization equipment with words or symbols designed to convey information or attract attention. (9)

electric welder. A piece of equipment that uses electricity to provide the heat necessary to join pieces of metal together. (9)

electrified truck parking space. A truck parking space that has been provided with an electrical system that allows truck operators to connect their vehicles while stopped and use off-board power sources to operate on-board systems without any engine idling. (9)

electrolytic cell. A tank or vat in which electrochemical reactions are caused by applying electric energy for the purpose of refining or producing usable materials. (9)

electroplating. The process of coating a metal object with another metal by the process of electrolytic deposition. (9)

emergency system. A system legally required and classed as emergency by municipal, state, federal, or other codes or by any governmental agency having jurisdiction. These systems are intended to automatically supply illumination, power, or both to designated areas and equipment in the event of failure of the normal supply or in the event of accident to elements of a system intended to supply, distribute, and control power and illumination essential for safety to human life. (10)

enclosure. The case or housing of apparatus, or the fence or walls surrounding an installation to prevent personnel from accidentally contacting energized parts or to protect the equipment from physical damage. (4, 6)

energy management system. A system consisting of any of the following: a monitor(s), communications equipment, a controller(s), a timer(s), or other device(s) that monitors and/or controls an electrical load or a power production of a storage source. (10)

energy storage system. One or more components assembled capable of storing energy and providing electrical energy into the premises wiring system or an electric power production and distribution network. (10)

equipment. A general term, including fittings, devices, appliances, luminaires, apparatus, machinery, and the like used as a part of, or in connection with, an electrical installation. (7)

equipment grounding conductor (EGC). A conductive path that is part of an effective ground-fault current path and connects normally non-current-carrying metal parts of equipment together and to the system grounded conductor or to the grounding electrode conductor, or both. (5, 19)

exception. A situation in which a requirement does not apply. These notes are placed immediately following the requirement for which the exception applies. (3)

extension ring. A component added to the top of a box to increase its volume. They are the same length and width of the box but have an open back to allow access to the original box. (14)

F

fault current. The amount of current available in the event of a short circuit or ground fault. (4)

fault-managed power system. A powering system that monitors for faults and controls current delivered to ensure fault energy is limited. (10)

feeder. The circuit conductors between the service equipment and the source of a separately derived system, or the other power supply source and the final branch-circuit overcurrent device. (5)

fire alarm system. A fire alarm control panel that monitors the building, connects to other building systems (such as elevators and HVAC), will signal the alarms in the case of a fire, and can notify the fire department. (10)

fire pump. An electric pump used to increase the water pressure to a fire suppression system. (9)

fire-resistive cable system. A cable and components used to ensure survivability of critical circuits for a specified time under fire conditions. (10)

first draft report. A report containing public inputs, CMP actions and statements, and first revisions. (1)

first revision. An edited portion of the *NEC* that has been rewritten to include the accepted public input. (1)

fixture stud. A device that is installed inside an electrical box to support a luminaire. (14)

fuel cell system. The complete aggregate of equipment used to convert chemical fuel into usable electricity and typically consisting of a reformer, stack, power inverter, and auxiliary equipment. (9)

G

general chapters. The first four chapters of the *NEC* that apply generally to all installations. (2)

general-purpose branch circuit. A branch circuit that supplies two or more receptacles or outlets for lighting and appliances. The most common type of branch circuit. (13)

general unit load. The value of expected energy use based on the useable square footage of a dwelling unit. (21)

general-use snap switch. A form of general-use switch constructed so that it can be installed in device boxes or on box covers, or otherwise used in conjunction with wiring systems recognized by the *Code*. (7)

general-use switch. A switch intended for use in general distribution and branch circuits. (7)

grade. The final elevation of the ground. (13)

ground. The earth. (19)

ground fault. An unintentional, electrically conducive connection between an ungrounded conductor of an electrical circuit and the normally non-current-carrying conductors, metallic enclosures, metallic raceways, metallic equipment, or earth. (5, 17, 19)

grounded. The state of being connected to the ground or a conductive body that extends the ground connection. (19)

grounded conductor. A system or circuit conductor that is intentionally grounded. (5, 19)

ground-fault circuit interrupter (GFCI). A device intended for the protection of personnel that de-energizes a circuit or portion thereof within an established period of time when a ground-fault current exceeds the values established for a Class A device. (13)

grounding. The act of connecting a system to ground or to a conductive body that extends the ground connection. (19)

grounding electrode. A conducting object through which a direct connection to earth is established. (5, 19)

grounding electrode conductor (GEC). A conductor used to connect the system grounded conductor or the equipment to a grounding electrode or to a point on the grounding electrode system. (5, 19)

grounding electrode system. A system of grounding electrodes and grounding electrode conductors that are connected together. (19)

H

habitable room. A room in a building for living, sleeping, eating, or cooking, but excluding bathrooms, toilet rooms, closets, hallways, storage or utility spaces, and similar areas. (5, 13)

hazardous location. An area that may contain combustible gasses, dusts, or fibers. (4)

hermetic refrigerant motor-compressor. A combination consisting of a compressor and motor, both of which are enclosed in the same housing, with no external shaft or shaft seals, with the motor operating in the refrigerant. (7)

hickey. A device that attaches a fixture nipple to the fixture stud. (14)

I

impedance. The total opposition to current flow in an ac circuit. The combination of reactance and resistance. (12)

individual branch circuit. A branch circuit that supplies only one utilization equipment. (5, 13)

induction heating, melting, and welding. The heating, melting, or welding of a nominally conductive material due to its own I^2R losses when the material is placed in a varying electromagnetic field. (9)

industrial control panel. An assembly of two or more components consisting of one of the following: 1) power circuit components only, 2) control circuit components only, 3) a combination of power and control circuit components. (7)

industrial machinery. A power-driven machine or system, not portable by hand while working, that is used to process materials by cutting; forming; pressure; electrical, thermal, or optical techniques; lamination; or a combination of these processes. (9)

information technology equipment (ITE). The equipment and systems rated 1000 volts or less, normally found in offices or other business establishments and similar environments classified as ordinary locations, that are used for creation and manipulation of data, voice, video, and similar signals that are not communications equipment. (9)

informational note. The explanatory material that immediately follows a section and provides the *Code* user additional information. (3)

informative annex. An appendix that contains additional information to help with understanding and application but is not an enforceable part of the *NEC*. (12)

insulated conductor. A conductor encased within material of composition and thickness that is recognized by the *Code* as electrical insulation. (15, 16)

interconnected electric power production source. A power supply that is installed in parallel with the primary source of electricity. (10)

intersystem bonding termination. A device that provides a means for connecting intersystem bonding conductors for communications systems to the grounding electrode system. (19)

intrinsically safe circuit. A circuit in which any spark or thermal effect is incapable of causing ignition of a mixture of flammable or combustible material in air under prescribed test conditions. (8)

irrigation machine. An electrically driven or controlled machine, with one or more motors, not hand-portable, and used primarily to transport and distribute water for agricultural purposes. (9)

J

junction box. A box where conductors are housed, splice, or terminate. (14)

K

keyword. An important term related to a topic. These are used to help locate information in the table of contents and the index of the *NEC*. (3)

knockout. An opening in a box or enclosures for the installation of cables and raceways. (12)

L

legally required standby system. A system required and so classed as legally required standby by municipal, state, federal, or other codes or by any governmental agency having jurisdiction. These systems are intended to automatically supply power to selected loads (other than those classed as emergency systems) in the event of a failure of the normal source. (10)

lighting outlet. An outlet intended for the direct connection of a lamp holder or luminaire. (14)

lighting outlet box. A box that is intended to connect and support a luminaire. (14)

listed. A label given by NRTLs that a product passes minimum safety requirements, and the product will be safe if it is installed under the conditions of use it was tested for. (4)

load. The portion of an electrical system that converts electrical energy into a useful form of energy, such as light, heat, and rotation. (20, 21)

location, damp. A location protected from weather and not subject to saturation with water or other liquids but subject to moderate degrees of moisture. (16)

location, dry. A location not normally subject to dampness or wetness. (16)

location, wet. An installation underground or in concrete slabs or masonry in direct contact with the earth; in locations subject to saturation with water or other liquids; and in unprotected locations exposed to weather. (16)

luminaire. A complete lighting unit consisting of a light source together with the parts designed to position the light source and connect it to the power supply. (7)

M

main bonding jumper (MBJ). A connection between the grounded circuit conductor and the equipment grounding conductor and/or supply-side bonding jumper at the service. (19)

mandatory language. The language used when a specific action is required or prohibited. Written using the words *shall* or *shall not*. (3)

manufactured building. Any building (other than manufactured homes, mobile homes, park trailers, or recreational vehicles) that is of closed construction and made or assembled in a manufacturing facility for installation, or for assembly and installation, on the building site. (8)

manufactured wiring system. A system containing component parts that are assembled in the process of manufacture and cannot be inspected at the building site without damage or destruction to the assembly. Used for the connection of luminaires, utilization equipment, continuous plug-in type busways, and other devices. (9)

meter socket. The enclosure that contains the utility meter that measures the amount of power consumed. (6)

mil. A unit of measurement that is equal to 1/1000 of an inch. (12)

modular data center. A prefabricated unit, rated 1000 volts or less, consisting of an outer enclosure housing multiple racks or cabinets of information technology equipment (ITE) and various support equipment. (9)

multiwire branch circuit. A branch circuit that consists of two or more ungrounded conductors that have a voltage between them, and a grounded conductor that has equal voltage between it and each ungrounded conductor of the circuit and that is connected to the neutral or grounded conductor of the system. (13)

N

National Fire Protection Association (NFPA). The organization which publishes the *NEC*, as well as many other standards. It is a nonprofit organization that is dedicated to the prevention of death, injury, and losses associated with fire, electricity, and other hazards. (1)

Nationally Recognized Testing Laboratory (NRTL). An organization recognized by OSHA to perform product safety testing. (4)

natural body of water. A body of water such as a lake, stream, pond, river, or other naturally occurring body of water, which may vary in depth throughout the year. (9)

NEC style manual. A writing tool for the *NEC* that is published by the NFPA. Its purpose is to ensure that the language and structure of the *NEC* are explicit, consistent, and promote uniform interpretation. (1)

Glossary

network interface unit. A device that converts a broadband signal into component voice, audio, video, data, and interactive services signals and provides isolation between the network power and the premises signal circuits. (11)

network terminal. A device that converts network-provided signlas into component signals and is considered a network device on the premises that is connected to a communications service provider and is powered at the premises. (11)

neutral conductor. The conductor connected to the neutral point of a system that is intended to carry current under normal conditions. (5)

NFPA standards council. A group of individuals that oversee NFPA standards development activities. They ensure compliance with NFPA rules and regulations and serve as the appeals body for standards development matters. (1)

nipple. A raceway that is 24″ or less in length. (12, 15)

nominal voltage. A value assigned to a circuit or system for the purpose of designating its voltage class. (4)

noncoincident loads. A series of loads that are unlikely to be used simultaneously. (21)

nonrenewable. A device that cannot be reset and must be replaced. (17)

O

office furnishing. A cubicle panel, partition, study carrel, workstation, desk, shelving system, or storage unit that may be mechanically and electrically interconnected to form an office furnishing system. (9)

one-shot bender. A conduit bending tool that makes a complete bend in one step rather in multiple steps. (12)

optional method (service calculation). An alternative method to the standard calculation of performing a service calculation. It can be used for dwellings, schools, restaurants, and existing installations. This method generally produces a smaller value than is found using the standard calculation. (21)

optional standby system. A system intended to supply power to public or private facilities or property where life safety does not depend on the performance of the system. These systems are intended to supply on-site generated or stored power to selected loads either automatically or manually. (10)

outlet. A point in the wiring system where current is taken to supply utilization equipment. (14)

outlet box. The box where power is taken to supply utilization equipment. (14)

oven. A permanently installed compartment with a door designed to cook food. It will often have a heating element on the top and bottom of the heating compartment. (20)

overcurrent. Any current in excess of the rated current of equipment, or the ampacity of a conductor. It may result from overload, short circuit, or ground fault. (5, 17)

overcurrent protective device (OCPD). A device capable of providing protection for service, feeder, and branch circuits and equipment over the full range of overcurrents between its rated current and its interrupting rating. (17)

overhead service conductor. The overhead conductors between the service point and the first point of connection to the service-entrance conductors at the building or other structure. (18)

overload. The operation of equipment in excess of normal, full-load rating, or of a conductor in excess of its ampacity that, when it persists for a sufficient length of time, would cause damage or dangerous overheating. (5, 17)

overvoltage. An increase in voltage beyond the maximum operating voltage of the system. (5)

P

panelboard. A single panel or group of panel units designed for assembly in the form of a single panel, including buses and automatic overcurrent devices, and equipped with or without switches for the control of light, heat, or power circuits. Designed to be placed in a cabinet or cutout box placed in or against a wall, partition, or other support, and accessed only from the front. (7)

parallel conductors. Two or more conductors that are electrically joined at both ends, making them equivalent to one large conductor. (16)

park trailer. A unit that is built on a single chassis mounted on wheels and has a gross trailer area not exceeding 37 m^2 (400 ft^2) in the set-up mode. (8)

part. The subdivisions of an article that divides the requirements/information into logical groupings. (2)

permissive language. The language used when a specific language is allowed, but not required. Written using the words *shall be permitted* or *shall not be required*. (3)

phase converter. An electrical device that converts single-phase power to 3-phase electric power. (7)

pipe organ. A musical instrument that produces sound by driving pressurized air (called wind) through pipes selected via a keyboard. (9)

plaster ring (mudring). A component that is installed on a box (typically 4″ square) to provide an opening in the finished wall surface for the installation of a device or luminaire. (14)

plug fuse. A screw-in type fuse that is no longer permitted to be installed in new construction. (17)

public comment. A comment submitted by the general public relating to the first draft report. (1)

public input. A proposed change to the current edition of the *NEC*. (1)

pull box. A box that is used as a pull point in the middle of the conduit run. (14)

R

raceway. An enclosed channel designed expressly for holding wires, cables, or busbars, with additional functions as permitted. (6, 15)

raised cover. A component that provides a means of mounting devices to square metallic boxes in areas where the raceway and boxes are exposed. (14)

range. A cooktop and an oven built in one unit. (20)

reactance. The nonresistive opposition to current flow that is present in ac circuits. May be capacitive, inductive, or a combination of the two. (12)

reactive load. A load with a current flow that is either leading or lagging the applied voltage. (21)

receptacle. A contact device installed at the outlet for the connection of an attachment plug, or for the direct connection of electrical utilization equipment designed to mate with the corresponding contact device. (7)

relocatable structure. A factory-assembled structure, or structures transportable in one or more sections, built on a permanent chassis, and designed to be used as other than a dwelling unit without a permanent foundation. (8)

remote-control circuit. A electrical circuit that controls any other circuit through a relay or an equivalent device. (10)

resistive load. A load that has a current flow that is in phase with the applied voltage. (21)

rough-in. The stage of the construction process where boxes, raceways, and cables are installed in the walls ahead of the finished wall surfaces. (2)

S

second draft report. A report containing the public comments, CMP actions and responses, and second revisions. (1)

second revision. An edited portion of the *NEC* that has been rewritten to reflect the changes from the second draft meeting. (1)

section. The divisions of an article for rules and requirements. (2)

separately derived system. An electrical source, other than a service, having no direct connection(s) to circuit conductors of any other electrical source other than those established by grounding and bonding connections. (19)

service. The conductors and equipment connecting the serving utility to the wiring system of the premises served. (5, 18)

service conductor. The conductors from the service point to the service disconnecting means. (18)

service drop. The overhead conductors between the serving utility and the service point. (18)

service equipment. The necessary equipment, consisting of a circuit breaker(s) or switch(es) and fuse(s) and their accessories, connected to the serving utility and intended to constitute the main control and disconnect of the serving utility. (18)

service head. A cap that prevents snow and rain from entering the raceway. (18)

service lateral. The underground conductors between the utility electric supply system and the service point. (18)

service mast. A raceway that extends through the roof for the attachment of the service-drop or overhead service conductors. (18)

service point. The point of connection between the facilities of the servicing utility company and the premises wiring. (18)

service-entrance conductor. The service conductors between the terminals of the service equipment to the service drop, overhead service conductors, service lateral, or underground service conductors. (18)

shore power. The electrical equipment required to power a floating vessel including, but not limited to, the receptacle and cords. (8)

short circuit. An excess current resulting from contact between two ungrounded conductors with a difference in potential, or an ungrounded conductor and a grounded (neutral) conductor. (17)

short radius conduit body. A conduit body that is too small to house anything but the conductors passing through. (14)

solar photovoltaic (PV) system. The total components, circuits, and equipment up to and including the PV system disconnecting means that, in combination, convert solar energy into electric energy. (9)

solidly grounded. The state of a system being connected to ground without inserting any resistor or impedance device. (19)

special chapters. The three chapters of the *NEC* that address special occupancies, special equipment, and special conditions that aren't common to every electrical installation. Contains requirements that will supplement or modify the rest of the *NEC*. (2)

stand-alone system. A system that is capable of supplying power independent of an electric power production and distribution network. (10)

standard method (service calculation). The methodical approach of calculating the loads in an electrical installation to determine the minimum size of the electric service. (21)

stranded conductor. A conductor consisting of multiple layers of strands, allowing for greater flexibility than a solid conductor. (12)

surge protective device. A protective device for limiting transient voltage by diverting or limiting surge current. (18)

switch leg. The conductor that is being controlled by a switch. (5)

switchboard. A large single panel, frame, or assembly of panels on which are mounted on the face, back, or both, switches, overcurrent and other protective devices, buses, and usually instruments. (7)

switchgear. An assembly completely enclosed on all sides and top with sheet metal (except ventilating openings and inspection windows) and containing primary power circuit switching, interrupting devices, or both, with buses and connections. (7)

system bonding jumper (SBJ). The connection between the grounded circuit conductor and the supply-side bonding jumper, or the equipment grounding conductor, or both, at a separately derived system. (19)

T

tail. A small length of wire that extends from a wire splice to a device or luminaire. (14)

technical correlating committee (TCC). A committee of individuals with varied backgrounds that correlate changes to the *NEC* brought forward by the code-making panels with other portions of the *NEC* as well as other standards. (1)

temporary installation. An installation with time constraints where the wiring is installed for a limited amount of time after which it will be removed. (8)

tentative interim amendment (TIA). A revision to the *NEC* that is implemented between *Code* cycles. (1)

torque. The amount of rotational force applied when tightening a fastener or termination. (12)

transformer. A device that changes the voltage of an ac system either up or down. (7)

trim. The stage of the construction process where electrical devices, such as lights and appliances are installed. This stage generally happens after the wall and ceiling finishes have been installed. (2)

tubing. A type of circular raceway that has a thinner wall and is therefore easier to bend and work with than some other circular raceways. (15)

Type S fuse. A type of fuse that screws into an adapter designed to restrict the size fuse it will accept. (17)

U

underground service conductors. The underground conductors between the service point and the first point of connection to the service-entrance conductors in a terminal box, meter, or other enclosure, inside or outside the building wall. (18)

ungrounded. The state of being not connected to ground or to a conductive body that extends the ground connection. (5)

W

wind turbine. A mechanical device that converts wind energy to electrical energy. (9)

wireway. A trough with a hinged or removable cover that houses and protects conductors and cables. (15)

Y

yoke. The mounting bracket portion of a device that is used to fasten it to the box. (14)

Z

Zone 0 location. A zone classification describing an area where flammable gases are present continuously. (8)

Zone 1 location. A zone classification describing an area where flammable gases are present under normal conditions. (8)

Zone 2 location. A zone classification describing an area where gases will not be present unless there is a malfunction. (8)

Zone 20 location. A zone classification describing an area where combustible dusts and ignitable fibers are present continuously. (8)

Zone 21 location. A zone classification describing an area where combustible dusts and ignitable fibers are present under normal conditions. (8)

Zone 22 location. A zone classification describing an area where dust and fibers are stored or handled but are not likely to be present unless there is a malfunction. (8)

Index of *NEC* References

Article 90: 22
   Section 90.2: 22
   Section 90.3: 22–23, 198, 268
   Section 90.4: 23–24
   Section 90.4(A): 23
   Section 90.4(B): 23
   Section 90.4(C): 23
   Section 90.4(D): 23
   Section 90.5: 24
   Section 90.6: 24
   Section 90.7: 24
   Section 90.9: 25
Chapter 1: 12, 22, 36, 38*f*
   Article 100: 13, 36, 58, 79, 264
   Article 110: 13, 37
      Part I: 37–40
      Part II: 41
      Part III: 41
      Part IV: 41
      Part V: 41–43
      Section 110.2: 37
      Section 110.3: 37
      Section 110.3(B): 37, 229
      Section 110.5: 282
      Section 110.12: 37
      Section 110.12(A): 37
      Section 110.13: 37
      Section 110.14: 39–40
      Section 110.14(C): 39, 294, 386
      Section 110.14(C)(1): 296, 318
      Section 110.14(D): 40
      Section 110.15: 40
      Section 110.16: 40
      Section 110.24: 40
      Section 110.26: 41, 99, 330
      Table 110.26(A)(1): 331, 332
      Section 110.26(C): 331
      Section 110.26(C)(2)(a): 332
      Section 110.26(C)(3): 332
      Section 110.26(D): 331
      Section 110.26(E): 331
      Section 110.28: 41
      Table 110.28: 41, 42*f*

Chapter 2: 13, 22, 50, 51*f*, 60, 60*f*
   Article 200: 50–52
      Section 200.1: 51
      Section 200.6: 51
      Section 200.10: 52
      Section 200.10(B): 52
      Section 200.10(C): 52
   Article 210: 52–55, 53*f*, 217, 218, 224
      Part I: 52
      Part II: 52–55
      Part III: 55, 224
      Section 210.2: 52
      Section 210.8: 52, 54, 220
      Section 210.8(A)(1): 228
      Section 210.8(A)(5): 226
      Section 210.8(B): 24
      Section 210.8(C): 233
      Section 210.11: 57
      Section 210.11(C): 224, 317
      Section 210.11(C)(1): 224, 398
      Section 210.11(C)(2): 224
      Section 210.11(C)(3): 224
      Section 210.11(C)(4): 224
      Section 210.12: 220, 226
      Section 210.18: 52, 53
      Section 210.19: 221–222
      Section 210.19(A): 222
      Section 210.19(C): 385, 387
      Section 210.20: 222
      Section 210.22: 54
      Section 210.50(C): 224
      Section 210.52: 224
      Section 210.52(A): 55, 225–226, 232
      Section 210.52(A)(1–4): 234
      Section 210.52(B)(3): 227
      Section 210.52(D): 228
      Section 210.52(E): 230
      Section 210.52(G)(1): 231
      Section 210.52(H): 231
      Section 210.52(I): 231
      Figure 210.52(C)(1): 227
      Section 210.63(A): 230
      Section 210.65: 234

Note: Page numbers followed by *f* indicate figures.

Section 210.70: 231
Section 210.70(8)(E): 234
Section 210.70(A)(1): 232
Section 210.70(A)(2)(1): 233
Section 210.70(A)(2)(2): 233
Section 210.70(C): 233
Article 215: 55f
　Section 215.2(A)(2): 221
Article 220: 55–57, 56f
　Section 220.82: 405
　Section 220.82(A): 406
　Section 220.82(C): 407
Article 225: 57f
　Section 225.22: 58
Article 230: 58, 59, 59f, 325, 326
　Part I: 58
　Part II: 58
　Part III: 58, 342
　Part IV: 58
　Part V: 58, 327
　Part VI: 58
　Part VII: 58, 334
　Figure 230.1: 59
　Section 230.24(B): 338
　Section 230.28: 338
　Section 230.30(B): 341, 342
　Section 230.32: 343
　Section 230.42: 336
　Section 230.43: 334
　Section 230.53: 334
　Section 230.54: 338
　Section 230.66(A): 58, 327
　Section 230.66(B): 327
　Section 230.67, 63: 327
　Section 230.70(A)(1): 329
　Section 230.70(A)(2): 329
　Section 230.71: 329–330
　Section 230.71(B): 329
　Section 230.77: 329
　Section 230.79: 328, 405, 408
　Section 230.79: 329
　Section 230.85(E): 332
Article 240: 58, 61, 61f, 310, 315
　Part I: 60–61
　Part II: 61
　Part III: 61
　Part IV: 62
　Part V: 62
　Part VI: 62, 310
　Part VII: 62
　Part VIII: 62
　Section 220.14(A): 316

Section 240.2: 61, 62
Table 240.6(A): 314
Section 240.20(A): 315
Section 240.21: 61
Section 240.24: 61
Section 240.24(A): 315
Section 240.24(C): 315
Section 240.24(D): 315, 329
Section 240.24(E): 315
Section 240.24(F): 315
Section 240.33: 315
Section 240.4: 318
Section 240.4(A): 318
Section 240.4(B): 318
Section 240.4(D): 318, 387
Section 240.6(A): 318
Section 240.6(C): 318
Section 240.60(D): 311
Section 240.81: 315
Section 240.83(C): 311
Section 240.83(D): 314
Article 242: 63, 63f
　Part II: 63
　Part III: 63
Article 250: 63, 64f, 65f, 349, 350
　Part I: 66
　Part II: 66
　Part III: 66, 352
　Part IV: 66
　Part V: 66
　Part VI: 66, 362
　Part VII: 66
　Figure 250.1: 350, 366
　Section 250.4: 66
　Section 250.4(A): 351
　Section 250.8: 352
　Section 250.8(A): 352
　Section 250.10: 352
　Section 250.12: 352
　Table 250.66: 358–360
　Section 250.24: 367
　Section 250.24(C): 367, 405
　Section 250.28: 367
　Section 250.28(D)(1): 367
　Section 250.30: 370
　Section 250.50: 352
　Section 250.52: 353
　Section 250.52(A)(1): 353
　Section 250.52(A)(2): 354
　Section 250.52(A)(3): 354
　Section 250.52(A)(4): 355
　Section 250.52(A)(5): 355

Section 250.53: 353
Section 250.53(A): 355
Section 250.53(D): 353
Section 250.53(F): 355
Section 250.64(B): 360
Section 250.64(B)(2): 360
Section 250.64(B)(3): 360
Section 250.64(C): 357
Section 250.64(E): 361
Section 250.66: 358, 367
Section 250.66(A): 358–359
Section 250.66(B): 358, 360
Section 250.66(C): 358, 360
Section 250.8: 352
Section 250.8(A): 352
Section 250.92: 366
Section 250.94: 369
Section 250.96: 366
Section 250.102(C)(1): 367, 370, 405
Section 250.104(A): 353, 354, 369
Section 250.104(B): 370
Section 250.104(C): 354, 370
Section 250.109: 362
Section 250.110: 362
Section 250.112: 362
Section 250.114: 362
Section 250.116: 362
Section 250.118: 363
Section 250.119: 364
Section 250.122: 365, 370
Section 250.122(B): 366
Article 250.53(A): 355
Chapter 3: 13, 22, 74, 76f, 77f, 82f, 83f, 84f, 85f
Article 300: 75, 77, 78
Section 300.4: 24, 77
Section 300.5: 343
Section 300.5(B): 283, 336
Section 300.5(D)(4): 434
Section 300.5(J): 343
Table 300.5(A): 342
Section 300.7(A): 334
Section 300.7(B): 335
Section 300.9: 283, 334, 336
Section 300.10: 366
Section 300.14: 247
Article 300.5(D)(3): 341
Article 305: 75
Article 310: 78f, 206
Table 310.4(1): 284, 286
Table 310.4(A): 267–268
Section 310.5: 317
Section 310.10: 283, 336

Section 310.10(G): 296
Section 310.12: 336, 405, 408
Section 310.12(A): 405, 408
Table 310.12(A): 337f
Section 310.14: 285
Section 310.15: 285–286
Section 310.15(A): 294
Section 310.15(B)(1): 289
Section 310.15(B)(2): 291
Table 310.15(B)(1)(1): 289, 290f, 291f
Section 310.15(C): 288
Section 310.15(C)(1): 288
Section 310.15(C)(1)(b): 288
Section 310.15(E): 288
Section 310.15(F): 288
Section 310.16: 285–287, 336
Table 310.16: 289, 317–318, 336, 386, 405
Section 310.17: 285–286
Article 312: 79, 80f
Section 312.6: 255
Table 312.6(A): 255
Article 314: 81, 81f, 241
Section 34.16: 244
Table 314.16(A): 244
Section 314.16(B): 242, 247
Section 314.16(B)(1): 247
Section 314.16(B)(2): 248
Section 314.16(B)(3): 249
Section 314.16(B)(4): 250
Section 314.16(B)(5): 250
Section 314.17(B)(2): 245
Section 314.28: 253
Section 314.28(A)(2): 253, 255
Table 314.16: 245
Table 314.16(B)(1): 250, 251f
Article 315: 80
Article 320: 74, 75f, 80
Article 334: 80, 83f
Section 334.80: 296
Article 340: 80
Article 342: 82
Article 344
Section 344.30(A): 339
Article 348
Section 348.22: 268
Table 348.22: 268
Article 352: 335
Table 352.44: 335
Article 358: 83, 84f, 264
Section 258.22: 270
Section 358.22: 198
Article 360: 264

Article 362: 75f, 82, 264
Article 366: 75, 84
Article 376: 265
Article 378: 265
Article 386: 84, 85f
Article 392: 84
Article 396: 74, 75, 75f
Chapter 4: 13, 22, 93f, 94f, 95f, 100f, 110f
   Article 400: 92, 94
      Table 400.4: 93
      Section 400.10: 93
      Section 400.12(1): 93
   Article 402: 92, 94f, 95
   Article 404: 94–96, 95f
      Section 404.2(C): 232
      Section 404.7: 329
      Section 404.8(A): 329
      Section 404.8(B): 246
   Article 406: 96f
      Section 406.5(G)(1): 227
      Section 406.5(G): 228
      Section 406.9(A): 230
      Section 406.9(B): 230
      Section 406.9(C): 228
      Section 406.12: 97
   Article 408: 97, 98f
   Article 409: 99f
   Article 410: 100, 101f, 102f
   Article 411: 102
   Article 422: 102–103, 103f
      Section 422.12: 219
      Section 422.60: 316
   Article 424: 103, 104, 104f, 105, 105f
   Article 430: 104, 107f, 108f, 397, 401
      Part XIV: 106
      Table 430.248: 401
   Article 440: 106, 109f, 397
      Part IV: 402
      Section 440.14: 109
      Section 440.22(C): 316
   Article 445: 106, 110f
   Article 450: 106, 110f, 111f
   Article 455: 106, 111f
   Article 480: 109, 112f, 172
Chapter 5: 14, 23, 119f
   Article 500: 118, 119
   Article 501: 119f, 120
   Article 502: 119, 121, 121f
   Article 503: 121, 122f
   Article 504: 121, 122f
   Article 505: 122, 123f
   Article 506: 119, 123f
   Article 511: 119, 123–24, 123f
   Article 512: 124f
   Article 513: 124f
   Article 514: 125f
   Article 515: 125f
   Article 516: 126f
   Article 517: 118, 119, 126, 127f
   Article 518: 128f
      Section 518.2(A): 128
   Article 520: 129f
   Article 522: 129–130, 130f
   Article 525: 130f
   Article 530: 130f
   Article 540: 128, 131, 132f
   Article 545: 131, 132f
   Article 547: 133f
   Article 550: 134f
   Article 551: 135f
   Article 552: 136f
   Article 555: 131, 136–137, 137f
   Article 590: 137–138, 138f
Chapter 6: 14, 23, 144, 145f
   Article 600: 144, 145f
      Section 600.5: 235
   Article 604: 145, 146f
   Article 605: 146f
   Article 610: 146–147, 147f
   Article 620: 148f
   Article 625: 149f
   Article 626: 149f, 150
   Article 630: 149f, 150
   Article 640: 149, 151f
   Article 645: 151–152, 152f
   Article 646: 152–153, 152f
   Article 647: 153f
   Article 650: 153f
   Article 660: 153f, 154
   Article 665: 153f, 154
   Article 668: 153, 155f
   Article 669: 155f
   Article 670: 155f
   Article 675: 155–156, 156f
   Article 680: 156, 157f, 220
      Section 680.26: 368
   Article 682: 156, 158f
   Article 685: 156, 158f
   Article 690: 159f
   Article 691: 159f
   Article 692: 160f
   Article 694: 161f
   Article 695: 161f
Chapter 7: 15, 23, 168f
   Article 700: 169–170, 169f
   Article 701: 170, 170f

Article 702: 171, 171*f*
Article 705: 171–172, 171*f*
Article 706: 172, 173*f*
Article 708: 172–173, 173*f*
Article 710: 174*f*
Article 722: 175*f*
Article 724: 175–176, 175*f*
   Section 724.40: 175
Article 725: 176*f*
   Section 725.60(C): 176
Article 726: 177*f*
Article 728: 177*f*
Article 750: 177*f*
Article 760: 177–178, 178*f*
Article 770: 178–179, 179*f*
Chapter 8: 15, 23, 186*f*
   Article 800: 186, 187*f*
      Section 800.44(C): 339
   Article 805: 186, 187*f*
      Section 805.156: 187
   Article 810: 187, 188*f*
   Article 820: 189*f*
   Article 830: 189*f*, 190*f*
   Article 840: 189–190, 190*f*
Chapter 9: 16, 23
   Table 1: 198–199, 198*f*, 268–269
   Table 2: 199*f*
   Table 4: 200*f*, 269, 270*f*
   Table 5: 200–201, 201*f*, 265, 271*f*
   Table 5A: 201
   Table 8: 201–202, 202*f*, 274*f*
   Table 9: 202, 203*f*
   Table 10: 202, 203*f*
   Table 11(A): 203, 204*f*
   Table 11(B): 203, 204*f*
   Table 12(A): 203, 204*f*
   Table 12(B): 203, 204*f*
   Table 13: 203–204, 205*f*
Informative Annexes
   Informative Annex A: 204–205
   Informative Annex B: 206
   Informative Annex C: 206, 269–270
      Table C.1.(A): 270
   Informative Annex D: 206, 397
   Informative Annex E: 206
   Informative Annex F: 207
   Informative Annex G: 207
   Informative Annex H: 208
   Informative Annex I: 208
   Informative Annex J: 208
   Informative Annex K: 208

Index

A

ADA. *See* Americans with Disabilities Act
adjustable speed drives, 106
adjustment factors, 287–288
AFCI. *See* arc-fault circuit interrupter
allowance table, 250–252
alter articles, *NEC*, 106–112
aluminum, 282–283
Americans with Disabilities Act (ADA), 208
ampacity
    adjustment factors, 287–288
    cable ampacity, 296–299
    correction factor, 289–291
    high temperature, 291–294
    more than three conductors, 291–294
    parallel conductor, 296
    provisions, 285
    rooftop raceways and cables, 291
    tables, 286
    weakest link, 294–296
angle pulls, 253–255
appliance, 102
appliance branch circuit, 218
application provisions, 285
arc blast, 40
arc-fault circuit interrupter (AFCI), 52, 220, 312–313
arc flash, 40
articles, 12
artificially made bodies of water, 156
assembly occupancies, 128
attics, 233
audio signal processing equipment, 149
auxiliary gutter, 265, 275–276

B

backup system, 168
bare conductor, 265
barrier, 246
bathroom, 224, 228–229
bonding, 66
    communication systems, 369
    definition, 366
    enclosures, 366
    main bonding jumper, 367
    piping systems, 369–370
    service equipment, 366
    structural metal, 369–370
    system bonding jumper (SBJ), 370
box
    angle pull leaving back of the box, 255
    box fill calculations, 244–252
    device, 243
    electrical boxes, 242–243
    fill calculations, 247–250
    junction, 243
    lighting outlet, 242
    metallic, 244
    nonmetallic, 244–245
    outlet, 242
    pull, 243
box fill calculations, 247–250
    allowance table, 250–252
    box volume calculations, 244–247
    clamps, 248–249
    conductors, 247–248
    devices, 250
    equipment grounding conductor (EGC), 250
    support fittings, 249
box volume
    calculations, 247–250
    items that add to overall volume, 245
    items that subtract from volume, 246–247
    metallic boxes, 244
    nonmetallic boxes, 244–245
branch circuit, 218–220
    appliance, 218
    arc-fault circuit interrupter (AFCI), 52, 220
    conductor sizing, 385–387
    definition, 52, 218
    dwelling, 224–34
    dwelling receptacle requirements, 225–231
    general purpose, 218
    ground-fault circuit interrupter (GFCI), 52, 220
    habitable room, 55
    individual, 219
    individual branch circuit, 54
    lighting, 231–234
    multiwire, 219
    non-dwelling, 234–235
    overcurrent protective device (OCPD), 387
    rating, 220–223

C

cabinet, 79
cable, 283
cable ampacity, 296–299
cable articles, *NEC*, 80–82
cartridge fuse, 310–311
chapters, 12
circuit breaker, 311
    arc-fault circuit interrupter, 312–313
    dual-function circuit, 313
    ground-fault circuit interrupter (GFCI), 312
    ratings, 313–314
circular mil, 202
circular raceway, 266
    conductor area, 270
    fill calculation, 268–272
    maximum number of conductors, 269–270
    minimum size raceway, 270
    permitted conduit and tubing fill, 268–269
    raceway area, 270
    raceway size, 266
    special conduit fill scenarios, 273–275
clamps, 248–249
Class 3 circuit, 176
Class 2 circuit, 176
Class I locations
    Class I, Division 1 locations, 119
    Class I, Division 2 locations, 119
Class II locations
    Class II, Division 1 locations, 121
    Class II, Division 2 locations, 121
Class III locations
    Class III, Division 1 locations, 121
    Class III, Division 2 locations, 121
CMPs. *See* code-making panels
code-change process, *NEC*
    first draft meeting, 6–7
    NFPA standards council, 8
    NFPA technical meeting, 7–8
    public comment stage, 7
    public input stage, 5–6
    tentative interim amendment, 8
code-making panels (CMPs), 5
combustible dust, 119, 121
common format, *NEC*, 74–75
communication circuit, 186
compact conductor, 201, 265
conductor, 265, 267
    ampacities, 285–299
    bare, 265
    box fill calculations, 247–248
    compact, 265
    covered, 265
    equipment grounding conductor, 52
    grounded conductor, 51
    insulated conductor, 265
    insulation, 265, 267–268
    materials, 282–283
    neutral conductor, 52
    specifications, 284–285
    switch leg, 52
    uses permitted, 283–284
conductor articles, *NEC*, 78–79
conductor sizing, 222
conduit, 264
conduit bodies, 80, 252
confined space, 42
consensus standard, 4
consume articles, *NEC*, 94–100
continuous load, 222, 315
control articles, *NEC*, 94–100
controller, 104
cooktop. *See* counter-mounted cooking unit
copper, 282
copper-clad aluminum, 283
COPS. *See* critical operations power systems
correction factor, 289
counter-mounted cooking unit (cooktop), 380
covered conductor, 265
crawl spaces, 233
create articles, *NEC*, 106–112
critical operations power systems (COPS), 172
cutout box, 79

D

demand factor, 380, 398
depth, 331
design articles, *NEC*, 50–58
device, 243
device box, 243
devices, 250
direct burial, 284
direct sunlight, 283–284
dot, 17
dual-function circuit, 313
dwelling branch circuit, 224–234
    outlet requirements and provisions, 224–225
    receptacle requirements, 225–231
    required branch circuit, 224
dwelling emergency disconnect, 332
dwelling receptacle requirements, 225–231
    bathroom, 228–229
    foyers, 231
    garages, 231
    hallways, 231
    kitchens, 227–228
    laundry, 231
    outdoor outlets, 230
    receptacle spacing, 225–226
    tamper-resistant receptacles, 225
dwelling scenario, 398

Index

E

Edison-base fuse, 310
effective ground-fault current path, 361
EGC. *See* equipment grounding conductor; equipment grounding system
egress, 42, 331
electrical boxes, 242–243
    device box, 243
    junction box, 243
    lighting outlet box, 242
    outlet box, 242
    pull box, 243
Electrical Safety in the Workplace (NFPA 70E), 5
electric sign, 144
electric vehicle supply equipment (EVSE), 403
electric welders, 149
electrified truck parking space, 149
electrolytic cell, 153
electroplating, 155
emergency system, 169
enclosure, 41, 79
enclosure articles, *NEC*, 79–80
energy management system, 177
energy storage system, 172
equipment, 92
    adjustable speed drives, 106
    appliance, 102
    controller, 104
    general-use snap switch, 96
    general-use switch, 96
    hermetic refrigerant motor-compressor, 106
    industrial control panel, 99
    luminaire, 100
    panelboard, 97
    phase converter, 106
    switchboard, 98
    switchgear, 99
    transformer, 106
equipment connection articles, *NEC*, 92–94
equipment grounding, 361–362
    effective ground-fault current path, 361
equipment grounding conductor (EGC), 52, 250
equipment grounding system (EGC), 363
    raceway system, 363
    sizing, 365–366
    wire type, 364
equipment requiring servicing, 234
EVSE. *See* electric vehicle supply equipment
exception, 24
extension rings, 245

F

fault current, 40
fault-managed power system, 177
feeder, 55
fire alarm system, 177
fire pumps, 160
fire-resistive cable system, 177
first draft ballot, 7
first draft meeting, 6–7
first draft report, 7
first revisions, 7
fixture stud, 249
foyers, 231
fuel cell system, 160
fuses, 309
    cartridge, 310–311
    Edison-base, 310
    nonrenewable, 309
    plug, 310
    Type S, 310

G

garages, 224, 231
general articles, *NEC*, 75–77
general chapters, 12
general-purpose branch circuit, 218
general-use snap switch, 96
general-use switch, 96
GFCI. *See* ground-fault circuit interrupter
grade, 230
ground, 350
grounded conductor, 51, 350
grounded service entrance conductor, 405
ground fault, 60, 308, 350
ground-fault circuit interrupter (GFCI), 220, 312
grounding
    and bonding connections, 352
    definition, 350
    electrode system, 352–361
    equipment grounding, 361–366
    grounding electrode conductor, 357–361
    purpose, 351–352
grounding electrode, 66
grounding electrode conductor, 66, 357
    protection, 360–361
    sizing, 358–360
grounding electrode system, 352–361
    concrete encased electrode, 354–355
    definition, 352
    ground ring, 355
    metal in-ground support structure, 354
    metal underground water pipe, 353–354
    not permitted as grounding electrodes, 357
    other local metal underground systems, 357
    plate electrodes, 356–357
    road and pipe electrodes, 355–356
gutter. *See* wireway

H

habitable bathroom, 232
habitable kitchen, 232
habitable room, 55, 225, 232
hallways, 231
hazardous locations, 118–127
height, 331
hermetic refrigerant motor-compressor, 106
hickey, 249
highlighting, 27–28
horsepower, 397
household cooking equipment
    demand factors, 380–385

I

illumination, 331
impedance, 202
index, 27
individual branch circuit, 54, 219
induction heating, 153
industrial control panel, 99
industrial machinery, 155
information, *NEC*, 25
    index, 27
    keywords, 25–26
    memorization, 27
    table of contents (TOC), 26–27
informational note, 24
information technology equipment (ITE), 151
Informative Annexes, 16, 197
    Informative Annex A, 204, 205*f*
    Informative Annex B, 206*f*
    Informative Annex C, 206, 207*f*, 269–270, 274
    Informative Annex D, 55, 206, 207*f*
    Informative Annex E, 206, 208*f*
    Informative Annex F, 207
    Informative Annex G, 207
    Informative Annex H, 208*f*
    Informative Annex I, 208, 209*f*
    Informative Annex J, 208, 210*f*
    Informative Annex K, 208
inrush current, 307
insulated conductor, 265, 282
interconnected electric power production source, 172
intersystem bonding termination, 369
intrinsically safe circuit, 121
irrigation machine, 155
ITE. *See* information technology equipment

J

junction box, 243, 252–255

K

keywords, 25–26
kitchens, 227–228
knockouts, 200

L

laundry, 224, 231
layout, *NEC*
    general chapters, 12–13
    Informative Annexes, 16
    NEC Chapter 9 tables, 16
    special chapters, 14–15
legally required standby system, 170
light fixture. *See* luminaire
lighting, 231–234
    attics, 233
    crawl spaces, 233
    equipment requiring servicing, 234
    habitable room, kitchen, and bathroom, 232
    outside lights, 233
    stairway, 233
    unfinished basements, 233
lighting outlet, 232
lighting outlet box, 242
load, 380
    continuous, 315
    determining, 316
    noncontinuous, 315
location, damp, 283
location, dry, 283
location, wet, 283
luminaire, 100

M

main bonding jumper, 367
mandatory language, 24
manufactured building, 132–133
manufactured wiring system, 145
meeting rooms, 234–235
melting, 153
memorization, 27
meter socket, 79
mil, 202
modular data center, 152
mudring, 245
multiwire branch circuit, 219

N

National Electrical Code (*NEC*), 3
    alter articles, 106–112
    autonomous power production systems, 174
    cable articles, 80–82
    changing, 4–5

code-change process, 5–8
common format, 74–75
communication systems, 186–191
conductor articles, 78–79
consume articles, 100–106
control articles, 94–100
control circuits, 174–178
create articles, 106–112
design articles, 50–58
enclosure articles, 79–80
energy backup and storage systems, 169–173
equipment connection articles, 92–94
finding information in, 25–27
gatherings and entertainment, 128–131
general articles, 75–77
hazardous locations, 118–127
history, 4
informative annexes, 204–210
layout, 12–16
managed power systems, 174–178
manufactured structures and agricultural buildings, 131–137
online subscription, 28
optical fiber cables, 178–179
outline format, 12
power-limited circuits, 174–178
protection articles, 58–66
revisions, 16–18
special conditions, 168
special equipment, 144–161
special occupancies, 118
store articles, 106–112
style manual, 6
tables, 198–204
temporary installations, 137–138
tools, 27–28
National Fire Protection Association (NFPA), 4
NFPA 70E. See *Electrical Safety in the Workplace*
standards council, 8
technical meeting, 7–8
Nationally Recognized Testing Laboratory (NRTL), 37
natural bodies of water, 156
NEC Handbook, 18
NEC. See *National Electrical Code*
network interface unit, 189
network terminal, 189
neutral conductor, 52
NFPA. See National Fire Protection Association
nipple, 199, 264
noncircular raceway, 275–276
noncoincidence leads, 402
noncontinuous load, 315
non-dwelling branch circuit, 234–235
nonmetallic boxes, 244–245
nonrenewable, 309
NRTL. See Nationally Recognized Testing Laboratory

O

Occupational Safety and Health Administration (OSHA), 37
OCPD. See overcurrent protective device
office furnishing, 146
one-shot bender, 199
online subscription, 28
operating temperature, 285
optional standby system, 171
OSHA. See Occupational Safety and Health Administration
outdoor outlets, 230
outlet, 242
outlet box, 242
outlet requirements and provisions, 224–225
outline format, *NEC*, 12
outside lights, 233
oven, 380
overcurrent, 58, 306
overcurrent protection, 222
    device sizing factors, 316–318
    location, 315
    protection devices, 308–315
    types, 306–308
overcurrent protective device (OCPD), 308–315
    branch circuits, 387
overhead service, 336–337
    conductors, 337–338
    raceway, 338–339
    service drop, 337
overload, 58, 306
overvoltage, 53

P

panelboard, 97
parallel conductor, 296
park trailer, 136
part, 12, 74–75
permissive language, 24
phase converter, 106
pipe organ, 153
plaster rings, 245
plug fuse, 310
protection articles, *NEC*, 58–66
public comment, 7
public input, 5
pull boxes, 243, 252–255
PV. See solar photovoltaic system

R

raceway, 82, 264
    circular raceway, 266, 268–275
    definitions, 264–265
    noncircular raceway, 275–276

raceway fill
　　auxiliary gutter, 265
　　bare conductor, 265
　　circular raceway, 266
　　compact conductor, 265
　　conductors, 265, 267–268
　　conduit, 264
　　covered conductor, 265
　　definition, 264
　　insulated conductor, 265
　　nipple, 264
　　tubing, 264
　　wireway, 265
raised covers, 245
range, 380
rating, 220–221
　　conductor sizing, 222
　　overcurrent protection, 222
reactance, 202
receptacle, 96
receptacle spacing, 225–226
relocatable structure, 133
remote-control circuit, 176
required branch circuit, 224
revisions, *NEC*, 16–18
revision symbols, 16–17
rounding, 397

S

SBJ. *See* system bonding jumper
SCADA. *See* supervisory control and data acquisition
second draft meeting, 7
second draft report, 7
second revision ballot, 7
second revisions, 7
sections, 12, 75
separately derived system, 370
service, 58
　　definition, 326
　　types of, 336–343
service calculations
　　demand factor, 398
　　dwelling scenario, 398
　　general requirements, 396–398
　　minimum size, 405, 408
　　optional method, 405–407
　　samples problems in annex, 397
　　setting up, 397–398
　　standard method, 398–404
　　values, 396–397
service conductor, 336
service disconnect
　　clearance, 330–332
　　dwelling emergency disconnect, 332
　　location, 329
　　number of, 329
　　overcurrent protection, 334
　　rating, 328–329
service drop, 337
service-entrance conductor, 336
service equipment, 327
service head, 338
service lateral, 340
service mast, 338
service point, 326
shaded N, 17
shaded text, 16
shaded triangle Δ, 16
shore power, 137
short circuit, 307
short conduit bodies, 252
sign circuit, 235
significant changes class, 18
small appliances, 224
solar photovoltaic system (PV), 158
solidly grounded, 350
SPD. *See* surge protection device
special chapters, 14
splice, 253–255
stairway, 233
stand-alone system, 174
standard method, service calculations
　　dryer, 401
　　electric vehicle supply equipment, 403–404
　　fastened in place appliances, 400–401
　　general lighting load, 398–400
　　heat/AC, 402–403
　　household cooking appliances, 401–402
　　other loads, 404
　　putting it together, 404
store articles, *NEC*, 106–112
straight pulls, 253
stranded conductor, 201
supervisory control and data acquisition (SCADA), 207
support fittings, 249
surface raceway, 275
surge protection device (SPD), 63, 327
switchboard, 98
switchgear, 99
switch leg, 52
system bonding jumper (SBJ), 370

T

table of contents (TOC), 26–27
tabs, 28
tail, 248
tamper-resistant receptacles, 225
TCC. *See* technical correlating committee

technical correlating committee (TCC), 5
tentative interim amendment (TIA), 8
termination, 255
TIA. *See* tentative interim amendment
TOC. *See* table of contents
torque, 208
trade name, 284–285
transformer, 106
tubing, 264
Type S fuses, 310

U
underground service, 339–343
    conductors, 340–341
    raceway, 342–343
unfinished basements, 233
ungrounded, 51
U pull, 253–255

V
voltage, 397
volt-ampere, 397

W
watt, 397
weakest link, 294–296
welding, 153
width, 331
wind turbine, 160
wireway, 265, 275
wiring methods, 334–336
    cable articles, 80–82
    circular raceways, 82–83
    conductor articles, 78–79
    enclosure articles, 79–80
    general articles, 75–77
    noncircular raceways, 84–85
    other wiring methods, 85

Y
yoke, 250

Z
Zone 0 location, 122
Zone 1 location, 122
Zone 2 location, 122
Zone 20 location, 123
Zone 21 location, 123
Zone 23 location, 123